Visible Light Photocatalysis in Organic Chemistry

Visible Light Photocatalysis in Organic Chemistry

Edited by Corey R. J. Stephenson, Tehshik P. Yoon, and David W. C. MacMillan

WILEY-VCH

Editors

Prof. Corey R. J. Stephenson
University of Michigan
Department of Chemistry
930 N University Avenue
Ann Arbor
MI 48109
USA

Prof. Tehshik P. Yoon
University of Wisconsin-Madison
Department of Chemistry
1101 University Avenue
Madison
WI 53706
USA

Prof. David W. C. MacMillan
Princeton University
Merck Center for Catalysis at Princeton
NJ 08544
USA

Library of Congress Card No.: applied for

British Library Cataloguing-in-Publication Data
A catalogue record for this book is available from the British Library.

Bibliographic information published by the Deutsche Nationalbibliothek
The Deutsche Nationalbibliothek lists this publication in the Deutsche Nationalbibliografie; detailed bibliographic data are available on the Internet at <http://dnb.d-nb.de>.

Print ISBN: 978-3-527-33560-2
ePDF ISBN: 978-3-527-67417-6
ePub ISBN: 978-3-527-67416-9
Mobi ISBN: 978-3-527-67415-2
oBook ISBN: 978-3-527-67414-5

Cover Design Adam-Design, Weinheim, Germany
Typesetting SPi Global, Chennai, India
Printing and Binding
CPI Group (UK) Ltd, Croydon, CR0 4YY

Printed on acid-free paper

C9783527335602_111225

The manufacturer's authorized representative according to the EU General Product Safety Regulation is Wiley-VCH GmbH, Boschstr. 12, 69469 Weinheim, Germany, e-mail: Product_Safety@wiley.com.

Contents

1

An Overview of the Physical and Photophysical Properties of $[Ru(bpy)_3]^{2+}$

Daniela M. Arias-Rotondo and James K. McCusker

Michigan State University, Department of Chemistry, 578 S Shaw Lane, East Lansing, MI 48824, United States

1.1 Introduction

The photophysics and photochemistry of transition-metal coordination compounds have been studied for over half a century [1, 2]. In particular, metal polypyridyl complexes – especially those that possess visible charge transfer absorptions – have played a central role in efforts to understand fundamental aspects of excited-state electronic structure and dynamics, as well as efforts to develop a wide range of solar energy conversion strategies [3, 4]. Their footprint in the area of synthetic organic chemistry was largely nonexistent until 2008 [5], when MacMillan and coworkers [6] reported the first example of a transition-metal-based charge transfer compound, $[Ru(bpy)_3]^{2+}$ (where bpy is 2,2′-bipyridine), acting as a photocatalyst (PC) in an asymmetric alkylation of aldehydes; simultaneously, Yoon and coworkers [7] reported [2+2] enone cycloadditions photocatalyzed by $[Ru(bpy)_3]^{2+}$. Following those initial reports, several groups have explored the use of coordination compounds as photocatalysts for a variety of organic transformations [8]. These compounds engage in single-electron transfer (SET) processes with organic substrates, generating organic radicals, which play a major role in organic synthesis. This new kind of catalysis has opened the door to synthetically useful reactions that could not be performed otherwise.

The majority of the photocatalysts used nowadays are polypyridyl complexes of either Ru(II) or Ir(III) [8]. The large number of examples using $[Ru(bpy)_3]^{2+}$ might make this compound look like a "one size fits all" photocatalyst, when in reality, the best photocatalyst for a reaction is determined by the kinetics and thermodynamics of the system of interest. The purpose of this chapter is to provide the necessary tools to understand the different factors that come into play when choosing a photocatalyst. To this end, we will use $[Ru(bpy)_3]^{2+}$ as an example; it is important to note that the concepts we will discuss apply to most transition-metal polypyridyl compounds.

* An expanded discussion of these topics can be found in Chem. Soc. Rev. 2016, 45, 5803–5820.

Visible Light Photocatalysis in Organic Chemistry, First Edition.
Edited by Corey R. J. Stephenson, Tehshik P. Yoon and David W. C. MacMillan.

Scheme 1.1 shows two examples of catalytic cycles using Ru(II)-based photoredox catalysts: in both cases, the first step is the absorption of a photon by the photocatalyst to generate an excited state that then engages in redox reactions. The first cycle in Scheme 1.1, reported by Zheng and coworkers [9], is called reductive, because the excited photocatalyst is reduced. The second one, reported by Cano-Yelo and Deronzier [10], is an oxidative cycle; the photocatalyst is first oxidized and then reduced to reform its resting state.

As shown in Scheme 1.1, most steps in a catalytic cycle are bimolecular reactions. In a very general way, for any catalytic cycle involving $[Ru(bpy)_3]^{2+}$, we can write the series of reactions in Scheme 1.2 [11, 12]. The first step is the absorption of a visible light photon by the photocatalyst in its ground state and its consequent promotion to an electronic excited state (PC*); the backward reaction is the ground-state recovery (this process can be radiative (i.e., emission) and/or nonradiative, as will be discussed in Section 1.3). For the excited photocatalyst to react with a molecule (R), both species must diffuse toward each other, forming a "precursor complex." Then, the reaction takes place; of the many kinds of reactions that could happen, only electron and energy transfer are relevant for our discussion. After the reaction, the products must diffuse away from each other; if they cannot escape the solvent cage fast enough, a back reaction may take place.

This relatively simple scheme allows us to outline the main points that need to be considered when choosing a photocatalyst:

1) Photocatalytic reactions make use of the enhanced reactivity of the photocatalyst in its excited state; for this reason, a photocatalyst must possess a good absorption cross section, preferably over a broad range of wavelengths that the other species in the reaction mixture do not absorb.[1]
2) The quantum yield of formation of the reactive excited state should be as high as possible (preferably, near unity); that state must persist long enough to undergo the desired reaction with the substrate, and then cleanly regenerate in order to maintain its viability as part of a catalytic cycle. In the context of Scheme 1.2, these latter criteria mean that k_d and k_q must be larger than k_0, so that the PC* can diffuse toward the appropriate molecule and react with it before going back to the ground state [13].
3) If the catalytic cycle involves electron transfer, the excited- and ground-state redox potentials of the photocatalyst must provide for an exothermic (or at worst weakly endothermic) reaction; reversible electrochemistry is also desirable as an indicator of the stability of the photocatalyst over multiple turnovers.[2]
4) Synthetic accessibility and, more importantly, tunability are critical in order to tailor the excited-state reactivity of the photocatalyst to the reaction of interest.

1 Strictly speaking, it is only necessary for the photocatalyst to absorb light of one wavelength that the other species present in the reaction mixture do not absorb; having the photocatalyst absorb over a wider range of wavelengths makes it more versatile.

2 This is not necessary in the case of an energy-transfer photocatalyst, but those are far less common (see Prier, C. K.; Rankic, D. A.; MacMillan, D. W. *op. cit.* and references therein).

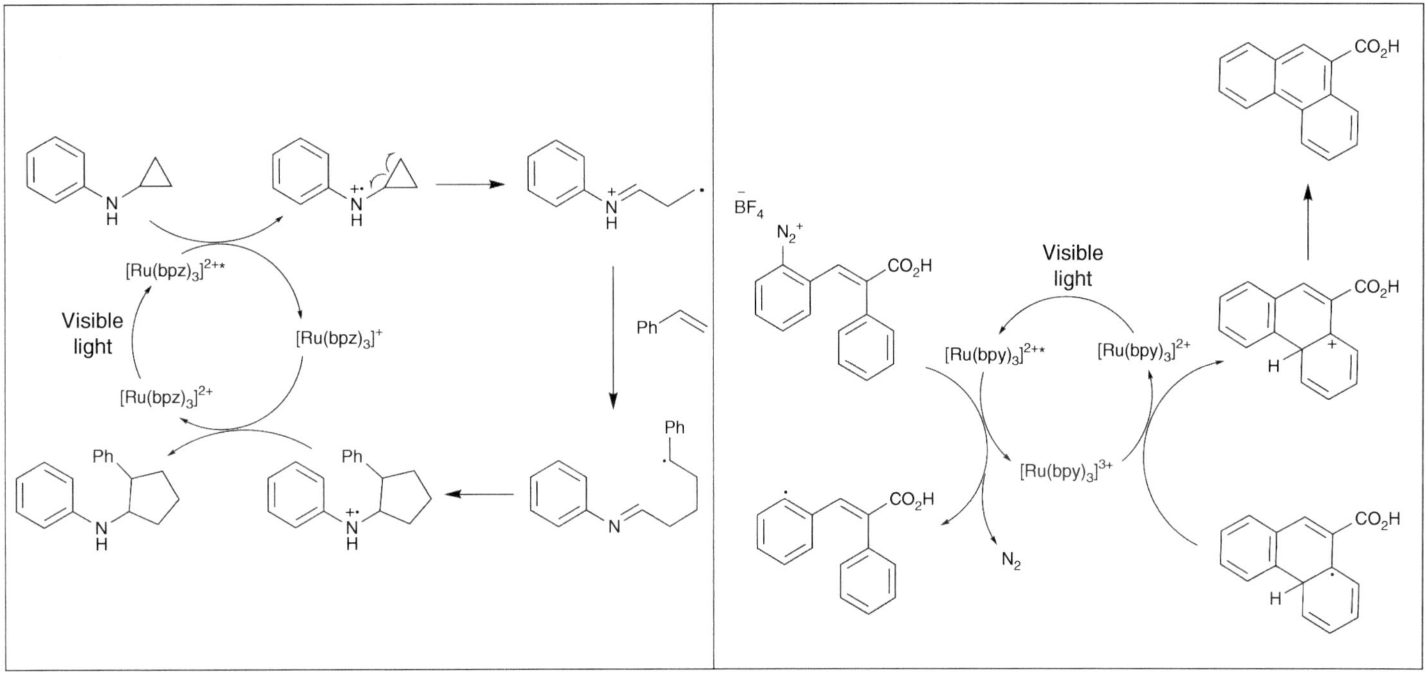

Scheme 1.1 Examples of reductive catalytic cycle (left; see also [9]) and oxidative catalytic cycle (right; see also [10]) involving Ru(II)-based photoredox catalysts; bpz is 2,2′-bipyrazine.

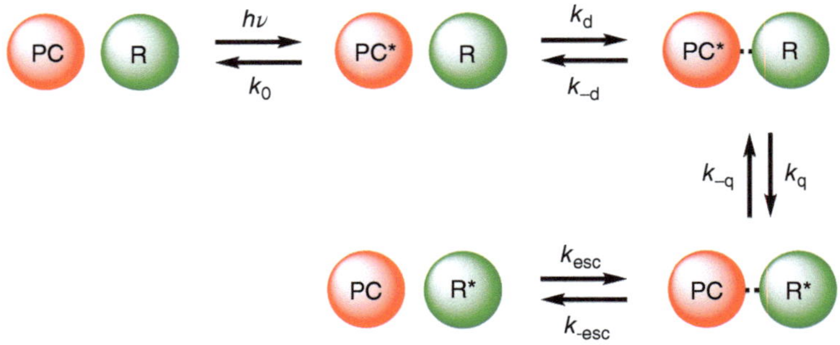

Scheme 1.2 Simplified kinetic scheme for a general quenching process (see also [11, 12]).

Given the various criteria just enumerated, it is no surprise that polypyridyl complexes of Ru(II) and Ir(III) have proved useful as photoredox catalysts. These compounds strongly absorb visible light, which makes it easy to selectively excite them relative to the organic substrates for typical reactions of interest. Their excited states are formed with ~100% efficiency [14] and their lifetimes range from 300 ns to 6 μs, which is long enough for them to engage in bimolecular reactions [3, 15]. As a class, these compounds are generally stable with respect to decomposition (both photochemical and thermal) and typically exhibit reversible redox behavior. They are also emissive, which facilitates mechanistic studies (as discussed in Sections 1.7 and 1.8); however, it is not a requirement. The synthesis of transition-metal polypyridyl complexes has been studied in great detail [4, 16], as well as the effect that different ligands have on the properties of the ground and excited states [17]. All these properties make these compounds the preferred choice for photocatalysts.

As mentioned above, we will discuss the properties of the ground and excited states of [Ru(bpy)$_3$]$^{2+}$, as a prototype for photoredox catalysis, describing the necessary experiments to fully understand their properties. Using this as a foundation, we will then focus on the processes that take place during a photocatalytic cycle and the experiments that allow for discriminating between various mechanistic possibilities (the main question being energy transfer versus reductive/oxidative electron transfer). In so doing, our goal is to provide a basic blueprint for how to identify, characterize, and ultimately design photocatalysts for use in a wide variety of chemical transformations.

1.2 [Ru(bpy)$_3$]$^{2+}$: Optical and Electrochemical Properties

1.2.1 Optical Properties

The electronic absorption spectrum of [Ru(bpy)$_3$](PF$_6$)$_2$ in acetonitrile is shown in Figure 1.1. The intense absorption at 285 nm corresponds to a ligand-centered transition ($\pi_L \rightarrow \pi_L{}^*$), which has been assigned by comparison with the absorption spectrum of the protonated ligand [18]. The band in the visible region

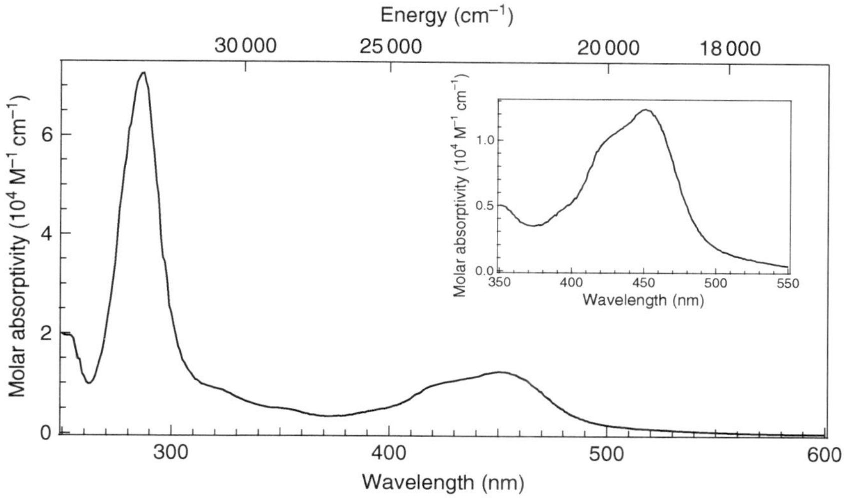

Figure 1.1 Electronic absorption spectrum of $[Ru(bpy)_3](PF_6)_2$ in acetonitrile at room temperature. The inset shows the metal-to-ligand charge transfer (MLCT) band.

(λ_{max} = 452 nm) corresponds to a metal-to-ligand charge transfer (MLCT) transition. As the name implies, this type of excited state can be viewed as the promotion of an electron from a metal-based orbital to a ligand-based one. Because of this spatial redistribution of electron density, this transition is responsible for the enhanced redox activity of the excited state relative to what is observed in the ground state, and makes the compound an efficient photocatalyst. Charge transfer transitions are typically very intense, with extinction coefficients in the range of 10^3 to $10^4\,M^{-1}\,cm^{-1}$ [19] (in acetonitrile at room temperature, $\varepsilon \sim 15\,000\,M^{-1}\,cm^{-1}$ for $[Ru(bpy)_3]^{2+}$).

Two additional features can be seen in the absorption spectrum of $[Ru(bpy)_3]^{2+}$. The origin(s) of the weaker features at 330 and 350 nm are less clear-cut and have been the subject of considerable debate over the years. They are most likely due to ligand–field (so-called "d–d") transitions within the d-orbital manifold of the metal. The inferred intensity belies this assignment to a certain extent (the symmetry-forbidden nature of d–d bands typically limits their absorptivities to the range of 10–100 $M^{-1}\,cm^{-1}$) [19] but the proximity of both the ligand-centered and MLCT features influences these values in the present case. These metal-centered transitions put electronic density in orbitals that are antibonding with respect to the metal–ligand bonds and are therefore responsible for ligand loss reactions [3]. These three types of transitions are schematized in the simplified molecular orbital diagram in Scheme 1.3.

It is worth noting that most organic substrates, with the exception of highly conjugated systems, do not absorb visible light (cf. ligand-based transition in Figure 1.1). Thus, the use of visible light allows the selective excitation of the photocatalyst and not the organic reactants, which prevents the uncontrolled formation of organic radicals that could lead to unwanted side reactions.

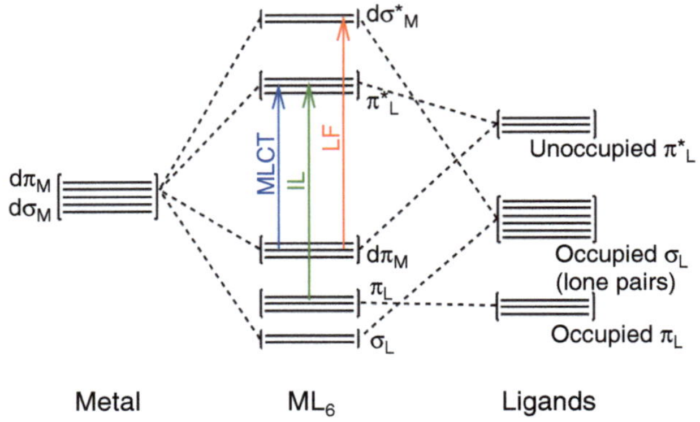

Scheme 1.3 Simplified molecular orbital diagram for an octahedral compound with π-acceptor ligands. The three types of electronic transitions discussed in the text are indicated by the arrows.

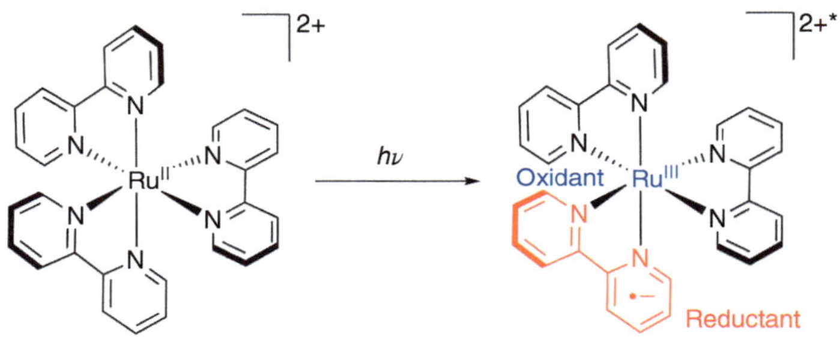

Scheme 1.4 A qualitative representation of a metal-to-ligand charge transfer state in $[Ru(bpy)_3]^{2+}$. The spatial separation of charge within the molecule following light absorption is critical for the redox activity of the excited state.

A metal-to-ligand charge transfer transition can be thought of as the simultaneous oxidation of the metal center and reduction of the ligand [20] that yields $[Ru^{III}(bpy^{\bullet-})(bpy)_2]^{2+*}$ (see Scheme 1.4). Unlike ligand- or metal-based electronic transitions (where the electron stays in the same spatial region before and after excitation), the MLCT results in the separation of charges within the compound, which confers a special reactivity to the resulting state: the oxidized metal (Ru^{III}) can act as an oxidant, gaining an electron to form Ru^{II}; likewise, the reduced ligand ($bpy^{\bullet-}$) can donate its extra electron, acting as a reductant. In its excited state, $[Ru(bpy)_3]^{2+}$ is both a stronger oxidant and reductant than in its ground state. Moreover, both the reductant and oxidant are simultaneously present in the same molecule, making this class of compounds very versatile for applications in photocatalysis.

1.2.2 Electrochemical Properties

Most of the examples using transition-metal photocatalysts take advantage of their ground- and excited-state redox properties. It is thus important to

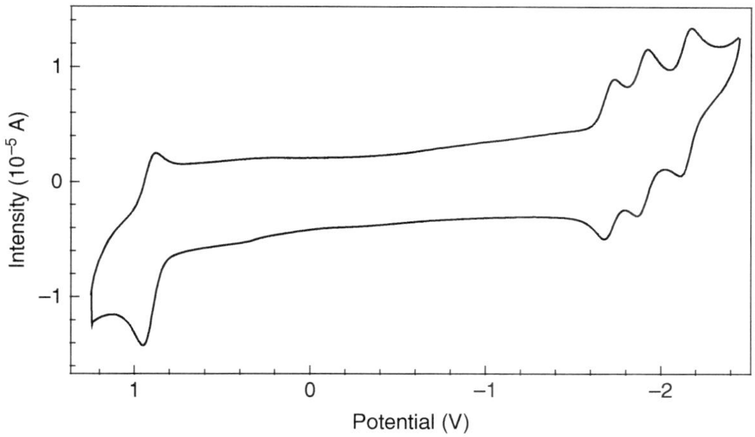

Figure 1.2 Cyclic voltammogram of $[Ru(bpy)_3](PF_6)_2$ in CH_3CN solution, using 0.1 M tetrabutylammonium hexafluorophosphate ($TBAPF_6$) as supporting electrolyte. Potentials are referenced to the ferrocene/ferrocenium couple, added as an internal standard.

understand those properties and how they affect the behavior of $[Ru(bpy)_3]^{2+}$ as a photocatalyst. The redox potentials for a coordination compound such as $[Ru(bpy)_3]^{2+}$ can be measured using cyclic voltammetry. The cyclic voltammogram for $[Ru(bpy)_3](PF_6)_2$ is shown in Figure 1.2. The oxidation of the metal center (Eq. (1.1)) is reversible and takes place around 1.00 V (vs. ferrocene/ferrocenium).

$$[Ru(bpy)_3]^{2+} \rightarrow [Ru(bpy)_3]^{3+} + e^- \tag{1.1}$$

Three reductions are also observed in the −1.50 to −2.30 V range, all of which correspond to one-electron reductions of each of the three ligands in succession (Eqs. (1.2a–1.2c)).

$$[Ru(bpy)_3]^{2+} + e^- \rightarrow [Ru(bpy^{\bullet-})(bpy)_2]^+ \tag{1.2a}$$

$$[Ru(bpy^{\bullet-})(bpy)_2]^+ + e^- \rightarrow [Ru(bpy)(bpy^{\bullet-})_2] \tag{1.2b}$$

$$[Ru(bpy)(bpy^{\bullet-})_2] + e^- \rightarrow [Ru(bpy^{\bullet-})_3]^- \tag{1.2c}$$

The first two reductions are reversible, whereas the last one (Eq. (1.2c) is quasi-reversible at best. In terms of photoredox reactions, only the first reduction (i.e., Eq. (1.2a)) will be relevant for one-electron processes, but the reversibility of these redox processes is an important consideration when these compounds are used as photocatalysts, since the compound must be stable enough in its oxidized or reduced form in order to be viable over the course of multiple turnovers of a given reaction.

Using the description above, the energy of the MLCT band can be thought of as the amount of energy necessary to reduce the ligand and oxidize the metal, as shown in Eq. (1.3).

$$E(\text{MLCT}) \approx |E(\text{Ru}^{\text{III}}/\text{Ru}^{\text{II}})| + |E(\text{bpy}/\text{bpy}^{\bullet-})| \tag{1.3}$$

Several aspects of Eq. (1.3) are worth noting: (i) this is an approximation: energetics associated with solvation as well as electron correlation effects are not accounted for in this simplified expression [21]; (ii) the fact that there are two contributions to the MLCT energy – the oxidation potential of the metal and the reduction potential of the ligand – implies that the value of E(MLCT) alone is not sufficient to determine whether a chromphore's energetics are suitable for a given reaction. One can observe MLCT bands at roughly the same energy where one is a very strong reductant but a very weak oxidant (i.e., very negative ligand reduction potential), or vice-versa. The electrochemical data on the compound (in addition to other details to be discussed later) is the means by which these specifics can be deconvolved.

1.3 Excited State Kinetics

We are ultimately interested in bimolecular reactions between an excited photocatalyst and an organic molecule. Before we can discuss these bimolecular reactions, however, it is necessary to understand the properties of the excited state in the absence of a substrate, since the presence (or absence) of a reaction will ultimately be determined by referring back to the photocatalysts' intrinsic excited-state behavior.

1.3.1 Steady-State Emission

Visible light excites $[Ru(bpy)_3]^{2+}$ into an ^{1}MLCT state; this short-lived state relaxes to an ^{3}MLCT state within ~100 fs via intersystem crossing (ISC, with rate constant k_{isc}) [22]. The ^{3}MLCT state can relax back to the ground state either nonradiatively (with rate constant k_{nr}) or via phosphorescence (a radiative pathway; its rate constant is k_r). Equations (1.4)–(1.6) illustrate these processes. Photoinduced reactions, such as the coordination of a solvent molecule or ligand loss, can also take place. However, these are not usually observed for $[Ru(bpy)_3]^{2+}$ and related compounds [14], so they will not be discussed here.

$$[Ru^{II}(bpy)_3]^{2+} \xrightarrow{h\nu_{abs}} {}^1[Ru^{III}(bpy^{\bullet-})(bpy)_2]^{2+*} \xrightarrow{k_{ISC}} {}^3[Ru^{III}(bpy^{\bullet-})(bpy)_2]^{2+*} \tag{1.4}$$

$$[Ru^{III}(bpy^{\bullet-})(bpy)_2]^{2+*} \xrightarrow{k_r} [Ru^{II}(bpy)_3]^{2+} + h\nu_{em} \tag{1.5}$$

$$[Ru^{III}(bpy^{\bullet-})(bpy)_2]^{2+*} \xrightarrow{k_{nr}} [Ru^{II}(bpy)_3]^{2+} + \text{heat} \tag{1.6}$$

The solution-phase steady-state emission spectrum of $[Ru(bpy)_3]^{2+}$ at room temperature is shown in Figure 1.3: the emission maximum is at 620 nm. The same spectrum is obtained regardless of the excitation wavelength, consistent with the near-unit quantum yield of formation of the emissive ^{3}MLCT state. The emission maximum can be used as a first-order approximation of the energy difference between the triplet excited state (^{3}MLCT) and the ground state (the zero point energy, E_0).

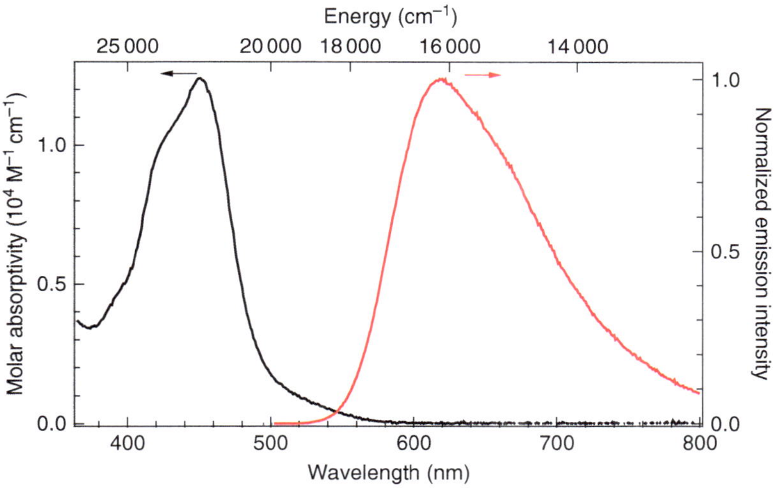

Figure 1.3 Electronic absorption spectrum (black) and steady-state emission spectrum (red) of $[Ru(bpy)_3](PF_6)_2$ in acetonitrile at room temperature.

For an emissive substance, the simplest definition of the quantum yield (Φ) of emission (also called the radiative quantum yield) is the ratio between the number of photons emitted by a sample and the number of photons absorbed, as shown in Eq. (1.7).

$$\Phi = \frac{\#\text{ photons emitted}}{\#\text{ photons absorbed}} = \frac{I_{\text{em}}}{I_{\text{abs}}} \tag{1.7}$$

For every photon absorbed, one molecule is promoted to the excited state. The radiative quantum yield can also be described in terms of a kinetic competition, specifically the relative rate(s) of processes giving rise to emission versus the rates of all processes that serve to deplete the population of that emissive state. Referring to Eqs. (1.5) and (1.6), for $[Ru(bpy)_3]^{2+}$ in the absence of any other species, Φ can be expressed as

$$\Phi_0 = \frac{k_r}{k_r + k_{nr}} = \frac{k_r}{k_0} \tag{1.8}$$

Radiative quantum yields can be measured as absolute values (i.e., Eq. (1.7)) or relative to some standard. To measure an absolute quantum yield it is necessary to detect every photon that is emitted by the sample, which tends to be quite labor intensive. Although instrumentation has recently become commercially available to allow for (relatively) facile measurement of absolute radiative quantum yields,[3] most of the quantum yields in literature are determined and reported relative to a standard with a known absolute quantum yield [23]. The choice of the standard depends on the characteristics of the molecule of interest; it is best if the standard and the molecule are dissolved in the same solvent and have similar absorption and emission spectra. $[Ru(bpy)_3]^{2+}$ is commonly used as a standard for relative

3 http://www.hamamatsu.com.

quantum yields of transition-metal complexes. In deoxygenated[4] acetonitrile at room temperature its quantum yield is 0.095 [24]. The relative quantum yield of a sample can be calculated using Eq. (1.9),

$$\Phi_x = \Phi_{\text{std}} \left(\frac{I_x/A_x}{I_{\text{std}}/A_{\text{std}}}\right)\left(\frac{\eta_x}{\eta_{\text{std}}}\right)^2 \tag{1.9}$$

where x refers to the molecule of interest and *std* to the standard; I_x and I_{std} are the integrated areas of the corrected emission spectra,[5] A_x and A_{std} are the absorbances at the excitation wavelength, and η_x and η_{std} are the indexes of refraction of the solutions, taken to be equal to those of the neat solvents. For relative quantum yield determinations, it is crucial for the experimental conditions for both the sample and the standard to be exactly the same. A more detailed discussion of methodology for measuring and quantifying emission data is beyond the scope of this chapter, but a number of excellent resources are readily available [25, 26].

As will be discussed later, observing a change (specifically, an attenuation) in the quantum yield of emission of a photocatalyst in the presence of a quencher is an important initial indicator that a reaction is occurring between the excited state of the photocatalyst and one or more substrate(s).

1.3.2 Time-Resolved Emission

Both the radiative and nonradiative decay processes (Eqs. (1.5) and (1.6)) are of first order with respect to the excited state (ES) and give rise to the following rate expression for the loss of the excited state:

$$-\frac{d[\text{ES}]}{dt} = k_{\text{r}}\,[\text{ES}] + k_{\text{nr}}\,[\text{ES}] = (k_{\text{r}} + k_{\text{nr}})[\text{ES}] = k_0\,[\text{ES}] \tag{1.10}$$

where $k_0 = k_{\text{nr}} + k_{\text{r}}$. Equation (1.10) can be integrated to yield the known rate law for a first-order reaction, shown in Eq. (1.11).

$$[\text{ES}] = [\text{ES}]_0\, e^{-k_0 t} \tag{1.11}$$

The inverse of the observed rate constant, k_0^{-1}, is the lifetime (τ_0) of the excited state; experimentally, this can be measured with time-resolved emission or absorption spectroscopy.

In a time-resolved emission experiment, the (emissive) sample is excited at a wavelength close to its absorption maximum, with the emission collected at 90° with respect to the excitation beam in order to minimize scatter. A typical time-resolved emission trace for $[Ru(bpy)_3]^{2+}$ in acetonitrile is shown in Figure 1.4. By fitting the trace to an exponential decay, τ_0 can be found. For $[Ru(bpy)_3]^{2+}$, the lifetime ranges from 500 to 1000 ns, depending on a number of variables including solvent, oxygen concentration in the sample, temperature, and so on [3].

---

4 This is necessary because O_2 can quench the ^{3}MLCT excited state of $[Ru(bpy)_3]^{2+}$.
5 Spectra refer to emission spectra that have been properly corrected for the fluorimeter's instrument response characteristics. References on emission spectroscopy can be consulted for further information on this point.

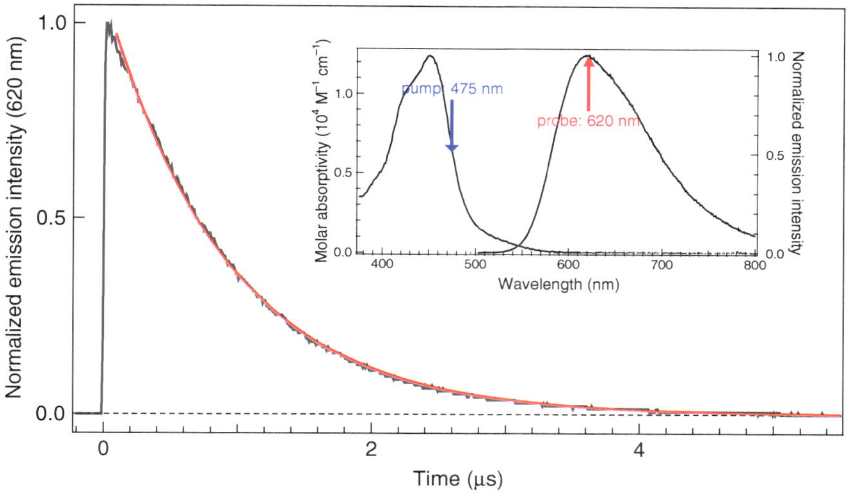

Figure 1.4 Time-resolved emission data (grey line) for $[Ru(bpy)_3]^{2+}$ in acetonitrile at room temperature. The sample was excited at 475 nm and emission was detected at 620 nm (as shown in the inset). The red trace shows the fit to a single exponential decay with $\tau = 930$ ns.

Combining the excited-state lifetime and the quantum yield, it is possible to calculate k_r and k_{nr}. Rearranging Eq. (1.8), we obtain Eqs. (1.12) and 1.13).

$$k_r = \Phi_0 \times k_0 \tag{1.12}$$

$$k_{nr} = k_0 - (\Phi_0 \times k_0) \tag{1.13}$$

It is important to remark that k_r is an intrinsic property of the molecule, and as such, it remains constant no matter what reactions the excited state engages in. On the other hand, k_{nr} varies when quenching processes (such as energy or electron transfer) take place. All the information that we will be interested in for a photocatalytic cycle (in other words, the information about any processes competing with the emission) is contained in k_{nr}; in this regard, k_r can be viewed as a probe, providing insight into the dynamics of the system manifesting in k_{nr}. This concept is discussed in greater detail in Section 1.7.

1.4 Excited-State Reactivity of $[Ru(bpy)_3]^{2+}$

In its excited state, $[Ru(bpy)_3]^{2+}$ can act as an energy donor, an electron acceptor, or an electron donor; which of these processes dominates is determined by thermodynamic and kinetic factors associated with a given reaction [27].

The inherent competition that exists among these various reaction pathways is depicted in Eqs. (1.14a–1.14c); the energy transfer route can furthermore be subdivided according to the specific mechanism of that process. As a result, although determining whether the excited state of the chromophore is reacting can be as straightforward as observing emission quenching, mechanistic discrimination as to the nature of that reaction generally requires considerably more work. The next sections will discuss these different processes and describe experiments that are typically employed in order to distinguish among them.

$$[Ru^{III}(bpy^-)(bpy)_2]^{2+*} \xrightarrow[k_{EnT}]{+Q} [Ru^{II}(bpy)_3]^{2+} + Q^* \quad (1.14a)$$

$$[Ru^{III}(bpy^-)(bpy)_2]^{2+*} \xrightarrow[k_{ox}]{+A} [Ru^{III}(bpy)_3]^{3+} + A^{\bullet -} \quad (1.14b)$$

$$[Ru^{III}(bpy^-)(bpy)_2]^{2+*} \xrightarrow[k_{red}]{+D} [Ru^{II}(bpy^-)(bpy)_2]^{+} + D^{\bullet +} \quad (1.14c)$$

1.5 Energy Transfer: Förster and Dexter Mechanisms

Energy transfer is a process by which excess energy contained in one molecule (the donor) is transferred to another molecule (the acceptor). In the context of the chemical systems being discussed herein, that excess energy comes from the absorption of a photon by the donor to create an electronic excited state. The product of the reaction is an electronically excited acceptor molecule concomitant with reformation of the electronic ground state of the donor, as shown in Eq. (1.15).

$$D^* + A \xrightarrow{k_{EnT}} D + A^* \quad (1.15)$$

Although energy transfer can occur as the result of emission from the donor and subsequent absorption of that emitted light by the acceptor (the so-called "trivial" mechanism), energy transfer more typically occurs via nonradiative processes (i.e., the emission and reabsorption of light do not occur). The two most common mechanisms of nonradiative energy transfer are known as Förster (through-space) and Dexter (through-bond or "exchange") energy transfer. These mechanisms are depicted in Scheme 1.5. It should be noted that both Förster and Dexter transfers yield the same products (i.e., ground-state donor and excited-state acceptor), although the physical origins of the reaction are fundamentally different [28].

Förster energy transfer [29] is a dipolar mechanism that takes place through space: the transition moment dipole of the donor couples nonradiatively with the transition moment dipole of the acceptor. Because of the dipolar nature of this mechanism, no orbital overlap is necessary between the donor and the acceptor. This makes Förster energy transfer operational at long distances that can range from 1 to 10 nm [30]. In the photosynthetic apparatus, the energy absorbed by the antenna complex is shuttled to the reaction center via Förster energy transfer [31].

An overlap between the emission spectrum of the donor and the electronic absorption spectrum of the acceptor is necessary for the energy transfer to occur: for this reason, Förster transfer is often referred to as fluorescence resonance energy transfer, or FRET. A schematic representation of this resonance condition (which in reality is simply a reflection of energy conservation for the energy transfer process) is shown in Scheme 1.6. The organic reactants usually involved

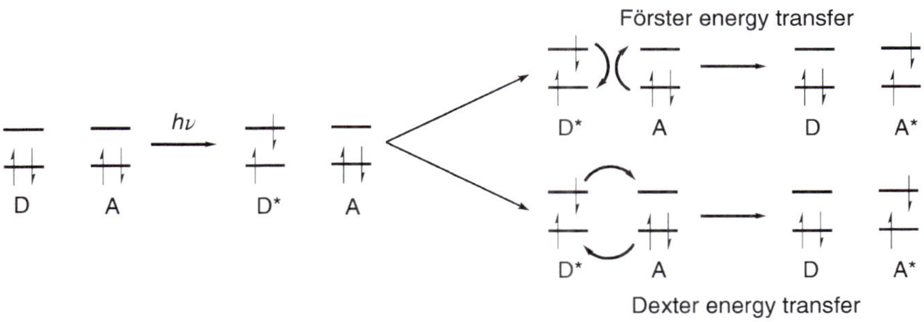

Scheme 1.5 Förster and Dexter energy transfer mechanisms.

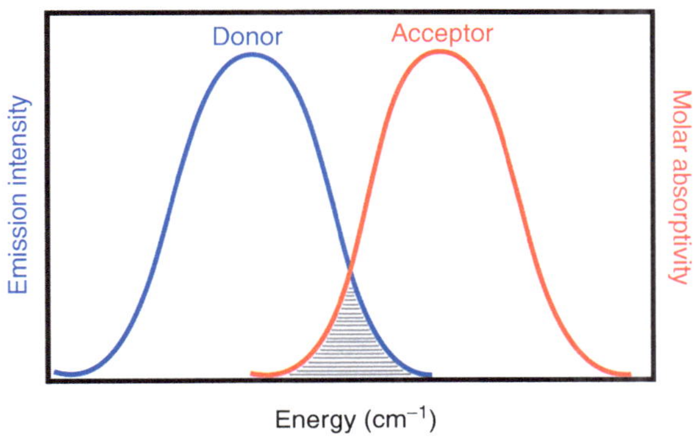

Scheme 1.6 Schematic emission spectrum of the donor and absorption spectrum of the acceptor. The shaded region is the spectral overlap.

in photocatalyzed reactions do not readily absorb light in the visible region of the spectrum, so the spectral overlap between their absorption spectrum and the emission spectrum of $[Ru(bpy)_3]^{2+}$ is usually negligible. As a consequence, Förster energy transfer is not a common reaction pathway for the systems that are discussed in this chapter.

The Dexter mechanism [32, 33], on the other hand, is best thought of as two concomitant electron transfer reactions (see Scheme 1.5). Except in rare cases, electron transfer is a through-bond process, meaning that Dexter transfer requires orbital overlap between the donor and the acceptor in order for the energy transfer process to proceed: this limits its occurrence to shorter distances than the Förster mechanism (typically no more than 10 Å). In other words, for a bimolecular reaction the Dexter process requires physical contact between the excited donor and the acceptor. On the plus side, since it is an exchange process (as opposed to a resonance one), no spectral overlap is required.

Molecular oxygen can quench the excited state of many transition-metal polypyridyl compounds via Dexter energy transfer [34, 35]. For this reason, most photophysical measurements involving $[Ru(bpy)_3]^{2+}$ and other transition-metal complexes must be carried out in deoxygenated solutions.

1.6 Electron Transfer

A generic electron transfer process is represented in Eq. (1.16). In a simple electron transfer reaction (the kind that we are interested in), no bonds are formed or broken; this is known as an outer sphere electron transfer.

$$D + A \xrightarrow{k_{eT}} D^{+} + A^{-} \tag{1.16}$$

The kinetics of outer sphere electron transfer can be described using Marcus theory [36]. In bimolecular systems (such as the ones of interest in organic synthesis), the distance between the donor and the acceptor (as well as their relative orientations) can vary, affecting the rate of electron transfer. For simplicity, we will consider the donor and the acceptor to be at a fixed distance and orientation. Under those conditions, the rate constant for outer sphere electron transfer can be written as shown in Eq. (1.17),

$$k_{eT} = \frac{2\pi}{\hbar}|H_{AB}|^2 \frac{1}{\sqrt{4\pi\lambda k_B T}} \exp\left[\frac{(-\Delta G^\circ + \lambda)^2}{4\lambda k_B T}\right] \tag{1.17}$$

where ΔG° is the driving force for electron transfer (which depends on the redox potentials of the donor and the acceptor), H_{AB} represents the electronic coupling between the donor and the acceptor, and λ is the reorganization energy. This latter term reflects energetics associated with the structural changes in going from reactants to products as well as the reorganization of the solvent molecules around them. The magnitude of the electronic coupling (H_{AB}) depends on the distance and orientation of the donor and the acceptor and therefore tends to be difficult to specify for bimolecular reactions in solution.

Even though electron transfer and Dexter energy transfer are closely related, two important differences should be noted. First, because two electrons are exchanged instead of one, Dexter energy transfer has a stronger distance dependence than electron transfer (typically e^{-2r} as opposed to e^{-r} for electron transfer) [33]. Second, since electron transfer leads to a new charge distribution, the reorganization energy (especially the solvent contribution) is much larger than that associated with Dexter energy transfer [37].

The product of a Dexter energy transfer differs from that of an electron transfer because no charge-separated species is formed. This turns out to be an extremely important distinction that helps differentiate these two reaction pathways, as will be discussed later.

1.7 Probing the Mechanism, Stage I: Stern–Volmer Quenching Studies

The simplest experiment that can be performed is a Stern–Volmer quenching study. With this experiment, it is possible to determine whether a bimolecular reaction is taking place. While this is extremely useful information, it is important to stress that this experiment does not provide any mechanistic information by itself. As will become apparent in the discussion to follow, both energy and electron transfer reactions involving the excited state of the chromophore will yield experimentally indistinguishable results from a Stern–Volmer quenching study. It is only through the application of additional experiments (most notably time-resolved absorption spectroscopy) that further insight into the nature of the reaction responsible for the quenching can be gleaned.

In Section 1.3, the radiative and nonradiative pathways for the excited state were described. When a species other than $[Ru(bpy)_3]^{2+}$ is present in solution, the possibility of one or more additional reactions, such as electron and/or energy transfer, is introduced. When this happens, the excited state is quenched (the ground state is recovered). In a very general way, when a quencher is present, we can write the reaction shown in Eq. (1.18).

$$[Ru^{II}(bpy)_3]^{2+} \xrightarrow{h\nu} [Ru^{III}(bpy^{-})(bpy)_2]^{2+*} \xrightarrow[k_q]{+Q} \text{products}$$
$$[Ru^{III}(bpy^{-})(bpy)_2]^{2+*} \xrightarrow{k_r} [Ru^{II}(bpy)_3]^{2+}; \quad [Ru^{III}(bpy^{-})(bpy)_2]^{2+*} \xrightarrow{k_{nr}} [Ru^{II}(bpy)_3]^{2+} \qquad (1.18)$$

In this scheme, the rate at which the excited state disappears is given by Eq. (1.19).

$$-\frac{d[ES]}{dt} = k_0\,[ES] + k_q\,[ES][Q] \qquad (1.19)$$

For bimolecular quenching to take place, $k_q\,[Q]$ must be larger than k_0. This condition is usually met if the excited-state lifetime is on the nanosecond to microsecond time scale. Otherwise, the excited state relaxes back to the ground state before it can diffuse to and react with the substrate (Q) [20]. The goal of Stern–Volmer studies is to determine whether the excited state reacts with the quencher. Quantifying k_q is most easily done by carrying out the study under pseudo first-order conditions: the concentration of the quencher must be at least two orders of magnitude larger than that of the photocatalyst,[6] so that $[Q]$ can be assumed to be constant throughout the experiment. This collapses Eq. (1.19) to Eq. (1.20) and allows for the determination of k_q (Eq. (1.21)). The observed rate constant (k_{obs}) varies with the concentration of the quencher.

$$-\frac{d[ES]}{dt} = (k_0 + k_q[Q])[ES] = k_{obs}\,[ES] \qquad (1.20)$$

$$k_{obs} = (k_0 + k_q[Q]) \qquad (1.21)$$

k_{obs} may be directly determined using time-resolved spectroscopy. If the sensitizer is emissive, this is most easily done via time-resolved emission spectroscopy: by measuring the decay rate constant at several quencher concentrations, the quenching constant k_q can be found when fitting the results to Eq. (1.22).

$$\frac{k_{obs}}{k_0} = \frac{k_0 + k_q[Q]}{k_0} = 1 + \frac{k_q[Q]}{k_0} \qquad (1.22)$$

6 Strictly, it must be $[Q] \gg [ES]$, but since evaluating the concentration of the excited state is not trivial, it is simpler to make $[Q] \gg$ [photocatalyst].

Alternatively, steady-state emission spectroscopy can be employed. In the absence of contributions from static quenching [38], the radiative quantum yield of the photocatalyst in the presence of a quencher depends on k_q, as shown in Eq. (1.23).

$$\Phi_q = \frac{k_r}{k_0 + k_q[Q]} \tag{1.23}$$

Provided that the radiative quantum yields in the presence and absence of the quencher are measured under identical experimental conditions, their ratio can be related to the Stern–Volmer expression (Eq. (1.24)).

$$\frac{\Phi_0}{\Phi_q} = \frac{I_0}{I_q} = \frac{k_0 + k_q[Q]}{k_0} = 1 + \frac{k_q[Q]}{k_0} \tag{1.24}$$

Assuming that the rate constant for excited-state decay of the chromophore (k_0) is known, k_q can be determined by measuring the radiative quantum yield as a function of quencher concentration.

Stern–Volmer studies are helpful because the excited-state lifetime is shortened if a reaction between the photocatalyst and a quencher takes place. However, the only information these studies can provide is whether or not the excited state is being quenched; ***they do not in any way provide mechanistic insights because energy and electron transfer quenching pathways will yield qualitatively indistinguishable results for this experiment.***

1.8 Probing the Mechanism, Stage II: Electron Versus Energy Transfer

The discussion above underscored the extremely important point that a Stern–Volmer study does not provide any insight into the actual reaction the excited state of a sensitizer is engaging in. A simple analogy can be drawn with, for example, a Schiff base condensation. If one used NMR to probe this reaction, the disappearance of the aldehyde proton resonance would never be used as proof that the imine had formed, only that a reaction involving the aldehyde had taken place. In the same way, the observation of quenching of emission from the sensitizer from a Stern–Volmer quenching study is nothing more than evidence that the starting material (i.e., the excited state) is being consumed. In order to determine what reaction actually occurred, one must identify the product(s) of the reaction.

As mentioned previously, the two dominant excited-state reaction pathways available in most systems are electron and energy transfer from the excited state to the substrate; in the case of the former, oxidative and reductive quenching are both possible, with each leading to distinctly different products. In the case of energy transfer, the photocatalyst will go back to the ground state (see Eq. (1.14a)), whereas electron transfer will result in the oxidation or reduction of the photocatalyst (with corresponding reduction or oxidation of the substrate, Eqs. (1.14b and 1.14c). Direct detection of one (or more) of these products is the gold standard by which mechanistic pathways in these reactions must be established.

Time-resolved absorption spectroscopy, also known as transient absorption (TA), is a very useful tool in these cases. This technique uses a laser pulse to excite the sample and a white light source to probe the absorption of the transient species formed due to excitation, using the absorption of the ground state as the blank. The TA signal is the change in absorbance of the sample before and after excitation. This renders the technique more versatile than time-resolved emission, because non-emissive molecules can be studied as well. Depending on the instrumentation available, difference spectra can be acquired at single wavelengths (yielding kinetic traces, see Figure 1.5) or a full spectrum can be obtained.

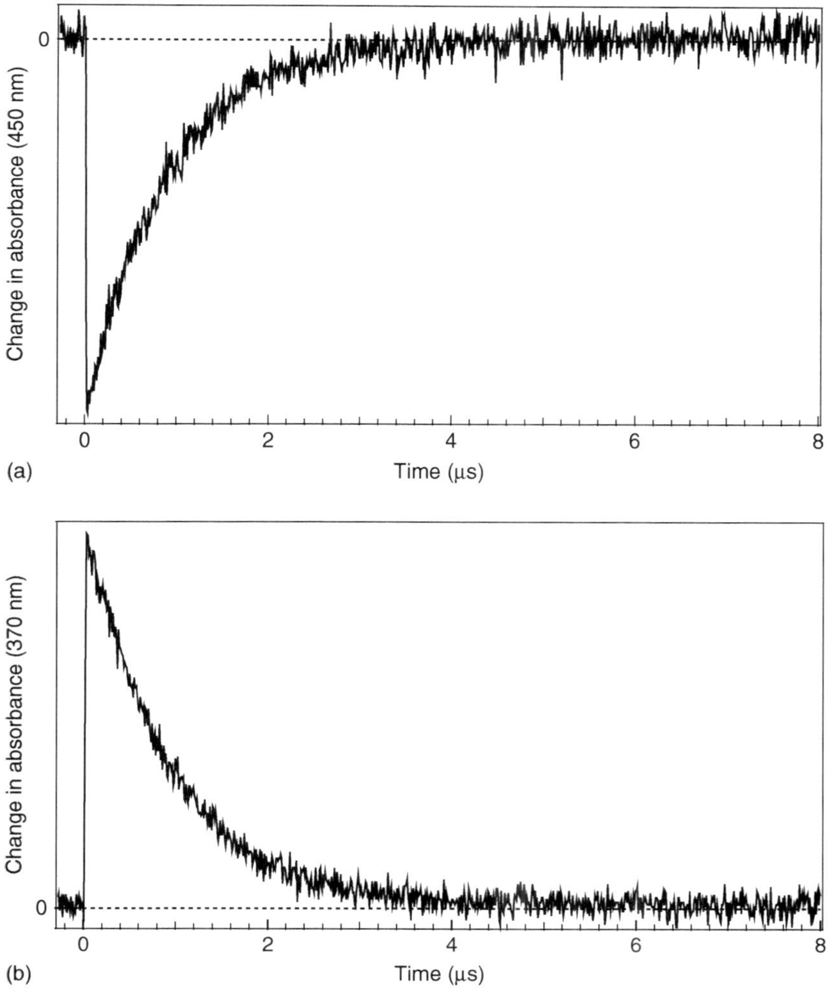

Figure 1.5 Kinetic traces for $[Ru(bpy)_3]^{2+}$ in acetonitrile; $\lambda_{pump} = 475$ nm. (a) $\lambda_{probe} = 450$ nm; the bleach is due to the presence of Ru^{III}. (b) $\lambda_{probe} = 370$ nm; this positive feature arises from the reduced ligand. Both traces go back to zero with the same time constant.

For a TA experiment, an expression derived from Beer's law can be written, as shown in Eq. (1.25).

$$\Delta A = \Delta\varepsilon \cdot b \cdot [\text{GS}] \cdot \eta^{\text{ex}} \tag{1.25}$$

where ΔA is the change in absorbance before and after excitation (i.e., excited state minus ground state), $\Delta\varepsilon$ is the change in molar absorptivity (the difference between the ground state and the excited state), b is the optical path length, [GS] is the concentration of the ground state (the concentration of the sample), and η^{ex} is the fraction of molecules that are excited from the ground state to the excited state ($0 < \eta^{\text{ex}} < 1$). For a given experiment, b and [GS] are constant. η^{ex} depends, among other factors, on the cross section between the pump and probe beams, but remains constant as long as the experimental conditions are not changed. When that is the case, any changes in the sign of ΔA are a direct reflection of the changes in $\Delta\varepsilon$.

Scheme 1.7 is a cartoon of a full spectrum obtained in a TA experiment. If at a certain wavelength the excited state absorbs more strongly than the ground state, a positive feature is observed. Conversely, if the ground state absorbs more than the excited state, a negative feature ("bleach") is obtained. The points where the excited and ground states have the same absorbance are called isosbestic points.

For $[\text{Ru(bpy)}_3]^{2+}$, it is known that the main feature for the oxidized species is a bleach centered around 450 nm; [39] this signal is diagnostic of the presence of Ru(III). In the case of $[\text{Ru(bpy)}_3]^{+}$, the absorption centered around 370 nm indicates the presence of a bpy radical anion. The kinetic traces at 450 and 370 nm are shown in Figure 1.5: as the excited molecules relax back to the ground state, the features of $[\text{Ru(bpy)}_3]^{2+}$ are recovered and the kinetic trace goes back to zero.

Now let us consider what happens to the TA traces upon adding a quencher. To illustrate the different scenarios, several simulated TA traces are shown in Figure 1.6. For the unquenched photocatalyst, a lifetime of 700 ns was used. To make comparisons easier, a lifetime of 300 ns was assumed for the quenched photocatalyst, regardless of the reaction taking place. Irrespective of the type of quenching, $[\text{Ru}^{\text{III}}(\text{bpy}^-)(\text{bpy})_2]^{2+*}$ is formed after excitation, leading to a positive feature at 370 nm (due to bpy^-) and a bleach at 450 nm (diagnostic of Ru^{III}).

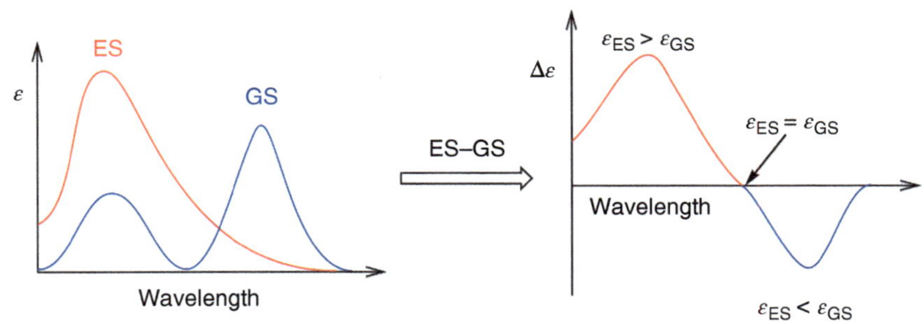

Scheme 1.7 (b) Schematic representation of a transient absorption plot. The positive feature is shown in red, while the bleach is in blue. (a) Schematic absorption spectra of the ground and excited states.

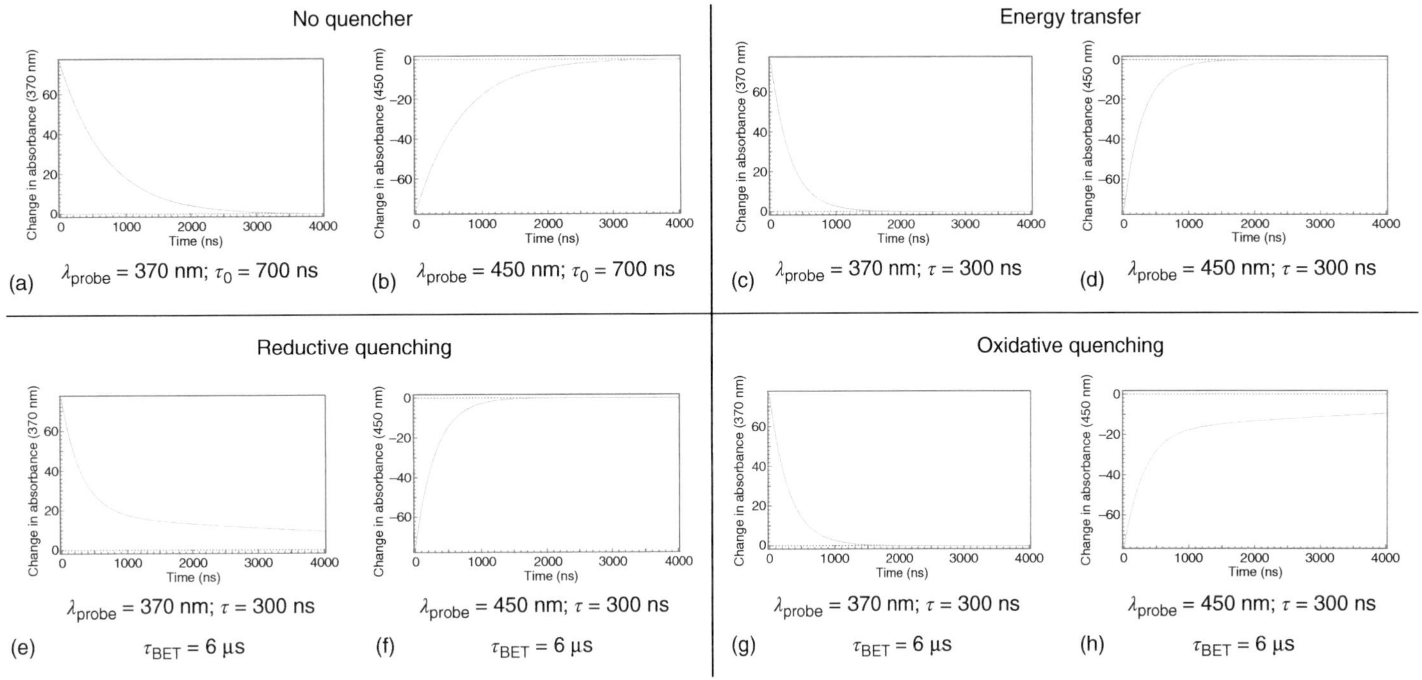

Figure 1.6 Simulated TA traces for $[Ru(bpy)_3]^{2+}$ with no quencher (a, b), in the presence of an energy transfer acceptor (c, d); in the presence of an electron donor (e, f); in the presence of an electron acceptor (g, h).

In the presence of an energy acceptor the product of the quenching reaction is $[Ru(bpy)_3]^{2+}$, the same as that before excitation (see Eq. (1.14a)), so both the Ru^{III} and the bpy^- signals are lost at the same time, with an observed rate constant k_{obs} that is larger than k_0 (see Eqs. (1.20)and (1.21)). As the excited photocatalyst goes back to the ground state, the kinetic trace goes back to zero as a single exponential, regardless of the probe wavelength (370 or 450 nm). To reiterate, the important diagnostic for this reaction pathway is the simultaneous, kinetically indistinguishable loss of both the bpy^- and Ru^{III} species. This occurs because both of these components comprise the reactive excited state, and therefore both are lost in an energy transfer process that returns the system to the ground state.

For an electron transfer, the products of the quenching reaction are chemically distinct from $[Ru(bpy)_3]^{2+}$ (see Eqs. (1.14b) and (1.14c)). In the case of a reductive quenching, the excited photocatalyst is reduced to $[Ru(bpy^-)(bpy)_2]^+$ owing to the oxidation of the substrate. This has two consequences: (i) persistence of the absorption feature at 370 nm, concomitant with (ii) a partial recovery of the bleach at 450 nm. The recovery of the bleach signal is only partial because, although reduction converts the Ru^{III} species present in the excited state to Ru^{II}, the original intensity of the ground-state MLCT absorption has three contributions (i.e., MLCT transitions to each of the three bpy ligands): the product of reductive quenching only recovers ⅔ of this intensity because of the presence of bpy^-. This is illustrated in Figure 1.6. Oxidative quenching, on the other hand, results in the formation of $[Ru^{III}(bpy)_3]^{3+}$. This will result in the mirror image of the observables just described for reductive quenching wherein the bleach persists concomitant with the loss of the bpy^- signal at 370 nm. The key qualitative differences between an electron and an energy transfer quenching process, then, lie in the wavelength dependence of the observed kinetics: for energy transfer one observes wavelength-independent kinetics, whereas electron transfer results in qualitatively different kinetic traces depending on probe wavelength and the nature (i.e., oxidative or reductive) of the reaction.

We have focused on the spectroscopic signatures of the excited state of $[Ru(bpy)_3]^{2+}$ for this discussion because its reduced and oxidized forms have quite different electronic absorption spectra. This does not exclude the possibility of monitoring one of the substrates via TA spectroscopy, provided that its reduced and oxidized forms absorb light at different wavelengths so that their spectra can be clearly distinguished.

1.9 Designing Photocatalysts: $[Ru(bpy)_3]^{2+}$ as a Starting Point

As it was mentioned in the introduction, there are some desirable characteristics for a photocatalyst:

1) Strong absorption of light over a wide spectral region
2) Stability in solution
3) Excited-state lifetime longer than 10^{-9} s

4) Reversible redox behavior (with certain values for ground- and excited-state redox potentials)
5) Ease of synthesis and tunability of ground- and excited-state properties

We have used $[Ru(bpy)_3]^{2+}$ as a convenient example to discuss the relevant properties of a photocatalyst as well as the processes it may be involved in. Scheme 1.8 illustrates both types of catalytic cycles that $[Ru(bpy)_3]^{2+}$ can take part in, where D and A represent a generic electron donor and acceptor, respectively. Two steps in these cycles are redox reactions, so the redox potentials of the photocatalyst must be such that each reaction is favorable.

Using the reductive quenching cycle as an example, two reactions have to be favorable for the cycle to proceed, as shown in Eqs. (1.26) and (1.27).

$$[Ru^{III}(bpy^{\bullet -})(bpy)_2]^{2+*} + D \rightarrow [Ru^{II}(bpy^{\bullet -})(bpy)_2]^{+} + D^{+} \tag{1.26}$$

$$[Ru^{II}(bpy^{\bullet -})(bpy)_2]^{+} + A \rightarrow [Ru^{II}(bpy)_3]^{2+} + A^{-} \tag{1.27}$$

For (Eq. (1.26)) to be a favorable reaction, $\Delta E_{M/D}$ (defined in Eq. (1.28)) must be positive; analogously, for (Eq. (1.27)) to be spontaneous, $\Delta E_{A/L}$ (Eq. (1.29)) must be positive. The relevant potentials for the photocatalyst are defined in Scheme 1.9.

$$\Delta E_{M/D} = E(M^*/M^+) - E(D^+/D) \tag{1.28}$$

$$\Delta E_{A/L} = E(A/A^-) - E(L/L^-) \tag{1.29}$$

It is easy to see that the identities of A/A^- and D/D^+ (and therefore their redox potentials) determine which compounds can act as photocatalysts for a given reaction. The redox potentials of the A/A^- and D/D^+ couples can be determined using electrochemistry. If either the donor or the acceptor is formed in situ during the reaction, that redox potential will be harder (if at all possible) to determine.

It was mentioned before that the redox activity of $[Ru(bpy)_3]^{2+}$ is enhanced in the excited state; that is the reason why this compound can be used as a photocatalyst. It is the extra energy of the excited state that makes both its oxidation and

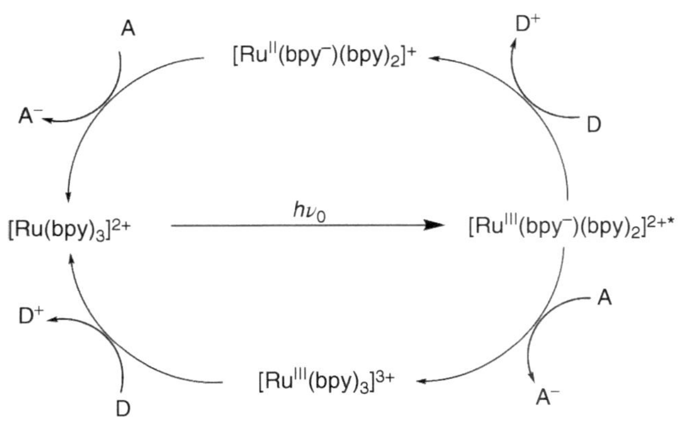

Scheme 1.8 Generic catalytic cycles via reductive quenching (top half) and oxidative quenching (bottom half).

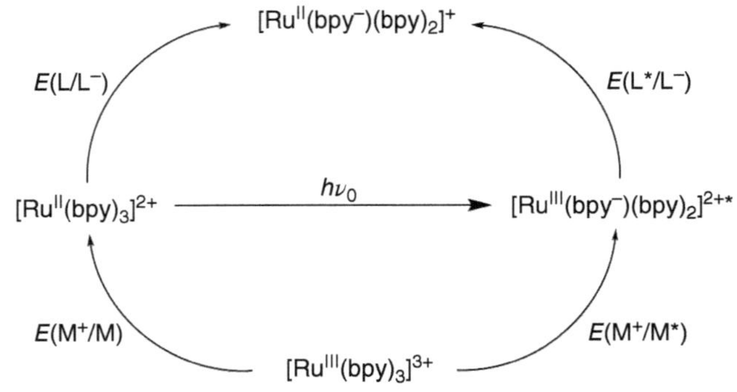

Scheme 1.9 Thermodynamic cycle relating the excited- and ground-state potentials.

reduction more favorable than those of the ground state. Whether a compound is a suitable photocatalyst for a given reaction depends on the redox potentials of both the ground and excited states. The latter potentials cannot be directly measured, but can be calculated using the redox potentials for the ground state and the energy of the excited state. The relationship between these quantities is presented in Scheme 1.9. Assuming that all the excited-state energy is available as free energy (i.e., the entropic contribution is neglected) [40], the excited state redox potentials can be calculated using Eqs. (1.30) and (1.31) [27].

$$E(M^+/M^*) = E(M^+/M) - E_0 \tag{1.30}$$

$$E(L^*/L^-) = E(L/L^-) + E_0 \tag{1.31}$$

We have mentioned that an advantage of transition-metal polypyridyl complexes is that their properties can be modified by changing the ligands coordinated to the metal center or the metal itself. As a general rule, electron-withdrawing substituents on the ligands make them easier to reduce (thus making the compound a better oxidant) and the metal harder to oxidize; conversely, electron-donating substituents make the compound a better reductant. In the case of the Ir(III) compounds, the metal is less electron rich than Ru(II), which makes it harder to oxidize [15, 41, 42] and also makes the bpy ligand harder to reduce. In the case of the cyclometalated compounds (those containing 2-phenylpyridine (ppy)), the reduction of ppy is much less favorable than that of bipyridine, because ppy is formally an anionic ligand. These anionic ligands also make the metal easier to oxidize [15].

Modifying the ligands or the metal affects not only the redox potentials of the compound but also the electronic absorption and emission spectra (which means that the properties of the excited state change as well). However, changes in the ligands (such as the presence of substituents) tend to have a greater impact on the electrochemical properties of the compound than on E_0 [17, 40].

1.10 Conclusion

The best photocatalyst for a given organic reaction is determined by factors such as the solvent used, the redox potentials of the reactants, their electronic absorption spectra, and the type of mechanism of the reaction. Being able to study all these factors is crucial to find the right photocatalyst for the transformation of interest.

Transition-metal polypyridyl compounds make a convenient choice for photocatalysts, partly because of the tunability of their properties. The purpose of this chapter has been to serve as a guide to better exploit that tunability. To this end, we have discussed the different kinds of reactions that the excited photocatalysts can engage in, and the necessary experiments to study those processes.

The use of photoredox catalysts in organic chemistry is a relatively young field that has already yielded very exciting results. We hope this chapter will help guide the choice of the right photocatalyst for a given reaction, thus helping advance this promising field.

References

1 Burstall, F.H. (1936) *J. Chem. Soc.*, 173.
2 Paris, J.P. and Brandt, W.W. (1959) *J. Am. Chem. Soc.*, **81**, 5001–5002.
3 Juris, A., Balzani, V., Barigelletti, F. *et al.* (1988) *Coord. Chem. Rev.*, **84**, 85–277.
4 Polo, A.S., Itokazu, M.K., and Murakami Iha, N.Y. (2004) *Coord. Chem. Rev.*, **248**, 1343–1361.
5 Hedstrand, D.M., Kruizinga, W.H., and Kellogg, R.M. (1978) *Tetrahedron Lett.*, **19**, 1255–1258.
6 Nicewicz, D.A. and MacMillan, D.W.C. (2008) *Science*, **322**, 77–80.
7 Ischay, M.A., Anzovino, M.E., Du, J., and Yoon, T.P. (2008) *J. Am. Chem. Soc.*, **130**, 12886–12887.
8 Prier, C.K., Rankic, D.A., and MacMillan, D.W.C. (2013) *Chem. Rev.*, **113**, 5322–5363.
9 Maity, S., Zhu, M., Shinabery, R.S., and Zheng, N. (2012) *Angew. Chem., Int. Ed.*, **51**, 222–226.
10 Cano-Yelo, H. and Deronzier, A. (1987) *J. Photochem.*, **37**, 315–321.
11 Rehm, D. and Weller, A. (1970) *Isr. J. Chem.*, **8**, 259–271.
12 Balzani, V., Bolletta, F., and Scandola, F. (1980) *J. Am. Chem. Soc.*, **102**, 2152–2163.
13 Clark, C.D. and Hoffman, M.Z. (1997) *Coord. Chem. Rev.*, **159**, 359–373.
14 Sutin, N. and Creutz, C. (1980) *Pure Appl. Chem.*, **52**, 2717–2738.
15 Flamigni, L., Barbieri, A., Sabatini, C. *et al.* (2007) *Top. Curr. Chem.*, **281**, 143–203.
16 Evans, I.P., Spencer, A., and Wilkinson, G. (1973) *J. Chem. Soc. Daltan Trans.*, 204–209.
17 Lin, C.T., Bottcher, W., Chou, M. *et al.* (1976) *J. Am. Chem. Soc.*, **98**, 6536–6544.
18 Lytle, F.E. and Hercules, D.M. (1969) *J. Am. Chem. Soc.*, **91**, 253–257.
19 Drago, R.S. (1992) *Physical Methods for Chemists*, Philadelphia, Saunders College Pub.
20 Sutin, N. and Creutz, C. (1978) *Adv. Chem. Ser.*, **168**, 1–27.

21 Vlcek, A.A., Dodsworth, E.S., Pietro, W.J., and Lever, A.B.P. (1995) *Inorg. Chem.*, **34**, 1906–1913.
22 Damrauer, N.H., Cerullo, G., Yeh, A. *et al.* (1997) *Science*, **275**, 54–57.
23 Crosby, G.A. and Demas, J.N. (1971) *J. Phys. Chem.*, **75**, 991–1024.
24 Suzuki, K., Kobayashi, A., Kaneko, S. *et al.* (2009) *Phys. Chem. Chem. Phys.*, **11**, 9850–9860.
25 Lakowicz, J.R. (2006) *Principles of Fluorescence Spectroscopy*, 3rd edn, Springer US, Boston.
26 Resch-Genger, U. and Rurack, K. (2013) *Pure Appl. Chem.*, **85**, 2005–2013.
27 Creutz, C. and Sutin, N. (1976) *Inorg. Chem.*, **15**, 496–499.
28 Arnaut, L., Formosinho, S., and Burrows, H. (2007) *Chemical Kinetics. From Molecular Structure to Chemical Reactivity*, Elsevier, Amsterdam.
29 Forster, T. (1959) *Discuss. Faraday Soc.*, **27**, 7–17.
30 Sahoo, H.J. (2011) *Photochem. Photobiol. C Photochem. Rev.*, **12**, 20–30.
31 Pullerits, T. and Sundström, V. (1996) *Acc. Chem. Res.*, **29**, 381–389.
32 Dexter, D.L.J. (1953) *Chem. Phys.*, **21**, 836–850.
33 Scandola, F., Indelli, M.T., Chiorboli, C., and Bignozzi, C.A. (1990) *Top. Curr. Chem.*, **158**, 73–149.
34 Brunschwig, B. and Sutin, N. (1978) *J. Am. Chem. Soc.*, **100**, 7568–7577.
35 Demas, J.N., Harris, E.W., and McBride, R.P. (1977) *J. Am. Chem. Soc.*, **99**, 3547–3551.
36 Marcus, R.A. (1993) *Angew. Chem., Int. Ed. Engl.*, **32**, 1111–1121.
37 Scandola, F. and Balzani, V. (1983) *J. Chem. Educ.*, **60**, 814–823.
38 Arias-Rotondo, D.M. and McCusker, J.K. (2016) *Chem. Soc. Rev.*, **45**, 5803–5820.
39 Damrauer, N.H. and McCusker, J.K. (1999) *J. Phys. Chem. A*, **103**, 8440–8446.
40 Balzani, V., Bolletta, F., Scandola, F., and Ballardini, R. (1979) *Pure Appl. Chem.*, **51**, 299–311.
41 Balzani, V., Bolletta, F., Gandolfi, M., and Maestri, M. (1978) *Top. Curr. Chem.*, **75**, 1–64.
42 Ladouceur, S., Fortin, D., and Zysman-Colman, E. (2011) *Inorg. Chem.*, **50**, 11514–11526.

2

Visible-Light-Mediated Free Radical Synthesis

Louis Fensterbank, Jean-Philippe Goddard, and Cyril Ollivier

Institut Parisien de Chimie Moléculaire (UMR 8232 CNRS), Université Pierre & Marie Curie – Sorbonne Universités, 4, Place Jussieu, 75252 Paris, France

2.1 Introduction

Radical chemistry has witnessed tremendous developments over the last three decades [1, 2]. The main pros of radical reactions are their mildness and high compatibility with functional groups. In contrast with ionic processes, the use of protecting groups is usually not necessary. A lot of radical reactions take place in salt-free conditions so that the influence of aggregation states and solvation is negligible. Moreover, owing to their early transition states, radical processes are ideal for the construction of densely functionalized centers, even quaternary ones, which is particularly suited for the synthesis of natural products. The vast amount of reliable kinetic data and number of stereoselectivity models [1–3] now render quite versatile the implementation of radical processes into a retrosynthetic analysis. Finally, the palette of radical transformations is wide and not limited to cyclizations: intermolecular additions, homolytic substitutions, and C—H functionalization *via* hydrogen transfers have equally gained wide acceptance. However, radical reactions also have suffered from a few cons that have thwarted their full recognition in mainstream chemistry. First, scale-up is often difficult because most of the reactions are run in dilute media. But, the most troublesome issue is related to the mediator generally used. The quite common tin (IV) derivatives are toxic and difficult to eliminate and alternative systems that have been proposed have so far not found a comparable utilization.

An interesting track is redox chemistry. The simple idea that a radical can be generated by single electron transfer (SET) from an anion by oxidation, or a cation or cation-like species by reduction is of course very attractive. Nevertheless, a lot of these processes rely on the stoichiometric use of metals that pose sustainability issues. Thus, there is a need to go further. The development of catalytic processes and more precisely photocatalytic redox processes using visible light certainly appears quite appealing. For that purpose, a lot of works have relied on the use of the $Ru(bpy)_3^{2+}$ complex, which is introduced in catalytic quantities. Well known for decades in the inorganic chemistry community as a photoredox catalyst for diverse applications such as water splitting and carbon dioxide reduction,

Visible Light Photocatalysis in Organic Chemistry, First Edition.
Edited by Corey R. J. Stephenson, Tehshik P. Yoon and David W. C. MacMillan.

it has witnessed only sporadic uses in organic synthesis until recently in 2008 when MacMillan used it to merge organocatalysis with photoredox catalysis and provide an asymmetric α-alkylation of aldehydes [4]. Since this renaissance, an impressive number of research groups have embarked on this topic, providing new ways to conduct ancestral radical organic chemistry. In this chapter, we will concentrate on the photocatalytic generation of radical intermediates that have already been encountered in the past and generated through other pathways. Their utilization will be addressed. We will particularly emphasize how photocatalysis can really improve the reaction conditions and outcomes.

2.2 Basics of the Photocatalytic Cycle

Visible-light photoredox catalysis requires the participation of a suitable photocatalyst M that absorbs light in the visible range. It can be an organic dye such as eosin Y or a polypyridine complex of ruthenium(II), iridium(IV), or copper(I) with different redox potentials that can be selected according to the substrate to reduce or oxidize (see Chapter 1). If we consider a photoredox-catalyzed transformation of a substrate R, there are two scenarios that have a common part, namely, the transition of the complex M from the ground state to its excited state by irradiation at a wavelength λ_{max} to form M*. In the first case, if a sacrificial electron donor is present in the reaction medium and is reductive enough (Scheme 2.1, green path), it can act as a reductive quencher (Qred) of M* and generate M^-. The reduced complex can then transfer an electron to the substrate R and regenerate the catalyst from M in a so-called "reductive quenching cycle." The reduction of the substrate R provides an anion radical $R^{•-}$, which can undergo subsequent transformations. Similarly, if the photoexcited complex reacts with a sacrificial electron acceptor acting as an oxidative quencher (Qox), it is oxidized to M^+ and can be reduced by a substrate R to regenerate the starting catalytic species (oxidative quenching cycle) and liberate a radical cation $R^{•+}$, which can react with a radical trap (Scheme 2.1, orange path). In the second case, the substrate R is also the quencher. If it acts as Qred, it can transfer an electron to the excited-state photocatalyst M* to form M^-. The oxidized substrate $R^{•+}$ can then participate in radical transformations. To regenerate the catalyst, a sacrificial electron acceptor A must be introduced into the medium (reductive

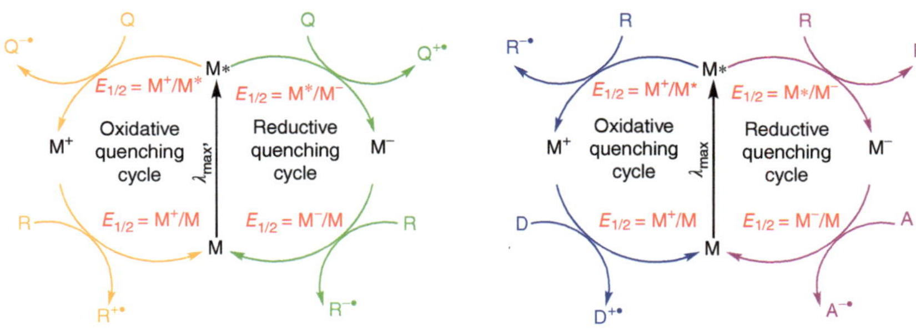

Scheme 2.1 Oxidative quenching cycle versus reductive quenching cycle.

quenching cycle) (Scheme 2.1, violet path). The opposite can be achieved. If the complex transfers an electron to the substrate (or oxidative quencher (Qox)), it is oxidized to M^+. The reduced substrate $R^{\bullet -}$ undergoes radical transformations and the presence of a sacrificial electron donor D allows the regeneration of the photocatalyst M (oxidative quenching cycle) (Scheme 2.1, blue path). Many contributions in this area have been reported in the literature for photoreduction and photooxidation of various functions and very interesting monographs and reviews have been published [5–17].

2.3 Generation of Radicals

This section will focus on the photocatalytic generation of radical intermediates from prefunctionalized substrates bearing a homolytically cleavable C—X bond followed by H-abstraction. The later functionalization of the radical intermediates will also be covered in some cases.

2.3.1 Formation of C-Centered Radicals

2.3.1.1 Dehalogenation (I, Br, Cl)

Traditional radical chemistry has heavily relied on the use of prefunctionalized substrates that bear an easily scissible bond in radical reaction conditions. Treatment of halogenated substrates with various mediators has formed a great part of this approach. It is thus not surprising that efforts have been made to study this reactivity under photocatalytic conditions.

Activated halides such as malonyl, α-carbonyl substrates have been engaged successfully. Pioneering works on this type of transformation were reported by Fukuzumi who described the reduction of phenacyl (*para*-$YC_6H_4COCH_2X$, X = Br or Cl; Y = CN, Br, Me, MeO) by 9,10-dihydro-10-methylacridine with $[Ru(bpy)_3]^{2+}$ as the photocatalyst [18]. It has to be noted that for these keto precursors there is an ambiguity about the exact mechanism of the dehalogenation step. Electron capture could indeed take place in the C—X antibonding orbital followed by fragmentation of the corresponding radical anion. Alternatively, formation of a ketyl radical intermediate followed by β-elimination of the halide radical could also provide the same final product.

More recently, Stephenson focused initially on the reductive debromination of bromopyrroloindoline [19]. The best set of conditions relied on the use of 2.5 mol% $Ru(bpy)_3Cl_2 \cdot H_2O$, 10 equiv of Hünig base and formic acid in DMF at room temperature, or 2.5 mol% $Ru(bpy)_3Cl_2 \cdot H_2O$, 2 equiv Hünig base and 1.1 equiv of Hantzch ester. The latter conditions proved to be compatible with vinyl iodide as well as bromo- and iodoaryl functions, which constitute a nice example of chemoselectivity. At this stage, it was proposed that the radical cation of the Hünig base is the main source of hydrogen for the homolytic reduction of the intermediate radical. The same group also examined the reduction of bromo diethylmalonate [20]. Other activated substrates such as perhalogenated derivatives [21, 22], benzyl bromides [23], cyclopropyl bromides [24], and glycosyl halides [25] have also been studied.

Zeitler reexamined the dehalogenation of phenacyl derivatives and activated α-carbonyl substrates with a series of dyes [26]. Eosin Y was selected as the most reliable photocatalyst for this series. It is also interesting that the reported protocol proved to be compatible with an aryliodide function. However, on using the highly reductive *fac*-Ir(ppy)$_3$, efficient generation of radical intermediates from alkyl, alkenyl, and aryl iodides could be observed as shown by Stephenson, which pushed the frontiers of this reaction [27].

Some of these representative findings are summarized in Scheme 2.2.

Quite recently, König has brought in the concept of consecutive PET (conPET), which relies on the use of perylene bisimide dye as the photocatalyst. In the presence of Et_3N as Qred, the first visible light excitation provides the radical anion of the perylene bisimide. The resulting radical anion is sufficiently stable to undergo a second excitation to reach an excited state possessing sufficient energy to trigger

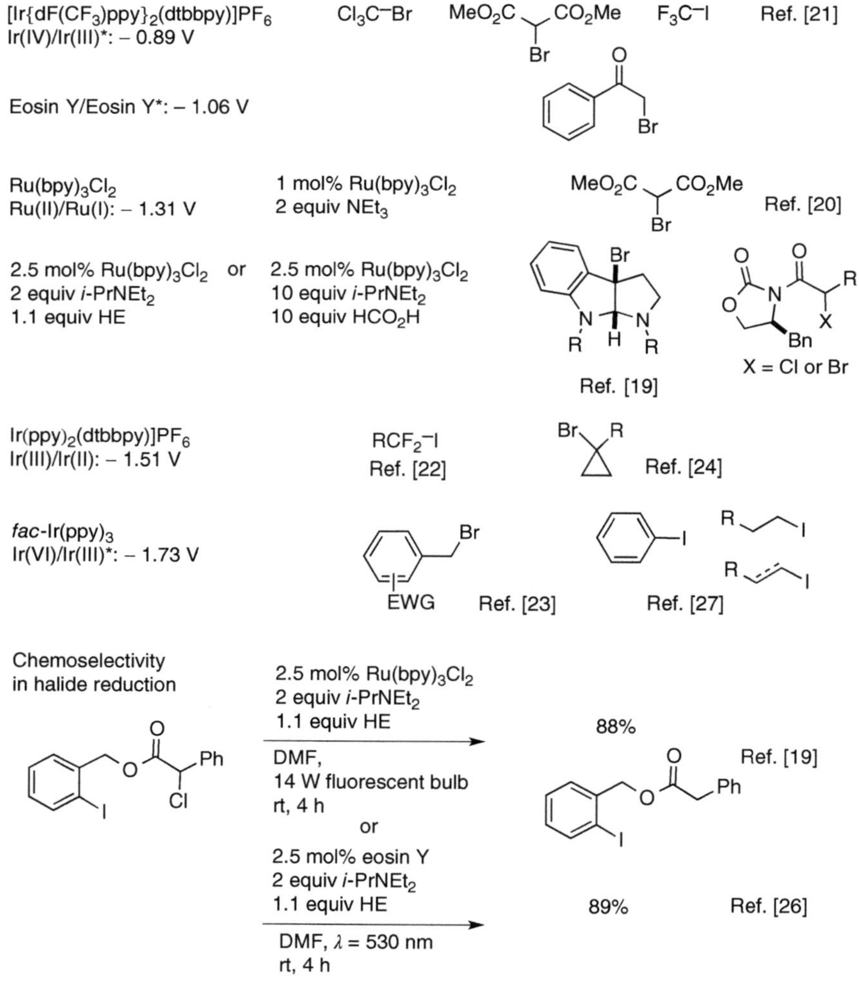

Scheme 2.2 Representative photocatalyzed dehalogenative processes.

the reductive dehalogenation of arylchlorides [28]. Finally, vicinal dibromides can be transformed into alkenes [29]. For a lot of these reactions, the implementation of flow processes has resulted in greatly optimized processes [30, 31]. All these dehalogenative processes have been applied in intermolecular and cyclization processes as shown below.

2.3.1.2 Other C-Heteroatom Cleavage

Fluoride is rarely thought of as a potential leaving group. Nevertheless, Weaver has recently reported a photocatalytic hydrodefluorination of polyfluorinated (mainly pentafluoro) aromatics using $Ir(ppy)_3$ as photocatalyst (Scheme 2.3). Generation of a radical aryl anion followed by release of a fluoride anion is invoked. Regioselectivity of the process is controlled by the electronics of the aryl system [32].

The cleavage of carbon–sulfur bonds has also been examined. Pioneering works involved α-ketosulfones [33] and α-ketosulfoniums [34]. The photocatalytic reduction of sulfoniums has recently been applied to C—C bond formation (see below).

More recently, the relatively strong C—O bond has been cleaved using adequate settings. Epoxides are versatile targets, but a series of works have shown that α-activation by a carbonyl group is required [35]. Using Hantzsch ester as the Qred and presumably a H donor, good yields of aldol products or β-amino keto derivatives could be obtained. Interestingly, the generated radical intermediate from keto epoxides or keto aziridines can be used for further diastereoselective C—C bond formation [36].

The Barton–McCombie deoxygenation reaction now fully belongs to mainstream chemistry and has found a lot of applications in total synthesis. However, this reaction is typically operated with stoichiometric amounts of tin hydride at warm temperatures. Finding a catalytic system resulting in tin-free and mild conditions is quite appealing. This was achieved by using imidazole-based thiocarbamates as adequate precursors and $Ir(ppy)_3$ as catalyst in the presence of Hünig base (Scheme 2.4). Stern–Volmer studies established that the thiocarbamate was the Qox of the excited state of $Ir(ppy)_3$.

2.3.1.3 C—C Bond Cleavage

Finally, C—C bonds can also be cleaved through reductive decarboxylation. Okada, Oda, and coworkers notably showed that *N*-(acyloxy)phthalimides were adequate precursors of alkyl radicals under $Ru(bpy)_3Cl_2$ visible-light photocatalysis in the presence of 1-benzyl-1,4-dihydronicotamide as sacrificial donor and *t*BuSH as radical scavenger [37]. The latter could be engaged in conjugate addition to activated olefins [38].

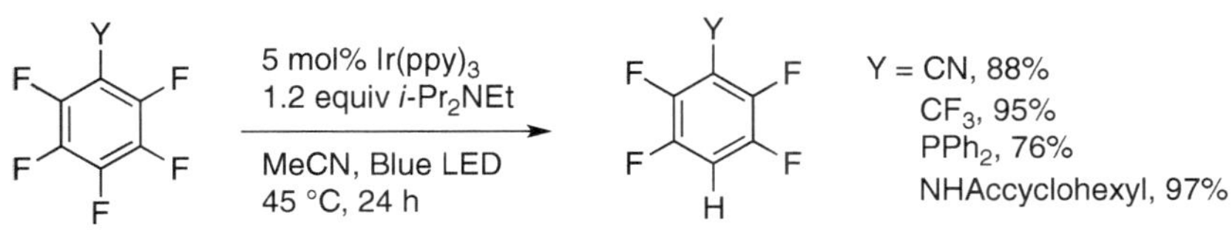

Scheme 2.3 Photocatalyzed selective hydrodefluorination.

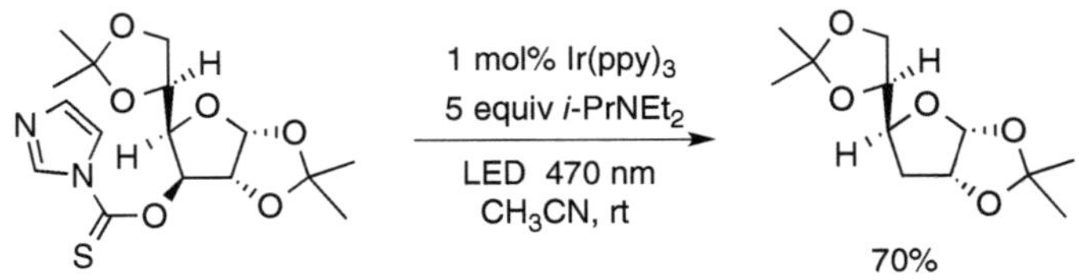

Scheme 2.4 Photocatalytic Barton–McCombie deoxygenation.

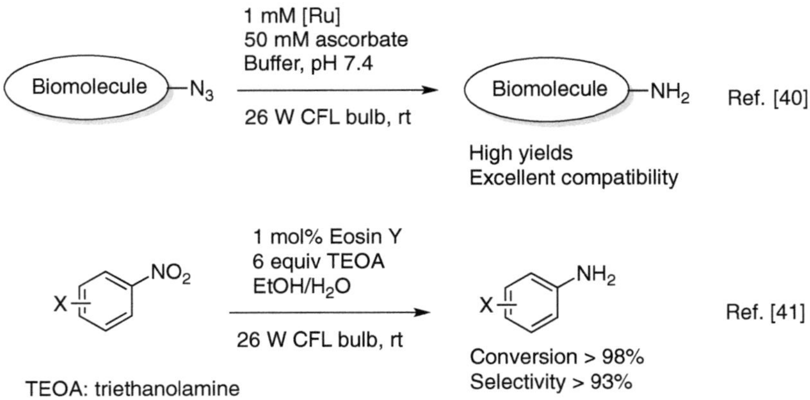

Scheme 2.5 Reduction of azido and nitro functions via photocatalysis-generated N-centered radicals.

Recently also, vicinal diamines such as TMEDA could be cleaved under $Ru(bpy)_3Cl_2$ photocatalysis through the C—C bond to generate an iminium cation and an alpha-amino radical intermediate and give as final products *N*,*N*-dialkylformamides [39].

2.3.2 Formation of N-Centered Radicals

N-Centered radicals have been generated using various nitrogenated functions. Azides can be reduced to primary amines using the $Ru(bpy)_3^{2+}$ catalysis. Importantly, the azide reduction using sodium ascorbate as sacrificial donor was shown to be compatible with alcohols, phenols, acids, alkenes, alkynes, aldehydes, alkyl halides, alkyl mesylates, and disulfides functions so that the reaction can be applied to DNA, oligosaccharides, and protein enzymes [40]. In the same line, arylnitro derivatives have been engaged in photocatalytic reduction to provide anilines using eosin as photocatalyst and triethanolamine as reducing agent [41] (Scheme 2.5).

2.4 C—X Bond Formation

2.4.1 C—O Bond

Different oxygenated moieties can be introduced by photocatalysis in place of an initial C—H bond. The simplest oxidation of benzene into phenol has been devised by Fukuzumi. Thus, the 3-cyano-1-methylquinolinium ion acts as an efficient photocatalyst for the selective oxygenation of benzene into phenol using both oxygen and water under homogeneous conditions (2.5 mol% of photocatalyst) at room temperature. A gram-scale preparation of phenol (41%) was disclosed (Scheme 2.6) [42].

Benzoyloxylation of electron-rich arenes and heteroaromatics can be achieved via Ru(II) photocatalysis using benzoyl peroxide and in the presence of a base. The reaction exhibits some regioselectivity and some compatibility with various functional groups so that it could be applied to the direct benzoyloxylation of fluorescein-OMe (Scheme 2.6) [43].

α-Oxyamination of 1,3-dicarbonyl derivatives has been developed by Koike and Akita. With acyclic diketones, ketoesters, and malonates, the reaction proceeds smoothly upon $[Ir(ppy)2(dtbbpy)]^+$ catalysis in the presence of TEMPO but without any additive. Under these conditions, TEMPO exhibits a dual reactivity as Qox and Qred to respectively generate TEMPO-, acting as a base that deprotonates the dicarbonyl derivative, and TEMP=O+, acting as an electrophile. In the case of cyclic 1,3-dicarbonyl derivatives, oxidatively dimerized products are formed [44]. Tan developed a related oxyamination of acyclic 1,3-dicarbonyl derivatives using Rose Bengal as photocatalyst. The method could be nicely extended to the synthesis of quaternary α-fluorinated α-hydroxy acid derivatives. [45] Alkyl, aryl, allyl, and benzylorganoborates can also be oxidized by iridium catalysis so that the resulting radicals can be trapped by TEMPO. At least 2 equiv of TEMPO are required since one is needed as a spin-trapping agent and the second acts as a sacrificial oxidant to regenerate the ground-state iridium(III) catalyst [46].

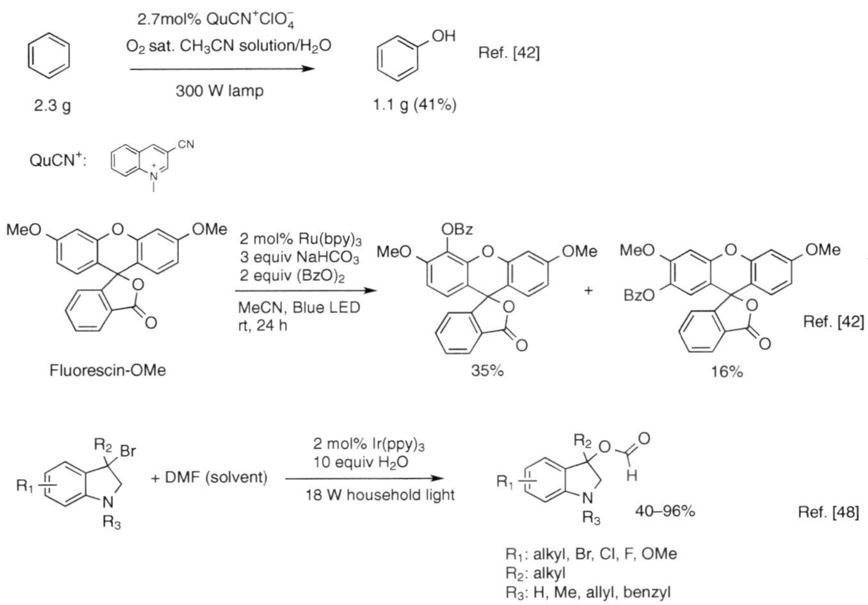

Scheme 2.6 Representative examples of photocatalyzed C—O bond formations.

Jiao disclosed a quite original aerobic oxidation of benzyl halides leading to carbonyl derivatives. The reaction relies on a combination of photocatalysis with $Ru(bpy)_3^{2+}$ and organocatalysis with 4-methoxypyridine, which forms an ammonium intermediate that is reduced and then fragments. The resulting radical in then oxidized by air in basic medium ensured by the presence of a carbonate salt. Ketones, 1,2-diketones, and 1,2-ketoesters could be obtained in high yields [47].

A quite versatile photolytic formyloxylation of 3-bromooxindoles with DMF in the presence of water has been reported by Chen and Xiao. As seen before (see Section 2.3.1.1) and confirmed by Stern–Volmer studies, the photoexcited state of fac-$Ir(ppy)_3$ is quenched by the bromo precursor liberating a C-centered radical, which homolytically adds on the oxygen atom of DMF. The resulting α-amino radical is readily oxidized to provide an iminium salt, which upon reacting with water provides the final product and dimethylamine ammonium. The transformation proved to be quite versatile in terms of yields and substrate scope (Scheme 2.6) [48].

The following examples of C—O bond formations do not result from a homolytic process but a polar one. Thus, a well-established alternative way to introduce an oxygenated group is through nucleophilic addition of an alcohol to a carbocationic center. The anomeric cation of glycosyl substrates is well disposed for this purpose. As illustrated by Bowers, S-PMP (*para*-methoxyphenyl) thioglycosides could be oxidized to a sulfur-centered radical cation intermediate using iridium and ruthenium photocatalysis with X_3CBr as Qox. The transient intermediate then fragments into the arylthiyl radical and the valuable cationic species that can be trapped by a series of alcohols, with no diastereoselectivity though [49]. Similarly, a phenyselenium group can also be oxidized, and photocatalyzed glycosylation of alcohols with selenoglycosides could be worked out [50].

Somewhat related is the 5-*exo*-trig cyclization of a phenolate anion to the electrophilic carbon atom of an amidinium radical cation of an amidine moiety that is generated through $Ru(bpy)_3Cl_2$ catalysis. The resulting 2-aminobenzoxazoles are obtained in good yields (>80 %) [51]. Finally, aldehydes can be smoothly oxidized to carboxylic acids with a broad scope of substrates. Upon $Ir(dFppy)_3$ photocatalysis under oxygen atmosphere, an intermediate acyl radical is formed, which combines with hydroperoxyl radical ($HOO^•$). The latter originates from hydrogen abstraction on the aldehyde by singlet oxygen [52].

2.4.2 C—N Bond

α-Aminoalkyl radicals, generated through the oxidation of the corresponding amine, readily react with a diazocarboxylate by radical addition on the N=N double bond or by radical–radical coupling involving the radical anion of the diazocarboxylate derivative. The corresponding aminals were obtained in fair to good yields. They proved to be valuable building blocks because of their high reactivity toward nucleophilic substitution reactions at the α-position (Scheme 2.7) [53].

Lee has concentrated on the photocatalytic imidation of arenes with *N*-chlorophthalimide. The easily generated N-centered radical undergoes homolytic aromatic substitution. The scope proved to be broad and the phthalimido group is easily removed to provide a free amino group (Scheme 2.7) [54].

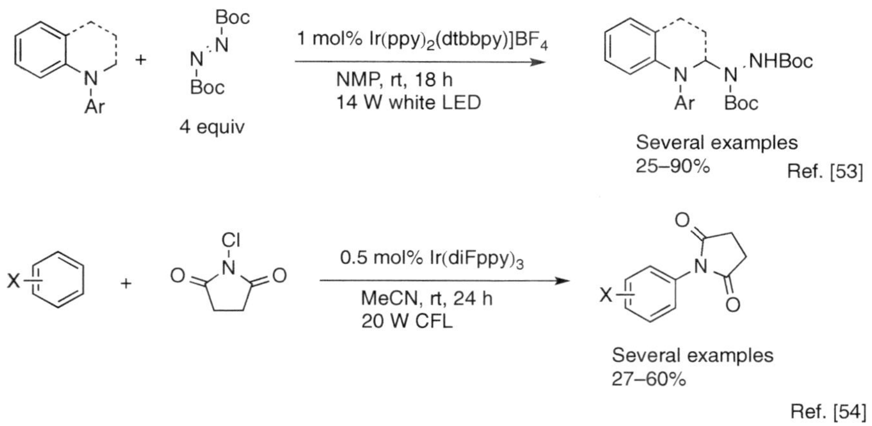

Scheme 2.7 Representative examples of photocatalyzed C—N bond formations.

Facile photocatalytic oxidation of the nitrogen atom of styrylaniline derivatives has been exploited by Zheng. The resulting N-centered radical cation intermediate then undergoes unusual 5-*endo*-trig radical cyclization, which can be rationalized by the presence in the tether of sp^2 carbon and a nitrogen atom. The resulting indoles are formed in high yields [55].

Lan and Lei have examined a photocatalytic decarboxylation of α-ketoacids in the presence of a primary amine to provide amides. An acyl radical has been shown to act as the key intermediate. As suggested by modeling, the latter would be trapped by nucleophilic addition of the amine partner to form an amide radical anion, which then loses the electron through SET [56].

2.4.3 C—S and C—Se Bonds

Aryldiazoniums have been shown by van Wangelin to be excellent precursors of arylsulfides upon photoreduction with eosin Y. The thiolating agents are disulfides. Extension to selenyl analogs was also reported [57].

In the same line, Ranu showed that heteroarylamines could be smoothly converted in a one-pot reaction to arylselenides and even tellurides upon eosin Y photocatalysis. The previously unknown sequence relies on the astucious *in situ* generation of a diazonium intermediate thanks to the presence of *t*-BuONO in the reaction medium [58].

Thiol–ene “click” addition has been imposed over the last decade as one of the privileged methods of ligation of diverse partners and it has found applications in biology and materials sciences. Different groups have worked out a photocatalytic version of this reaction. Stephenson and workers reported a photocatalytic initiation of this reaction. They first studied the background reaction taking place without photocatalysis. This process proved to be extremely substrate dependent and was presumably initiated by residual traces of oxygen. Nevertheless, irradiation in the presence of $Ru(bpy)_3^{2+}$ catalyst, sodium ascorbate as Qred, $BrCCl_3$, which liberates a CCl_3 radical, and a threefold excess of thiol provides an efficient initiation of the thiol–ene process. In the same study, the authors discovered a very efficient solvent-free and non-photocatalyzed amidation/hydrothiolation tandem that could be applied to the preparation of pharmacologically relevant N-substituted-6-methylthiomorpholin-3-ones [59].

In the same line, Yoon focused his study on developing a quite robust catalytic system of the thiol–ene reaction. While the photooxidation of thiols with $Ru(bpz)_3{}^{2+}$ alone proves to be rather sluggish, the addition of *p*-toluidine, presumably serving as Qred, efficiently initiates the reaction, so that on using 0.25 mol% of $Ru(bpz)_3PF_6$ and 0.5 equiv of *p*-toluidine, thiols such as glutathione can be added to a series of biologically relevant alkenes in water as solvent. An interesting feature of this approach is that the thiol is the limiting reagent [60].

An intriguing sulfenylation of *N*-methylindoles using arylsulfonylchlorides and $Ru(bipy)_3{}^{2+}$ photocatalyst has been reported by Zheng. While *N*-methylindole serves as Qred to generate the Ru(I) reducing agent, the arylsulfonylchloride partner would be reduced to the corresponding arylsulfenyl chloride, which would add to the indole nucleus through electrophilic aromatic substitution [61].

Thioanilides can serve as convenient precursors of benzothiazoles through a photocatalyzed C—H radical under dioxygen. A base such as DBU would deprotonate the starting anilide and the corresponding anion would get oxidized by Ru(III) originating from the oxidative quenching of excited Ru(II) by molecular oxygen [62].

2.4.4 C—Br Bond

Fukuzumi has developed a photocatalytic bromination of electron-rich aromatic hydrocarbons with HBr under oxygen and based on 9-mesityl-10-methylacridinium ion as photocatalyst. The reaction is limited by the oxidation potential of the substrates but has proved to be quite selective, with no polybromination. The catalytic cycle has been thoroughly studied [63].

Thanks to a $Ru(bpy)_3{}^{2+}$ photocatalyzed generation of Br_2 from CBr_4, a mild bromination of electron-rich aromatics has been devised. The generated bromine can also be used in all typical electrophilic bromination reactions: bromoetherification, bromination of keto derivatives, dibromination of alkenes [64].

Electron-rich aryl epoxides can be smoothly transformed into α-aryl bromoketones using CBr_4 as the brominating agent. A detailed mechanistic study suggests that the reaction proceeds first by an oxidation of the aromatic ring. The resulting aryl radical cation then retrieves its aromaticity, triggering the oxidative homolytic opening of the epoxide ring. Further reaction with the $CBr_3{}^{\bullet}$ radical and the Br^- anion provides the final bromoketo derivative [65].

2.4.5 C—F Bond

Photocatalytic approaches have also been worked out for efficient C—F formation on benzylic substrates, even if the C—F critical bond formation event is not *per se* homolytic. Cheap and readily available diarylketones were used as the photocatalysts with low loadings (5 mol%) in the presence of selectfluor as the source of fluorine. Remarkably, 9-fluorenone catalyzed selective benzylic C—H monofluorination and xanthone *gem*-difluorination [66].

Sammis and Paquin have developed a photocatalytic decarboxylative fluorination of aryloxyacetic derivatives. The required ingredients include $Ru(bpy)_3{}^{2+}$

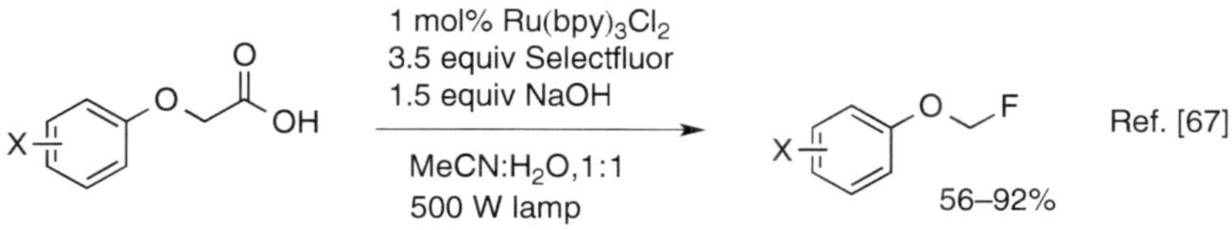

Scheme 2.8 Photocatalyzed C—F bond formation.

as photocatalyst, NaOH to provide a more easily oxidized carboxylate salt, and selectfluor as Qox, as supported by quite insightful transient absorption spectroscopy studies (Scheme 2.8) [67].

2.4.6 C—B Bond

Using eosin Y as photocatalyst, aryldiazoniums can be homolytically reduced and the resulting aryl radical undergoes homolytic substitution on the fluoride-activated bis-pinacol boronate [68].

2.5 C—C Bond Formation

2.5.1 Formation and Reactivity of Aryl Radicals

Aryl radical formation has been intensively explored by using the classical $n\text{Bu}_3\text{SnH}$/AIBN system and the related processes through the reduction of aryl halides. The synthetic potential of such a reactive intermediate has been proved by the huge number of applications in total synthesis, methodological developments, and material sciences [1]. As an efficient alternative and as seen above, the SET reduction of such aryl halides has been investigated under photoredox catalysis conditions using visible light. This represents a challenge because of their reduction potential being lower than −2 V (SCE). Thus, highly reductive catalysts have to be used to allow the reduction. Stephenson [27] and Lee [69] reported the generation of aryl radicals from the corresponding halides by using *fac*-Ir(ppy)_3 and $[\text{Ir(ppy)}_2\text{(dtbbpy)}]\text{PF}_6$, respectively (Scheme 2.9).

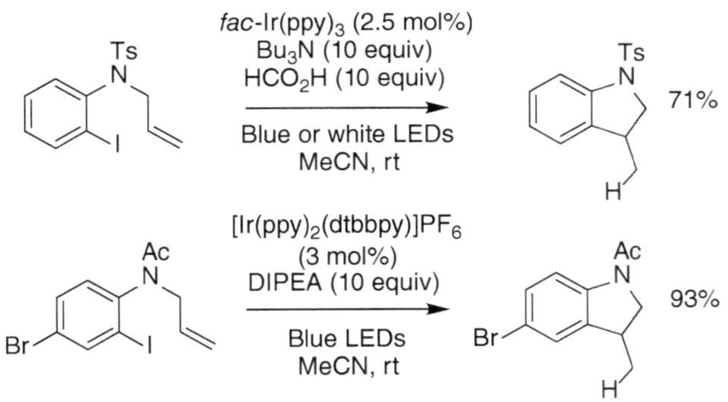

Scheme 2.9 Aryl radical cyclization under photoredox conditions.

Both transformations involved a tertiary amine as the Qred of iridium catalysts. Stephenson used an excess of tributylamine in the presence of formic acid while diisopropylethylamine was preferred by Lee. Under those conditions, aryl iodides are selectively reduced to afford the corresponding aryl radicals that cyclize in a 5-*exo*-trig manner. Then, hydrogen transfer occurs onto the resulting primary alkyl radicals. These protocols have been applied to the photocatalytic reduction of alkyl, alkenyl, and aryl iodides. In the study reported by Lee, an alteration of the reaction rates has been highlighted, depending on the light source. The direct reduction of aromatic halides has also been investigated and showed high selectivities in favor of the C—I bond in the presence of C—Br and C—Cl bonds.

Intermolecular arylation has been reported under visible light irradiation in the presence of *fac*-Ir(ppy)$_3$ as catalyst (Scheme 2.10) [70]. Screening of reaction parameters proved that DMSO is the best solvent. Different additives such as tertiary alkylamines, sodium and potassium hydroxide, and lithium, sodium, and potassium *tert*-butoxide have been tested. Conditions involving only three equivalents of *t*-BuOK allowed efficient transformation from aryl iodides and bromides. Without the iridium catalyst, the arylation was performed in 20% yield from iodobenzene and benzene while 0.5 mol% of *fac*-Ir(ppy)$_3$ increased the yield up to 85%. The arylations of *p*-iodo and *p*-bromo anisole with benzene gave 76% and 80% yields respectively of the corresponding biaryl compounds. Aryl iodides needed 0.5 mol% of iridium catalyst at room temperature while bromides reacted at 70 °C with a catalyst loading of 2.5 mol%. Potassium *tert*-butoxide is known to promote SET reduction of aryl halides [71]. So, the role of the iridium catalyst is not clear in the present reaction but allows a significant improvement.

Milder reductive conditions have been developed by Barriault for the reduction of aryl bromide to the corresponding aryl radical (Scheme 2.11) [72]. Iridium catalysts are advantageously replaced by [Au$_2$(μ-dppm)$_2$]Cl$_2$, which has been previously reported by Che, who determined the photophysical properties and applied it to the photoinduced reduction of unactivated alkyl

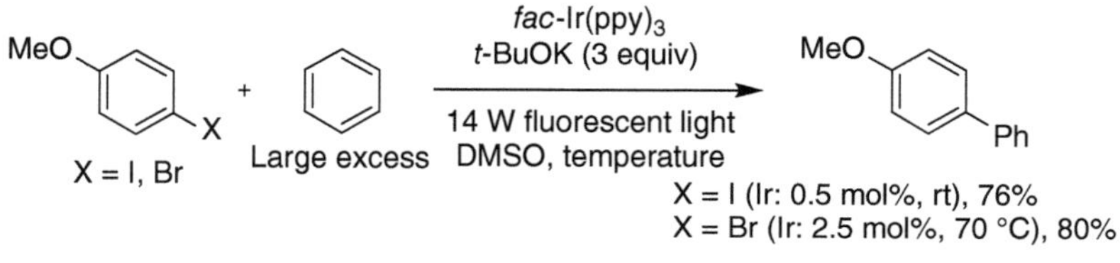

Scheme 2.10 Intermolecular biaryl formation.

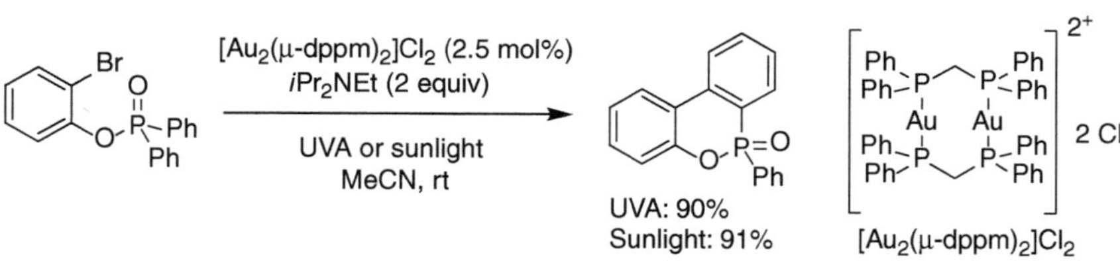

Scheme 2.11 Intramolecular biaryl formation.

halides [73]. This luminescent gold catalyst has a high reduction potential ($E°Au_2^{3+}/Au_2^{2+*} = -1.6$ to -1.7 V vs. SCE) and its excited state can be quenched by a tertiary amine. Thus, the reduction of aryl and alkyl bromides has been investigated and 2–5 equiv of Hünig base (DIPEA) with UVA or sunlight irradiation appeared to be the best conditions. For example, the *o*-bromophosphenic ester allowed generating the corresponding aryl radical by SET reduction, which cyclized onto a phenyl moiety. Then, rearomatization occurred to afford the cyclic biaryl compound in very high yield. This transformation has been used for radical cyclization of alkyl bromides and cascade reactions following the SET reduction/intermolecular addition/cyclization/hydrogen transfer in moderate to good yields.

Historically, the reduction of onium salts represents the early development of visible-light photoredox catalysis [74]. Indeed, Kellog was a pioneer in the application of such transformations to organic synthesis by the SET photoreduction of sulfonium salts catalyzed by $Ru(bpy)_3Cl_2$ and Hantzch ester to generate carbon-centered radicals [75, 76]. This transformation was applied to the generation of alkyl radicals and the corresponding mechanism was investigated. Subsequently, a number of other onium salt derivatives such as diazonium, iodonium, or sulfonium were studied for their ability to generate radicals in a compatible manner with organic synthesis. In continuity with Kellog's seminal work, Fensterbank, Goddard, and Ollivier reported the visible-light photocatalytic reduction of triarylsulfonium salts, which proved to be a source of aryl radicals (Scheme 2.12) [77].

The triarylsulfonium salts are difficult to reduce because of their high reduction potential (−1.2 V vs SME for triphenylsulfonium). Nevertheless, they can produce aryl radicals under mild conditions with the use of $Ru(bpy)_3Cl_2$ as catalyst and DIPEA as Qred. Fluorescent light bulb lamp has been used as the visible light source. Counterion effect, aryl substitution, and acceptor modulation have been screened. The radical so generated added onto olefins to afford alkylation or allylation products in good yields.

Since more than one century, the SET reduction of diazonium salts has been reported by famous chemists such as Sandmeyer [78, 79], Pschorr [80], Gomberg and Bachmann [81], or Meerwein *et al.* [82]. In the 1950s, Kochi improved the

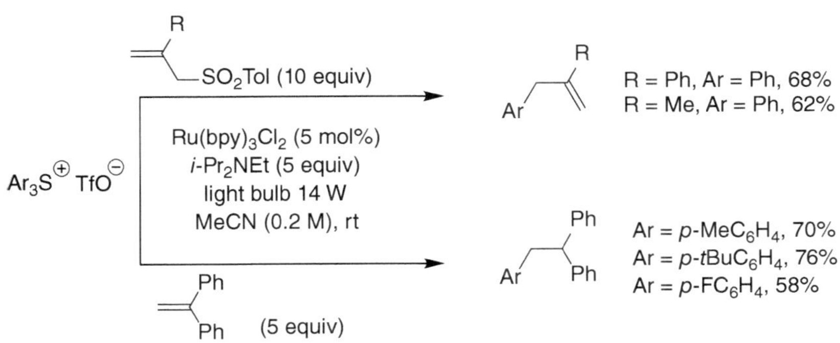

Scheme 2.12 Formation and reaction of aryl radicals by photoreduction of sulfonium salts.

knowledge around these reactions by investigating their kinetics parameters [83]. Indeed, diazonium salts, especially the arene derivatives, proved to be a valuable source of carbon-centered radicals that can be engaged in carbon–carbon bond formation, promoted by a stoichiometric or catalytic reductant [84]. In the context of photoredox catalysis under visible light activation, Deronzier was a pioneer in the reactivity of diazonium salts with transition-metal-based photocatalysts such as $Ru(bpy)_3Cl_2$. The ability of arendiazonium salts to be the Qox of the ruthenium complex excited state has been reported and applied to a photocatalytic version of the Pschorr reaction [85, 86]. Fluorenone, fluorene, dibenzofuran, and phenanthrene derivatives were obtained by the reduction of the corresponding salts through the cyclization of aryl radicals. This concept has also been extended to the oxidation of primary and secondary alcohols into aldehydes and ketones [87, 88]. More recently, König extended this process to the Meerwein arylation catalyzed by transition metal complexes or organic dyes (Scheme 2.13) [89, 90].

Arylation of unsaturated compounds has been reported with $Ru(bpy)_3Cl_2$ or eosin Y. *p*-Bromodiazonium salt was reduced in the presence of 5 equiv of styrene to afford 94% of the stilbene derivative from the reaction catalyzed by the ruthenium complex and a lower yield of 66% with that catalyzed by eosin Y [89]. After oxidative quenching of the photocatalyst excited state, the reduced diazonyl radical decomposed to liberate the corresponding aryl radical. Addition onto olefin followed by oxidation of the resulting radical allowed the formation of stilbenyl derivatives and the recovery of the catalyst. The same transformation was extended to biaryl synthesis in a metal-free manner with furan, thiophene, and pyrrol acceptors in good yields [90]. Enol acetates have also been investigated as aryl radical acceptors with a comparison between ruthenium catalyst and organic dyes [91].

Diaryliodonium salts are also an important source of aryl radicals through single electron transfer processes and have been recently submitted to photoredox-catalyzed transformations. They are versatile reagents for organic transformation as suggested by their applications in cationic photopolymerization of epoxides [92, 93]. Moreover, their reduction is easier than that of the corresponding triarylsulfonium salts because of their lower reduction potentials ($E_{1/2}{}^{red}(Ph_2I^+) = -0.2$ V vs. SCE). Symmetric and dissymmetric diaryliodonium salts played the role of Qox of $Ru(bpy)_3Cl_2$ and the resulting aryl radical can be

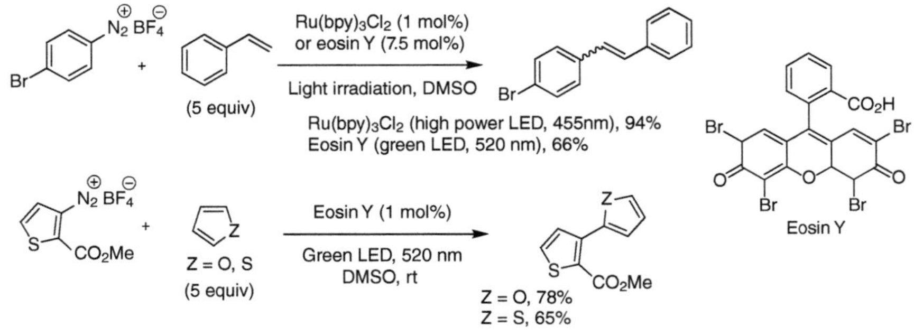

Scheme 2.13 Formation and reaction of aryl radical by photoreduction of diazonium salts.

engaged in arylation processes. Xio and Xue reported the arylation or arenes and heteroarenes under mild conditions in good yields [94]. A radical mechanism has been highlighted on the basis of preliminary experiments with TEMPO as radical trap.

Fensterbank, Goddard, and Ollivier worked on the application of iodonium salts as aryl radical sources. This transformation gave the opportunity to evaluate cheaper and more sustainable alternatives to ruthenium and iridium catalysts [95]. For this purpose, photoactive copper complexes appeared a valuable field of investigation. Copper(I)-biphenanthroline $Cu(dap)_2^+$ (dap = 2, 9-bis(*para*-anisyl)-1,10-phenanthroline) and $Cu(dpp)_2^+$ (dpp = 2,9-diphenyl-1, 10-phenanthroline) have been reported by McMillin [96] to have promising photoredox catalysis and photophysical and redox properties (Scheme 2.14) [97]. Indeed, reduction potentials of $Cu(dap)_2^{+*}$ and $Cu(dpp)_2^{+*}$ are lower than that of $Ru(bpy)_3^{2+*}$ (−1.43 V and −1.11 V respectively vs. −0.85 V vs. SCE). Sauvage [98] and Reiser [99] explored the Cu(II)/Cu(I) redox system for the reduction of benzyl bromides, ATRA (atom transfer radical addition), and allylation of α-haloketones.

The allylation of phenyl radical liberated by the reduction of diphenyliodonium hexafluorophosphate has been used as a model reaction in order to compare ruthenium, iridium, and copper photoactive catalysts (Scheme 2.14). Irradiation with a fluorescent lamp in the presence of 2 equiv of DIPEA and 0.5 mol% of $Ru(bpy)_3Cl_2$ or *fac*-$Ir(ppy)_3$ gave respectively 81% and 80% yields. At the same low level of catalyst loading the two copper catalysts had comparable efficiency and the allylation product was isolated in 82% yield with $[Cu(dpp)_2]PF_6$. ^1H NMR investigation allowed proposing a reasonable mechanism and gave evidence about the modification of the copper oxidation state. The development of cheaper and environmentally compatible photocatalysts constitutes certainly one of the most promising perspectives of this domain.

Photoredox catalysis under visible light irradiation has been proposed to promote oxidative C—C bond formation. Collins reported the preparation of carbazole derivatives by an intramolecular arylation process, leading to the formation aryl–aryl moiety from triarylamine substrates (Scheme 2.15) [100]. Ruthenium and copper complexes have been studied on model substrates and extended to substituted triarylamines and alkyldiarylamines. Iodine was used as Qox of the photocatalysts. The classical $Ru(bpy)_3Cl_2$ did not allow promoting the oxidative coupling reaction of diphenylaniline, but replacing the hydrogen atom by a phenyl group increases the yield to 27%. On the same substrate, copper complex $[Cu(Xantphos)(DMP)]BF_4$ (either used as isolated compound

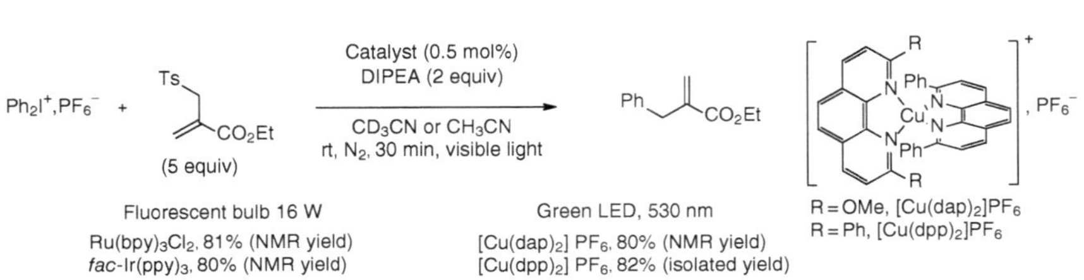

Scheme 2.14 Reduction of iodonium salts by photoactive copper complexes.

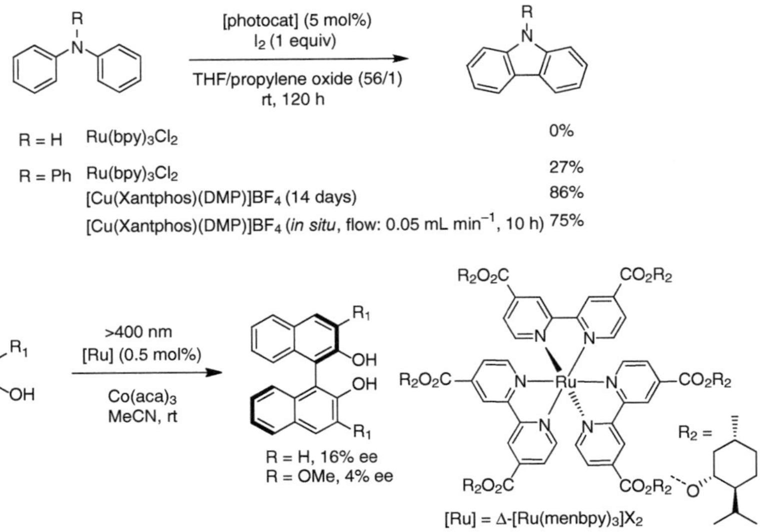

Scheme 2.15 Carbazol and enantioenriched bis-naphthol synthesis by oxidative coupling.

or generated *in situ*) proved to be a significantly better catalyst in the formation of the corresponding *N*-phenyl carbazole. Modification of copper ligands did not improve the yield of this transformation. These conditions allowed full conversion but with very slow reaction rates (full conversion was obtained in 14 days). Then, the flow process appeared very attractive because it permits increasing the irradiation surface and accelerating the reaction rate. With a flow of 0.05 mL min^{-1} and a residence time of 10 h, *N*-phenyl carbazole was obtained in 75% yield.

Oxidative C—C bond formation has been nicely illustrated by Ohkubo in the enantioselective preparation of binaphthol derivatives by intermolecular reaction (Scheme 2.15) [101, 102]. This study is the first application of photoredox catalysis under visible light in an enantioselective manner. For this, chiral ruthenium trispyridine complex $\Delta[Ru(menbpy)_3]X_2$ has been used. The chirality is brought about by the bipyridine-modified ligands bearing two menthyl residues. One electron oxidation of naphthol led to the corresponding radical cation that can add onto a second molecule of naphthol. The resulting dearomatized intermediate was then oxidized to form the binaphthol products in 16% and 4% ee depending on the substitution. The origin of the enantioselectivity was not clearly reported but was suspected to come from the second oxidation step based on a kinetic resolution.

2.5.2 Formation and Reactivity of Trifluoromethyl and Related Radicals

2.5.2.1 Photocatalyzed Reduction of Perfluorohalogen Derivatives

In 2011, MacMillan reported the synthesis of α-trifluoromethyl carbonyl compounds through a potential radical addition of trifluoromethyl radical onto silylated enol ethers [103]. Owing to the high lability of CF_3I, even weak light was enough to initiate the reaction. To extend this transformation to less activated substrates, trifluoromethylation reaction has been studied by Stephenson [21, 104] (reported in details in Chapter 3) and Cho (*vide infra*) in the context of photoredox catalysis. Iodotrifluoromethane has been considered as a source of trifluoromethyl radical and engaged in different transformations (Scheme 2.16). In the presence of heteroarenes, $Ru(bpy)_3Cl_2$ and TMEDA, the reduction of CF_3I occurred and the resulting trifluoromethyl radical added onto the aromatic system [105]. Consecutive oxidation allowed formation of the corresponding aromatic system. This transformation is efficient and the trifluoromethyl-substituted rings were obtained in good to high yields, 90% in the case of 3-methylindole. Alkynes have also been investigated as radical addition partners under similar photoredox conditions [106]. It has been shown that depending on the additives, the radical addition of a trifluoromethyl radical generated by SET can follow multiple pathways and results in different products.

The $Ru(phen)_3Cl_2$/TMEDA (2 equiv) system favored ATRA to give the iodotrifluoromethylation product in high yield and with a good control of the stereoselectivity. With *fac*-$Ir(ppy)_3$/DBU (10 equiv) and considerable amount of THF, radical addition occurred but the intermediate vinyl radical was reduced faster by hydrogen atom transfer (HAT) than by iodine transfer. Formal hydrotrifluoromethylation products were obtained. Finally, trifluoromethylation of alkynes has been reported by replacing DBU or TMEDA by potassium *tert*-butoxide in DMF. In the same vein, the trifluoromethylation of alkenes has been reported by Cho following an ATRA/elimination sequence [107].

The difluoroalkylation of olefins and heteroarenes has been reported by Cho and coworkers [108] and Qing and coworkers [109], respectively, by using visible-light photoredox catalysis and ethyl bromodifluoroacetate. Substituted alkanes, alkenes, and heteroaromatic derivatives were obtained by controlling the reaction parameters. Such incorporation of $—CF_2R$ moiety is synthetically useful because it opened the possibility for further transformation of the appended ester group.

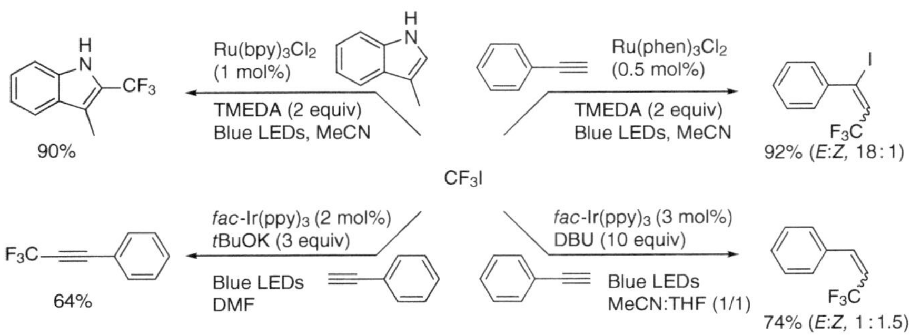

Scheme 2.16 Photoredox-catalyzed trifluoromethylation by reduction of trifluoromethyl iodide.

2.5.2.2 Photocatalyzed Reduction of Perfluoroalkyl-Substituted Onium Salts

As a valuable alternative to perfluorohalogen derivatives, hypervalent iodine and sulfonium compounds bearing a trifluoromethyl group proved to be highly efficient sources of the trifluoromethyl radical (Scheme 2.17).

Togni's and Umemoto's reagents have been usually used as electrophilic sources of the trifluoromethyl group but they also proved to be prompted to form the corresponding stabilized radical. Hydrotrifluoromethylation of alkenes has been investigated by Gouverneur and Médebielle under photoredox conditions (Scheme 2.18) [110, 111].

A set of reagents has been tested for the trifluoromethylation of a model olefin in the presence of classical $Ru(bpy)_3Cl_2$. The best transformation was obtained with the Umemoto triflate salt. Hydrogen donor was also a key parameter in this reaction. Mainly Hantzsch ester and H-donor solvents were evaluated and the best compromise was given by methanol. This transformation was adapted to the formal addition of fluoroform to an alkyne to yield the corresponding trifluoromethylated alkenes. The proposed mechanism emphasized the action of the Umemoto reagent as Qox of the Ru(II) excited state. The so-formed sulfuranyl radical decomposed and liberated the trifluoromethyl radical. Addition onto a double or triple C—C bond occurred and the resulting C-centered radical abstracted a hydrogen atom from methanol. The Ru(II) catalyst was recovered by oxidation of the ketyl radical to formaldehyde. Based on this reported procedure, the authors proposed the trifluoromethylation of allylsilanes directed toward the preparation of branched allylic CF_3 derivatives [112]. The stereoselectivity of the reaction has been highlighted by the formation of enantioenriched compounds.

Free anilines have been engaged in a trifluoromethylation reaction under visible-light photoredox conditions by Zhu and Ma (Scheme 2.19) [113]. Diverse sources of trifluoromethyl radical have been tested and the best results were obtained with the Togni'reagent T1/*fac*-$Ir(ppy)_3$. High functional group tolerance was observed and this clearly indicates that the method can be used in

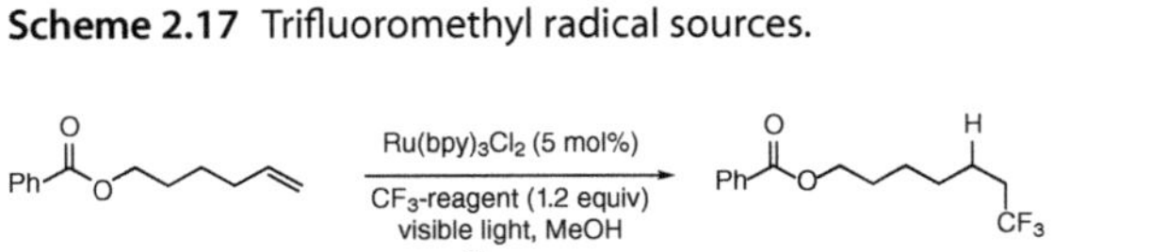

Scheme 2.17 Trifluoromethyl radical sources.

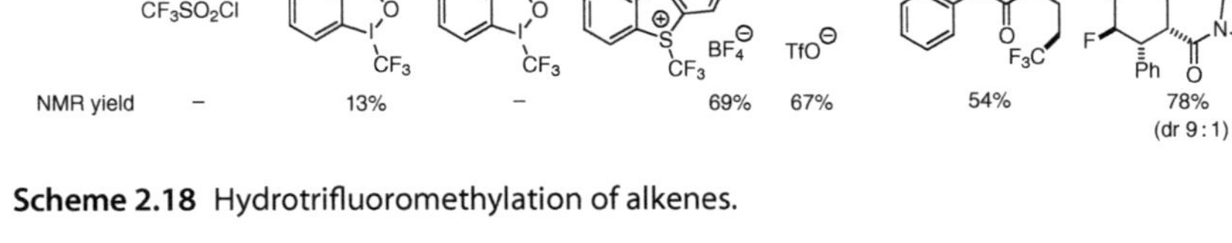

Scheme 2.18 Hydrotrifluoromethylation of alkenes.

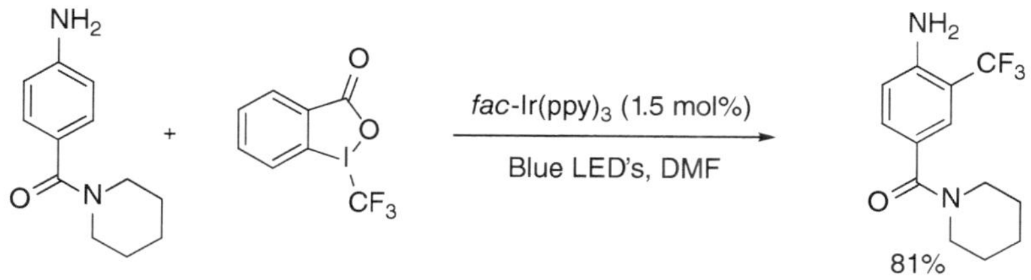

Scheme 2.19 Trifluoromethylation of free anilines.

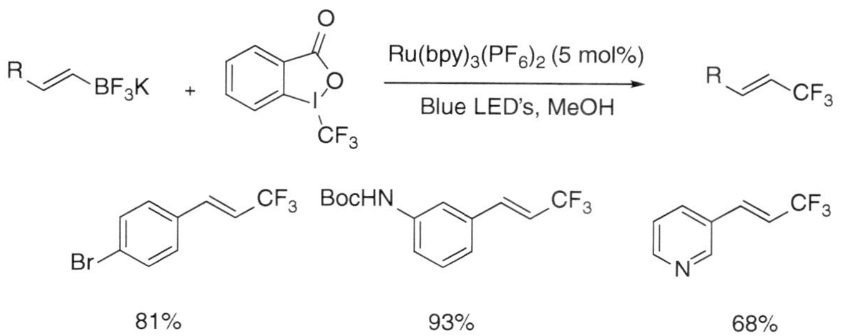

Scheme 2.20 Synthesis of trifluoromethylated alkenes from vinyltrifluoroborate salts.

advanced steps of a total synthesis for the late incorporation of trifluoromethyl group onto an aniline moiety.

A synthetically useful method for the preparation of trifluoromethylated alkenes from vinyltrifluoroborate salts has been reported by Koike and Akita (Scheme 2.20) [114]. The Togni's reagent T1 was superior in terms of efficiency. Styrenyl and alkyl olefins gave good to high yields under such mild conditions. Togni's reagent acted as Qox of the excited ruthenium catalyst and generated the trifluoromethyl radical. This transformation is very attractive because it constitutes an alternative to transition-metal-catalyzed cross-coupling reactions.

2.5.2.3 Photocatalyzed Formation of Perfluoroalkyl Radicals from Sulfonyl and Sulfinyl Derivatives

Among the different known sources of trifluoromethyl radical, trifluoromethanesulfonyl chloride (triflyl chloride) appeared a very attractive reagent mainly because of its low cost and low molecular weight side-products. In 2011, MacMillan reported the trifluoromethylation of arenes and heteroarenes through the formation of trifluoromethyl radical formation by visible-light photoreduction of triflyl chloride (Scheme 2.21) [115]. The SET reduction of triflyl chloride affords the corresponding radical anion, which decomposes to the sulfonyl radical. The release of SO_2 generates the trifluoromethyl radical that added onto five and six arenes and heteroarene compounds. The oxidative rearomatization of the ensuing cyclohexadienyl-type radical is driven by the reduction of Ru(III) and recovers the Ru(II) catalyst. Good to high yields were obtained and this transformation has been applied to the preparation of a set of trifluoromethylated drug analogs.

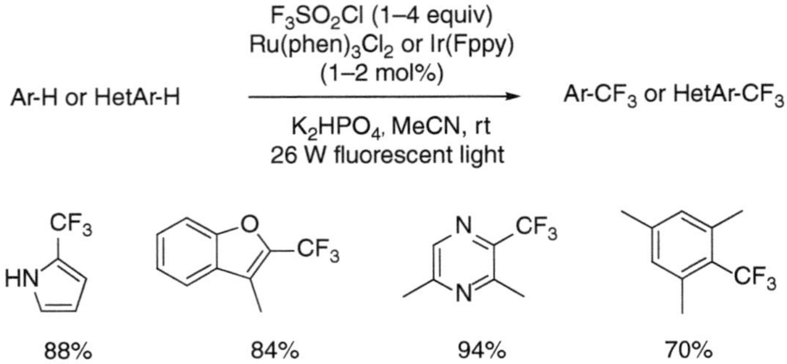

Scheme 2.21 Radical trifluoromethylation of arenes and heteroarenes.

Following a similar procedure, Yu and Zhang reported the reduction of triflyl chloride by Ir(ppy)$_2$(dtbbpy)]PF$_6$ [116]. In this work, the trifluoromethyl radical added onto enol esters to afford the corresponding α-trifluoromethyl ketones. A one-pot synthesis from ketones has been reported by Kappe in a metal-free version with eosin Y as the photocatalyst and extended to continuous flow synthesis [117].

In 2013, Nicewicz [118] and Itoha [119] reported in the same time the photooxidation of the trifluoromethanesulfinate salt (Langlois reagent) and analogs as a way of generating perfluoroalkyl radicals. Moreover, these transformations involved photoredox organocatalysts, mesityl acridinium salt for Nicewicz, and anthraquinone-2-carboxylic acid for Itoha (Scheme 2.22).

Hydrotrifluoromethylation of alkenes has been performed in the presence of a catalytic amount of tetrafluoroborate *N*-methyl mesitylacridinium as the photocatalyst and thiophenol derivatives as the hydrogen atom donor. A broad range of aliphatic and styrenyl alkenes with different substitution patterns have been reported and the anti-Markovnikov regioisomers were exclusively isolated in moderate to good yields. Trifluoro- and perfluoroalkyl sulfinate salts have also been engaged in radical addition onto aromatic and heteroaromatic compounds under organic photooxidation conditions with anthraquinone-2-carboxylic acid as catalyst. Trifluoromethyl, pentafluoroethyl, heptafluoropropyl, and perfluorohexyl and octyl radicals were generated and successfully added onto aromatic rings in good to high yields.

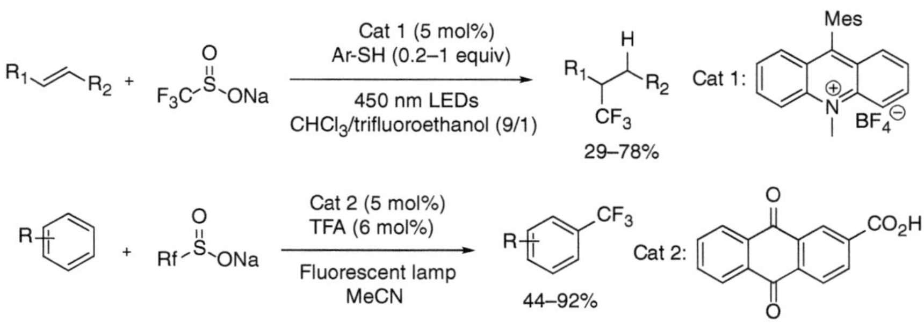

Scheme 2.22 Photoredox organocatalyzed trifluoromethylation of olefins and arenes.

2.5.3 Formation and Reactivity of Alkyl and Related Radicals

2.5.3.1 C—C Bond Formation Through Photocatalyzed Reduction of Halogen Derivatives and Analogs

In the early stage development of photoredox catalysis dedicated to organic synthesis, carbon–carbon bond formation was already a stimulating axis of investigation. In 1984, Tanaka reported a dramatic effect of $Ru(bpy)_2Cl_2$ on an NADH model reduction of benzyl bromide [120]. Indeed, the photoreduction of benzyl bromide mediated by the sole 1-benzyl-1,4-dihydronicotinamide yielded toluene as the major product. When ruthenium photosensitizer was added in catalytic amounts, 1,2-diphenylethane was isolated almost exclusively, indicating an SET sequence. A few years later, Sauvage proposed the use of the $Cu(dap)_2^+$/triethylamine system to achieve the same transformation from nitro-substituted benzyl bromides to the corresponding dibenzyl products [98].

Then, the photoreduction of activated halogen compounds became an efficient way to generate carbon-centered radicals localized at benzylic, allylic, or α-carbonyl positions. In 2009, Stephenson reported a tin-free dehalogenation that proceeded through the SET reduction of bromide and chloride [19]. This transformation has been applied by the same group to the total synthesis of (+)-gliocladin C by extension to C—C bond formation (Scheme 2.23) [121]. In this case, the radical was not reduced by HAT but added onto the indole derivative to stereoselectively substitute the benzylic position. Among others, this reaction has been adapted by Stephenson for visible-light photoredox catalysis in flow [122].

Intra- and intermolecular additions of electrophilic radicals based on photoreduction of the corresponding halogen derivatives have been reported (Scheme 2.24). Prior to the total synthesis of (+)-gliocladin C, malonyl radicals were engaged in functionalization of electron-rich heterocycles [20]. A reductive quenching of the ruthenium excited state was proposed to generate

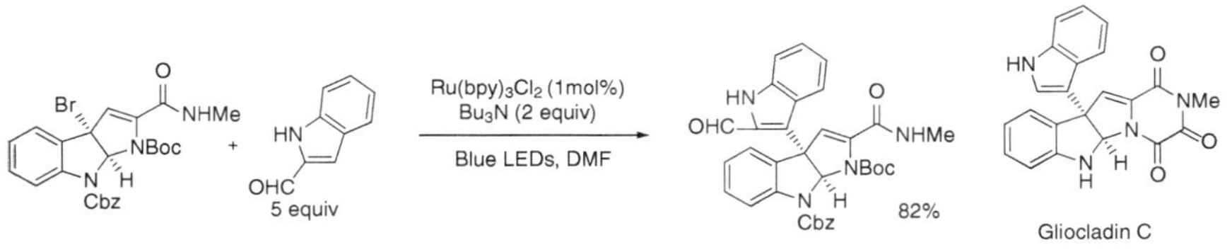

Scheme 2.23 Photoredox-catalyzed step in the total synthesis of (+)-gliocladin C.

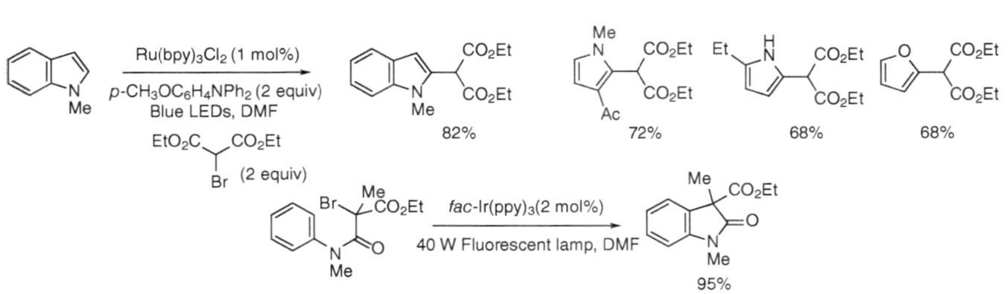

Scheme 2.24 Reduction of activated bromides and radical additions.

a highly reductive Ru(I) complex that was able to reduce the C—Br bond of ethyl bromomalonate. The so-formed radical added onto a series of furan, pyrrole, and indole derivatives. The yields are moderate to high. Triethylamine and anisyldiphenylamine were tested as sacrificial donors of electrons and the latter proved to be superior. A similar transformation in an intramolecular version has been published by Yu for the synthesis of oxindoles (Scheme 2.24) [123]. In this case, no additive had to be used and a different mechanism has been proposed. Oxidation of radical as the last step of the photocatalytic process has been applied to an alkenylation reaction. In such an event, this sequence allowed recovering the catalyst and converting the resulting radical into a cation, generally followed by elimination. In 2013, Yu reported a formal C—H functionalization of enamides and enecarbamates [124]. At the same time, Masson reported α-alkylation of imines through radical/cationic tandem reaction [125].

In 1991, OKada and Oda reported the decarboxylation of *N*-(acyloxy) phthalimide derivatives under mild conditions and visible light activation (Scheme 2.25) [38]. $Ru(bpy)_3Cl_2$ was used as photosensitizer in the presence of 1-benzyl-1,4-dihydronicotinamide, which plays the role of a sacrificial donor of electron. The phthalimide moiety was reduced by SET to generate the carboxylic radical after the cleavage of the labile N—O bond. Decarboxylation allowed obtaining the alkyl radical that added on to the electron-poor olefin. This transformation has been applied to the decarboxylation of primary, secondary, and tertiary carbon centers, generally in good yields. The same authors showed that in the presence of a hydrogen atom donor such as *tert*-butylthiol, the reaction became a highly efficient reductive decarboxylation to afford the corresponding alkanes in high yields [37].

In 2012, Overman applied Okada's decarboxylation to a straightforward total synthesis of (−)-aplyviolene [126]. A key step of the synthesis involved the construction of a quaternary center through tertiary radical conjugate addition from the decarboxylation of a tertiary *N*-(acyloxy)-phthalimide derivative. The same group reported an extension of this methodology dedicated to the formation of a tertiary alkyl radical from the corresponding tertiary alcohols (Scheme 2.26) [127]. Derivatization of alcohols into *N*-phthalimidoyl oxalates was followed by photocatalytic reduction of the phthalimidoyl moiety. Then, two successive decarboxylations generated the tertiary alkyl radical, which added onto diverse electrodeficient olefins to create a quaternary center. After optimization, the best catalytic system involved $Ru(bpy)_3(PF_6)_2$, Hantzsch ester, and *i*-$Pr_2NEt.HBF_4$ as additive under blue LED irradiation. Based on the acceptor quantity, good to high yields were obtained.

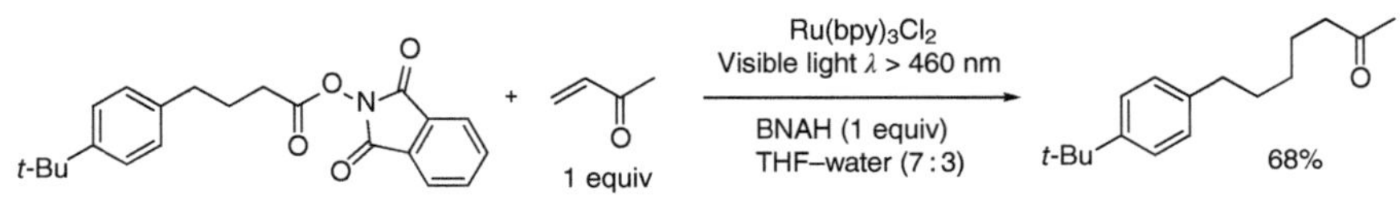

Scheme 2.25 Giese reaction from decarboxylation of *N*-(acyloxy)phthalimide derivatives.

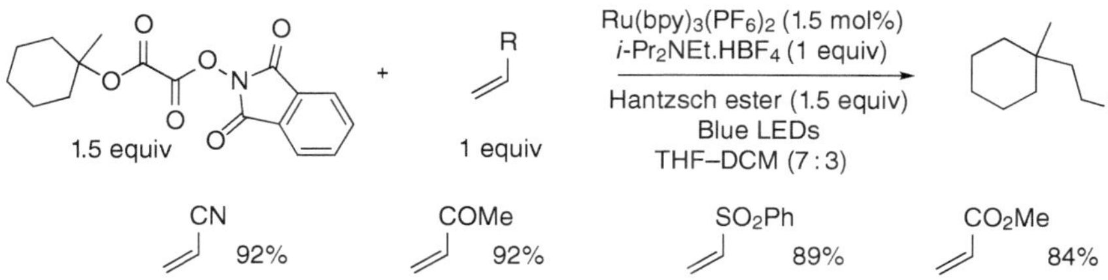

Scheme 2.26 Quaternary carbon center formation by decarboxylation of N-(acyloxy)phthalimide derivatives.

2.5.3.2 C—C Bond Formation Through Photocatalyzed Oxidation of Electron-Rich Functional Group

Since decades, neutral boron derivatives were known to be valuable alkyl radical sources, mainly due to the ability of the C—B bond to be homolytically cleaved under oxidative conditions. Oxygen-centered radicals were generally used for this purpose. Recently, boronate complexes also proved to be precursors of alkyl radicals through SET using stoichiometric amounts of the oxidant. In 2012, Akita and Koike extended this concept to the use of photoredox catalytic systems (Scheme 2.27) [46].

Organotrifluoroborates and organo(triol)borates were investigated, first in the formation of C—O bond through radical trapping experiments with TEMPO as mentioned in Section 2.4.1 and second in C—C bond formation processes with electron-poor olefins. The latest were more effective than trifluoroborate salts for the formation of secondary and primary alkyl and aryl radicals. The most efficient catalyst among a set of ruthenium and iridium complexes was $[Ir(dF(CF_3)ppy)_2(bpy)](PF_6)$, which has the highest oxidation potential at the photoexcited state (+0.91 V vs. Fc^+/Fc). This is consistent with the fact that no additional Qox is needed for this transformation and the organoborate itself is oxidized by the [Ir(III)] excited state to generate the corresponding alkyl radical and the reduced [Ir(II)] complex. The same authors applied this methodology to the hydroalkoxymethylation of electron-deficient olefins [128]. Chen took advantage of organoborate oxidation to develop a deboronative alkynylation under visible photoredox catalysis [129]. Alkyl radicals were subsequently

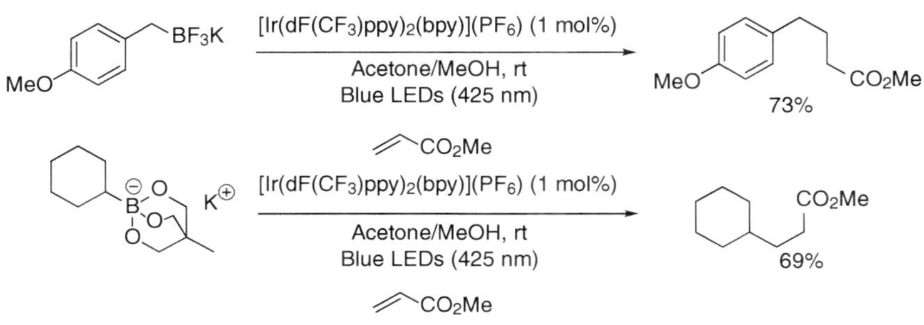

Scheme 2.27 Radical formation by photooxidation of tetravalent boron compounds.

formed after trifluoroborate salt oxidation by $Ru(bpy)_3(PF_6)_2$ and added onto alkynyl benziodoxoles. This transformation was conducted in neutral aqueous medium and appeared to be compatible with a series of biomolecules such as DNA, proteins, or free amino acids.

SET oxidation of enols and enol ethers has been extensively studied for the formation of carbon-centered radicals that were involved in intermolecular additions or cyclizations. These processes were generally based on iron(III), cerium(IV), or manganese(III) and used overstoichiometric amounts of metal salts. Photoredox catalysis has been successfully applied to the oxidation of 1,3-dicarbonyl derivatives by Tan [45] and Akita [44] but limited to carbon–oxygen bond formation from a TEMPO radical-trapping sequence. In 2014, Fensterbank, Goddard, and Ollivier reported versatile conditions for carbon–carbon bond formation in either an intramolecular or an intermolecular manner (Scheme 2.28) [130]. The intramolecular process will be discussed in the cascade reaction section. Under photoredox catalysis, 1,3-dicarbonyl compounds such as ethyl benzoylacetate dimerized through the formation of intermediate radical species. In the presence of a radical acceptor such as allylsulfone, allylation was observed in good yields. Formally resulting from C—H activation, this transformation proceeded through the oxidation of enolate to the corresponding stabilized radical. Electron-rich enamines were engaged in photoredox oxidation through the formation of the corresponding α-iminium radicals [131]. Addition of such radicals onto silyl enol ethers led to the synthesis of γ-diketones. In 2009, Fagnoni reported an efficient way for C—H activation based on hydrogen atom abstraction [132]. Tetrabutylammonium decatungstate was used as photocatalyst under solar-light activation. The so-formed carbon-centered radicals were engaged in Giese-type reactions.

Photoredox catalysis under visible light has also been reported for the oxidation of aliphatic, α-oxy, and α-amino carboxylic acids as a method to generate the corresponding alkyl radicals subsequent to the extrusion of carbon dioxide [133]. The synthetic potential of this transformation was demonstrated by the radical additions onto a set of Michael acceptors and applied to the three-steps synthesis of the medicinal agent pregabalin. Oxidative decarboxylation of aryl acetic acids has been investigated for the formation of benzyl radicals and used in a similar transformation [134]. An amino group at the *para*-position of the benzene ring is necessary for the decarboxylation/addition sequence.

2.5.3.3 C—C Bond Formation Through Photocatalyzed Oxidation of Amino Group

Tertiary amines are easily oxidized by single electron transfer into radical cations that can be transformed into the corresponding iminium salts by the loss of a hydrogen atom or into the α-aminoalkyl radicals by the loss of a

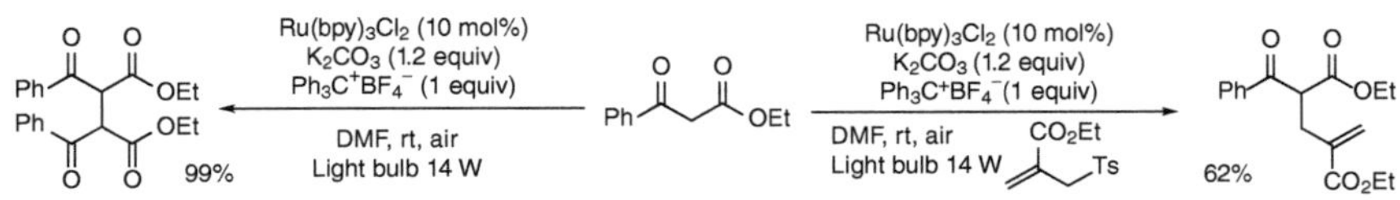

Scheme 2.28 Photocatalyzed oxidation of 1,3-dicarbonyl substrates.

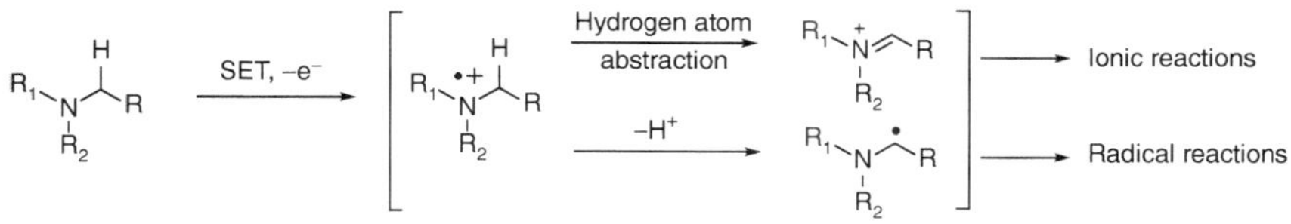

Scheme 2.29 Behavior of aliphatic tertiary amines under oxidative conditions.

proton (Scheme 2.29). Number of groups investigated this dual reactivity under photoredox catalysis mediated by visible light [135]. This section will give a summary of the reactivity of α-aminoalkyl radicals generated in this manner. More details are provided in Chapter 4.

The formation of α-aminoalkyl radicals by oxidation can be seen as a C—H activation process from amines and has been used for α-functionalization of amino groups. Arylation and heteroarylation were developed in *ipso* radical substitution leading to the formation of benzylamine derivatives [136, 137]. Alkylation [138, 139] and vinylation [140] proved to be efficient in the presence of electron-deficient olefins. The loss of proton following the oxidation of amino group has been replaced by the heterolytic bond cleavage of a C—Si bond to generate α-aminoalkyl radicals from α-silylamine [141]. By this way, radical addition onto α,β-unsaturated carbonyl allowed obtaining the corresponding alkylated silyl enol ether by trapping the released silylium residue. α-Amino amide and thioamide were also prepared by addition of α-aminoalkyl radicals onto isocyanates and isothiocyanates [142]. The photooxidation of alkylamines was also investigated in intramolecular processes for the synthesis of *N*-aryltetrahydroisoquinoline [143] and 3-acylindole [144] derivatives.

2.6 Radical Cascade Applications

Radical cascades have emerged as a convenient strategy for elaborating complex architectures in a one-pot process. Arguably, they represent one of the most elegant and efficient approaches to form multiple carbon–carbon and carbon–heteroatom bonds from unsaturated precursors in a single operation [145, 146]. Either radical chain or non-chain redox processes have largely participated in this assembly. As mentioned above, the development of visible-light photoredox catalysis for radical synthesis has demonstrated high efficiency, selectivity, versatility, and functional group tolerance in various transformations, becoming of great interest in cascade processes. Starting from simple alkenes to polyenes/polyenynes, various strategies involving intramolecular polycyclizations, sequential inter- and intramolecular radical steps, or both radical and polar sequences will be discussed. However, atom transfer radical reactions (ATRA) will not be covered in this chapter (see Chapter 5) and limited information on cycloaddition will be given (see Chapter 5).

2.6.1 Intramolecular Polycyclization Processes

The notion of cascade reactions becomes fully apparent with the construction of polycyclic molecules from polyunsaturated precursors upon a combination of

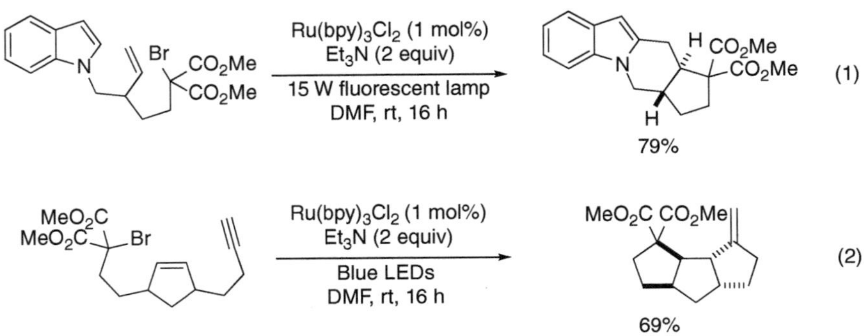

Scheme 2.30 Polycyclization of unsaturated bromomalonates.

intramolecular radical steps. In the example depicted in Scheme 2.30, Eq. (1), Stephenson reported the radical cyclization of a bromomalonate bearing indole and alkene moieties under $Ru(bpy)_3Cl_2/Et_3N$ (Q_{red}) photoreductive conditions. The malonyl radical undergoes 5-*exo*-trig cyclization to a terminal alkene followed by 6-*endo* intramolecular homolytic aromatic substitution to an indole, affording a tetracyclic alkaloid compound in 79% yield as the sole product [147]. In the same manner and inspired by the strategy developed by Curran and Tanabe for the synthesis of Hirsutene and Capnellene [148, 149], they applied these photocatalytic conditions to a double 5-*exo*-trig/5-*exo*-dig cyclization from a polyunsaturated bromomalonate, reaching the linear triquinane core structure of the natural product (Scheme 2.30, Eq. (2)) [150].

Intramolecular [2 + 2] enone cycloadditions have been investigated by Yoon for the construction of bicyclic systems with cyclobutane units. Photoreduction of lithium-activated enone with the $Ru(bpy)_3Cl_2$/Jünig base (Q_{red}) system liberates a radical enolate intermediate, which triggers a 5-*exo*-trig/4-*exo*-trig cyclization sequence in a diastereoselective manner (Scheme 2.31, Eq. (1)) [151]. One-carbon chain extension between both enone units led to the formation of the dihydropyrane ring in place of the cyclobutane ring (Scheme 2.31, Eq. (2)) [152]. Cyclopropylmethyl-radical-mediated rearrangement was also exploited in cascade processes for the generation of reactive homoallylic radical intermediates. The photocatalytic system developed has been successfully implemented to promote reductive intramolecular [3 + 2] cycloaddition reactions from aryl cyclopropyl ketones with activated alkenes and alkynes under $La(OTf)_3$ Lewis acid activation. A radical cyclopropyl ring opening and two successive 5-*exo*-trig cyclizations gave rise to the substituted cyclopentane derivatives, and particularly 5,5-membered fused bicyclic compounds in high yield (Scheme 2.31, Eq. (3)) (see Chapter 5) [153].

Based on Snider's manganese chemistry, Fensterbank, Goddard, and Ollivier developed a new and unique photocatalytic oxidation process of 1,3-dicarbonyl compounds to generate the corresponding radical, which can be involved in cascade reactions. The method proved to be complementary to existing methods using stoichiometric amounts of manganese triacetate. Upon treatment with the $Ru(bpy)_3Cl_2/O_2$ (Q_{ox}) oxidative system in the presence of 2 equiv of tritylium

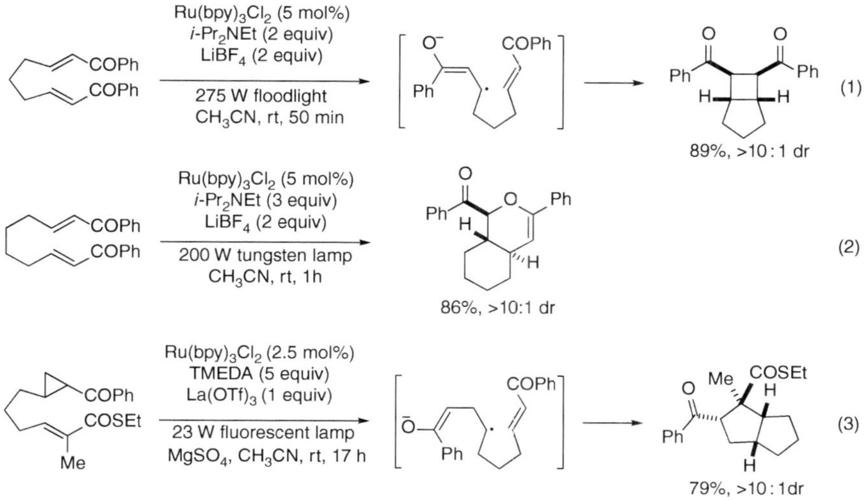

Scheme 2.31 Intramolecular [2 + 2] and [3 + 2] cycloadditions.

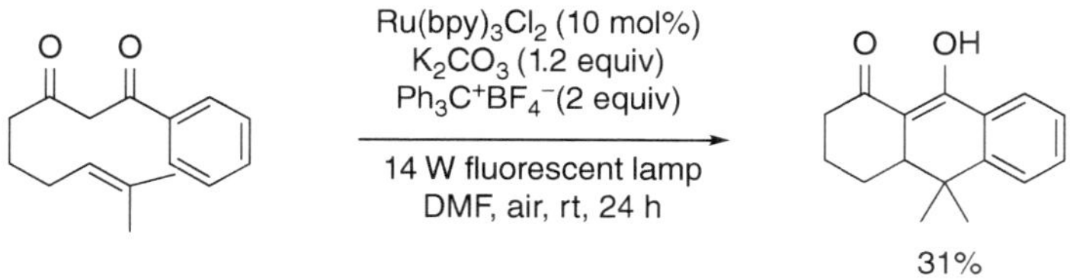

Scheme 2.32 Intramolecular radical cascade reaction involving an unsaturated 1-phenyl 1,3-diketone.

Ph_3CBF_4 as a sacrificial acceptor, two examples of cascade reactions have been realized, starting from unsaturated phenyldiketones to obtain the tricyclic products with moderate yields (Scheme 2.32) [130].

2.6.2 Sequential Inter- and Intramolecular Processes

One of the most developed cascade sequence is tandem intermolecular radical addition/homolytic aromatic substitution, which first comprises radical addition onto activated double bond followed by cyclization onto arene and rearomatization by subsequent oxidation of the radical adduct. To illustrate, aryl radicals generated from aryl diazoniums by reduction with a photoactivated ruthenium complex [154] reacted with *N*-arylacrylamides and led to the formation of substituted indoles. Other examples have been reported with alkyl or CF_3 radicals generated from hypervalent iodide reagents (RCOOH/DIB [155] or Togni's reagent [156]) and fluoroalkyl radicals from fluoroalkylsulfonyl chlorides (Scheme 2.33) [157].

The methodology was also carried out using isocyanides. Addition of alkyl, fluoroalkyl, and particularly CF_3 radicals obtained respectively by the photoreduction of the corresponding bromides [158] and iodoniums (Umemoto's

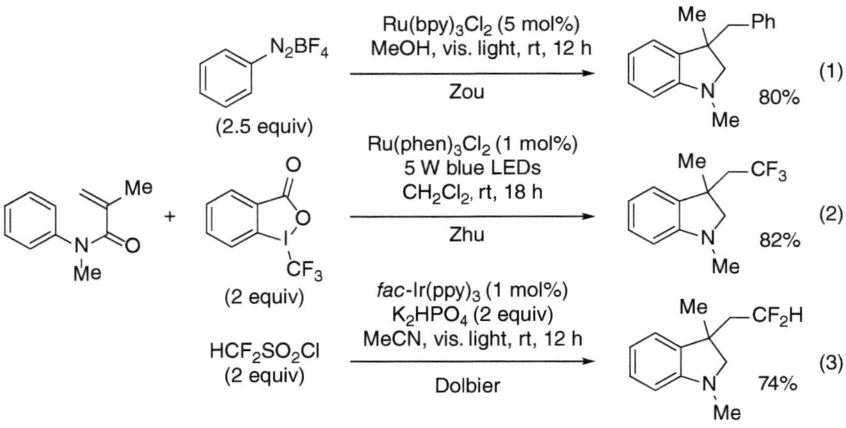

Scheme 2.33 Tandem radical cyclization of *N*-arylacrylamides.

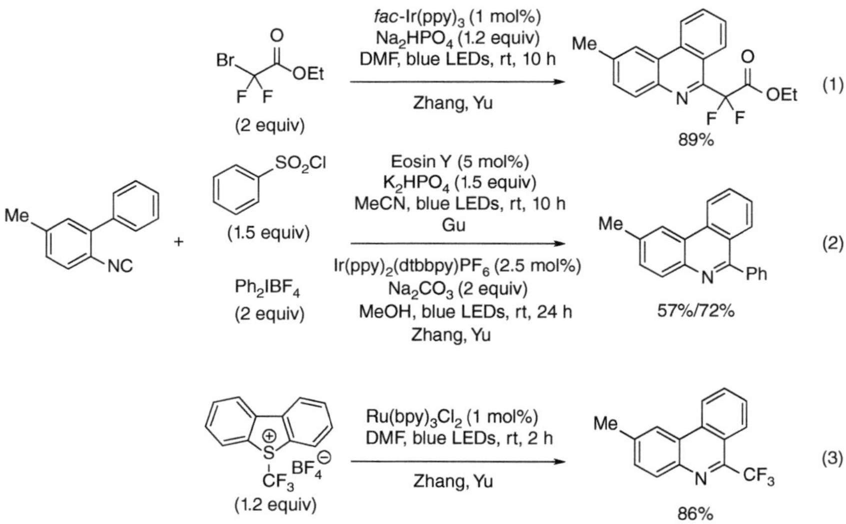

Scheme 2.34 Synthesis of 6-substituted phenanthridines from a 2-isocyanobiphenyl.

reagent) [159] to 2-isocyanobiphenyls gave rise to the formation of 6-alkylated phenanthridines (Scheme 2.34, Eqs. (1) and (3)). Moreover, further explorations revealed that the strategy can be applied to aryl radicals derived from arylsulfonyl chlorides [159] or diaryliodonium salts [160] to furnish 6-arylated phenanthridines (Scheme 2.34, Eq. (2)). Starting from vinyl isocyanides, addition of phenyl [161] and trifluoromethyl radicals [162] allowed the synthesis of isoquinolines (Scheme 2.35).

The tandem sequence was extended by Li to the addition of aryl radicals, derived from arylsulfonylchlorides, onto the triple bond of *ortho*-alkyl arylalkynes followed by H-abstraction to secondary benzylic carbon atom/5-*endo*-trig cyclization and oxidation of the radical adduct to give substituted 1*H*-indenes

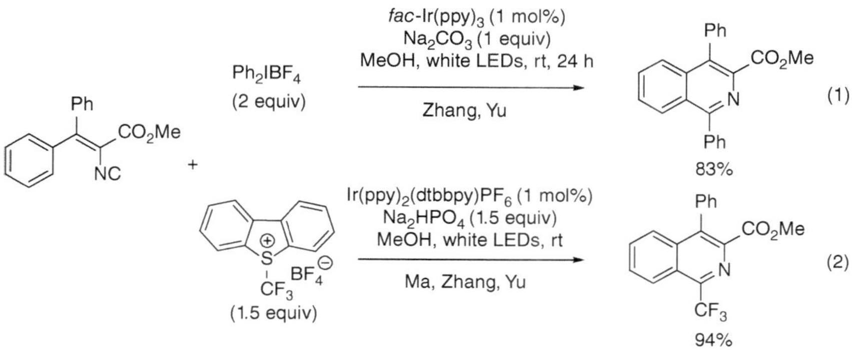

Scheme 2.35 Synthesis of isoquinolines from a vinyl isocyanide.

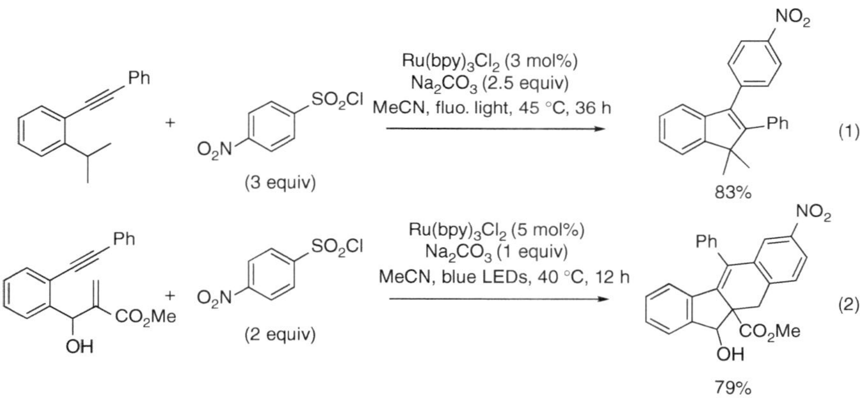

Scheme 2.36 Tandem cyclizations of *ortho*-isopropyl phenylacetylene and a 1,6-enyne.

(Scheme 2.36, Eq. (1)) [163]. Similarly, 1,6-enynes underwent aryl radical addition onto triple bond and successive 5-*exo*-trig and 6-*endo* cyclizations onto alkene and arene rings respectively. Under photoreductive conditions, a range of 10a,11-dihydro-10*H*-benzo[*b*]fluorenes have been synthesized (Scheme 2.36, Eq. (2)) [164].

[2 + 2] Cycloaddition of acyclic enones mediated by visible-light photoredox catalysis was reported by Yoon [165]. Under photoreductive conditions ($Ru(bpy)_2Cl_2$/Hünig base (Q_{red})/visible light) and in the presence of $Eu(OTf)_3$ and a chiral ligand L*, aryl enone was reduced to radical enolate, added to another enone, and the radical adduct underwent a 4-*exo*-trig cyclization to give the cyclobutane with an excellent enantioselectivity for the major diastereoisomer (Scheme 2.37, see Chapter 5) [166]. However, the reduction of chalcones afforded cyclopentanol derivatives resulting from sequential reductive coupling and aldol cyclization (Scheme 2.38) [167].

Other ambitious strategies based on [3 + 2] and [4 + 2] annulation reactions have been explored. To illustrate, König investigated the radical annulation of *o*-methylthioarene diazoniums and alkynes for the preparation of substituted

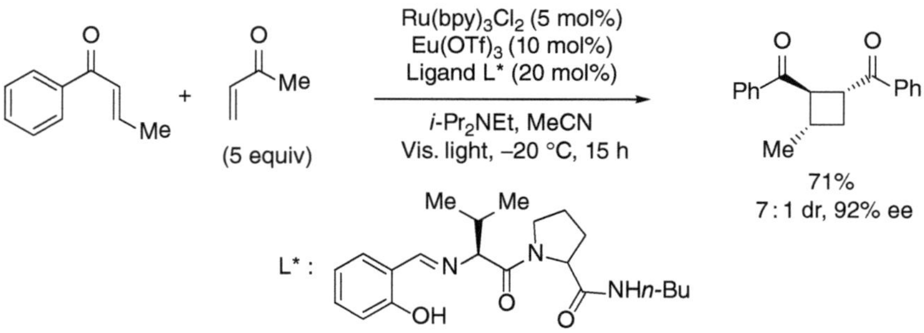

Scheme 2.37 Enantioselective [2 + 2] cycloaddition of acyclic enones.

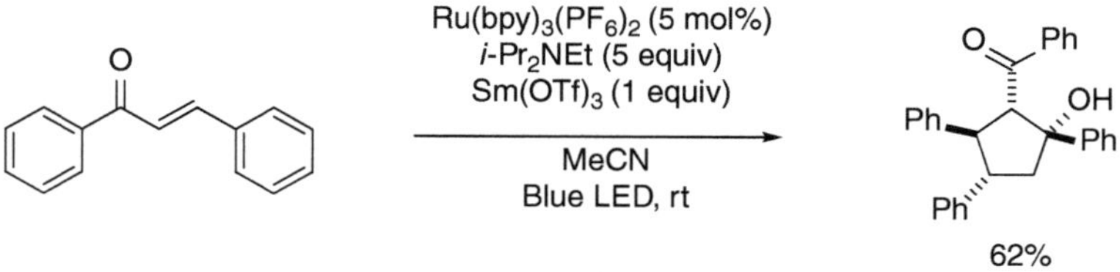

Scheme 2.38 Reductive dimerization of chalcones.

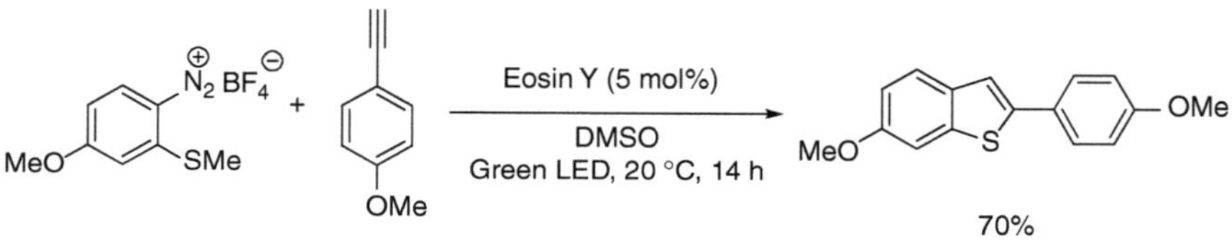

Scheme 2.39 Synthesis of benzothiophene from an *o*-methylthioarene diazonium and an alkyne.

benzothiophenes. Upon irradiation of eosin Y, photoreduction of a diazonium led to an aryl radical, which added to an alkyne. Then, the resulting vinyl radical cyclized onto the sulfur atom, leading to the benzothiophene (Scheme 2.39). A formal synthesis of raloxifene was also proposed [168].

In this area, Zheng has shown that cyclopropylamines can also participate in radical cascade reactions with activated olefins under photooxidative conditions. The β-carbon radical iminium ion, obtained by oxidation of the cyclopropylamine with the photoactivated ruthenium complex and ring opening, can add to the olefin and cyclize to form 5,5-fused bicyclic heterocycles (Scheme 2.40) [169]. Interestingly, the reaction was extended to alkynes, enynes, and diynes with success [170].

Zhang and Yu reported some examples of [3 + 2] and [4 + 2] cycloadditions between alkynes and 2-bromo-1,3-dicarbonyl compounds, the general reaction course of which depends on the nature on the dicarbonyl derivative. Reaction of phenylacetylene and 2-bromocyclohexane-1,3-dione with photoactivated $Ir(ppy)_2(dtbbpy)PF_6$ gave the furan as the [3 + 2] product, while changing the cyclohexane-1,3-dione to an aryl ketobromide furnished the

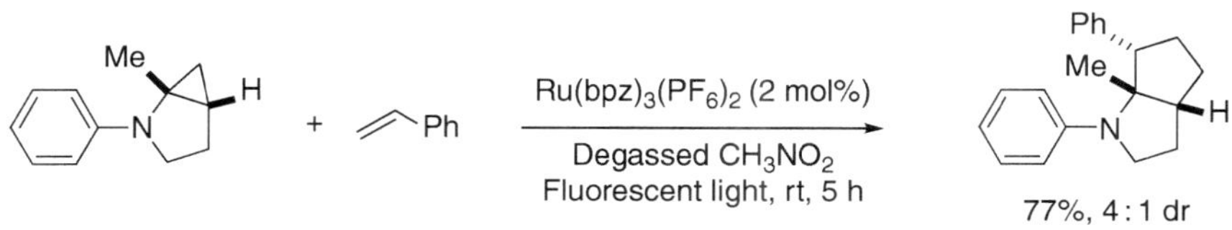

Scheme 2.40 Intermolecular [3 + 2] cycloaddition of a bicyclic cyclopropylamine and styrene.

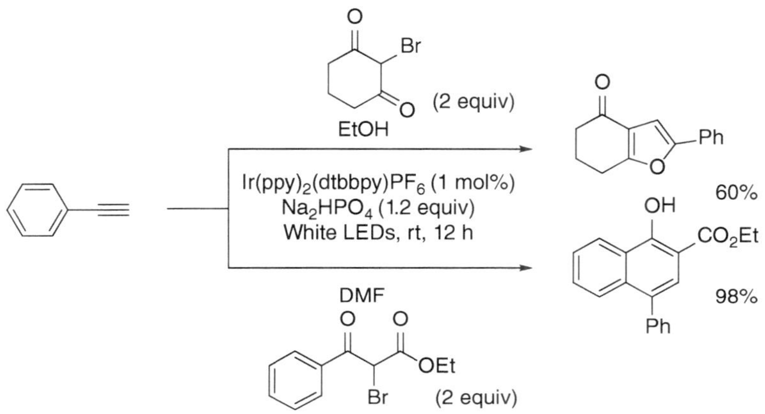

Scheme 2.41 Synthesis of furan and naphthol from 2-bromo-1,3-dicarbonyl compounds and phenylacetylene.

corresponding [4+2] naphthol (Scheme 2.41) [171]. Reactants and radical addition intermediates participated in an oxidative quenching cycle where the 2-bromo-1,3-dicarbonyl compounds were reduced by the excited state of the photocatalyst. The Ir(III) has been regenerated from Ir(IV) by oxidation of the radical adduct in each case.

A photocatalyzed [4+2] benzannulation reaction between biaryldiazonium salts and alkynes was described by Zhou. For instance, photosensitized reduction of a biphenyl diazonium salt with 1 mol% of eosin Y followed by inter-/intramolecular radical cascade reactions with methyl propiolate gave the 9-substituted phenanthrene (Scheme 2.42) [172]. Interestingly, Reiser explored a tandem vinyl radical addition onto heteroarene/cyclization onto arene ring/oxidation sequence to access polycyclic heteroarenes. As depicted in Scheme 2.43, the photoactivated iridium-complex-mediated reaction of α-bromochalcone and furan involving an oxidative quenching cycle of the photocatalyst yielded the polycyclic framework [173]. A methodology for the synthesis of 2-trifluoromethyl quinolines from trifluoroacetimidoyl chlorides and alkynes was proposed by Zhou. For example, the process involves a [4+2] annulation between a trifluoroacetimidoyl radical generated by photoreduction and phenylacetylene. An oxidative quenching cycle of the ruthenium complex was also evidenced (Scheme 2.44) [174]. As mentioned in Section 2.5.3.3 and Chapter 8, photooxidation of tertiary arylamine gave rise to the generation of α-aminoalkyl radical. The latter can be trapped by maleimides [175] or

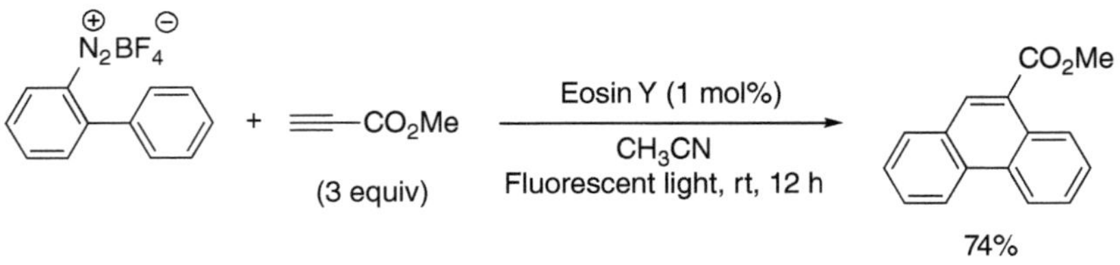

Scheme 2.42 Synthesis of phenanthrene from a biphenyl diazonium salt and methyl propiolate.

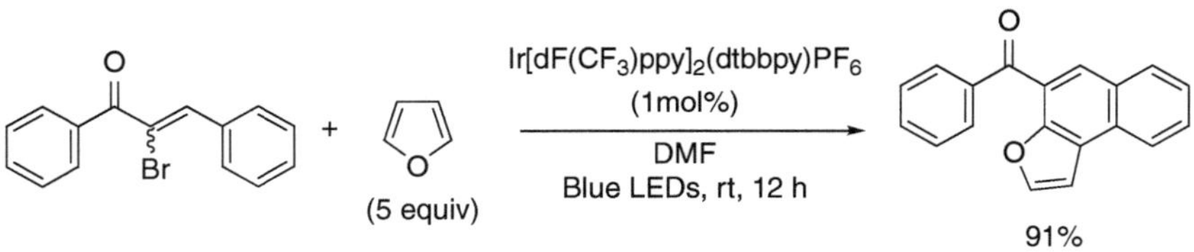

Scheme 2.43 Synthesis of a polycyclic heteroarene from α-bromochalcone and furan.

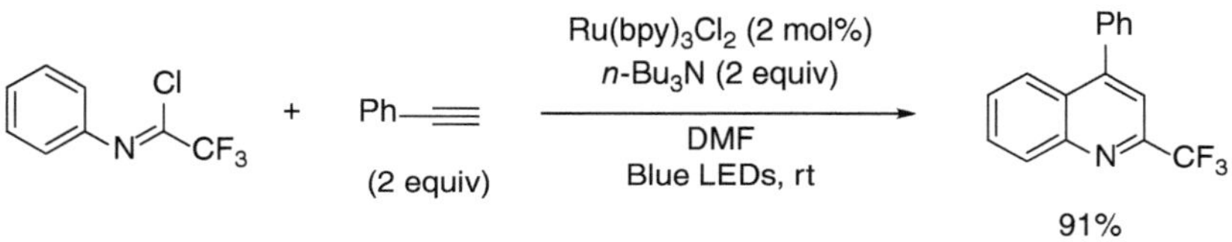

Scheme 2.44 Synthesis of 2-trifluoromethyl quinolines from trifluoroacetimidoyl chlorides and alkynes.

2-benzylidenemalononitrile [176]. Then, the radical adduct was converted to tetrahydroquinoleines via a sequential homolytic aromatic cyclization/oxidation pathway (Scheme 2.45).

2.6.3 Sequential Radical and Polar Processes

Radical multicomponent processes have emerged as a convenient and flexible strategy to elaborate complex molecular building blocks. Among them, visible-light photoredox catalysis proved to be efficient for promoting radical/cationic multicomponent reactions (MCRS) [177]. The sequence consists of a cascade of radical intermolecular addition, oxidation of the radical adduct by the photocatalyst, and nucleophile trapping. Examples of three-component radical/cationic oxyalkylation, and oxy- and aminoarylation of styrene-type olefins were described. For instance, Lei showed that photoreduction of bromoacetonitrile with $Ir(ppy)_3{}^*$ in the presence of 4-*tert*-butylstyrene and methanol gave the γ-alkoxynitrile (Scheme 2.46, Eq. (1)) [178]. The same process performed by Wu and Liu with α-bromo esters, styrenes, and water furnished the corresponding γ-lactones (Scheme 2.46, Eq. (2)) [179]. Earlier, Masson examined the reactivity of enecarbamates in combination with diethyl-2-bromomalonate and ethanol and observed the formation of the expected β-alkylated α-amino ethers (Scheme 2.46, Eq. (3)) [125]. Finally, oxy- and aminoarylation of styrene-type

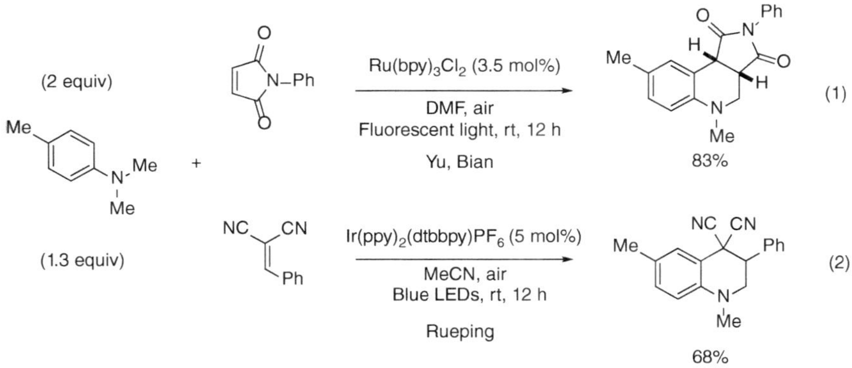

Scheme 2.45 Synthesis of tetrahydroquinoleines from *N,N*-dimethyl-4-methylaniline and activated alkenes.

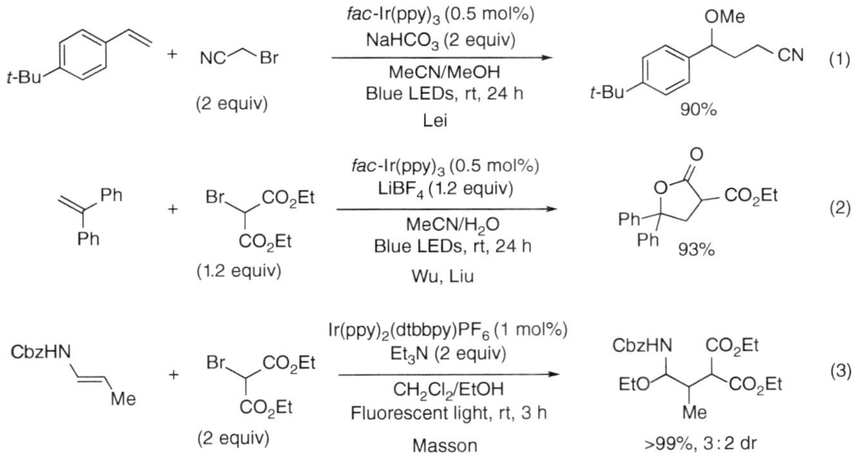

Scheme 2.46 Three-component radical/cationic oxyalkylation of alkenes.

olefins can be achieved by photoreduction of diphenyliodonium salts and phenyl radical addition to the alkene. Then, the intermediate carbocation generated by oxidation can be trapped either by an alcohol [180] or by a nitrile, which undergoes a Ritter-type reaction (Scheme 2.47, Eq. (1)). An example of aminoarylation from diazonium salts was also reported (Scheme 2.47, Eq. (2)) [181].

As mentioned before, the trifluoromethyl radical can be generated by photoreduction of Umemoto's or Togni's reagent with photoactivated ruthenium(II) or iridium(III) photocatalyst and added onto alkenes (Section 2.5.2). Extensions to radical-cationic MCRs have been reported recently for oxy- and aminotrifluoromethylation of styrene-type olefins [182–187] and enecarbamates [188] as depicted in Schemes 2.48 and 2.49. In these cases, the intermediate benzylic carbocation can be trapped intermolecularly by an alcohol [182–188], DMSO [183, 184], a nitrile [186], an amine [187], or an azide [187, 188] or intramolecularly by an acid moiety leading to CF_3-substituted lactones [185].

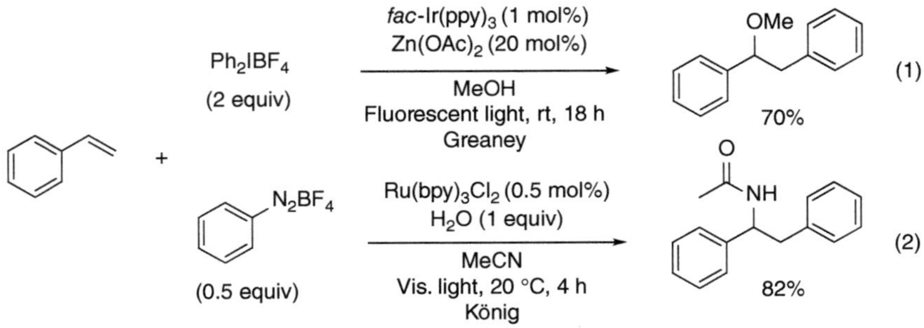

Scheme 2.47 Oxy- and aminoarylation of styrene.

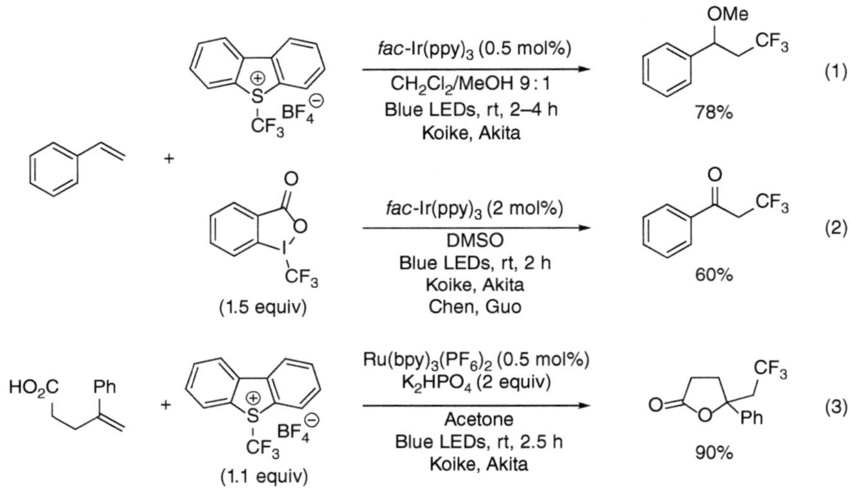

Scheme 2.48 Oxytrifluoromethylation of styrene-type olefin.

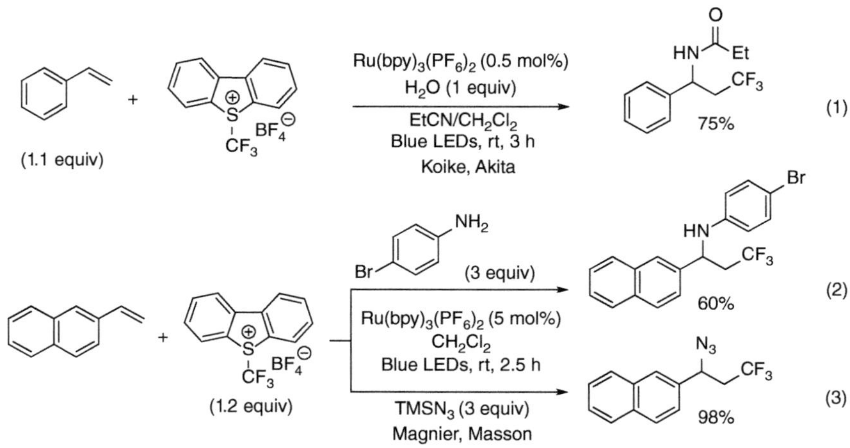

Scheme 2.49 Aminotrifluoromethylation of styrene-type olefin.

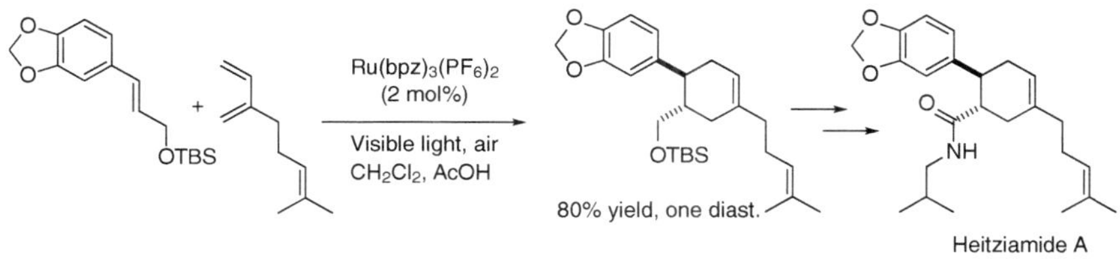

Scheme 2.50 [4 + 2] Cycloaddition between anethol and isoprene-type components.

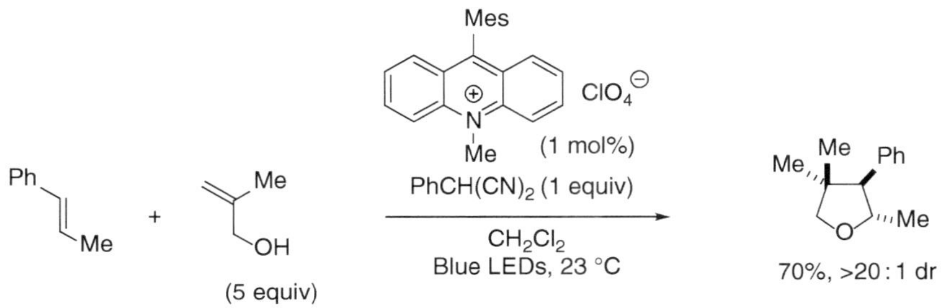

Scheme 2.51 [3 + 2] Cycloaddition reaction of methyl styrene and methallyl alcohol.

Phooxidation of methyl styrene-type alkene provides a transient radical cation, which can be trapped by an unsaturated carbon–carbon bond and a nucleophile. For instance, this represents a convenient way to realize radical cation Diels–Alder cycloadditions with electron-rich partners. A combination of ruthenium(II) tris-(bipyrazyl)/air (oxygen) quencher (Q_{ox}) enables the electronically mismatched [4 + 2] cycloaddition between anethol and isoprene-type components. This methodology reported by Yoon has been applied to the stereoselective synthesis of heitziamide A (Scheme 2.50) [189]. One year earlier, examples of intramolecular [2 + 2] cycloaddition between anethol-type alkene and styrene moiety were reported by the same author and exposed in Chapter 5 [190]. In 2013, Nicewicz examined the [3 + 2] cycloaddition reaction of methyl styrene-type alkenes and allylic or propargylic alcohols for the synthesis of substituted tetrahydrofurans (Scheme 2.51). This transformation involved oxidation of the alkene by the excited acridinium as a visible light organic photocatalyst, reaction of the cation with the unsaturated alcohol, 5-*exo*-trig -or dig- cyclization, and radical hydrogen abstraction to $PhCH(CN)_2$. Finally, the resulting phenyl malononitrile radical participated in the regeneration of the photocatalyst [191].

References

1 (a) Studer, A. and Chatgilialoglu, C. (eds) (2012) *Encyclopedia of Radicals in Chemistry, Biology and Materials*, vol. **1 and 2**, John Wiley & Sons, Chichester; (b) Renaud, P. and Sibi, M.P. (eds) (2001) *Radicals in Organic Synthesis*, vol. **1** and **2**, Wiley-VCH, New York.

2 Gansäuer, A. (ed.) (2006) Radicals in Synthesis I & II, in *Topics in Current Chemistry*, vol. **263** and **264**, Springer, Heidelberg.

3 Curran, D.P., Porter, N.A., and Giese, B. (1995) *Stereochemistry of Radical Reactions*, VCH, Weiheim.
4 Nicewicz, D. and McMillan, D.W.C. (2008) Merging photoredox catalysis with organocatalysis: the direct asymmetric alkylation of aldehydes. *Science*, **322**, 77–80.
5 Fagnoni, M., Dondi, D., Ravello, D., and Albini, A. (2007) Photocatalysis for the formation of the C—C bond. *Chem. Rev.*, **107** (6), 2725–2756.
6 Ravelli, D., Dondi, D., Fagnoni, M., and Albini, A. (2009) Photocatalysis: a multi-faceted concept for green chemistry. *Chem. Soc. Rev.*, **38** (7), 1999–2011.
7 Zeitler, K. (2009) Photoredox catalysis with visible light. *Angew. Chem., Int. Ed.*, **48** (52), 9785–9789.
8 Yoon, T.P., Ischay, M.A., and Du, J. (2010) Visible light photocatalysis as a greener approach to photochemical synthesis. *Nature Chem.*, **2** (7), 527–532.
9 Teplý, F. (2011) Photoredox catalysis by $[Ru(bpy)_3]^{2+}$ to trigger transformations of organic molecules: organic synthesis using visible-light photocatalysis and its 20th century roots. *Collect. Czech. Chem. Commun.*, **76** (7), 859–917.
10 Narayanaman, J.M.R. and Stephenson, C.R.J. (2011) Visible light photoredox catalysis: applications in organic synthesis. *Chem. Soc. Rev.*, **40** (1), 102–113.
11 Tucker, J.W. and Stephenson, C.R.J. (2012) Shining light on photoredox catalysis: theory and synthetic applications. *J. Org. Chem.*, **77** (4), 1617–1622.
12 Xuan, J. and Xiao, W.-J. (2012) Visible-light photoredox catalysis. *Angew. Chem., Int. Ed.*, **51** (28), 6828–6838.
13 Ischay, M.A. and Yoon, T.P. (2012) Accessing the synthetic chemistry of radical ions. *Eur. J. Org. Chem.*, (18), 3359–3372.
14 Wallentin, C.-J., Nguyen, J.D., and Stephenson, C.R.J. (2012) Radical carbon–carbon bond formations enabled by visible light active photocatalysts. *Chimia*, **66** (6), 394–398.
15 Prier, C.K., Rankic, D.A., and MacMillan, D.W.C. (2013) Visible light photoredox catalysis with transition metal complexes: applications in organic synthesis. *Chem. Rev.*, **113** (7), 5322–5363.
16 Xi, Y., Yi, H., and Lei, A. (2013) Synthetic applications of photoredox catalysis with visible light. *Org. Biomol. Chem.*, **11** (15), 2387–2403.
17 (a) Jahn, U. (2012) Radicals in transition metal catalyzed reactions? Transition metal catalyzed radical reactions? A fruitful interplay anyway: Part 1. Radical catalysis by group 4 to group 7 elements. *Top. Curr. Chem.*, **320**, 121–189; (b) Jahn, U. (2012) Radicals in transition metal catalyzed reactions? Transition metal catalyzed radical reactions? A fruitful interplay anyway: Part 2. Radical catalysis by group 8 and 9 elements. *Top. Curr. Chem.*, **320**, 191–322; (c) Jahn, U. (2012) Radicals in transition metal catalyzed reactions? Transition metal catalyzed radical reactions? A fruitful interplay anyway: Part 3. Catalysis by group 10 and 11 elements and bimetallic catalysis. *Top. Curr. Chem.*, **320**, 323–451.
18 Fukuzumi, S., Mochizuki, S., and Tanaka, T. (1990) Photocatalytic reduction of phenacyl halides by 9,l0-dihydro-10-methylacridine: control between the reductive and oxidative quenching pathways of tris(bipyridine)ruthenium complex utilizing an acid catalysis. *J. Phys. Chem.*, **94**, 722–726.

19 Narayanam, J.M.R., Tucker, J.W., and Stephenson, C.R.J. (2009) Electron-transfer photoredox catalysis: development of a tin-free reductive dehalogenation reaction. *J. Am. Chem. Soc.*, **131**, 8756–8757.
20 Furst, L., Matsuura, B.S., Narayanam, J.M.R. *et al.* (2010) Visible C—H functionalization of electron-rich heterocycles with malonates. *Org. Lett.*, **12**, 3104–3107.
21 Wallentin, C.-J., Nguyen, J.D., Finkbeiner, P., and Stephenson, C.R.J. (2012) Visible light-mediated atom transfer radical addition via oxidative and reductive quenching of photocatalysts. *J. Am. Chem. Soc.*, **134**, 8875–8884.
22 Nagib, D.A., Scott, M.E., and MacMillan, D.W.C. (2009) Enantioselective α-trifluoromethylation of aldehydes via photoredox organocatalysis. *J. Am. Chem. Soc.*, **131**, 10875–10877.
23 Shi, H.-W., Vander Wal, M.N., Grange, R.L., and McMillan, D.W.C. (2010) Enantioselective α-benzylation of aldehydes via photoredox organocatalysis. *J. Am. Chem. Soc.*, **132**, 13600–136003.
24 Tucker, J.W. and Stephenson, C.R.J. (2011) Tandem visible light-mediated radical cyclization–divinylcyclopropane rearrangement to tricyclic pyrrolidinones. *Org. Lett.*, **13**, 5468–5471.
25 Andrews, R.S., Becker, J.J., and Gagné, M.R. (2010) Intermolecular addition of glycosyl halides to alkenes mediated by visible light. *Angew. Chem., Int. Ed.*, **49**, 7274–7276.
26 Neumann, M., Füldner, S., König, B., and Zeitler, K. (2011) Metal-free, cooperative asymmetric organophotoredox catalysis with visible light. *Angew. Chem., Int. Ed.*, **50**, 951–954.
27 Nguyen, J.D., D'amato, E.M., Narayanam, M.R., and Stephenson, C.R.J. (2012) Engaging unactivated alkyl, alkenyl and aryl iodides in visible-light-mediated free radical reactions. *Nature Chem*, **4**, 854–859.
28 Ghosh, I., Ghosh, T., Bargadi, J.I., and König, B. (2014) Reduction of aryl halides by consecutive visible light-induced electron transfer processes. *Nature*, **346**, 725–728.
29 Maji, T., Karmakar, A., and Reiser, O. (2011) Visible-light photoredox catalysis: dehalogenation of vicinal dibromo-, α-halo-, and α,α-dibromocarbonyl compounds. *J. Org. Chem.*, **76**, 736–739.
30 Neuman, M. and Zeitler, K. (2012) Application of microflow conditions to visible light photoredox catalysis. *Org. Lett.*, **14**, 2658–2661.
31 Bou-Hamdan, F.R. and Seeberger, P.H. (2012) Visible-light-mediated photochemistry: accelerating $Ru(bpy)_3{}^{2+}$-catalyzed reactions in continuous flow. *Chem. Sci.*, **3**, 1612–1616.
32 Senaweera, S.M., Singh, A., and Weaver, J.D. (2014) Photocatalytic hydrodefluorination: facile access to partially fluorinated aromatics. *J. Am. Chem. Soc.*, **136**, 3002–3005.
33 Nakamura, K., Fujii, M., Mekata, H. *et al.* (1986) Desulfonylation of β-keto sulfones with the hantzsch ester, an Nad(P)H model. *Chem. Lett.*, **15**, 87–88.
34 van Bergen, T.J., Hedstrand, D.M., Kruizinga, W.H., and Kellog, R.M. (1979) Hydride transfer from 1,4-dihydropyridines to sp^3-hybridized carbon in sulfonium salts and activated halides: studies with NAD(P)H models. *J. Am. Chem. Soc.*, **44**, 4953–4962.

35 Hasegawa, E., Takizawa, S., Seida, T. *et al.* (2006) Photoinduced electron-transfer systems consisting of electron-donating pyrenes or anthracenes and benzimidazolines for reductive transformation of carbonyl compounds. *Tetrahedron*, **62**, 6581–6588.

36 Larraufie, M.-H., Pellet, R., Fensterbank, L. *et al.* (2011) Visible-light-induced photoreductive generation of radicals from epoxides and aziridines. *Angew. Chem., Int. Ed.*, **50**, 4463–4466.

37 Okada, K., Okubo, K., Morita, N., and Oda, M. (1992) Reduction decarboxylation of *N*-(acyloxy)phthalimides via redox-initiated radical chain mechanism. *Tetrahedron Lett.*, **33**, 7377–7380.

38 Okada, K., Okamoto, K., Morita, N. *et al.* (1991) Photosensitized decarboxylative michael addition through *N*-(acyloxy)phthalimides via an electron-transfer mechanism. *J. Am. Chem. Soc.*, **113**, 9401–9402.

39 Zhao, Y., Cai, S., JLi, J., and Wang, D.Z. (2013) Visible-light photo-catalytic C—C bond cleavages: preparations of *N,N*-dialkylformamides from 1,2-vicinal diamines. *Tetrahedron*, **69**, 8129–8131.

40 Chen, Y., Kamlet, A.S., Steinman, J.B., and Liu, D.R. (2011) A biomolecule-compatible visible-light-induced azide reduction from a DNA-encoded reaction-discovery system. *Nat. Chem.*, **3**, 146–153.

41 Yang, X.-J., Chen, B., Zheng, L.-Q. *et al.* (2014) Highly efficient and selective photocatalytic hydrogenation of functionalized nitrobenzenes. *Green Chem.*, **16**, 1082–1086.

42 Ohkubo, K., Kobayashi, T., and Fukuzumi, S. (2011) Direct oxygenation of benzene to phenol using quinolinium ions as homogeneous photocatalysts. *Angew. Chem., Int. Ed.*, **50**, 8652.

43 Rao, H., Wang, P., and Li, C.-J. (2012) Visible-light-triggered direct benzoyloxylation of electron-rich arenes at room temperature without chelation assistance. *Eur. J. Org. Chem.*, 6502–6507.

44 Koike, T., Yasu, Y., and Akita, M. (2012) Visible-light-driven oxidation of 1,3-dicarbonyl compounds via catalytic disproportionation of TEMPO by photoredox catalysis. *Chem. Lett.*, **41**, 999.

45 Liu, H., Feng, W., Kee, C.H. *et al.* (2010) Organic dye photocatalyzed α-oxyamination through irradiation with visible light. *Green Chem.*, **12**, 953.

46 Yasu, Y., Koike, T., and Akita, M. (2012) Visible light-induced selective generation of radicals from organoborates by photoredox catalysis. *Adv. Synth. Catal.*, **354**, 3414.

47 Su, Y., Zhang, L., and Jiao, N. (2011) Utilization of natural sunlight and air in the aerobic oxidation of benzyl halides. *Org. Lett.*, **13**, 2168.

48 Zou, Y.-Q., Guo, W., Liu, F.-L. *et al.* (2014) Visible-light-induced photocatalytic formyloxylation reactions of 3-bromooxindoles with water and DMF: the scope and mechanism. *Green Chem.*, **16**, 3787.

49 Wever, W.J., Cinelli, M.A., and Bowers, A.A. (2013) Visible light mediated activation and O-glycosylation of thioglycosides. *Org. Lett.*, **15**, 30.

50 Spell, M., Wang, X., Wahba, A.E. *et al.* (2013) An α-selective, visible light photocatalytic glycosylation of alcohols with selenoglycosides. *Carbohydr. Res.*, **369**, 42.

51 Srivastava, V.P. and Yadav, L.D.S. (2013) Visible-light-triggered oxidative C–H aryloxylation of phenolic amidines; photocatalytic preparation of 2-aminobenzoxazoles. *Synlett*, **24**, 2758.

52 Iqbal, N., Choi, S., You, Y., and Cho, E.J. (2013) Aerobic oxidation of aldehydes by visible light photocatalysis. *Tetrahedron Lett.*, **54**, 6222.

53 Miyake, Y., Nakajima, K., and Nishibayashi, Y. (2012) Direct sp^3 C–H amination of nitrogen-containing benzoheterocycles mediated by visible-light-photoredox catalysts. *Chem. Eur. J.*, **18**, 16473.

54 Kim, H., Kim, T., Lee, D.G. *et al.* (2014) Nitrogen-centered radical-mediated C–H imidation of arenes and heteroarenes via visible light induced photocatalysis. *Chem. Commun.*, **50**, 9273.

55 Maity, S. and Zheng, N.A. (2012) Visible-light-mediated oxidative C–N bond formation/aromatization cascade: photocatalytic preparation of *N*-arylindoles. *Angew. Chem., Int. Ed.*, **51**, 9562.

56 Liu, J., Liu, Q., Yi, H. *et al.* (2014) Visible-light-mediated decarboxylation/oxidative amidation of α-keto acids with amines under mild reaction conditions using O_2. *Angew. Chem., Int. Ed.*, **53**, 502.

57 Majek, M. and von Wangelin, A.J. (2013) Organocatalytic visible light mediated synthesis of arylsulfides. *Chem. Commun.*, **49**, 5507.

58 Kundu, D., Ahmammed, S., and Ranu, B.C. (2014) Visible light photocatalyzed direct conversion of aryl-/heteroarylamines to selenides at room temperature. *Org. Lett.*, **16**, 1814.

59 Keylor, M.H., Park, J.E., Wallentin, C.-J., and Stephenson, C.R.J. (2014) Photocatalytic initiation of thiol–ene reactions: synthesis of thiomorpholin-3-ones. *Tetrahedron*, **70**, 4264.

60 Tyson, E.L., Niemeyer, Z.L., and Yoon, T.P. (2014) Redox mediators in visible light photocatalysis: photocatalytic radical thiol–ene additions. *J. Org. Chem.*, **79**, 1427.

61 Chen, M., Huang, Z.-T., and Zheng, Q.-Y. (2012) Visible light-induced 3-sulfenylation of *N*-methylindoles with arylsulfonylchlorides. *Chem. Commun.*, **48**, 11686.

62 Cheng, Y., Yang, J., Qu, Y., and Li, P. (2012) Aerobic visible-light photoredox radical C–H functionalization: catalytic synthesis of 2-substituted benzothiazoles. *Org. Lett.*, **14**, 98.

63 Ohkubo, K., Mizushima, K., Iwata, R., and Fukuzumi, S. (2011) Selective photocatalytic aerobic bromination with hydrogen bromide via an electron-transfer state of 9-mesityl-10-methylacridinium ion. *Chem. Sci.*, **2**, 715.

64 Zhao, Y., Li, Z., Yang, C. *et al.* (2014) Visible-light photoredox catalysis enabled bromination of phenols and alkenes. *Beilstein J. Org. Chem.*, **10**, 622.

65 Guo, L., Yang, C., Zheng, L., and Xia, W. (2013) Visible light-mediated oxidative quenching reaction to electron-rich epoxides: highly regioselective synthesis of α-bromo (di)ketones and mechanism study. *Org. Biomol. Chem.*, **11**, 5787.

66 Xia, J.-B., Zhu, C., and Chen, C. (2013) Visible light-promoted metal-free C–H activation: diarylketone- catalyzed selective benzylic mono- and difluorination. *J. Am. Chem. Soc.*, **135**, 17494.

67 Rueda-Becerril, M., Mahé, O., Drouin, M. *et al.* (2014) Direct C−F bond formation using photoredox catalysis. *J. Am. Chem. Soc.*, **135**, 17494.

68 Yu, J., Zhang, L., and Yan, G. (2012) Metal-free, visible light-induced borylation of aryldiazonium salts: a simple and green synthetic route to arylboronates. *Adv. Synth. Catal.*, **354**, 2625–2628.

69 Kim, H. and Lee, C. (2012) Visible-light-induced photocatalytic reductive transformations of organohalides. *Angew. Chem., Int. Ed.*, **51**, 12303–12306.

70 Cheng, Y., Gu, X., and Li, P. (2013) Visible-light photoredox in homolytic aromatic substitution: direct arylation of arenes with aryl halides. *Org. Lett.*, **11**, 2664–2667.

71 Sun, C.-L. and Shi, Z.-J. (2014) Transition-metal-free coupling reactions. *Chem. Rev.*, **114**, 9219–9280.

72 Revol, G., McCallum, T., Morin, M. *et al.* (2013) Photoredox transformations with dimeric gold complexes. *Angew. Chem., Int. Ed.*, **52**, 13342–13345.

73 Li, D., Che, C.-M., Kwong, H.-L., and Yam, V.W.-W. (1992) Photoinduced C–C bond formation from alkyl halides catalysed by luminescent dinuclear gold(I) and copper(I) complexes. *J. Chem. Soc., Dalton Trans.*, 3325–3329.

74 Fensterbank, L., Goddard, J.-P., Malacria, M., and Ollivier, C. (2012) Homolytic reduction of onium salts. *Chimia*, **66**, 425–432.

75 Hedstrand, D.M., Kruizinga, W.H., and Kellogg, R.M. (1978) Light induced and dye accelerated reductions of phenacyl onium salts by 1,4-dihydropyridines. *Tetrahedron Lett.*, **19**, 1255.

76 van Bergen, T.J., Kruizinga, W.H., and Kellogg, R.M. (1979) Chemistry of dihydropyridines. 9. Hydride transfer from 1,4-dihydropyridines to sp3-hybridized carbon in sulfonium salts and activated halides: studies with NAD(P)H models. *J. Org. Chem.*, **44**, 4953.

77 Donck, S., Baroudi, A., Fensterbank, L. *et al.* (2013) Visible-light photocatalytic reduction of sulfonium salts as a source of aryl radicals. *Adv. Synth. Catal.*, **355**, 1477–1482.

78 Sandmeyer, T. (1884) Ueber die Ersetzung der Amidgruppe durch Chlor in den aromatischen Substanzen. *Ber. Dtsch. Chem. Ges.*, **17**, 1633–1635.

79 Sandmeyer, T. (1884) Ueber die Ersetzung der Amid-gruppe durch Chlor, Brom und Cyan in den aromatischen Substanzen. *Ber. Dtsch. Chem. Ges.*, **17**, 2650–2653.

80 Pschorr, R. (1896) Neue Synthese des Phenanthrens und seiner Derivate. *Chem. Ber.*, **29**, 496.

81 Gomberg, M. and Bachmann, W.E. (1924) The synthesis of biaryl compounds by means of the diazo reaction. *J. Am. Chem. Soc.*, **42**, 2339–2343.

82 Meerwein, H., Buchner, E., and van Emsterk, K. (1939) *J. Prakt. Chem.*, **152**, 237–245.

83 Kochi, J.K. (1956) The kinetics of the meerwein reaction. *J. Am. Chem. Soc.*, **78**, 1228–1233.

84 Heinrich, M.R. (2009) Intermolecular olefin functionalisation involving aryl radicals generated from arenediazonium salts. *Chem. Eur. J.*, **15**, 820–833.

85 Cano-Yelo, H. and Deronzier, A. (1984) Photocatalysis of the Pschorr reaction by tris-(2,2′-bipyridyl)ruthenium(II) in the phenanthrene series. *J. Chem. Soc., Perkin Trans. 2*, 1093–1098.

86 Cano-Yelo, H. and Deronzier, A. (1987) Photocatalysis of the Pschorr reaction by $Ru(bpy)_3{}^{2+}$. *J. Photochem.*, **37**, 315–321.
87 Cano-Yelo, H. and Deronzier, A. (1984) Photo-oxidation of some carbinols by the Ru(II) polypyridyl complex-aryl diazonium salt system. *Tetrahedron Lett.*, **25**, 5517–5520.
88 Cano-Yelo, H. and Deronzier, A. (1984) Photo-oxidation of tris(2,2′-bipyridine)ruthenium(II) by *para*-substituted benzene diazonium salts in acetonitrile: two-compartment photoelectrochemical cell applications. *J. Chem. Soc., Faraday Trans. 1*, **80**, 3011–3019.
89 Schroll, P., Hari, D.P., and König, B. (2012) Photocatalytic arylation of alkenes, alkynes and enones with diazonium salts. *ChemistryOpen*, **1**, 130–133.
90 Hari, D. P., Schroll, P., König, B. (2012) Metal-free, visible-light-mediated direct C–H arylation of heteroarenes with aryl diazonium salts *J. Am. Chem. Soc.* **134**, 2958–2961.
91 Hering, T., Hari, D. P., König, B. (2012) Visible-light-mediated α-arylation of enol acetates using aryl diazonium salts. *J. Org. Chem.* **77**, 10347–10352.
92 Lalevée, J., Blanchard, N., Tehfe, M.-A. *et al.* (2010) Green bulb light source induced epoxy cationic polymerization under air using tris(2,2′-bipyridine)ruthenium(II) and silyl radicals. *Macromolecules*, **43**, 10191.
93 Lalevée, J., Peter, M., Dumur, F. *et al.* (2011) Subtle ligand effects in oxidative photocatalysis with iridium complexes: application to photopolymerization. *Chem. Eur. J.*, **17**, 15027–15031.
94 Liu, Y.-X., Xue, D., Wang, J.-D. *et al.* (2013) Room temperature arylation of arenes and heteroarenes with diaryliodonium salts by photoredox catalysis. *Synlett*, **24**, 507–513.
95 Baralle, A., Fensterbank, L., Goddard, J.-P., and Ollivier, C. (2013) Aryl radical formation by copper(I) photocatalyzed reduction of diaryliodonium salts: NMR evidence for a CuII/CuI mechanism. *Chem. Eur. J.*, **19**, 10809–10813.
96 McMillin, D.R., Buckner, M.T., and Ahn, B.T. (1977) A light-induced redox reaction of bis(2,9-dimethyl-1,10-phenanthroline)copper(I). *Inorg. Chem.*, **16**, 943–945.
97 Armaroli, N., Accorsi, G., Cardinali, F., and Listorti, A. (2007) Photochemistry and photophysics of coordination compounds: copper. *Top. Curr. Chem.*, **280**, 69–115.
98 Kern, J.-M. and Sauvage, J.-P. (1987) Photoassisted C–C coupling via electron transfer to benzylic halides by a bis(di-imine) copper(I) complex. *J. Chem. Soc., Chem. Commun.*, 546–548.
99 (a) Pirtsch, M., Paria, S., Matsuno, T. *et al.* (2012) [Cu(dap)2Cl] as an efficient visible-light-driven photoredox catalyst in carbon–carbon bond-forming reactions. *Chem. Eur. J.*, **18**, 7336–7340; (b) For a review, see:Paria, S. and Reiser, O. (2014) Copper in photocatalysis. *ChemCatChem*, **6**, 2477–2483.
100 Augusto, C., Hernandez-Perez, C., and Collins, S.K. (2013) A visible-light-mediated synthesis of carbazoles. *Angew. Chem,. Int. Ed.*, **52**, 12696–12700.

101 Hamada, T., Ishida, H., Usui, S. *et al.* (1993) A novel photocatalytic asymmetric synthesis of (*R*)-(+)-1,1′-bi-2-naphthol derivatives by oxidative coupling of 3-substituted-2-naphthol with Δ-$[Ru(menbpy)_3]^{2+}$[menbpy = 4,4′-di(1*R*,2*S*,5*R*)-(−)-menthoxycarbonyl-2,2′-bipyridine], which posseses molecular helicity. *J. Chem. Soc., Chem. Commun.*, 909–911.

102 Hamada, T., Ishida, H., Usui, S. *et al.* (1994) Enantioselective and photocatalytic oxidation of 1,1′-bi-2-naphthol with a chiral ruthenium complex which includes molecular helicity. *J. Mol. Catal.*, **88**, L1–L5.

103 Pham, P.V., Nagib, D.A., and MacMillan, D.W.C. (2011) Photoredox catalysis: a mild, operationally simple approach to the synthesis of a-trifluoromethyl carbonyl compounds. *Angew. Chem., Int. Ed.*, **50**, 6119–6122.

104 Nguyen, J.D., Tucker, J.W., Konieczynska, M.D., and Stephenson, C.R.J. (2011) Intermolecular atom transfer radical addition to olefins mediated by oxidative quenching of photoredox catalysts. *J. Am. Chem. Soc.*, **133**, 4160–4163.

105 Iqbal, N., Choi, S., Ko, E., and Cho, E.J. (2012) Trifluoromethylation of heterocycles via visible light photoredox catalysis. *Tetrahedron Lett.*, **53**, 2005–2008.

106 Iqbal, N., Jung, J., Park, S., and Cho, E.J. (2013) Controlled trifluoromethylation reactions of alkynes through visible-light photoredox catalysis. *Angew. Chem., Int. Ed.*, **52**, 539–542.

107 Iqbal, N., Choi, S., Kim, E., Cho, E. J. (2012) Trifluoromethylation of alkenes by visible light photoredox catalysis. *J. Org. Chem.* **77**, 11383–11387.

108 Yu, C., Iqbal, N., Park, S., Cho, E. J. (2014) Selective difluoroalkylation of alkenes by using visible light photoredox catalysis. *Chem. Commun.* **50**, 12884-12887.

109 Lin, Q., Chu, L., and Qing, F.-L. (2013) Direct introduction of ethoxycarbonyldifluoromethyl-group to heteroarenes with ethyl bromodifluoroacetate via visible-light photocatalysis. *Chin. J. Chem.*, **31**, 885–891.

110 Mizuta, S., Verhoog, S., Engle, K. M., Khotavivattana, T., O'Duill, M., Wheelhouse, K., Rassias, G., Médebielle, M., Gouverneur, V. (2013) Catalytic hydrotrifluoromethylation of unactivated alkenes. *J. Am. Chem. Soc.* **135**, 2505–2508.

111 Mizuta, S., Verhoog, S., Wangb, X. *et al.* (2013) Redox chemistry of trifluoromethyl sulfonium salts as CF_3 radical sources. *J. Fluorine Chem.*, **155**, 124–131.

112 Mizuta, S., Engle, K.M., Verhoog, S. *et al.* (2013) Trifluoromethylation of allylsilanes under photoredox catalysis. *Org. Lett.*, **15**, 1250–1253.

113 Xie, J., Yuan, X., Abdukader, A., Zhu, C., Ma, J. (2014) Visible-light-promoted radical C–H trifluoromethylation of free anilines. *Org. Lett.* **16**, 1768–1771.

114 Yasu, Y., Koike, T., and Akita, M. (2013) Visible-light-induced synthesis of a variety of trifluoromethylated alkenes from potassium vinyltrifluoroborates by photoredox catalysis. *Chem. Commun.*, **49**, 2037–2039.

115 Nagib, D.A. and MacMillan, D.W.C. (2011) Trifluoromethylation of arenes and heteroarenes by means of photoredox catalysis. *Nature*, **480**, 224–228.

116 Jiang, H., Cheng, Y., Zhang, Y., and Yu, S. (2013) Sulfonation and trifluoromethylation of enol acetates with sulfonyl chlorides using visible-light photoredox catalysis. *Eur. J. Org. Chem.*, 5485–5492.

117 Cantillo, D., de Frutos, O., Rincón, J. A., Mateos, C., Kappe, C. O. (2014) Continuous flow α-trifluoromethylation of ketones by metal-free visible light photoredox catalysis. *Org. Lett.* **16**, 896–899.

118 Wilger, D.J., Gesmundo, N.J., and Nicewicz, D.A. (2013) Catalytic hydrotrifluoromethylation of styrenes and unactivated aliphatic alkenes via an organic photoredox system. *Chem. Sci.*, **4**, 3160–3165.

119 Cui, L., Matusaki, Y., Tada, N. *et al.* (2013) Metal-free direct C–H perfluoroalkylation of arenes and heteroarenes using a photoredox organocatalyst. *Adv. Synth. Catal.*, **355**, 2203–2207.

120 Fukuzumi, S. and Tanaka, T. (1984) Tris(bipyridyl)ruthenium(II)-photosensitized reaction of 1-benzyl-1,4-dihydronicotinamide with benzyl bromide. *J. Chem. Soc., Perkin Trans*, **2**, 1705–1709.

121 Furst, L., Narayanam, J.M.R., and Stephenson, C.R.J. (2011) Total synthesis of (+)-gliocladin C enabled by visible-light photoredox catalysis. *Angew. Chem., Int. Ed.*, **50**, 9655–9659.

122 Tucker, J.W., Zhang, Y., Jamison, T.F., and Stephenson, C.R.J. (2012) Visible light photoredox catalysis in flow. *Angew. Chem., Int. Ed.*, **51**, 4144.

123 Ju, X., Liang, Y., Jia, P. *et al.* (2012) Synthesis of oxindoles via visible light photoredox catalysis. *Org. Biomol. Chem.*, **10**, 498–501.

124 Jiang, H., Huang, C., Guo, J. *et al.* (2012) Direct C—H functionalization of enamides and enecarbamates by using visible-light photoredox catalysis. *Chem. Eur. J.*, **18**, 15158–15166.

125 Courant, T. and Masson, G. (2012) Photoredox-initiated α-alkylation of imines through a three-component radical/cationic reaction. *Chem. Eur. J.*, **18**, 423–427.

126 Schnermann, M.J. and Overman, L.E. (2012) A concise synthesis of (−)-aplyviolene facilitated by a strategic tertiary radical conjugate addition. *Angew. Chem., Int. Ed.*, **51**, 9576–9580.

127 Lackner, G. L., Quasdorf, K. W., Overman, L. E. (2013) Direct construction of quaternary carbons from tertiary alcohols via photoredox-catalyzed fragmentation of *tert*-alkyl *N*-phthalimidoyl oxalates. *J. Am. Chem. Soc.* **135**, 15342–15345.

128 Miyazawa, K., Yasu, Y., Koike, T., and Akita, M. (2013) Visible-light-induced hydroalkoxymethylation of electron-deficient alkenes by photoredox catalysis. *Chem. Commun.*, **49**, 7249–7251.

129 Huang, H., Zhang, G., Gong, L. *et al.* (2014) Visible-light-induced chemoselective deboronative alkynylation under biomolecule-compatible conditions. *J. Am. Chem. Soc.*, **136**, 2280–2283.

130 Daniel, M., Fensterbank, L., Goddard, J.-P., and Ollivier, C. (2014) Visible-light photocatalytic oxidation of 1,3-dicarbonyl compounds and carbon–carbon bond formation. *Org. Chem. Front.*, **1**, 551–555.

131 Yasu, Y., Koike, T., and Akita, M. (2012) Sunlight-driven synthesis of c-diketones via oxidative coupling of enamines with silyl enol ethers catalyzed by $[Ru(bpy)_3]^{2+}$. *Chem. Commun.*, **48**, 5355–5357.

132 Protti, S., Ravelli, D., Fagnoni, M., and Albini, A. (2009) Solar light-driven photocatalyzed alkylations. *Chem. Commun.*, 7351–7353.

133 Chu, L., Ohta, C., Zuo, Z., and MacMillan, D.W.C. (2014) Carboxylic acids as a traceless activation group for conjugate additions: a three-step synthesis of (±)-lyrica. *J. Am. Chem. Soc.*, **136**, 10886–10889.

134 Miyake, Y., Nakajima, K., and Nishibayashi, Y. (2013) Visible light-mediated oxidative decarboxylation of arylacetic acids into benzyl radicals: addition to electron-deficient alkenes by using photoredox catalysts. *Chem. Commun.*, **49**, 7854–7856.

135 Hu, J., Wang, J., Nguyen, T.H., and Zheng, N. (2013) the chemistry of amine radical cations produced by visible light photoredox catalysis. *Beilstein J. Org. Chem.*, **9**, 1977–2001.

136 McNally, A., Prier, C.K., and MacMillan, D.W.C. (2011) Discovery of an α-amino C—H arylation reaction using the strategy of accelerated serendipity. *Science*, **334**, 1114–1117.

137 Singh, A., Arora, A., and Weaver, J.D. (2013) Photoredox-mediated C—H functionalization and coupling of tertiary aliphatic amines with 2-chloroazoles. *Org. Lett.*, **15**, 5390–5393.

138 Miyake, Y., Nakajima, K., Nishibayashi, Y. (2012) Visible-light-mediated utilization of α-aminoalkyl radicals: addition to electron-deficient alkenes using photoredox catalysts. *J. Am. Chem. Soc.* **134**, 3338–3341.

139 Dai, X., Cheng, D., Guan, B. *et al.* (2014) The coupling of tertiary amines with acrylate derivatives via visible light photoredox catalysis. *J. Org. Chem.*, **79**, 7212–7219.

140 Noble, A. and MacMillan, D.W.C. (2014) Photoredox α-vinylation of α-amino acids and *N*-aryl amines. *J. Am. Chem. Soc.*, **136**, 11602–11605.

141 Miyake, Y., Ashida, Y., Nakajima, K., and Nishibayashi, Y. (2012) Visible-light-mediated addition of α-aminoalkyl radicals generated from a-silylamines to α,β-unsaturated carbonyl compounds. *Chem. Commun.*, **48**, 6966–6968.

142 Zhou, H., Lu, P., Gu, X., and Li, P. (2013) Visible-light-mediated nucleophilic addition of an α-aminoalkyl radical to isocyanate or isothiocyanate. *Org. Lett.*, **15**, 5646–5649.

143 Kohls, P., Jadhav, D., Pandey, G., and Reiser, O. (2012) Visible light photoredox catalysis: generation and addition of *N*-aryltetrahydroisoquinoline-derived R-amino radicals to michael acceptors. *Org. Lett.*, **14**, 672–675.

144 Zhang, P., Xiao, T., Xiong, S. *et al.* (2014) Synthesis of 3-acylindoles by visible-light induced intramolecular oxidative cyclization of O-alkynylated *N,N*-dialkylamines. *Org. Lett.*, **16**, 3264–3267.

145 Baralle, A., Baroudi, A., Daniel, M. *et al.* (2012) Radical cascade reactions, in *Encyclopedia of Radicals in Chemistry, Biology and Materials* (eds C. Chatgilialoglu and A. Studer), John Wiley & Sons, Chichester, UK, pp. 729–766.

146 Albert, M., Fensterbank, L., Lacôte, E., and Malacria, M. (2006) Tandem radical reactions. *Top. Curr. Chem.*, **264**, 1–62.

147 Tucker, J.W., Nguyen, J.D., Narayanam, J.M.R. *et al.* (2010) Tin-free radical cyclization reactions initiated by visible light photoredox catalysis. *Chem. Commun.*, **46** (27), 4985–4987.

148 Curran, D. P., Rakiewicz, D. M. (1985) The tandem radical approach to linear condensed cyclopentanoids: total synthesis of (±)-hirsutene. *J. Am. Chem. Soc.*, **107** (5), 1448–1449.

149 Tanabe, Y.; Nishii, Y., Wakimura, K.-I. (1994) A novel and regioselective radical cyclization of gem-dihalocyclopropyl substituted alkenes and alkynes using tributyltin hydride and catalytic AIBN *Chem. Lett.*, (9), 1757–1760.

150 Tucker, J. W., Narayanam, J. M. R., Krabbe, S. W., Stephenson, C. R. J. (2010) Electron transfer photoredox catalysis: intramolecular radical addition to indoles and pyrroles. *Org. Lett.*, **12** (2), 368–371.

151 Ischay, M.A., Anzovino, M.E., Du, J., and Yoon, T.P. (2008) Efficient visible light photocatalysis of [2 + 2] enone cycloadditions. *J. Am. Chem. Soc.*, **130** (39), 12886–12887.

152 Lin, S., Padilla, C.E., Ischay, M.A., and Yoon, T.P. (2012) Visible light photocatalysis of intramolecular radical cation Diels–Alder cycloadditions. *Tetrahedron Lett.*, **53** (24), 3073–3076.

153 Lu, Z., Shen, M., and Yoon, T.P. (2011) [3 + 2] Cycloadditions of aryl cyclopropyl ketones by visible light photocatalysis. *J. Am. Chem. Soc.*, **133** (5), 1162–1164.

154 Fu, W., Xu, F., Fu, Y. *et al.* (2013) Synthesis of 3,3-disubstituted oxindoles by visible-light-mediated radical reactions of aryl diazonium Salts with *N*-arylacrylamides. *J. Org. Chem.*, **78** (23), 12202–12206.

155 Xie, J., Xu, P., Li, H. *et al.* (2013) A room temperature decarboxylation/C–H functionalization cascade by visible-light photoredox catalysis. *Chem. Commun.*, **49** (50), 5672–5674.

156 Xu, P., Xie, J., Xue, Q. *et al.* (2013) Visible-light-induced trifluoromethylation of *N*-aryl acrylamides: a convenient and effective method to synthesize CF_3-containing oxindoles bearing a quaternary carbon center. *Chem. Eur. J.*, **19** (42), 14039–14042.

157 Tang, X.-J., Thomoson, C.S., and Jr. Dolbier, W.R. (2014) Photoredox-catalyzed tandem radical cyclization of *N*-arylacrylamides: general methods to construct fluorinated 3,3-disubstituted 2-oxindoles using fluoroalkylsulfonyl chlorides. *Org. Lett.*, **16** (17), 4594–4597.

158 Jiang, H., Cheng, Y., Wang, R. *et al.* (2013) Synthesis of 6-alkylated phenanthridine derivatives using photoredox neutral somophilic isocyanide insertion. *Angew. Chem., Int. Ed.*, **52** (50), 13289–13292.

159 Wang, R., Jiang, H., Cheng, Y. *et al.* (2014) Somophilic isocyanide insertion: synthesis of 6-arylated and 6-trifluoromethylated phenanthridines. *Synthesis*, **46** (20), 2711–2726.

160 Gu, L., Jin, C., Liu, J. *et al.* (2014) Transition-metal-free, visible-light-induced cyclization of arylsulfonyl chlorides with 2-isocyanobiphenyls to produce phenanthridines. *Chem. Commun.*, **50** (35), 4643–4645.

161 Jiang, H., Cheng, Y., Wang, R. *et al.* (2014) Synthesis of isoquinolines via visible light-promoted insertion of vinyl isocyanides with diaryliodonium salts. *Chem. Commun.*, **50** (46), 6164–6167.

162 Cheng, Y., Yuan, X., Jiang, H. *et al.* (2014) Regiospecific synthesis of 1-trifluoromethylisoquinolines enabled by photoredox somophilic vinyl isocyanide insertion. *Adv. Synth. Catal.*, **356** (13), 2859–2866.
163 Xia, J.-D., Deng, G.-B., Zhou, M.-B. *et al.* (2012) Reusable visible light photoredox catalysts; catalyzed benzylic C(sp^3)—H functionalization/carbocyclization reactions. *Synlett*, **23** (18), 2707–2713.
164 Deng, G.-B., Wang, Z.-Q., Xia, J.-D. *et al.* (2013) Tandem cyclizations of 1,6-enynes with arylsulfonyl chlorides by using visible-light photoredox catalysis. *Angew. Chem., Int. Ed.*, **52** (5), 1535–1538.
165 Du, J. and Yoon, T.P. (2009) Crossed intermolecular [2 + 2] cycloadditions of acyclic enones via visible light photocatalysis. *J. Am. Chem. Soc.*, **131** (41), 14604–14605.
166 Du, J., Skubi, K.L., Schultz, D.M., and Yoon, T.P. (2014) A dual-catalysis approach to enantioselective [2 + 2] photocycloadditions using visible light. *Science*, **344** (6182), 392–396.
167 Zhao, G., Yang, C., Guo, L., Sun, H., Lin, R., Xia, W. (2012) Reactivity insight into reductive coupling and aldol cyclization of chalcones by visible light photocatalysis. *J. Org. Chem.*, **77** (14), 6302–6306.
168 Hari, D.P., Hering, T., and König, B. (2012) Visible light photocatalytic synthesis of benzothiophenes. *Org. Lett.*, **14** (20), 5334–5337.
169 Maity, S., Zhu, M., Shinabery, R.S., and Zheng, N. (2012) Intermolecular [3 + 2] cycloaddition of cyclopropylamines with olefins by visible-light photocatalysis. *Angew. Chem., Int. Ed.*, **51** (1), 222–226.
170 Nguyen, T.H., Morris, S.A., and Zheng, N. (2014) Intermolecular [3 + 2] Annulation of cyclopropylanilines with alkynes, enynes, and diynes via visible light photocatalysis. *Adv. Synth. Catal.*, **356** (13), 2831–2837.
171 Jiang, H., Cheng, Y., Zhang, Y., and Yu, S. (2013) De novo synthesis of polysubstituted naphthols and furans using photoredox neutral coupling of alkynes with 2-bromo-1,3-dicarbonyl compounds. *Org. Lett.*, **15** (18), 4884–4887.
172 Xiao, T., Dong, X., Tang, Y., and Zhou, L. (2012) Phenanthrene synthesis by eosin Y-catalyzed, visible light-induced [4 + 2] benzannulation of biaryldiazonium salts with alkynes. *Adv. Synth. Catal.*, **354** (17), 3195–3199.
173 Paria, S. and Reiser, O. (2014) Visible light photoredox catalyzed cascade cyclizations of α-bromochalcones or α-bromocinnamates with heteroarenes. *Adv. Synth. Catal.*, **356** (2–3), 557–562.
174 Dong, X., Xu, Y., Liu, J.J. *et al.* (2013) Visible-light-induced radical cyclization of trifluoroacetimidoyl chlorides with alkynes: catalytic synthesis of 2-trifluoromethyl quinolines. *Chem. Eur. J.*, **19** (50), 16928–16933.
175 Ju, X., Li, D., Li, W. *et al.* (2012) The reaction of tertiary anilines with maleimides under visible light redox catalysis. *Adv. Synth. Catal.*, **354** (18), 3561–3567.
176 Zhu, S., Das, A., Bui, L., Zhou, H., Curran, D. P., Rueping, M. (2013) Oxygen switch in visible-light photoredox catalysis: radical additions and cyclizations and unexpected C—C-bond cleavage reactions. *J. Am. Chem. Soc.*, **135** (5), 1823–1829.

177 Godineau, E. and Landais, Y. (2009) Radical and radical-ionic multicomponent processes. *Chem. Eur. J.*, **15** (13), 3044–3055.
178 Yi, H., Zhang, X., Qin, C., Liao, Z., Liu, J., Lei, A. (2014) Visible light-induced γ-alkoxynitrile synthesis via three-component alkoxycyanomethylation of alkenes. *Adv. Synth. Catal.* **356** (13), 2873–2877.
179 Wei, X.-J., Yang, D.-T., Wang, L., Song, T., Wu, L.-Z., Liu, Q. (2013) A novel intermolecular synthesis of γ-lactones via visible-light photoredox catalysis. *Org. Lett.* **15** (23), 6054–6057.
180 Fumagalli, G., Boyd, S., and Greaney, M.F. (2013) Oxyarylation and aminoarylation of styrenes using photoredox catalysis. *Org. Lett.*, **15** (6), 1250–1253.
181 Prasad Hari, D., Hering, T., and Koenig, B. (2014) The photoredox-catalyzed Meerwein addition reaction: intermolecular amino-arylation of alkenes. *Angew. Chem., Int. Ed.*, **53** (3), 725–728.
182 Yasu, Y., Koike, T., and Akita, M. (2012) Three-component oxytrifluoromethylation of alkenes: highly efficient and regioselective difunctionalization of C=C bonds mediated by photoredox catalysts. *Angew. Chem., Int. Ed.*, **51** (38), 9567–9571.
183 Tomita, R., Yasu, Y., Koike, T., and Akita, M. (2014) Combining photoredox-catalyzed trifluoromethylation and oxidation with DMSO: facile synthesis of α-trifluoromethylated ketones from aromatic alkenes. *Angew. Chem., Int. Ed.*, **53** (28), 7144–7148.
184 Li, L., Chen, Q.-Y., Guo, Y. (2014) Synthesis of α-CF_3 ketones from alkenes and electrophilic trifluoromethylating reagents by visible-light driven photoredox catalysis. *J. Fluorine Chem.*, **167** (1), 79–83.
185 Yasu, Y., Arai, Y., Tomita, R. *et al.* (2014) Highly regio- and diastereoselective synthesis of CF_3-substituted lactones via photoredox-catalyzed carbolactonization of alkenoic acids. *Org. Lett.*, **16** (3), 780–783.
186 Yasu, Y., Koike, T., and Akita, M. (2013) Intermolecular aminotrifluoromethylation of alkenes by visible-light-driven photoredox catalysis. *Org. Lett.*, **15** (9), 2136–2139.
187 Dagousset, G., Carboni, A., Magnier, E., and Masson, G. (2014) Photoredox-induced three-component azido-, and aminotrifluoromethylation of alkenes. *Org. Lett.*, **16** (16), 4340–4343.
188 Carboni, A., Dagousset, G., Magnier, E., and Masson, G. (2014) Photoredox-induced three-component oxy-, amino-, and carbotrifluoromethylation of enecarbamates. *Org. Lett.*, **16** (4), 1240–1243.
189 Lin, S., Ischay, M.A., Fry, C.G., and Yoon, T.P. (2011) Radical cation Diels–Alder cycloadditions by visible light photocatalysis. *J. Am. Chem. Soc.*, **133** (48), 19350–19353.
190 Ischay, M.A., Lu, Z., and Yoon, T.P. (2010) [2 + 2] Cycloadditions by oxidative visible light photocatalysis. *J. Am. Chem. Soc.*, **132** (25), 8572–8574.
191 Grandjean, J.-M.M. and Nicewicz, D.A. (2013) Synthesis of highly substituted tetrahydrofurans by catalytic polar-radical-crossover cycloadditions of alkenes and alkenols. *Angew. Chem., Int. Ed.*, **52** (14), 396.

3

Atom Transfer Radical Addition using Photoredox Catalysis

Theresa M. Williams and Corey R. J. Stephenson

University of Michigan, Department of Chemistry, 930 North University Avenue, Ann Arbor, MI 48109, USA

3.1 Introduction

The mono- and difunctionalization of olefins is a critical paradigm in synthetic chemistry that has inspired solutions as simple as halohydrin formation, to complex ones such as transition metal-mediated C—C bond formation. Olefins are one of the most fundamental functional groups in organic chemistry and are often derived from simple chemical feedstock. The difunctionalization of alkenes is a rapid and straightforward strategy adopted to synthesize highly functionalized compounds.

Formally, atom transfer radical addition (ATRA) is the transposition of a σ bond of an atom transfer agent across a π bond of an alkene or an alkyne, via free radical intermediates, to afford two new sigma bonds, a C—X bond and a C—C bond [1]. The ATRA reaction was first investigated in great detail by Kharasch and coworkers in the mid-1900s [2]. The critical discovery afforded in these reports was that the use of a methyl free radical enabled the study of ester free radicals with olefins. It was previously known that the methyl free radical could abstract α-hydrogen atoms from various moieties, such as carboxylic acids, esters, and nitriles. However, the addition of these radicals to olefins could not be studied as the olefins could successfully compete with the acid or ester for the methyl free radical. Kharasch and coworkers utilized α-bromo esters, as these could produce the ester free radical and methyl free radical. Since an α-bromine atom is much more susceptible to attack by a methyl free radical than an α-hydrogen atom of an ester or any hydrogen atom in an olefin, these ester free radicals could be generated in olefinic solvents, such as 1-octene or propylene, and their reactions with these olefins could now be studied (Figure 3.1). The developed method was found to efficiently propagate if a sterically unencumbered bromo ester was used in combination with a terminal olefin.

Kharasch *et al.* were the first to demonstrate free radical chain reactions between ester free radicals and various olefins using a methyl free radical initiator [1]. The organic peroxide radical initiator undergoes thermal decomposition to generate an alkyl radical ($In^\bullet$), which is able to cleave the σ bond of the atom transfer agent, thus resulting in the alkyl radical ($R^\bullet$). The formed $R^\bullet$ can

Visible Light Photocatalysis in Organic Chemistry, First Edition.
Edited by Corey R. J. Stephenson, Tehshik P. Yoon and David W. C. MacMillan.

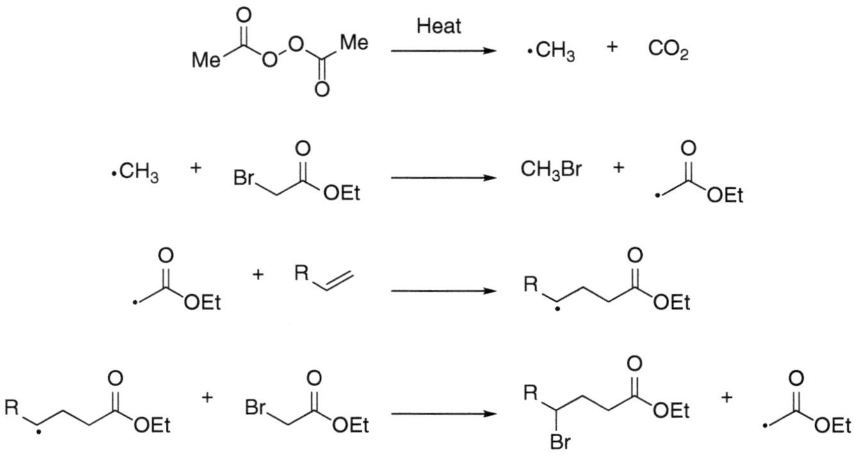

Figure 3.1 Kharasch and coworkers' seminal investigations of ATRA.

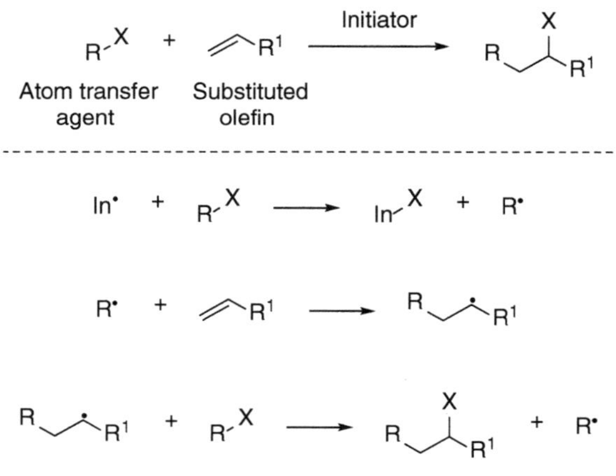

Figure 3.2 General ATRA mechanism.

subsequently add to the olefin, causing cleavage of the π-bond and creating a new carbon radical, which abstracts X• from the starting atom transfer agent [1] (Figure 3.2).

The seminal studies by Kharasch and coworkers were followed by investigations from the groups of Curran *et al.* [2], Oshima and coworkers [3], and Renaud *et al.* [4] who further developed ATRA into a useful tool for organic chemists. Curran and coworkers [2] used organotin reagents, such as hexabutylditin, as radical initiators to induce atom transfer radical cyclizations conducted using alkyl iodides and iodomalonates. They were able to show that the generated electrophilic radicals could undergo cyclization, thereby enabling the construction of five-, six-, and seven-membered rings. The size of the formed ring is dictated by the chain length and substituent effects. In principle, cyclization could be conducted using stoichiometric tin hydride; however, this method affords a reduced cyclized product. ATRA reactions using organotin reagents preserve the iodide and result in alkyl iodide products, allowing for subsequent synthetic transformations. Oshima and coworkers [3] employed triethylborane/O_2 as an efficient radical initiator in atom transfer radical cyclization reactions. Triethylborane-induced radical reactions can take place efficiently at low temperatures, even at −78 °C, in the presence of a trace amount of dioxygen. Furthermore, they can also be conducted in various solvents including alcohols and aqueous solutions due to the stability of triethylborane. The authors were able to show efficient atom transfer radical cyclization reactions at ambient temperatures in water.

Typical ATRA initiators include peroxides [2], organotin reagents [3], and triethylborane [4], which are toxic and hazardous to work with. Other less common initiators have also been used, including *p*-methoxybenzene–diazonium tetrafluoroborate in combination with $TiCl_3$ [5], dimanganese decacarbonyl [6], copper [7], iron [8], bimetallic Rh—Ru complexes [9], and chromium(II) acetate [10]. Initially, ATRA was limited to substrates that did not readily undergo free radical polymerization. However, it was found that transition metal complexes act as better halogen atom transfer reagents and can catalyze ATRA through a reversible redox process. These catalysts provided better selectivity for the monoadduct; however, these systems required large concentrations of the catalysts to be efficient [11]. To overcome this limitation, regeneration of the catalysts was employed through the addition of reducing agents such as, free radical diazo initiators, magnesium, or ascorbic acid [12]. Ruthenium complexes with reducing agents such as Mg led to powerful catalytic systems for ATRA reactions [13]. These were also advantageous as the Ru^{III} complexes were air stable and easy to synthesize. Copper complexes for transition metal-catalyzed ATRA reactions have also been effective using ascorbic acid as the reducing agent [14]. Although ATRA reactions allow for efficient difunctionalization of alkenes or alkynes, most methods unfortunately employ harsh reaction conditions and/or toxic reagents and lack broad functional group tolerance.

Given the propensity for photoredox catalysis to initiate radical reactions, it was the natural choice for the further development of ATRA reactions toward obviating the aforementioned shortcomings. Over the past decade the utility of photoredox catalysis to initiate organic transformations has been widely demonstrated [15]. The most commonly employed photoredox catalysts include ruthenium- and iridium-based polypyridyl complexes (Figure 3.3) due to their stability, their ability to produce a long-lived excited triplet state, and favorable photoredox properties. The use of visible light as a renewable energy source allows for simple reaction setups. In these methods, a photocatalyst is initially excited by visible light; this excited photocatalyst can now enter into either the reductive or the oxidative quenching cycle (Figure 3.4) While photoredox-catalyzed ATRA was considered to be a closed cycle, recent recognition of propagation pathways has provided a greater understanding as to the mechanisms and efficiencies of these processes. Utilizing visible light-mediated catalysts as initiators, the ATRA protocol can be performed under mild reaction conditions, have minimal side reactions, and operate with optimal catalytic efficiency and straightforward purification.

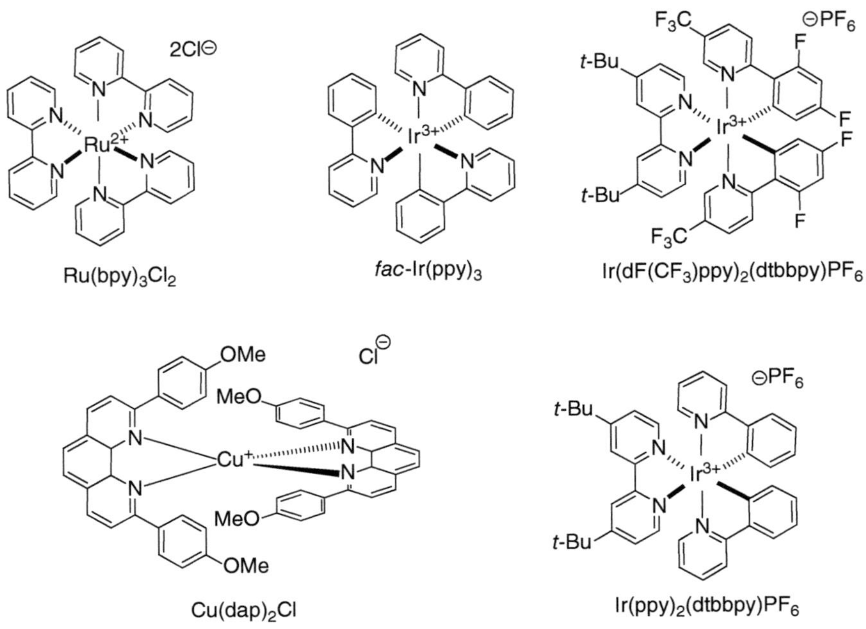

Figure 3.3 Common photoredox catalysts.

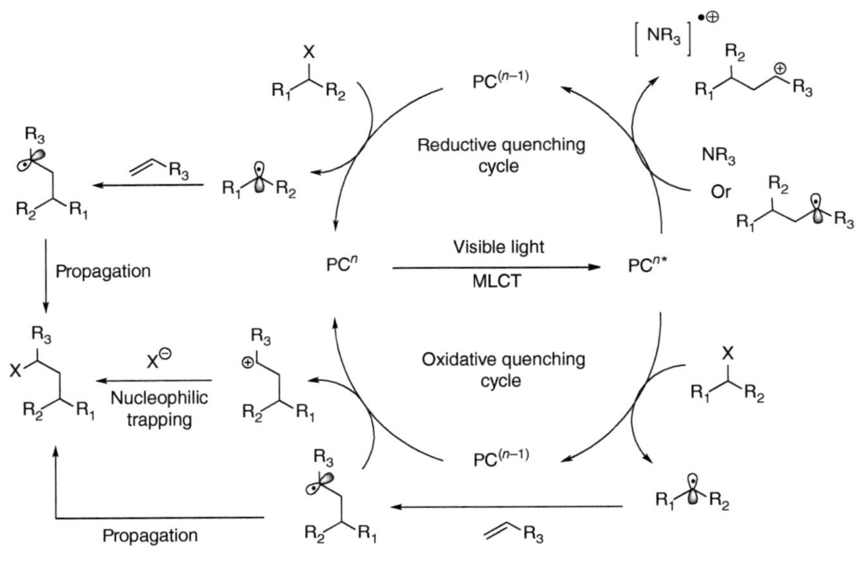

Figure 3.4 Reductive and oxidative photocatalytic cycle for ATRA.

3.2 Transition Metal-Catalyzed ATRA

3.2.1 Ruthenium- and Iridium-Based ATRA

An early demonstration of visible light-promoted ATRA was provided by Barton and coworkers (Scheme 3.1) in 1994 [16]. Using a $[Ru(bpy)_3]^{2+}$ photocatalyst, a probe was conducted on the transposition of phenylselenium sulfonates about a number of vinyl ether substrates and it was found that β-phenylselenosulfones could be produced in high yields. At that time, the proposed mechanism was through excitation of the $[Ru(bpy)_3]^{2+}$ photocatalyst, which would promote single-electron reduction of phenylselenium sulfonates that could subsequently be added to a vinyl ether substrate. The newly generated radical has the ability to react with the starting phenylselenium sulfonate in a propagative manner. Despite this early example, visible light-mediated ATRA did not receive major recognition until around 2011.

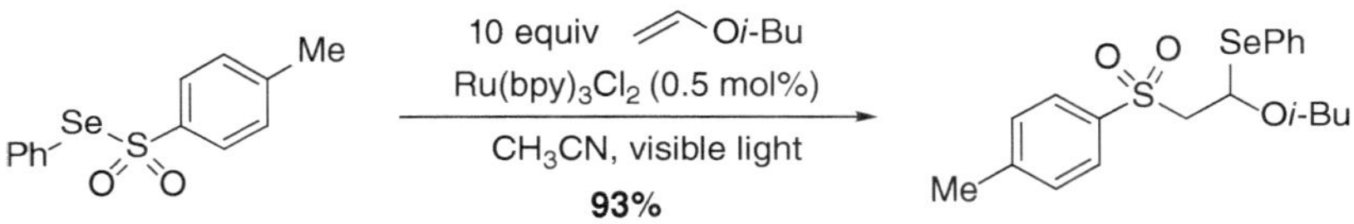

Scheme 3.1 Early example of photochemical ATRA.

The Stephenson group has shown the ability of these photoredox catalysts to both functionalize C—H bonds and reduce C—Br bonds through the reductive quenching cycle [17]. However, the reactive intermediates generated by the reductive quenching cycle often led to competitive side reactions such as hydrogen atom abstraction [17e]. The use of the substrate itself as the excited state quencher suppressed the side reactions and led to a net redox-neutral transformation. Stephenson and coworkers developed a redox-neutral reaction between bromomalonates and olefins, which invigorated interest in ATRA chemistry. Stephenson and coworkers provided a general method for visible light-mediated ATRA using either the oxidative or reductive quenching cycle of photocatalysts [18]. A broad range of atom transfer agents were utilized and successfully added across terminal alkenes, cyclic alkenes, and terminal alkynes (Figure 3.5). Two potential reaction pathways may be operative for this transformation. The ATRA product can be formed through either a propagative mechanism or a radical-polar crossover mechanism (Figure 3.6).

3.2.1.1 Mechanistic Investigations

Using the oxidative quenching cycle of $[Ir\{dF(CF_3)ppy\}_2(dtbbpy)]PF_6$ ($Ir^{IV}/Ir^{III*} = -0.89$ V vs. SCE) or $[Ru(bpy)_3]Cl_2$ ($Ru^{III}/Ru^{II*} = -0.81$ V vs. SCE) several ATRA products are obtained in excellent yields [18]. The postulated mechanism is initiated by the excitation of $[Ru(bpy)_3]^{2+}$ with blue light ($\lambda \approx 450$ nm). The excited photocatalyst can subsequently reduce the haloalkane substrate, formally oxidizing the photocatalyst. The generated carbon-centered radical is sufficiently nucleophilic to add to an alkene (Figure 3.6). The ATRA product can then be formed via two different reaction pathways. The radical addition intermediate

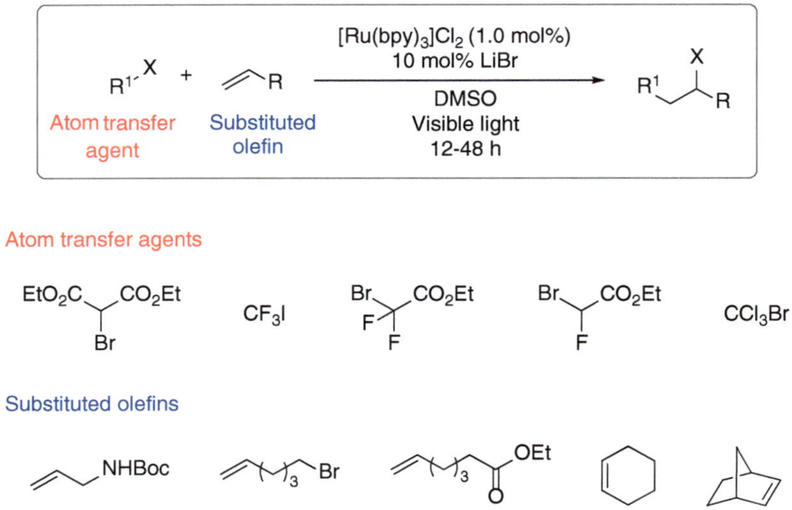

Figure 3.5 Stephenson select ATRA examples.

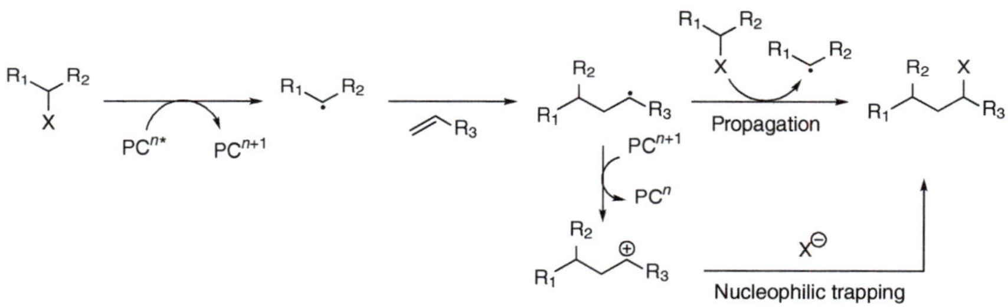

Figure 3.6 Photocatalytic ATRA mechanism.

can abstract a halogen from an equivalent of the starting material in a radical propagation step to give the ATRA product. Alternatively, the radical addition intermediate can be oxidized to the carbocation by the photocatalyst to complete the catalytic cycle and generate the ATRA product through nucleophilic trapping of a halide, which is a radical-polar crossover pathway [19].

Upon investigation of the mechanism of this ATRA reaction, evidence was found that supported both the radical-polar crossover (Figure 3.7a) and the radical propagation pathways (Figure 3.7b). When the ATRA of diethyl bromomalonate was performed with 4-penten-1-ol, the tetrahydrofuran by-product A was isolated in 10% yield. Re-subjecting the atom transfer product to the reaction conditions or refluxing it in toluene did not produce by-product A. These observations suggest that A is generated from nucleophilic trapping of a carbocation intermediate by the tethered alcohol, implicating a radical-polar crossover mechanism. An additional experiment was conducted in which the reaction was subjected to periods of visible light irradiation and periods of complete lack of visible light until the reaction was complete. It was observed that the

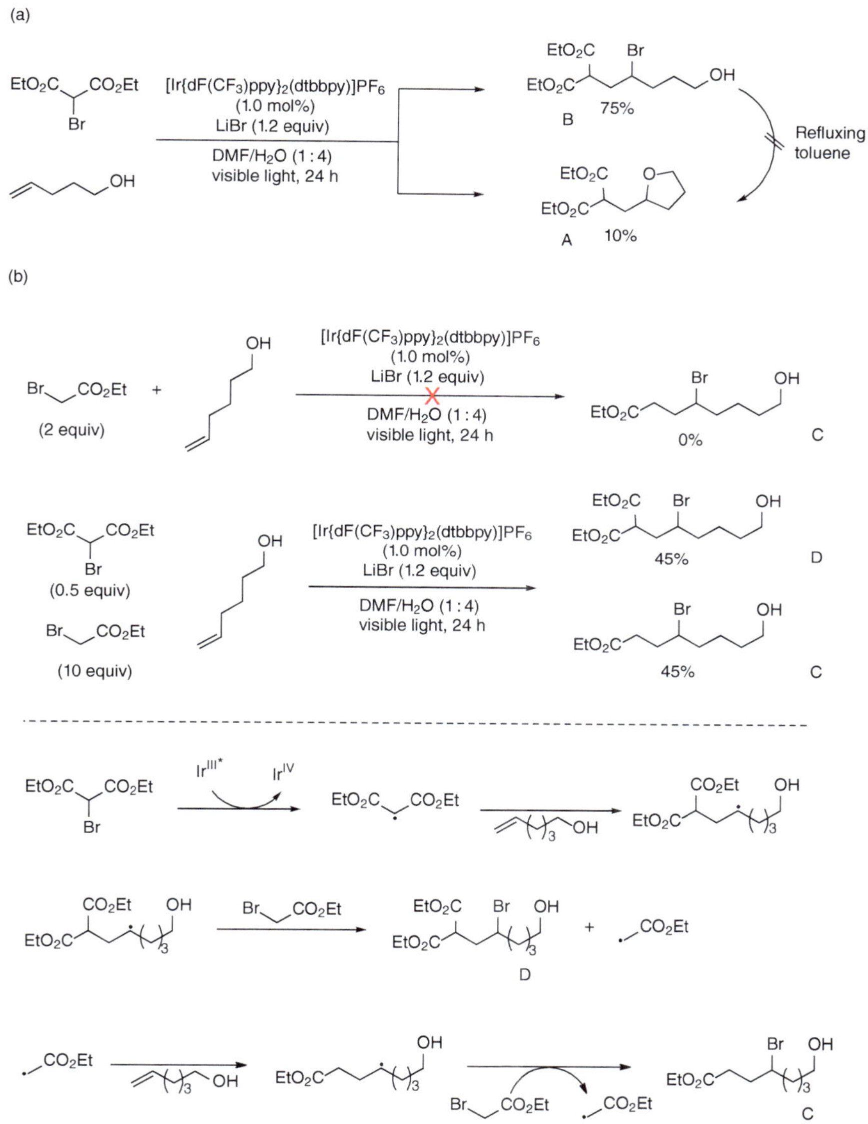

Figure 3.7 (a) Evidence for radical-polar crossover mechanism. (b) Evidence for propagation mechanism.

reaction progressed steadily with light irradiation; however, consumption of the alkene stalled in periods of darkness. While this does not negate a propagation mechanism it is evident that if propagation is an active component, the chains are short lived.

To assess if propagation is an active mechanistic pathway, ethyl bromoacetate was used as a probe. Ethyl bromoacetate does not undergo ATRA via oxidative quenching when [Ir{dF(CF_3)ppy}$_2$(dtbbpy)]PF_6 is used as a photocatalyst. If ethyl bromoacetate is present in a reaction with a functional substrate such as diethyl bromomalonate, it is possible that any product formed with ethyl bromoacetate originates from a propagative mechanism. In fact, the ATRA product of ethyl bromoacetate C was obtained as a mixture with the ATRA product of diethyl bromomalonate D in a 1 : 1 molar ratio. This suggests that the radical chain propagation mechanism is operative in the oxidative quenching cycle of the photocatalyst for ATRA. All these mechanistic experiments indicate that both the radical-polar crossover and the radical chain propagation pathways are accessible in the photoredox ATRA reaction. The extent to which one pathway is favored over the other depends on the nature of the substrates and the conditions used [20].

3.2.1.2 Ruthenium- and Iridium-Based ATRA

Addition reactions of perfluoroalkyl iodides to alkenes and alkynes are a common mode of synthesizing fluorous compounds [21]. Compounds with perfluorinated carbons are of particular interest for the synthesis of small molecules [21]. By attaching a perfluorinated alkyl chain, either temporarily or permanently, to a reagent or a reactant, a previously tedious product isolation is now considerably made easy [22]. While the iridium-based methodology was successfully used to carbobrominate a number of olefins, any attempt to apply this to the iodoperfluoroalkylation of alkenes and alkynes was unsuccessful. Due to the unsatisfactory results of iodoperfluoroalkylation of alkenes using the oxidative quenching pathway of $[Ir\{dF(CF_3)ppy\}_2(dtbbpy)]PF_6$, the reductive quenching cycle of $[Ru(bpy)_3]Cl_2$ ($Ru^{II}/Ru^{I} = -1.33$ V vs. SCE) was considered. Using sodium ascorbate as the reductant in a mixed solvent system gave excellent yields of the iodoperfluoroalkylation for a variety of substrates [18]. The development of ATRA via the oxidative and reductive quenching of photocatalysts has been shown to be a reliable and versatile method to form new carbon–carbon and carbon–halogen bonds. These early studies contributed to the growing interest in visible light-mediated ATRA reactions.

Further development in the field of ATRA chemistry was achieved by Yu and coworkers [23] who showed that *N*-chloroamines could be utilized as atom transfer agents to obtain 1,2-haloamines using photoredox catalysis. Vicinal haloamines can be versatile synthetic intermediates for the synthesis of functional materials by the substitution of the halogen atom with various nucleophiles. They reported the use of an iridium catalyst ($Ir(ppy)_2(dtbbpy)PF_6$) to affect the ATRA of *N*-chlorosulfonamides onto alkenes. The use of $Ir(ppy)_2(dtbbpy)PF_6$ in dichloroethane (DCE) with white LED strips was found to reduce *N*-chlorosulfonamides, thus enabling the addition to olefins (Figure 3.8). The substrate scope was broad with electron-rich benzenesulfonamides giving the highest yields compared to their electron-neutral and electron-poor counterparts. The alkyl groups of the *N*-chlorosulfonamides were well tolerated with decreases in yields resulting from bulky groups due to steric hindrance. A variety of styrene derivatives afforded satisfactory yields while aliphatic olefins were found to be less efficient. This work highlights the use of visible light photoredox catalysis to promote regioselective 1,2-chloroamination of olefins through ATRA (Figure 3.8).

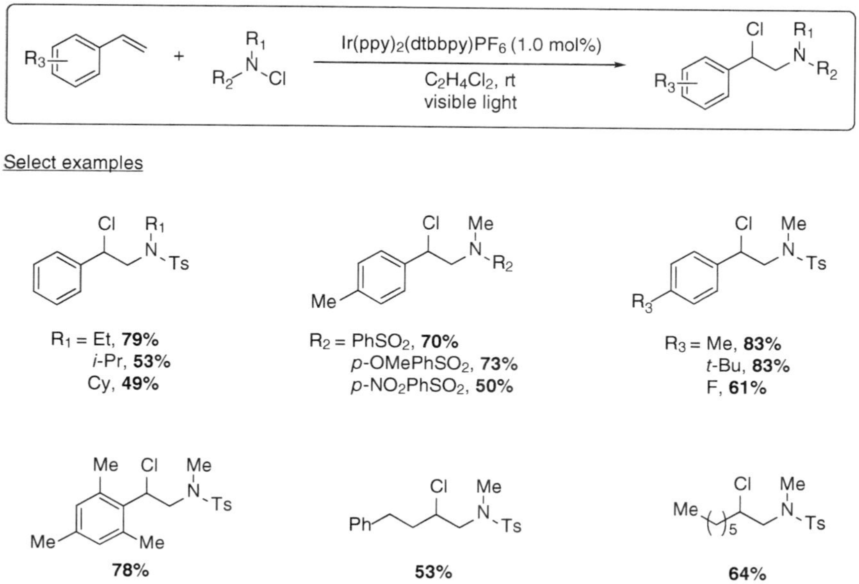

Figure 3.8 Method of forming 1,2-haloamine using ATRA.

3.2.2 Copper-Mediated ATRA

The examples provided thus far have demonstrated the utilization of ruthenium or iridium photocatalysts, which are robust, photochemically efficient, bench-stable solids. However, due to the scarcity of these metals they may be cost-prohibitive for potential large-scale applications. To address this concern Reiser and coworkers [24] reported the use of a copper(I) photoredox catalyst developed by Sauvage [25], ($[Cu(dap)_2]Cl$) for ATRA reactions (see copper chapter). $[Cu(dap)_2]Cl$ has a reduction potential of $Cu^{II}/Cu^{I*} = -1.43$ V vs. SCE, which is more negative than $[Ru(bpy)_3]Cl_2$ ($Ru^{II}/Ru^{I} = -1.33$ V vs. SCE) or $[Ir\{dF(CF_3)ppy\}_2(dtbbpy)]PF_6$ ($Ir^{III*}/Ir^{II} = -1.21$ V vs. SCE) and does not require a sacrificial electron donor making this an attractive alternative. Upon irradiation with green light ($\lambda = 530$ nm), $[Cu(dap)_2]Cl$ was able to promote the ATRA of alkyl halides to alkenes in good yields at room temperature and low catalyst loadings (Figure 3.9). The authors' proposed mechanism proceeds through photoexcited $[Cu(dap)_2]^{*+}$, which transfers an electron to the alkyl halide to generate a radical that can add to the alkene. The resulting radical then transfers an electron back to the oxidized copper catalyst, resulting in the generation of a carbocation that can be trapped by the halogen nucleophile. Similar to the mechanistic studies of Ru and Ir discussed previously, a propagative pathway cannot be discounted.

Reiser and coworkers also demonstrated the ability of $[Cu(dap)_2]Cl$ to reduce benzyl halides. The products of the ATRA reactions of benzyl halides and styrenes were used for the synthesis of 2-substituted tetrahydroquinolines (Scheme 3.2) [26]. The tetrahydroquinoline ring system is a common structural motif in biologically active natural products and is present in many therapeutic agents.

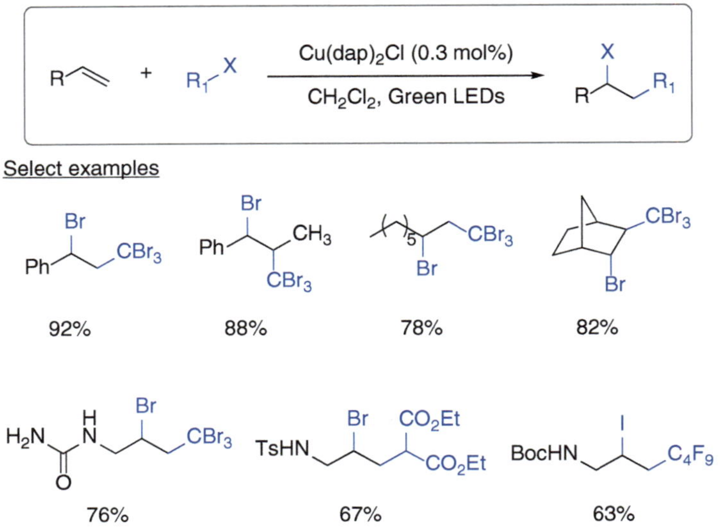

Figure 3.9 Copper catalyzed ATRA reaction.

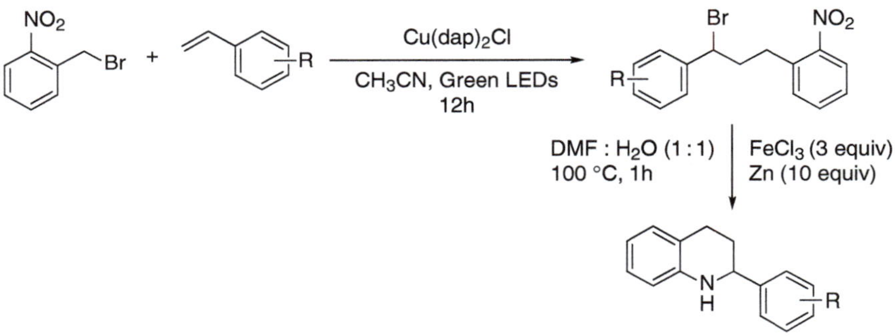

Scheme 3.2 Visible light ATRA reaction for the synthesis of substituted tetrahydroquinolines.

3.2.2.1 Trifluoromethylation

As shown previously by Stephenson and coworkers [18], visible light-mediated ATRA can be used to incorporate perfluorinated alkyl chains into organic compounds. The incorporation of fluorine or fluoroalkyl groups into organic molecules can dramatically improve the pharmacokinetic properties of drug molecules [27]. This phenomenon is supported by the presence of fluorine in a large proportion of recent drug candidates [28]. In particular, the incorporation of a trifluoromethyl (CF_3) group usually leads to increased chemical and metabolic stability of drug candidates [27]. The addition of fluorinated radicals to alkenes and alkynes is particularly useful for further synthetic manipulations [29].

Han and coworkers [30] developed a visible light-mediated method for the vicinal difunctionaliztion of alkenes using triflyl chloride, which incorporates a CF_3 group as well as a Cl into olefins. Using $[Ru(phen)_3]Cl_2$ as a photocatalyst the chlorotrifluoromethylation of several terminal alkenes, 1,1-disubstituted alkenes, and internal alkenes were successfully shown (Figure 3.10a). This protocol offers

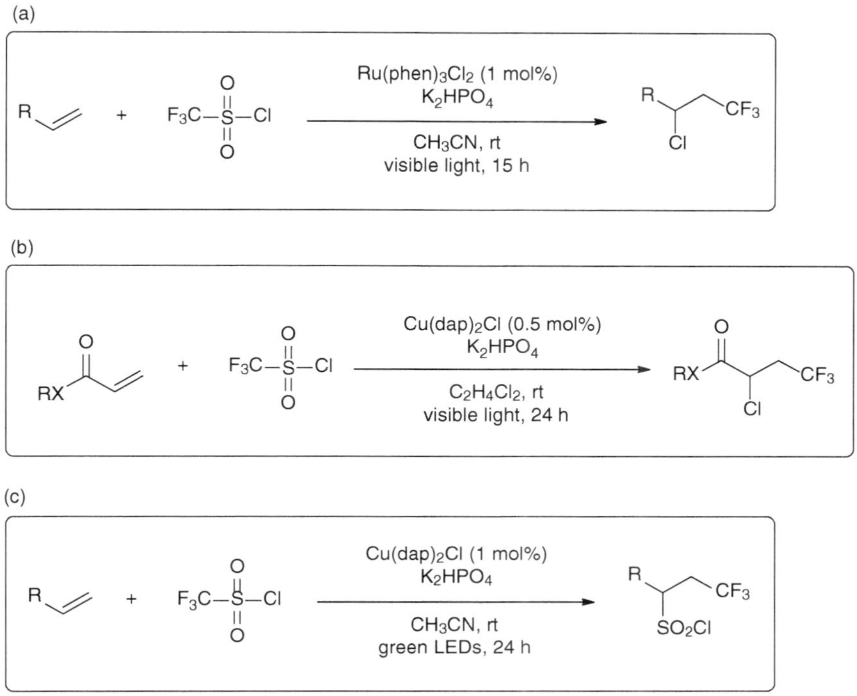

Figure 3.10 Trifluoromethylation using ATRA (a) Han *et al.* (b) Dolbier *et al.* (c) Reiser et al.

the possibility of late-stage incorporation of CF_3 groups into biologically active compounds that contain alkene functionalities.

Dolbier and Tang [31] also demonstrated the utility of photoredox ATRA for the incorporation of the trifluoromethyl group. Utilizing $[Cu(dap)_2]Cl$ as their photoredox catalyst, several fluoroalkylsulfonyl chlorides were added across electron-deficient alkenes. Several α-chloro-β-trifluoromethyl amides, esters, and ketones were obtained in moderate to excellent yields (Figure 3.10b). The reaction was also successfully extended to other fluoroalkylsulfonyl chlorides, such as $C_4F_9SO_2Cl$, HCF_2SO_2Cl, H_2CFSO_2Cl, and $CF_3CH_2SO_2Cl$. These products contain an activated C—Cl bond that can be further transformed into fluorinated amino acids, alkenes, and other derivatives.

Reiser and coworkers [32] similarly exploited the potential of $[Cu(dap)_2]Cl$ to form trifluoromethyl radicals from triflyl chloride for the addition to alkenes (Figure 3.10c). However, they reported an unprecedented trifluoromethylchlorosulfonylation of unactivated alkenes, which suggests an inner-sphere mechanism. In contrast, commonly used ruthenium and iridium complexes give rise to trifluoromethylchlorination products (as shown earlier) through the loss of SO_2 which is in agreement with the outer-sphere electron transfer mechanism commonly involved in photoredox-catalyzed ATRA reactions. Reiser and coworkers were able to demonstrate moderate substrate scope of substituted allyl benzenes and cyclic alkenes. It is noted that compounds with a heteroatom in close proximity to the alkenes resulted in trifluoromethylchlorination products. They also employed $[Ru(bpy)_3]Cl_2$ as a catalyst and observed exclusive formation of

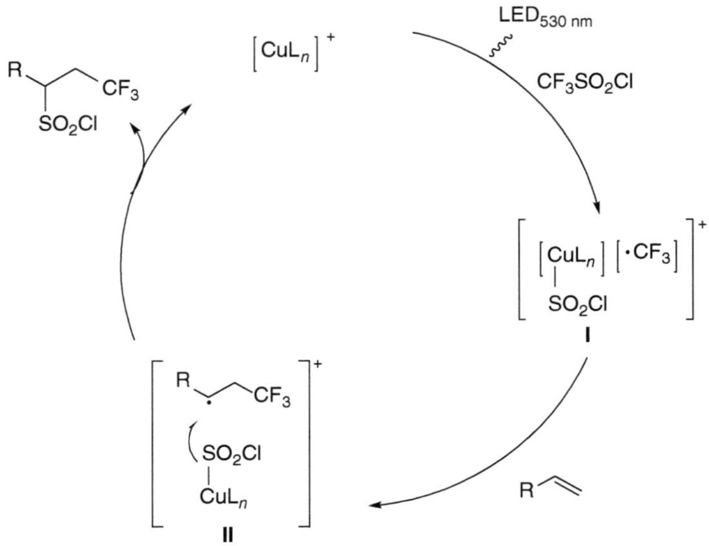

Figure 3.11 Proposed mechanism for $Cu(dap)_2Cl$ catalyzed trifluoromethyl-chlorosulfonylation.

the trifluoromethylchlorination products, which points toward the essential role of copper in obtaining the trifluoromethylchlorosulfonylated compounds. The authors' proposed mechanism (Figure 3.11) begins with the photoexcited copper catalyst reducing the triflyl chloride by a single-electron transfer (SET), giving rise to complex I with copper binding to the SO_2Cl^-. The trifluoromethyl radical is subsequently added to the alkene, and transfer of SO_2Cl from copper to the resulting radical intermediate affords the ATRA products. Given the well-known instability of free SO_2Cl^-, which rapidly decomposes to SO_2 and Cl^-, the formation of trifluoromethylchlorosulfonylated products (Figure 3.10c) demonstrate the role of copper as a ligand-binding entity. The binding of SO_2Cl^- to copper appears to be weak from the observation of trifluoromethylchlorination products with alkenes containing donor/heteroatoms, which could readily displace SO_2Cl^- resulting in SO_2 extrusion. This is consistent with the findings reported by Dolbier and coworkers who used heteroatom-containing alkenes in close vicinity to the double bond and exclusively produced trifluoromethylchlorinated products.

3.3 Other Photocatalysts for ATRA Transformations

3.3.1 *p*-Anisaldehyde

While most of these examples of visible light-mediated ATRA reactions have employed metal complexes as the photocatalyst, Melchiorre and coworkers [33] have demonstrated a metal-free photochemical ATRA process. Through the use of *p*-anisaldehyde and a household 23 W CFL bulb this aldehydic catalyst generates a reactive radical species by energy transfer to the haloalkane substrate. CFL emission was required to bring the catalyst to an excited state, which then enters an excited triplet state through intersystem crossing (ISC) (Figure 3.12). The excited triplet state of *p*-anisaldehyde has a relatively long lifetime (up to 3 μs in MeCN) [34] as well as a triplet energy (E_T) of 300 kJ mol^{-1} [35] . The authors propose that the excited catalyst sensitizes the alkyl halides to their excited triplet state, which results in the rapid homolytic dissociation of the C—X bond. The bond dissociation energies of the alkyl halides used were 260–300 kJ mol^{-1}, which is congruent with homolytic cleavage induced by an energy transfer mechanism. The carbon radical subsequently enters into the general ATRA propagation mechanism to produce the difunctionalized products. The authors were able to show the synthetic utility of this methodology through an expansive substrate scope. Different substituted alkyl halides were successfully added to terminal olefins, internal cyclic alkenes, olefins containing several sensitive functionalities, and alkynes (Figure 3.13). This ATRA process was also successfully performed on naturally occurring compounds (*R*)-limonene and

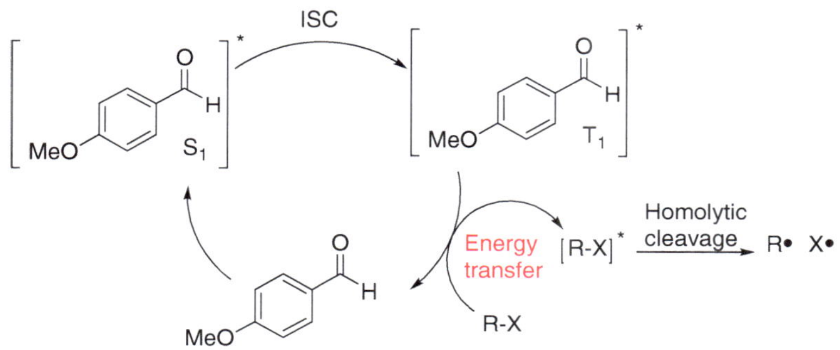

Figure 3.12 Proposed triple sensitization mechanism.

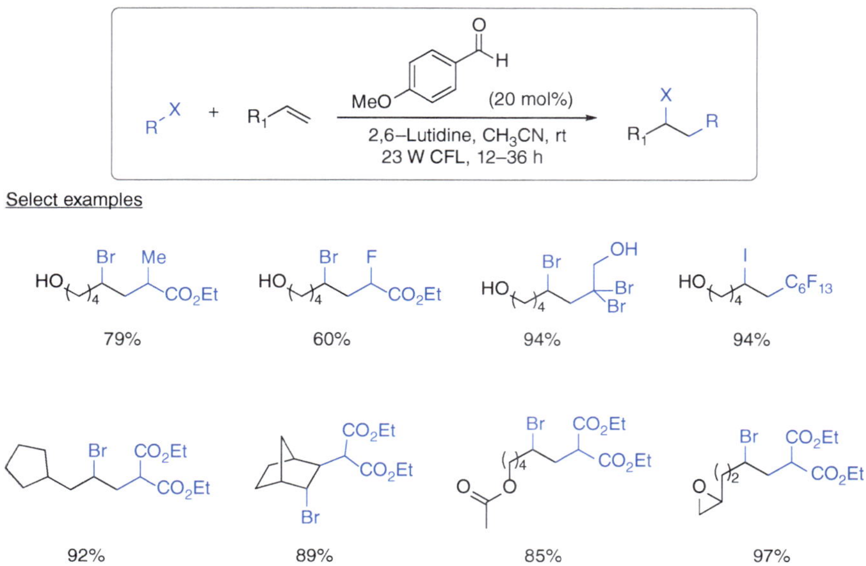

Figure 3.13 Metal free photochemical ATRA.

(−)-β-pinene. This work shows the ability of a simple organic molecule to induce intermolecular ATRA through illumination by a household light bulb.

3.4 Semiconductor

Ruthenium- and iridium-based photoredox catalysts have proven useful for ATRA reactions; however, the search for "green" and less expensive catalysts is always a priority for the scientific community. In this context, a number of semiconductors [36] and organic dyes [37] have been evaluated as photocatalysts. Pericàs and Riente [38] developed a photochemical ATRA process through the use of the inexpensive and non-toxic semiconductor bismuth oxide (Scheme 3.3). They utilized a 23 W CFL bulb to promote this reaction; however, they found that natural daylight could also be used to successfully catalyze this reaction. The authors propose that irradiation with light promotes the excitation of electrons on the surface of the semiconductor from the valence to the conduction

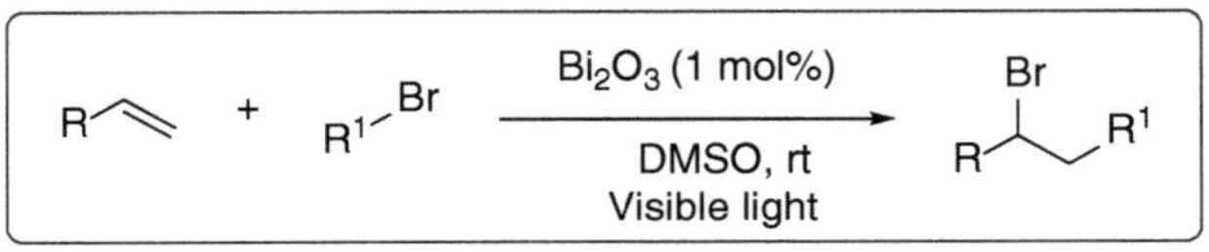

Scheme 3.3 Photochemical ATRA using semiconductor bismuth oxide.

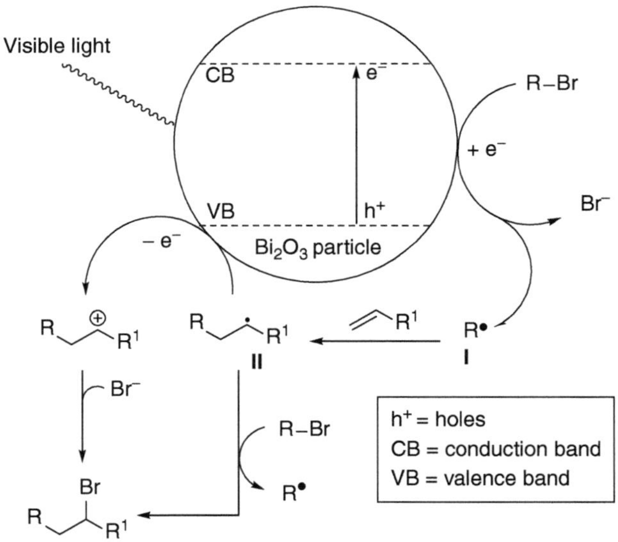

Figure 3.14 Proposed mechanism for Bi_2O_3 catalyzed, visible-light induced ATRA reaction.

band, which generates positive holes (h^+) (Figure 3.14). The photoexcited electrons induce the reductive cleavage of the organobromide to generate the carbon-centered radical 1. This radical then undergoes addition to the olefin to generate a second carbon radical 1, which can go through two possible pathways. This intermediate 2 can deliver an electron to the semiconductor to neutralize a positive hole and provide a carbocation intermediate, which will subsequently be trapped by a bromine anion. The second path involves the radical chain propagation pathway in which radical intermediate 2 abstracts a bromine atom from the starting material leading to the ATRA product and formation of another radical 1. Both routes may be operative for this transformation; however, no mechanistic studies have been performed to show the preference for one or the other. This simple catalytic system composed of commercially available bismuth oxide powder at low catalyst loadings displayed excellent performance in the ATRA reaction for a variety of olefins with different organobromides with isolated yields ranging from 45% to 95%.

3.5 Atom Transfer Radical Cyclization (ATRC)

In addition to the intermolecular ATRA reactivity shown thus far, ATRA can serve as an efficient cyclization mode for small molecule synthesis. In this variant of ATRA, C—C bonds are formed from *exo*-type cyclizations leaving the C—X bond formed on the exterior of the scaffold. This is significantly useful, as reactive functionality remains after ATRA cyclization, allowing for efficient multi-step synthetic plans. Radical cyclization is a desirable process because of the mild reaction conditions and high functional group compatibility. Atom transfer radical cyclization (ATRC) is a useful transformation for the formation of C—C bonds in the construction of cyclic frameworks. ATRC has been used for the formation of five-membered rings as 5-*exo*-trig cyclizations are among the fastest and most prominent radical cyclizations. The advances in visible light photoredox chemistry have shown significant improvements in radical chemistry, such as ATRA. However, photocatalytic ATRC is a challenge due to the preference for reduction products to form in which the halogen functionalities are lost. Through the use of the oxidative quenching cycle of iridium photocatalysts, Photocatalytic ATRC has been shown to be successful. Yao and coworkers [39] utilized $Ir(ppy)_3$ for a visible light-mediated radical intermolecular [3+2] ATRC reaction. Using dimethyl-2-(iodomethyl) cyclopropane-1,1-dicarboxylate, the authors were able to carry out intermolecular visible light [3+2] ATRC with a variety of alkenes or alkynes to form cyclopentane/cyclopentene derivatives (Scheme 3.4) [39]. The same group expanded the scope of this ATRC reaction to form tetrahydrofuran-containing oxabicycles as well as fused [6,5] and [5,5] systems of piperidine, pyridine, cyclohexane, or cyclopentane with tetrahydrofuran (Scheme 3.5) [40].

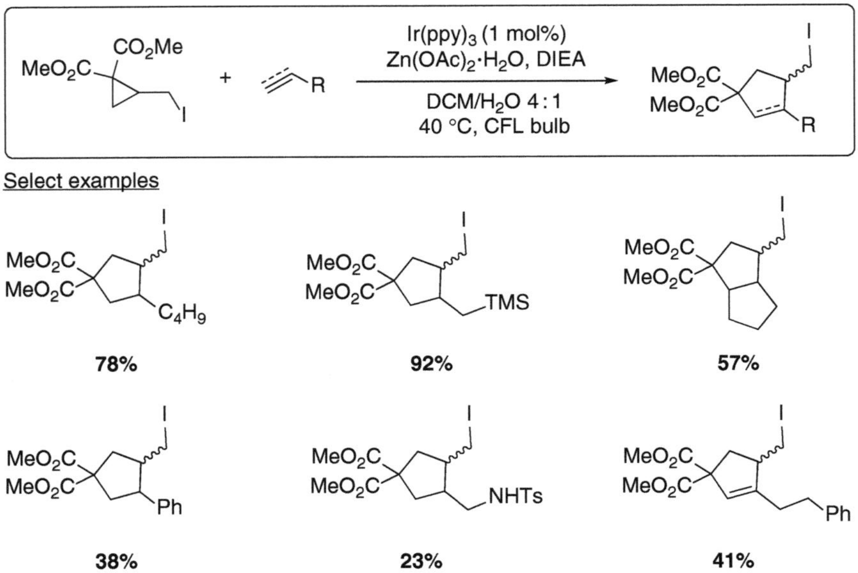

Scheme 3.4 ATRC to form 5 membered rings.

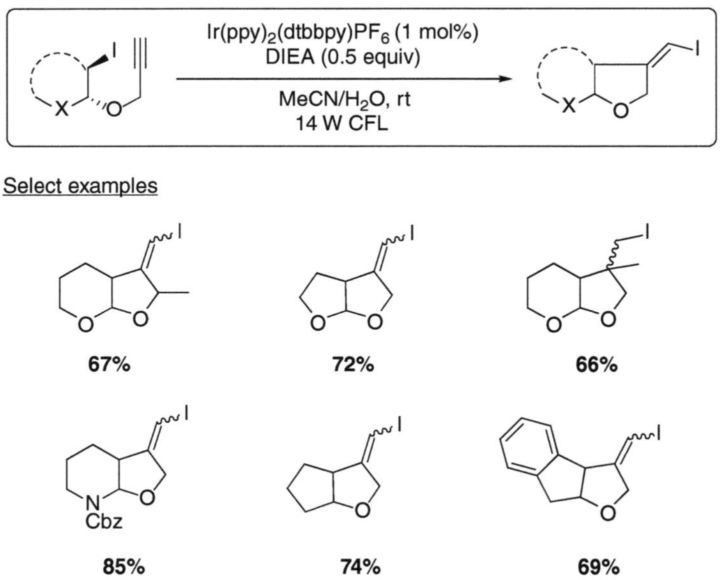

Scheme 3.5 ATRC to form fused [6, 5] and [5, 5] ring systems.

3.6 Atom Transfer Radical Polymerization (ATRP)

Atom transfer radical polymerization (ATRP) is among the most effective methods for controlled radical polymerization. ATRP allows for the synthesis of a wide variety of polymers with site-specific tailored functionalities targeting specific properties for high-value applications [41]. Photoredox ATRA reactions can also be applied to ATRPs. The use of light-activated catalysts to mediate ATRPs allows for excellent spatial and temporal control over the chain-growth process. The Hawker group developed a photocontrolled living radical polymerization procedure (Scheme 3.6) [42]. They employed $Ir(ppy)_3$ and proposed that this would reduce an alkyl bromide initiator, which would generate an alkyl radical that initiates polymerization of the monomer. The radical on the chain end of the polymer can then be oxidized by the iridium complex, completing the catalytic cycle and allowing for bromine attack of the carbocation. The polymer chain with a bromo end group can subsequently be reduced again and initiate another cycle and another addition of a monomer. This photocontrolled ATRP process allows for complete control of chain growth through activation of polymerization with visible light. In addition to its responsive nature to activation, the advantages of this process include facile reaction setup, use of only ppm levels of catalyst, and excellent functional group tolerance. The ATRP procedure offers a versatile platform for the preparation of functional materials with applications in electronics and health.

A limiting factor in using ATRP for applications such as microelectronics and biomaterials is metal contamination. A challenge for polymer chemists is therefore the lowering of catalyst loadings or removal of residual metals. To address this issue, Hawker and coworkers developed a metal-free catalyst system for ATRP [43]. Through the use of the highly reducing 10-phenylphenothiazine (PTH) ($PTH^{\bullet+}/PTH^* = -2.1$ V vs. SCE) the controlled polymerization of methyl methyacrylate could be achieved. A unique feature of PTH in contrast to $Ir(ppy)_3$ ($Ir^{III*}/Ir^{IV} = -1.73$ V vs. SCE) is the highly reducing excited state that may allow for a wider selection of functional groups which can greatly extend the scope of photomediated ATRP. This metal-free ATRP system uses the affordable organic catalyst PTH, which avoids the issue of metal contamination and allows for the production of a variety of functional materials.

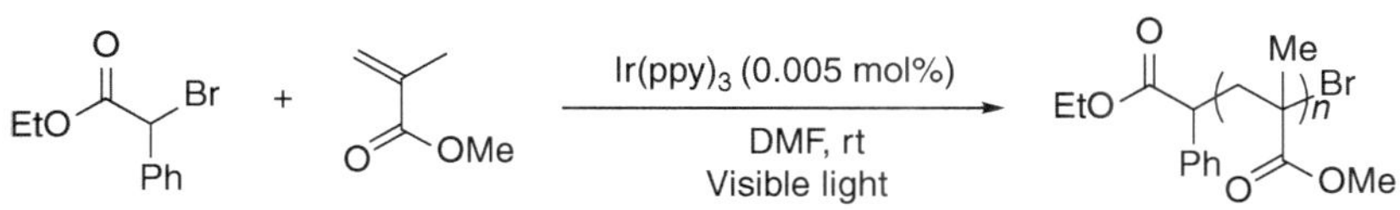

Scheme 3.6 Photocontrolled ATRP.

3.7 Conclusion

The difunctionalization of alkenes is an efficient means for the synthesis of complex compounds. ATRA began as a useful tool in the mid-1900s for the functionalization of alkenes as it affords two new sigma bonds, a C—X bond and a C—C bond. Typical ATRA reactions of that time employed harsh reaction conditions and/or toxic reagents and lacked broad functional group tolerance. With the reinvigoration of photoredox catalysis and its proclivity to initiate radical reactions it allowed for the development of ATRA as a more accessible tool. Utilizing visible light-mediated catalysts as initiators, the ATRA protocol can be performed under mild reaction conditions, have minimal side reactions, and operate with optimal catalytic efficiency and straightforward purification. Ruthenium, iridium, and copper complexes have all been successfully used for visible light-mediated ATRA. Other photocatalysts such as *p*-anisaldehyde and the semiconductor bismuth oxide have also shown to be efficient in catalyzing ATRA reactions. Visible light-mediated photocatalysts have also been applied to ATRC and ATRP, further developing this field for the synthesis of highly functionalized compounds. Utilizing visible light-mediated catalysts to perform ATRA reactions allows for effective difunctionalization of olefins in potentially complex systems, which was previously problematic. ATRA is a highly atom-economical process and through the use of photoredox catalysis has become a much milder method giving it the potential for use in natural product synthesis or for potential pharmaceuticals.

References

1 (a) Kharasch, M.S., Skell, P.S., and Fischer, P. (1948) *J. Am. Chem. Soc.*, **70**, 1055; (b) Kharasch, M.S., Jensen, E.V., and Urry, W.H. (1945) *Science*, **102**, 128.

2 (a) Curran, D.P., Bosch, E., Kaplan, J., and Newcomb, M. (1989) *J. Org. Chem.*, **54**, 1826; (b) Curran, D.P. and Chang, C.-T. (1989) *J. Org. Chem.*, **54**, 3140; (c) Curran, D.P., Chen, M.-H., Spleterz, E. *et al.* (1989) *J. Am. Chem. Soc.*, **111**, 8872; (d) Curran, D.P. and Seong, C.M. (1990) *J. Am. Chem. Soc.*, **112**, 9401; (e) Curran, D.P. and Tamine, J. (1991) *J. Org. Chem.*, **56**, 2746; (f) Curran, D.P. and Kim, D. (1991) *Tetrahedron*, **47**, 6171; (g) Curran, D.P., Kim, D., and Ziegler, C. (1991) *Tetrahedron*, **47**, 6189.

3 (a) Yorimitsu, H., Nakamura, T., Shinokubo, H., and Oshima, K. (1998) *J. Org. Chem.*, **63**, 8604; (b) Yorimitsu, H., Nakamura, T., Shinokubo, H. *et al.* (2000) *J. Am. Chem. Soc.*, **122**, 11041; (c) Yorimitsu, H., Shinokubo, H., Matsubara, S. *et al.* (2001) *J. Org. Chem.*, **66**, 7776.

4 (a) Renaud, P., Ollivier, C., and Panchaud, P. (2002) *Angew. Chem., Int. Ed.*, **41**, 3460; (b) Panchaud, P., Ollivier, C., Renaud, P., and Zigmantas, S. (2004) *J. Org. Chem.*, **69**, 2755; (c) Chabaud, L., Landais, Y., and Renaud, P. (2005) *Org. Lett.*, **7**, 2587;(d) Schar, P. and Renaud, P. (2006) *Org. Lett.*, **8**, 1569;

(e) Weidner, K., Giroult, A., Panchaud, P., and Renaud, P. (2010) *J. Am. Chem. Soc.*, **132**, 17511.

5 Cao, L. and Li, C. (2008) *Tetrahedron Lett.*, **49**, 7380.

6 Gilbert, B.C., Kalz, W., Lindsay, C.I. *et al.* (2000) *J. Chem. Soc., Perkin Trans. 1*, 1187.

7 (a) Yang, Z.-Y. and Burton, D.J. (1991) *J. Org. Chem.*, **56**, 5125; (b) Metzger, J.O. and Mahler, R. (1995) *Angew. Chem., Int. Ed.*, **34**, 902.

8 Forti, L., Ghelfi, F., Libertini, E. *et al.* (1991) *Tetrahedron*, **53**, 17761.

9 Quebatte, L., Scopelliti, R., and Severin, K. (2004) *Angew. Chem., Int. Ed.*, **43**, 1520.

10 Lübbers, T. and Schäfer, H.J. (1991) *Synlett*, 861.

11 (a) Iqbal, J., Bhatia, B., and Nayyar, N.K. (1994) *Chem. Rev.*, **94**, 519; (b) Gossage, R.A., Van De Kuil, L.A., and Van Koten, G. (1998) *Acc. Chem. Res.*, **31**, 423; (c) Severin, K. (2006) *Curr. Org. Chem.*, **10**, 217; (d) Clark, A.J. (2002) *Chem. Soc. Rev.*, **31**, 1.

12 (a) Quebatte, L., Thommes, K., and Severin, K. (2006) *J. Am. Chem. Soc.*, **128**, 7440; (b) Eckenhoff, W.T. and Pintauer, T. (2007) *Inorg. Chem.*, **46**, 5844.

13 Severin, K. (2012) *Chimia*, **66**, 386.

14 Taylow, M.J.W., Eckenhoff, W.T., and Pintauer, T. (2010) *Dalton Trans.*, **39**, 11475.

15 For selected reviews on visible light photocatalysis, see:(a) Yoon, T.P., Ischay, M.A., and Du, J. (2010) *Nat. Chem.*, **2**, 527; (b) Narayanam, J.M. and Stephenson, C.R.J. (2011) *Chem. Soc. Rev.*, **40**, 102; (c) Teplý, F. (2011) *Collect. Czech. Chem. Commun.*, **76**, 859; (d) Xuan, J. and Xiao, W.-J. (2012) *Angew. Chem., Int. Ed.*, **51**, 6828; (e) Shi, L. and Xiao, W.-J. (2012) *Chem. Soc. Rev.*, **41**, 7687; (f) Yoon, T.P. (2013) *ACS Catal.*, **3**, 895; (g) Xi, Y.-M., Yi, H., and Lei, A.-W. (2013) *Org. Biomol. Chem.*, **11**, 2387; (h) Hopkinson, M.N., Sahoo, B., Li, J.- L., and Glorious, F. (2014) *Chem. Eur. J.*, **20**, 3874; (i) Schultz, D.M. and Yoon, T.P. (2014) *Science*, **343**, 1239176.

16 Barton, D.H.R., Csiba, M.A., and Jaszberenyi, J.C. (1994) *Tetrahedron Lett.*, **35**, 2869.

17 (a) Narayanam, J.M.R., Tucker, J.W., and Stephenson, C.R.J. (2009) *J. Am. Chem. Soc.*, **131**, 8756; (b) Tucker, J.W., Narayanam, J.M.R., Krabbe, S.W., and Stephenson, C.R.J. (2010) *Org. Lett.*, **12**, 368; (c) Condie, A.G., Gonzalez-Gomez, J.C., and Stephenson, C.R.J. (2010) *J. Am. Chem. Soc.*, **132**, 1464; (d) Tucker, J.W., Nguyen, J.D., Narayanam, J.M.R. *et al.* (2010) *J. Chem. Commun.*, **46**, 4985; (e) Furst, L., Matsuura, B.S., Narayanam, J.M.R. *et al.* (2010) *Org. Lett.*, **12**, 3104; (f) Dai, C., Narayanam, J.M.R., and Stephenson, C.R.J. (2011) *Nature Chem.*, **3**, 140.

18 (a) Nguyen, J.D., Tucker, J.W., Konieczynska, M.D., and Stephenson, C.R.J. (2011) *J. Am. Chem. Soc.*, **133**, 4160; (b) Wallentin, C.-J., Nguyen, J.D., Finkbeiner, P., and Stephenson, C.R.J. (2012) *J. Am. Chem. Soc.*, **134**, 8875.

19 (a) Cismesia, M.A. and Yoon, T.P. (2015) *Chem. Sci.*, **6**, 5426;(b) Kärkäs, M.D., Matsuura, B.S., and Stephenson, C.R. (2015) *J. Science*, **349**, 1285.

20 Majek, M., Filace, F., and Wangelin, A.J. (2014) *Beilstein J. Org. Chem.*, **10**, 981.

21 Gladysz, J.A., Curran, D.P., and Horvath, I.T. (2004) *Handbook of Fluorous Chemistry*, Weinheim, Wiley-VCH Verlag GmbH & Co. KGaA.

22 (a) Curran, D.P. (2001) *Synlett*, 1488; (b) Curran, D.P. (1998) *Angew. Chem., Int. Ed.*, **37**, 1175; (c) Curran, D.P. and Luo, Z.Y. (1999) *J. Am. Chem. Soc.*, **121**, 9069.

23 Qin, Q., Ren, D., and Yu, S. (2015) *Org. Biomol. Chem.*, **13**, 10295.

24 Pirtsch, M., Paria, S., Matsuno, T. *et al.* (2012) *Chem. Eur. J.*, **18**, 7336.

25 Kern, J.-M. and Sauvage, J.-P. (1987) *J. Chem. Soc., Chem. Commun.*, 546.

26 Paria, S., Pirtsch, M., Kais, V., and Reiser, O. (2013) *Synthesis*, **45**, 2689.

27 (a) Müller, K., Faeh, C., and Diederich, F. (2007) *Science*, **317**, 1881; (b) Shimizu, M. and Hiyama, T. (2005) *Angew. Chem., Int. Ed.*, **44**, 214.

28 Hagmann, W.K. (2008) *J. Med. Chem.*, **51**, 4359.

29 For reviews on reactions of the trifluoromethyl radicals, see:(a) Xu, J., Liu, X., and Fu, Y. (2014) *Tetrahedron Lett.*, **55**, 585; (b) Merino, E. and Nevado, C. (2014) *Chem. Soc. Rev.*, **43**, 6598; (c) Koike, T. and Akita, M. (2014) *J. Fluorine Chem.*, **167**, 30; (d) Egami, H. and Sodeoka, M. (2014) *Angew. Chem., Int. Ed.*, **53**, 8294; (e) Koike, T. and Akita, M. (2014) *Top. Catal.*, **57**, 967.

30 Oh, S.H., Malpani, Y.R., Ha, N. *et al.* (2014) *Org. Lett.*, **16**, 1310.

31 Tang, X.-J. and Dolbier, W.R. (2015) *Angew. Chem., Int. Ed.*, **54**, 4246.

32 Bagal, D.B., Kachkovskyi, G., Knorn, M. *et al.* (2015) *Angew. Chem., Int. Ed.*, **54**, 6999.

33 Arceo, E., Montroni, E., and Melchiorre, P. (2014) *Angew. Chem., Int. Ed.*, **53**, 12064.

34 Gorner, H. and Kuhn, H.J. (1986) *J. Phys. Chem.*, **90**, 5946.

35 Koyanagi, M. and Goodman, L. (1971) *Phys. Chem.*, **55**, 2959.

36 For reviews, see:(a) Ravelli, D., Dondi, D., Fagnoni, M., and Albini, A. (2009) *Chem. Soc. Rev.*, **38**, 1999; (b) Kisch, H. (2013) *Angew. Chem., Int. Ed.*, **52**, 812.

37 (a) Fidaly, K., Ceballos, C., Falguières, A. *et al.* (2012) *Green Chem.*, **14**, 1293; For reviews, see:(b) Nicewicz, D.A. and Nguyen, T.M. (2014) *ACS Catal.*, **4**, 355; (c) Fukuzumi, S. and Ohkubo, K. (2014) *Org. Biomol. Chem.*, **12**, 6059.

38 Riente, P. and Pericàs, M.A. (2015) *ChemSusChem.*, **8**, 1841.

39 Gu, X., Li, X., Qu, Y. *et al.* (2013) *Chem Eur. J.*, **19**, 11878.

40 Gu, X., Lu, P., Fan, W. *et al.* (2013) *Org. Biomol. Chem.*, **11**, 7088.

41 (a) Matyjaszewski, K. (2010) *Macromolecules*, **45**, 4015; (b) Xia, J. and Matyjaszewski, K. (2001) *Chem. Rev.*, **101**, 2921.

42 Fors, B.P. and Hawker, C.J. (2012) *Angew. Chem., Int. Ed.*, **51**, 8850.

43 Treat, N.J., Sprafke, H., Kramer, J.W. *et al.* (2014) *J. Am. Chem. Soc.*, **136**, 16096.

4

Visible Light Mediated α-Amino C—H Functionalization Reactions

You-Quan Zou and Wen-Jing Xiao

Central China Normal University, Key Laboratory of Pesticide & Chemical Biology, Ministry of Education, College of Chemistry, 152 Luoyu Road, Wuhan, Hubei 430079, China

4.1 Introduction

The ubiquitous C—H bonds underlie the methods by which feedstock and intermediates are used in the elaboration of organic molecules, and the direct functionalization of C—H bonds represents a powerful strategy for the preparation of fine chemicals [1]. However, traditional methods are mainly based on prefunctionalizing the C—H bond of the starting materials (Figure 4.1a). Converting C—H bond into the corresponding C—C or C—X (X = O, N, S, P, etc.) bond without the prefunctionalization step is a formidable task resulting from the relatively high bond dissociation energies of the unactivated C—H bonds [2]. During the past few years, transition-metal catalysis has unlocked opportunities for direct C—H bond functionalization, and a considerable number of elegant methods have been established (Figure 4.1b) [3]. In this context, the C—H bonds were first activated by transition metals and the resulting highly active C-M species were then reacted with the other reactants to afford the functionalized compounds. Despite these advances, transition-metal catalyzed C—H bond functionalization is still suffering from some impediments: (1) a directing group is needed to control the chemoselectivity, regioselectivity, and stereoselectivity; (2) expensive transition metals Pd, Ru, Rh, and Ir are often used, and the more cheaper metals such as Fe, Co, and Zn are not suitable in most cases; (3) reactions are mainly based on sp^2 C—H bonds, and examples about sp^3 C—H bonds are rare; (4) most of the reactions are performed at high temperature. Therefore, the development of novel economical and environmentally friendly methods for the direct functionalization of C—H bonds at room temperature into the highly functionalized groups is highly desirable.

Visible light, especially sunlight, has often been regarded as a clean energy because of its high natural abundance, greenness, benign environmental impact, and sustainability. Nevertheless, the fact that a large number of organic molecules cannot absorb visible wavelengths of light has limited the applications of visible light in organic reactions for a long time. Recently, visible light photocatalysis has emerged as a popular and effective strategy for the activation of various

Visible Light Photocatalysis in Organic Chemistry, First Edition.
Edited by Corey R. J. Stephenson, Tehshik P. Yoon and David W. C. MacMillan.

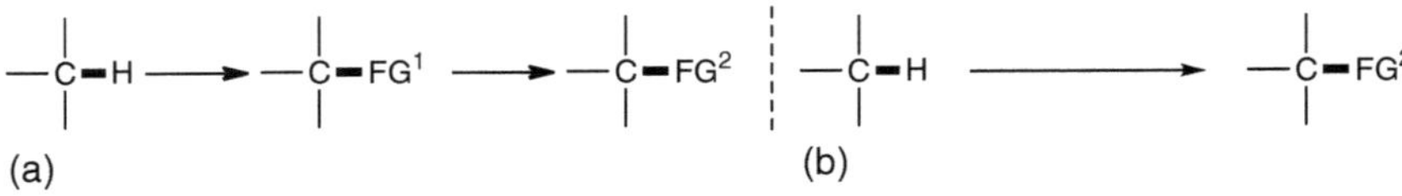

Figure 4.1 (a) Traditional method and (b) transition-metal catalysis for the C—H bond functionalization (FG = functional group).

C—H bonds, and many useful compounds that are not easily accessible under thermal conditions can be constructed under mild conditions [4]. In 2008, the asymmetric alkylation of aldehydes by the MacMillan group [5] and the [2+2] cycloaddition of enones by the Yoon group [6] constituted landmark achievements in this area. All the advances in this field should be attributed to the characteristics of metal–polypyridyl complexes and organic dyes, which serve as highly efficient photocatalysts or photosensitizers. Typically, electron- and energy-transfer processes were introduced to explain the mechanism of these transformations [7]. The former refers to the visible light irradiation of photocatalyst, for example, when $Ru(bpy)_3^{2+}$ produces its electronic excited state $Ru^*(bpy)_3^{2+}$ via metal-to-ligand charge transfer (MLCT), which serves as both a good electron donor (ED) and electron acceptor (EA). $Ru^*(bpy)_3^{2+}$ is subsequently either oxidized or reduced by organic quenchers through single-electron transfer (SET) mechanisms to initiate the subsequent organic reactions (Figure 4.2, Paths A and B). The latter refers to the energy of excited state $Ru^*(bpy)_3^{2+}$ that is transferred directly to the substrate for the following steps (Figure 4.2, Path C) [8], which is attributed to the relatively high triplet state energy of $Ru^*(bpy)_3^{2+}$ (ET = 46 kcal/mol).

Recent examples of visible light induced C—H functionalization reactions [9] feature the processes involving redox photocatalysis by metal–polypyridyl complexes and organic dyes. In this context, C—H bonds adjacent to a nitrogen atom were elegantly converted to various functional groups by means of visible light photoredox catalysis, which were corresponding to the transition metal-catalyzed cross-dehydrogenative coupling (CDC) reactions [10]. Firstly, visible light photocatalyst **PC** was sensitized to the excited state in the irradiation of visible light, which was then quenched by a tertiary amine to form the amino

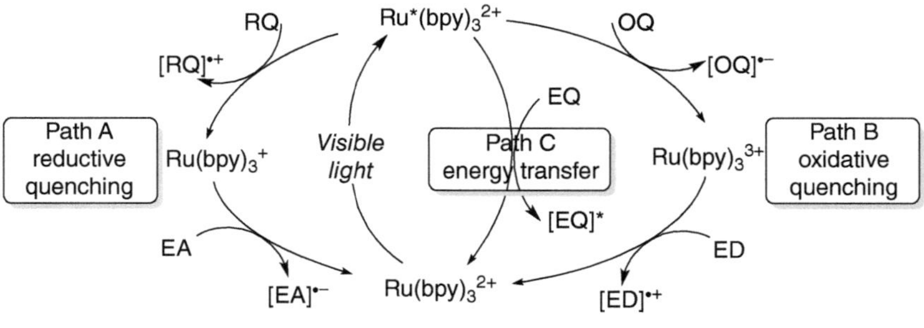

Figure 4.2 Electron- and energy-transfer processes for $Ru(bpy)_3^{2+}$ promoted by irradiation with visible light (RQ, reductive quencher; OQ, oxidative quencher; EQ, energy quencher; EA, electron acceptor; ED, electron donor).

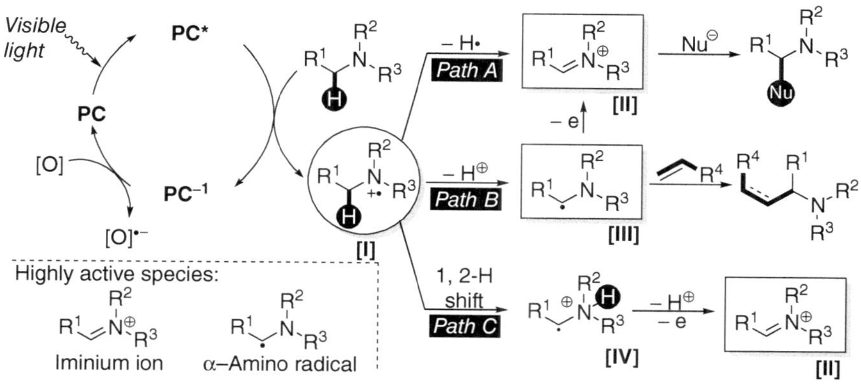

Figure 4.3 Possible reaction pathways for visible light photoredox catalytic C—H functionalization of α-amino C—H bonds (PC, photocatalyst; O, oxidant).

radical cation intermediate **[I]** with the release of low oxidation ground-state $\mathbf{PC^{-1}}$. $\mathbf{PC^{-1}}$ was then oxidized to the ground-state **PC** by the external oxidant **[O]** to finish the catalytic cycle accompanying radical anion $\mathbf{[O]^{•-}}$ as a highly active species. Subsequently, hydrogen atom abstraction occurred between the radical cation **[I]** and radical anion $\mathbf{[O]^{•-}}$ to afford useful and important electrophilic iminium ion **[II]**, which could be easily intercepted by various nucleophiles to generate the desired C—H bond functionalized compounds (Figure 4.3, path A). On the other hand, the amino radical cation intermediate **[I]** could also be converted to the α-amino radical **[III]** through a deprotonation process. This highly nucleophilic α-amino radical **[III]** could add to a large number of electron-deficient alkenes and aromatics to deliver the corresponding functionalized amines (Figure 4.3, path B). Meanwhile, further oxidation of the α-amino radical **[III]** also gave the iminium ion **[II]** to take part in the following nucleophilic addition reactions. The third way for the generation of iminium ion **[II]** might result from the 1, 2-hydrogen shift of intermediate **[I]**. As shown in Figure 4.3, **[IV]** was formed after the hydrogen shift followed by deprotonation and single-electron oxidation to give the iminium ion (Figure 4.3, path C). In this chapter, we will highlight the recent advances in visible light mediated α-amino C—H functionalization reactions involving iminium ions or α-amino radicals as the key intermediates.

4.2 Visible Light Mediated α-Amino C—H Functionalization Via Iminium Ions

4.2.1 Aza-Henry Reaction

In 2010, Stephenson and coworkers realized a highly effective C—H bond oxidation/aza-Henry reaction under visible light photocatalytic conditions [11]. By only using 1 mol% $[Ir(ppy)_2(dtbbpy)]PF_6$ as the catalyst, various *N*-aryltetrahydroisoquinolines **1** coupled with nitromethane or nitroethane

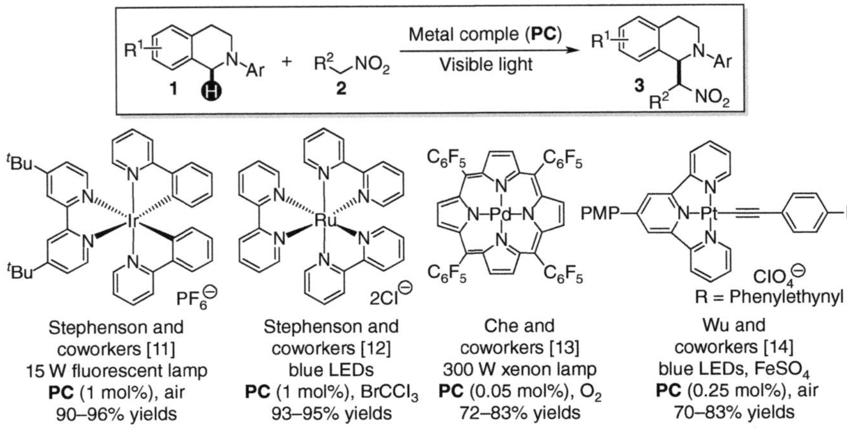

Figure 4.4 Homogeneous visible light induced aza-Henry reaction using metal complexes (PMP, 4-methoxyphenyl; PC, photocatalyst).

2 to give the desired aza-Henry products **3** in excellent yields (Figure 4.4). It is noteworthy that unactivated amine 1-phenylpyrrolidine was also suitable for this reaction with somewhat lower isolated yield (27% yield). Control experiments indicated that both visible light and photocatalyst were required for the high efficiency of this transformation. To get some insight into the mechanism of the photoredox aza-Henry reaction, the authors carried out fluorescence quenching experiments, and results showed that the fluorescence emission of the excited state iridium complex was indeed quenched by *N*-phenyltetrahydroisoquinoline. To shed some light on the catalyst turnover, they introduced (nitromethyl)benzene as the coupling partner to react with *N*-aryltetrahydroisoquinoline. The reductive product of (nitromethyl)benzene was not detected in the crude reaction mixture, which indicated that catalyst turnover may proceed via reaction with oxygen although the role of nitroalkane could not be absolutely ruled out.

Lately, the Stephenson group reported a new approach to aza-Henry products by using bromotrichloromethane as the stoichiometric oxidant to replace oxygen [12]. The iminium ion intermediate was first generated under the irradiation of visible light in the presence of $Ru(bpy)_3Cl_2$ and bromotrichloromethane (Figure 4.4). Interestingly, the subsequent nucleophilic addition proceeded smoothly without visible light, which could avoid the formation of some undesired products. Following the pioneering work by Stephenson, many kinds of metal complexes were well established for visible light initiated aza-Henry reaction. In 2013, Che and coworkers successfully applied palladium(II)–porphyrin complex $\mathbf{PdF_{20}TPP}$ in this kind of reaction [13]. They found that $\mathbf{PdF_{20}TPP}$ was an efficient and robust catalyst for photoinduced oxidative aza-Henry reaction, and the corresponding products could be obtained in good yields. The advantage of this methodology was that the catalyst loading could be reduced to 0.05 mol% without loss in reaction efficiency. Moreover, results of emission quenching, nanosecond time-resolved absorption (ns-TA) measurements, and cyclic voltammetry indicated that the mechanism of this process might involve a singlet-oxygen oxidation pathway. Shortly after, Wu and coworkers disclosed platinum(II)–terpyridyl complex mediated photocatalytic aza-Henry reaction [14]. Using $FeSO_4$ (2 equiv.) as an additive, amide, the by-product generally observed in the CDC transformation, was completely inhibited and the desired coupling products were obtained. The authors also proposed a possible mechanism involving a superoxide radical anion based on electron paramagnetic resonance (EPR) experiments.

Organic dyes, a new class of environmentally friendly, cost-effective, readily available, and easy to handle photocatalysts, were also widely applied in visible light photoredox catalysis [15]. The significance of this photo-organocatalysis was that all reactions were performed under metal-free conditions. In 2011, König and coworkers disclosed an elegant Eosin Y-catalyzed visible light oxidative aza-Henry reaction [16]. In the presence of 2 mol% Eosin Y, a variety of tetrahydroisoquinoline derivatives could react with nitroalkanes under the irradiation of green LEDs and provided results comparable to those of previously reported metal–polypyridyl complexes catalyzed reactions (Figure 4.5). Later, Tan's group [17] and Wu's group [18] realized the same reaction using Rose Bengal and TBA-eosin Y (TBA = tetrabutylammonium) as the photocatalyst, respectively, furnishing cross-coupling products in excellent yields. Although the precise mechanism of organic dyes-mediated aza-Henry reaction was unclear, they proposed a possible mechanism that was similar to Stephenson's model [11]. Under the irradiation of visible light, organic dye was sensitized to its excited state, which was then quenched by tetrahydroisoquinoline derivatives via a reductive quenching cycle. The generated nitrogen radical cation lost a hydrogen atom with the aid of oxygen or nitroalkane to generate an iminium ion. Subsequently, trapping of iminium ion with nitroalkanes delivered the final products.

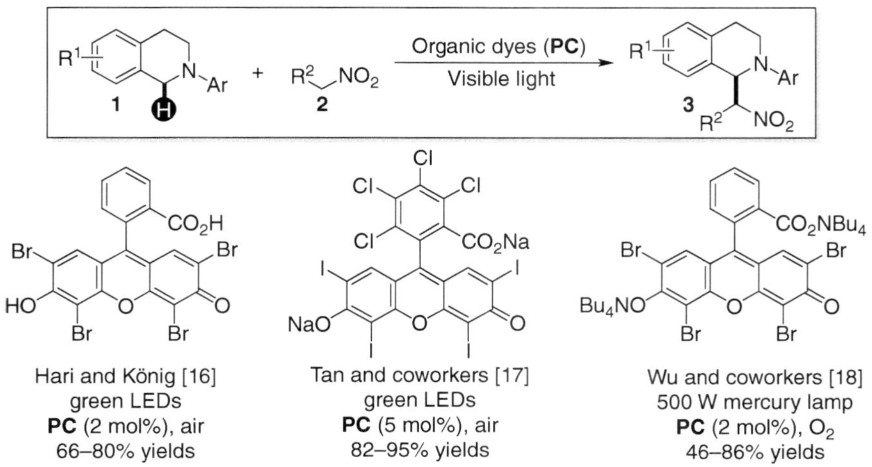

Figure 4.5 Homogeneous visible light induced aza-Henry reaction using organic dyes (PC, photocatalyst).

Heterogeneous catalysts feature high efficiency, simple recycle, and reuse, which makes the application of heterogeneous photocatalysis [19] into visible light photoredox catalysis more attractive. Recently, several elegant examples have been reported in this field. Notably, Lin and coworkers [20] synthesized novel porous cross-linked polymers (PCPs) with phosphorescent iridium–polypyridyl and ruthenium–polypyridyl complexes and successfully applied them into visible light induced aza-Henry reactions (Figure 4.6). The PCPs-catalyzed aza-Henry reactions gave higher conversions compared with their homogeneous counterparts, and no obvious deterioration of conversion was observed after recycling four times. Lately, the same group developed another kind of nonporous, phosphorescent cross-linked polymers (Ru-CP and Ir-CP), and these nonporous CPs also exhibited highly active and recyclable properties for visible light catalyzed aza-Henry reaction [23]. In 2012, the Rueping group [21] and the König group [22] utilized commercially available titanium dioxide and cadmium sulfide to achieve the same transformation under visible light conditions. Interestingly, Blechert and coworkers [24] reported the first example of organic heterogeneous photocatalyzed aza-Henry reaction by using their mesoporous graphitic carbon nitride (mpg-C_3N_4) semiconductor [25].

During the past few years, continuous flow technology has been well developed as a novel and popular methodology for the synthesis of important and useful compounds [26]. Compared to the batch method, continuous flow technology has many advantages including high surface-to-volume ratio, improved reproducibility, predictable reaction scale-up, and shortened reaction time as well as improved reaction efficiency. Therefore, applying this new technology into visible light photoredox catalysis is highly desirable (Figure 4.7). In 2012, Stephenson and coworkers designed a readily prepared and easily implemented flow reactor for the visible light initiated transformation, and coupling reaction between *N*-phenyl-tetrahydroisoquinoline and nitromethane

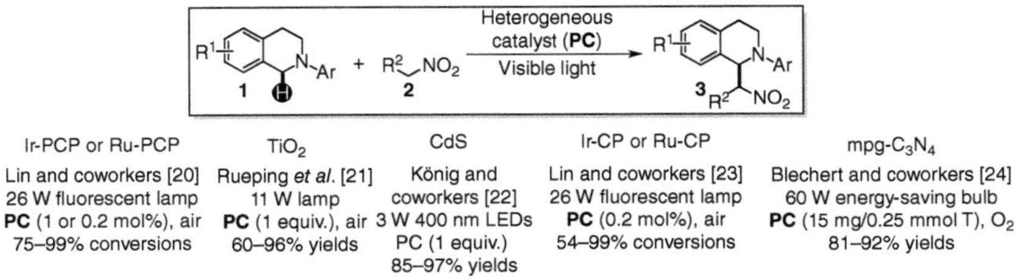

Figure 4.6 Heterogeneous visible light induced aza-Henry reaction (T, tetrahydroisoquinoline; PC, photocatalyst).

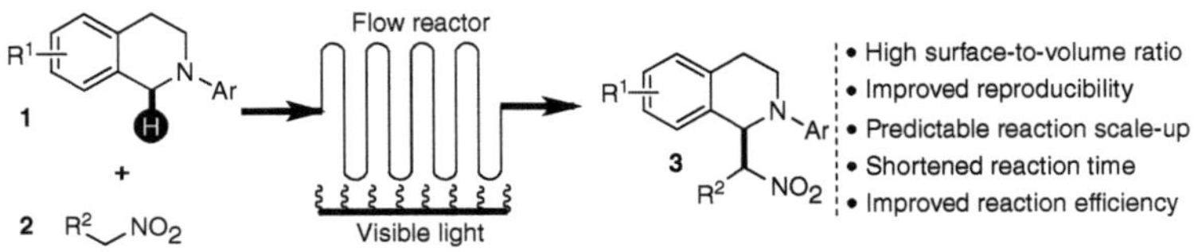

Figure 4.7 Continuous flow technology in visible light induced aza-Henry reaction.

could be completed within a short time [27]. Moreover, this protocol could be performed on a gram scale, which highlighted the synthetic potential of flow photochemistry. In addition, the Zeitler group developed a similar strategy by using iridium– or ruthenium–polypyridyl complex as the photocatalyst under microflow conditions [28]. Remarkably, unactivated substrates such as *N*-phenylpyrrolidine and 4,*N*,*N*,-trimethylaniline were also suitable in optimal conditions and afforded the corresponding products in moderate to good yields. Furthermore, Rueping and coworkers combined environmentally benign organic dyes and continuous flow technology into visible light aza-Henry reaction [29]. Importantly, all reactions were performed under metal-free conditions by using Rose Bengal as the photocatalyst.

In 2012, Wang, Li and coworkers reported a fascinating visible light promoted C—C bond cleavage of vicinal diamines [30]. Two highly reactive species, iminium ion and α-amino radical, were generated simultaneously in the reaction system, which were intercepted by nitroalkanes and 2-hydroxyethyl-acrylate (HEA), respectively. In the presence of 5 mol% $Ru(bpy)_3Cl_2 \cdot 6H_2O$ and 1 atm oxygen, various nitro substrates **5** could react with N^1,N^1,N^2,N^2-tetramethylethane-1,2-diamine (TMEDA) **4** to give the desired aza-Henry products **6** in moderate to good yields under the irradiation of visible light (Figure 4.8). β-Aryl-nitroethanes bearing electron-neutral, electron-donating, and electron-withdrawing groups on the aromatic ring tolerated well under the optimized conditions. Notably, indole- and furan-substituted substrates as well as purely aliphatic nitroalkane also proceeded smoothly to give the expected products with comparable yields (63–91% yields). More importantly, the vicinal diamines could be extended to 1,2-di(piperidin-1-yl)ethane and 1,2-dimorpholinoethane, albeit with lower reaction efficiencies. On the other hand, polymer **10** was obtained when HEA was added as the coupling partner, which demonstrated that α-amino radical did indeed exist during this process.

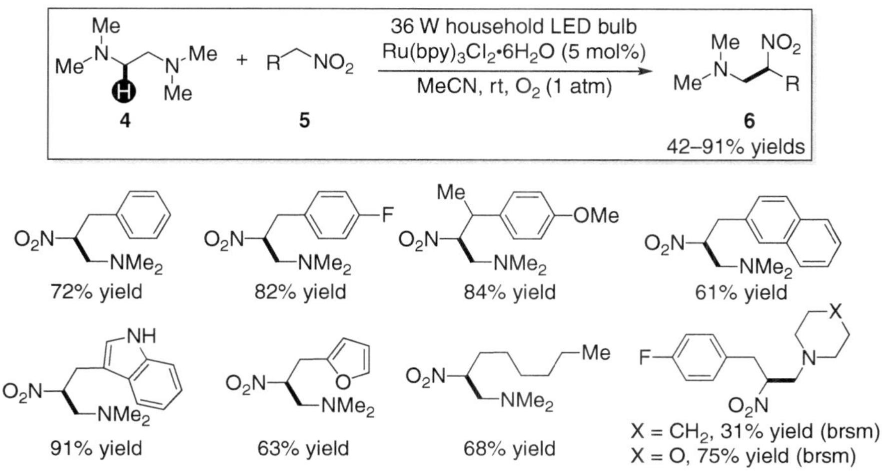

Figure 4.8 Visible light promoted C—C bond cleavage of vicinal diamines and application in aza-Henry reaction (brsm, based on recovered starting material).

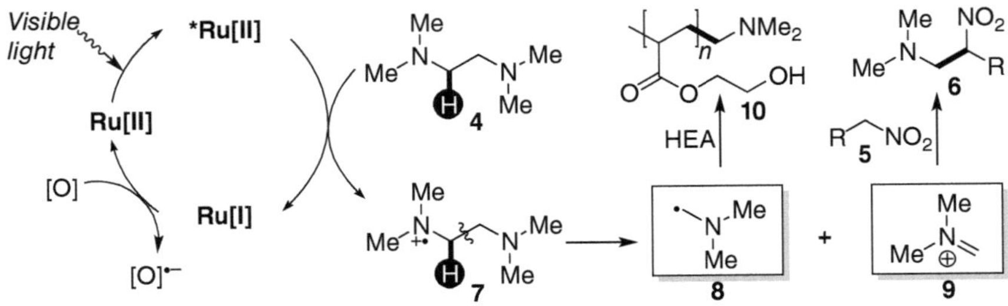

Figure 4.9 Possible mechanism for visible light promoted C—C bond cleavage of vicinal diamines (Ru[II], $Ru(bpy)_3Cl_2{\cdot}6H_2O$; HEA, 2-hydroxyethylacrylate).

Based on experimental data and previous reports, Wang, Li and coworkers also proposed a mechanism to explain the visible light promoted C—C bond cleavage of vicinal diamines (Figure 4.9). Initially, TMEDA **4** served as a reductive quencher (RQ) to reduce the excited state ***Ru[II]** to **Ru[I]** and deliver the amine radical cation **7**. Intermediate **7** then underwent a C—C bond cleavage process to deliver α-amino radical **8** and iminium ion **9**. The following radical photopolymerization and aza-Henry condensation would occur upon the reaction partner. This methodology not only broadens the substrate scope of photocatalytic aza-Henry reaction, but also gives some new insight into visible light photoredox catalysis and helps us to design new types of reactions.

4.2.2 Mannich Reaction

Mannich reaction generally refers to the addition of enolizable carbonyl compounds to iminium salts and imines, which is a powerful and straightforward strategy for the construction of β-amino carbonyls [31]. The first example of visible light photoredox catalytic Mannich reaction was reported by the Rueping group [32] in early 2011 (Figure 4.10). Using 1 mol% $Ru(bpy)_3(PF_6)_2$ as the photocatalyst and 10 mol% *L*-proline as the organocatalyst (OC), various *N*-aryl-tetrahydroisoquinolines **1** coupled with methyl-ketones **11** smoothly to give the desired β-amino ketones in moderate to good yields with careful fine tuning of the light source. In this reaction, the photoredox catalytic cycle oxidized *N*-aryl-tetrahydroisoquinolines to iminium ions, while the organocatalytic cycle converted methyl-ketones to the corresponding highly reactive enamines. The synergistic strategy made this transformation more attractive and demonstrated a further advancement in the area of visible light photoredox catalysis. In 2013, the Zhu group developed a novel and highly efficient gold(III)-complex catalyzed photo-oxidation-Mannich type reaction [33]. Remarkably, unmodified long-chain methyl-ketones, cycloketones, and linear amines could be subjected to the photo-oxidation-Mannich reaction providing the desired products in good to excellent yields. The broad substrate scopes, high yields, and mild reaction conditions made this protocol very practical. Later, Che and coworkers [13] examined palladium(II)–prophyrin complex $\mathbf{PdF_{20}TPP}$ which was highly effective for Mannich reaction, and good results were also observed.

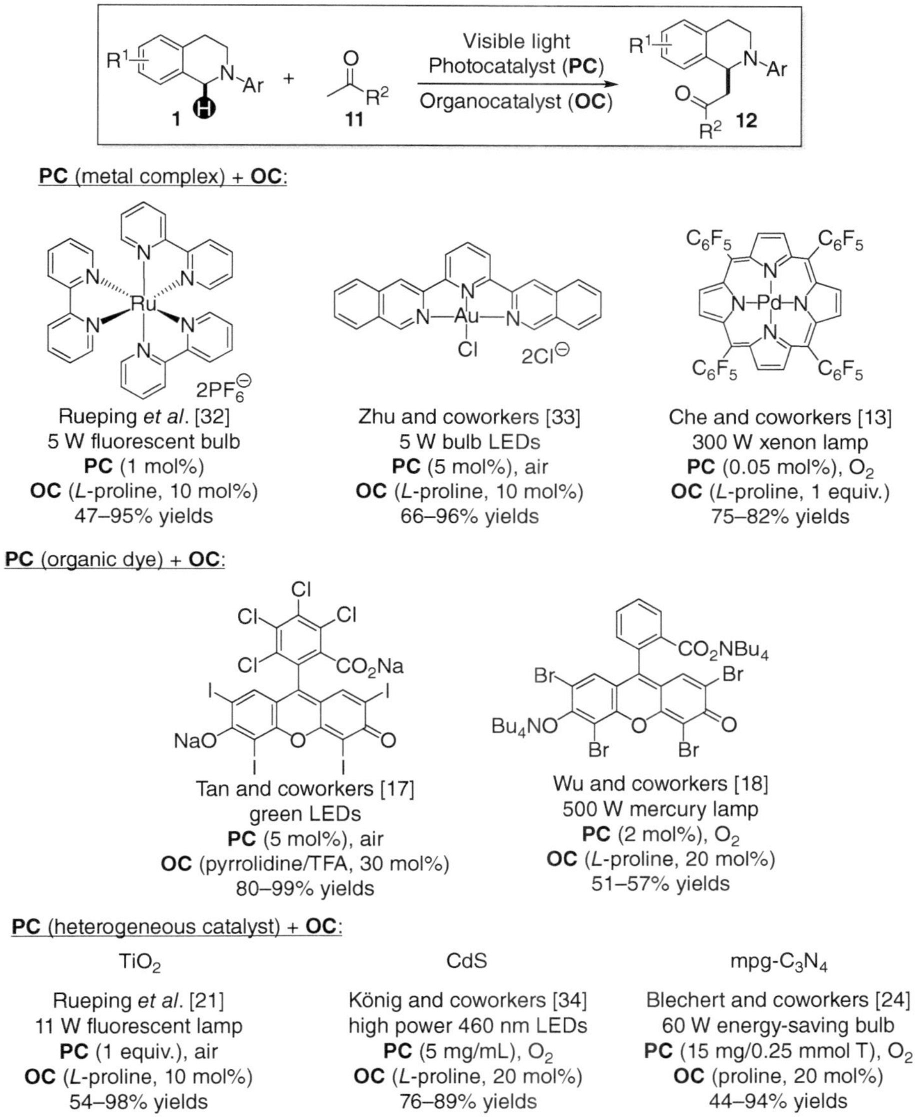

Figure 4.10 Visible light promoted Mannich reaction (TBA, tetrabutylammonium; T, tetrahydroisoquinoline).

In 2011, the Tan group [17] realized the Mannich reaction by using a cheaper organic dye, Rose Bengal, as the photocatalyst (Figure 4.10). When the organocatalyst pyrrolidine/TFA was replaced by (*S*)-2-(methoxymethyl)pyrrolidine, the ee value of the reaction between *N*-PMP-protected tetrahydroisoquinoline and acetone could be improved to 15%. Subsequently, TBA-eosin Y was successfully introduced into this transformation by Wu and coworkers [18]. Besides these metal complexes and organic dyes, heterogeneous materials were also found to be effective photocatalysts for the visible light promoted Mannich reactions (Figure 4.10). In 2012, the Rueping group [21] and the König group [34] examined the catalytic performances of commercially available TiO_2 and CdS, respectively. In the same year, Blechert and coworkers [24] demonstrated that organic heterogeneous photocatalyst mpg-C_3N_4 was also a robust and fascinating catalyst in this reaction. All the above examples enriched the kinds of photocatalysts in visible light photoredox catalysis and the chemistry of the Mannich reaction.

In 2012, Xia and coworkers [35] developed an alternative method for the visible light induced Mannich reaction by using enol silanes as the coupling partners. Under the optimized conditions, several enol silanes **14** could react with *N*-aryl tetrahydroisoquinolines **13** to give the Mannich adducts in good to excellent yields, and results showed that enol silanes bearing electron-donating groups gave better results than the electron-withdrawing ones (Figure 4.11). Meanwhile, a large number of *N*-aryl tetrahydroisoquinolines were subjected to the coupling reaction and delivered the desired β-amino ketones in excellent yields. More importantly, the authors found that *N*-acetic ester was also compatible with this condition and afforded the corresponding product in 92% yield upon isolation.

In order to get some insight into the mechanism of this reaction, the authors conducted some control experiments. Upon treatment of *N*-phenyl tetrahydroisoquinoline with 5 mol% $Ru(bpy)_3Cl_2$ in methanol under the irradiation of visible light, only hemiaminal **18** and methoxyaminal **19** were observed in the crude mixture. Then the terminal product was isolated in high yield after adding enol silane to the resulting mixture. These findings indicated that intermediates hemiaminal **18** and methoxyaminal **19** were beneficial for the following coupling reaction with enol silanes, which inhibited the formation of the amide byproduct (Figure 4.12).

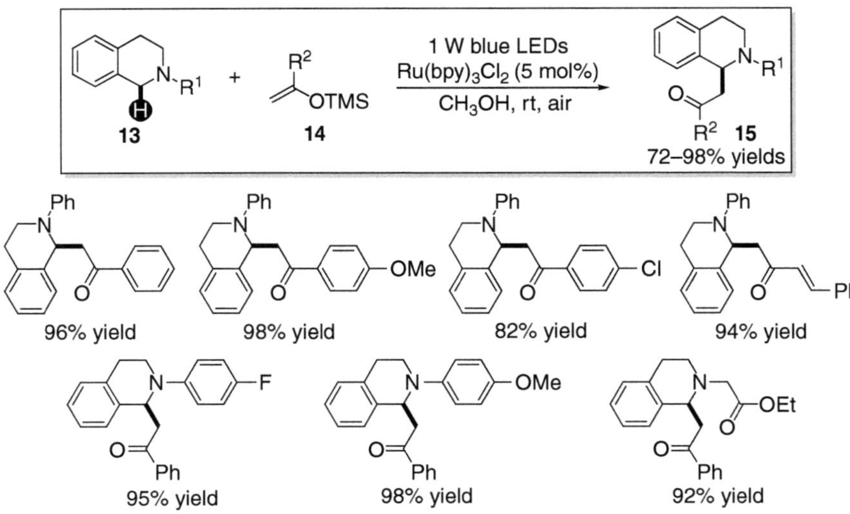

Figure 4.11 Visible light induced oxidative coupling reaction of amines with enol silanes.

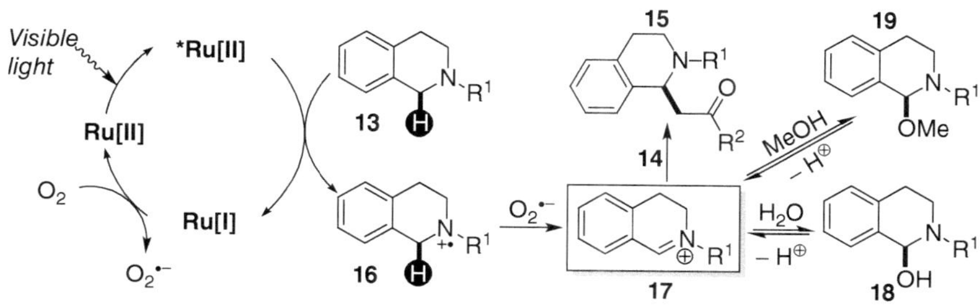

Figure 4.12 Possible mechanism for visible light induced oxidative coupling reaction of amines with enol silanes (Ru[II], $Ru(bpy)_3Cl_2$).

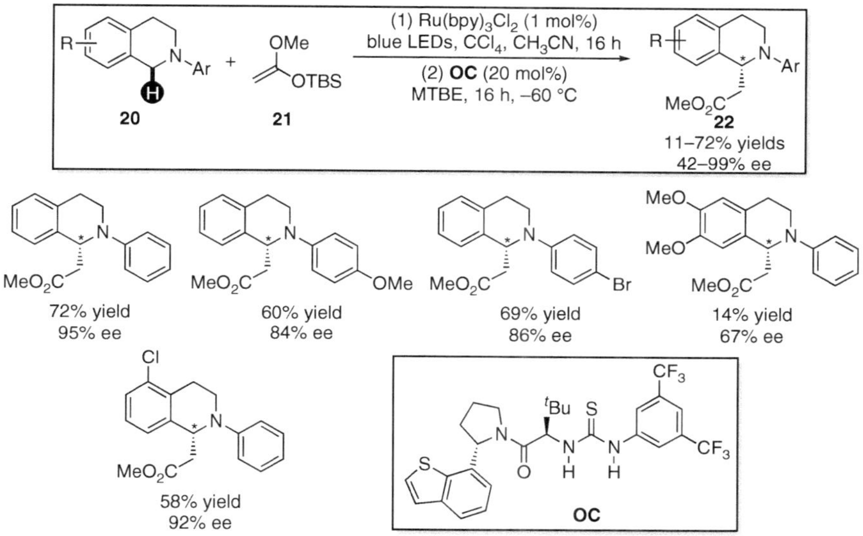

Figure 4.13 Photoredox activation and anion-binding catalysis in oxidative coupling reaction of amines with silyl ketene acetals (OC, organocatalyst; MTBE, methyl tert-butyl ether).

The Stephenson group [12] investigated a similar reaction under anerobic conditions. During the whole process, bromotrichloromethane was used as the oxidant. Recently, the asymmetric version of this transformation was achieved by Stephenson, Jacobsen and coworkers (Figure 4.13) [36]. Firstly, in the presence of 1 mol% $Ru(bpy)_3Cl_2$ as the photocatalyst and 4 equiv. of CCl_4 as the stoichiometric oxidant, iminium ions were generated *in situ* from *N*-aryl tetrahydroisoquinolines **20** by the irradiation of visible light in CH_3CN. Then switching the solvent to methyl tert-butyl ether (MTBE), a chiral thiourea-catalyzed Mannich-type reaction between the generated iminium ion and silyl ketene acetal **21** occurred and gave the final optically active β-amino esters **22** in good to excellent enantioselectivities. This rational combination of visible light photoredox catalysis and asymmetric anion-binding catalysis provides new opportunities for asymmetric transformations in oxidative photocatalysis [37].

4.2.3 Strecker Reaction

Strecker reaction is a highly effective and convenient strategy for the preparation of valuable α-amino nitriles that could be easily hydrolyzed to the useful α-amino acids and 1,2-diamines [38]. Developing clean and sustainable Strecker reactions for the construction of α-amino nitriles is of great importance for organic and medicinal chemistry. In 2011, Rueping and coworkers [39] reported a visible light induced aerobic oxidative Strecker reaction by using readily available potassium cyanide (KCN) as the CN^- source (Figure 4.14). The reaction was applicable to a variety of N-aryl tetrahydroisoquinolines with electron-donating and electron-withdrawing groups on the N-aryl groups. Interestingly, N-allyl-substituted tetrahydroisoquinoline was also compatible

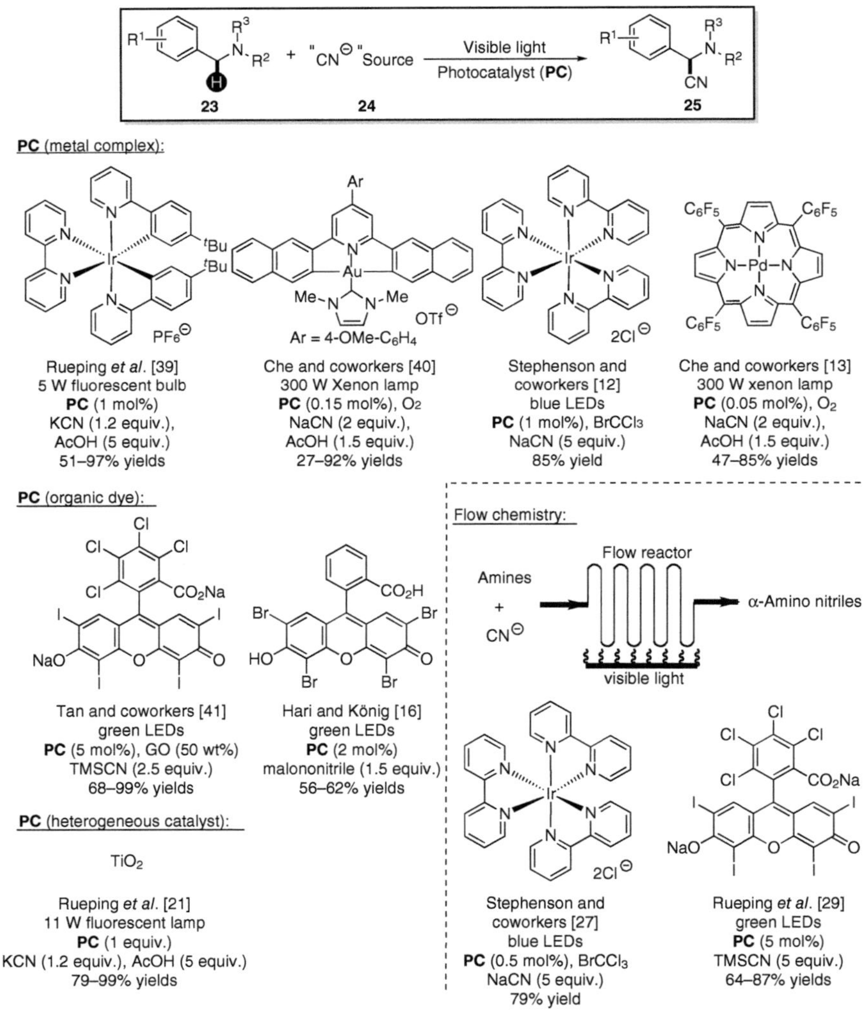

Figure 4.14 Visible light promoted Strecker reaction (GO, graphene oxide).

with the optimized reaction conditions and afforded the desired product with 51% yield. Significantly, acyclic tertiary amines could be efficiently subjected to this oxidative cyanation reaction, yielding the corresponding α-amino nitriles in good yields. In 2012, Che and coworkers [40] synthesized a novel luminescent organogold(III) complex with long-lived triplet excited state and demonstrated its high efficiency in the light-induced cyanation reaction of tertiary amines. The catalytic protocol that was further developed highlighted the richness of organogold(III) photochemistry and enriched the Strecker reaction. Lately, the same group used palladium(II)–prophyrin complex **$PdF_{20}TPP$** as the photocatalyst to achieve the same transformation [13]. Moreover, Stephenson's group also realized this transformation under oxygen-free conditions by using bromotrichloromethane as the oxidant [12].

In 2011, the Tan group [41] and the König group [16] independently developed an organic dye-catalyzed visible light photocatalytic Strecker reaction. It is noteworthy that graphene oxide (GO) could accelerate the reaction rates and increase the yields, as demonstrated in Tan's work (Figure 4.14). Additionally, the Rueping group reported the heterogeneous photocatalytic cyanation reaction using inexpensive TiO_2 as the photocatalyst [21]. Continuous flow technology was also introduced into this reaction by the groups of Stephenson [27] and Rueping [29].

Recently, visible light induced oxidative cyanation of primary and secondary amines was reported by Seeberger and coworkers [42]. Only in the presence of 0.1 mol% tetraphenylporphyrin (TPP), secondary amines were fully oxidized to imines in a continuous-flow photoreactor followed by trapping with trimethylsilylcyanide (TMSCN) affording the cyanation products with good to excellent yields (Figure 4.15). Meanwhile, using THF as the reaction media and tetrabutylammonium fluoride (TBAF) as the additive, many kinds of primary amines reacted with TMSCN smoothly to give the primary α-amino nitriles in good yields. Significantly, this protocol provides a concise and atom-economic way for the synthesis of unprotected amino acids. The authors proposed that a singlet-oxygen mediated pathway may be involved in this reaction. Firstly, triplet oxygen was sensitized to its singlet state by the excited state TPP under the irradiation of visible light. Then singlet-oxygen oxidized amines to imine intermediates for the following nucleophilic additions.

4.2.4 Friedel–Crafts Reaction

In 2012, Stephenson and coworkers developed an elegant method for the oxidative functionalization reactions of tetrahydroisoquinolines by using $BrCCl_3$ instead of oxygen [12]. Versatile nucleophiles including indoles could take part in the coupling reactions. When indole was used as the trapping reagent, the desired Friedel–Crafts product was isolated in 83% yield in the presence of 5 equiv. of potassium tert-butoxide (Figure 4.16). Lately, Wu and coworkers [14] reported an easily obtained platinum(II) terpyridyl complex with visible light absorption, long excited-state lifetime, and high luminescent quantum yield as well as good chemical stability and applied it in visible light promoted Friedel–Crafts-type reactions. They found that addition of 2 equiv. of $FeSO_4$ yielded the coupling products exclusively. Recently, Wu and coworkers [43]

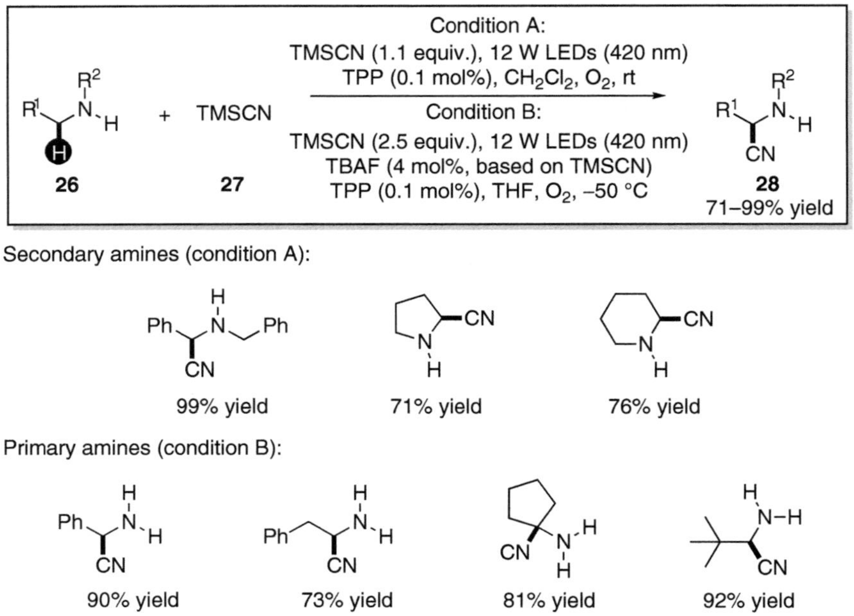

Figure 4.15 Visible light promoted oxidative cyanation of primary and secondary amines (TPP, tetraphenylporphyrin; TBAF, tetrabutylammonium fluoride).

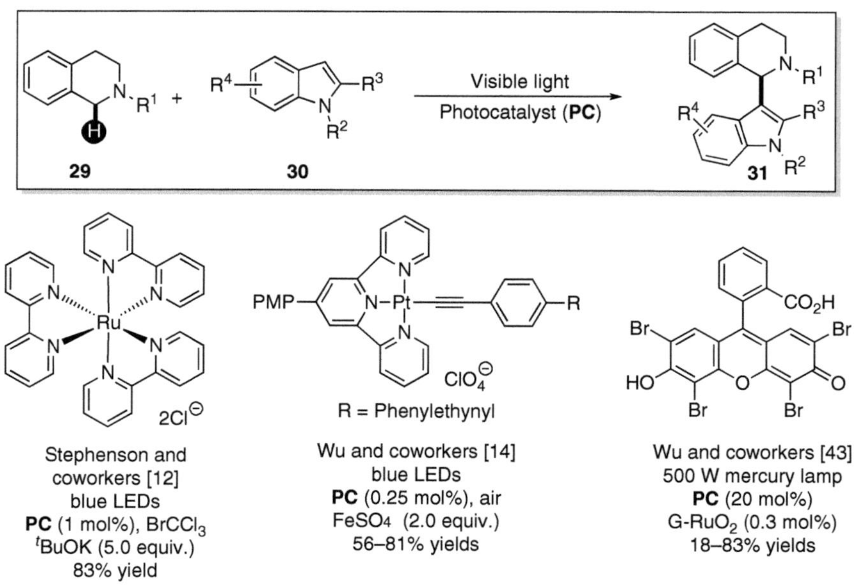

Figure 4.16 Visible light promoted Friedel–Crafts reactions (PMP, 4-methoxyphenyl).

disclosed a novel cross-coupling hydrogen evolution (CCHE) reaction by means of visible light photoredox catalysis. In this reaction, Eosin Y served as a photocatalyst to initiate cross-coupling of tetrahydroisoquinolines with indoles, and graphene-supported RuO_2 nanocomposite (G-RuO_2) played as an electron and proton capture. By combining these two kinds of catalysts, a variety of indoles and tetrahydroisoquinolines underwent the CCHE reactions smoothly to give the coupling products in good yields accompanying H_2 as the only by-product. Importantly, no external sacrificial oxidant was needed for this reaction. Mechanistic studies through flash-photolysis technology and product analysis indicated that G-RuO_2 did indeed capture the electron and proton eliminated from the C—H bonds of the substrates.

In 2012, Stephenson and coworkers [44] reported a mechanistically distinct visible light promoted Friedel–Crafts amidoalkylation via the oxidative quenching cycle. In the presence of 5 equiv. of $(NH_4)_2S_2O_8$, the excited state of ***Ru[II]** was oxidized to **Ru[III]**, resulting in the formation of reactive intermediate sulfate radical anion **34** via oxidative quenching (Figure 4.17). Subsequently, dialkylamides **32** interacted with **34** to give α-amido radical **35** via an H-atom abstraction process. The next step is the oxidation of **35** to iminium ion **37** by two possible pathways: (1) **35** was oxidized by **Ru[III]**, regenerating **Ru[II]** to complete the catalytic cycle; (2) **35** was oxidized by $(NH_4)_2S_2O_8$ through a chain-propagation step. Finally, nucleophilic addition of arenes to the generated iminium ions delivered the Friedel–Crafts amidoalkylation products.

Li and coworkers [45] demonstrated that α-carbonyl secondary amines could also be oxidized to imines when intercepted with indoles to give numerous functionalized 2-(1*H*-indol-3-yl)-2-amino-carbonyl compounds. A wide range of indoles could couple with α-amino carbonyls under the optimal conditions, affording the Friedel–Crafts-like adducts in moderate to good yields (Figure 4.18). Importantly, this protocol expanded the substrate scope of amines in visible light photoredox catalysis and provided a more clean and effective strategy for the synthesis of natural and non-natural α-amino carbonyl compounds. Rueping and coworker [46] also reported a highly effective oxidative

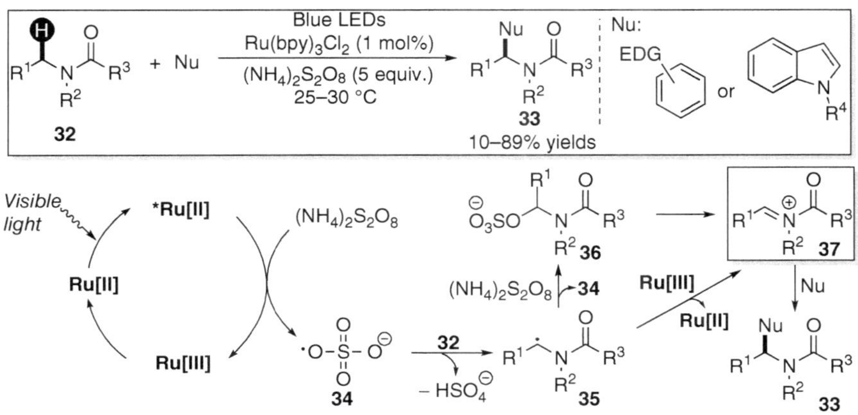

Figure 4.17 Visible light promoted Friedel–Crafts amidoalkylation reactions.

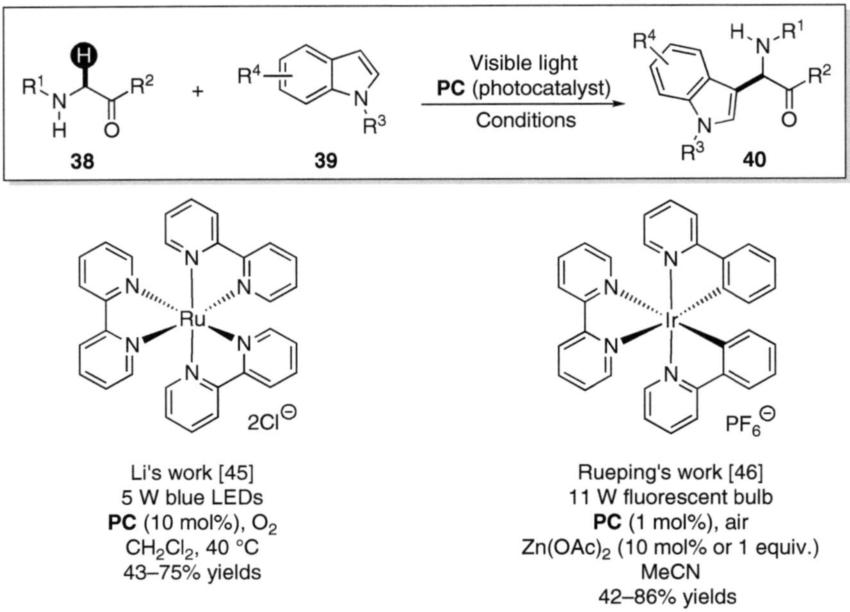

Figure 4.18 Visible light promoted α-arylation of α-amino carbonyls with indoles.

functionalization of glycine derivatives and peptides via combination of visible light photoredox catalysis and Lewis acid catalysis. They found that $Zn(OAc)_2$ played a crucial role in this transformation, which not only accelerated the formation of imine intermediate, but also served as a Lewis acid to activate the generated imine for the subsequent nucleophilic attack. In the presence of 1 mol% $Ir(ppy)_2(bpy)PF_6$ and 10 mol% $Zn(OAc)_2$, a variety of glycine derivatives and indoles reacted well and gave the desired products in good yields. More importantly, this method was also applicable to glycine-derived dipeptides by increasing the loading of $Zn(OAc)_2$ to 1 equiv.

4.2.5 Alkynylation Reaction

In 2012, Rueping and coworkers [47] developed a dual catalysis strategy for the aerobic oxidative α-alkynylation of tetrahydroisoquinoline derivatives by merging visible light photoredox catalysis with metal catalysis [48]. The authors found that photocatalyst $Ru(bpy)_2(dtbbpy)(PF_6)_2$ and metal salt $(MeCN)_4CuPF_6$ were compatible in this transformation and co-catalyzed the coupling reaction between tetrahydroisoquinolines and alkynes efficiently (Figure 4.19). They also proposed that iminium ions were generated in the photocatalytic cycle, and active copper acetylide **43** was formed *in situ* in the metal catalytic cycle for the subsequent addition step. This concept is a great advancement in visible light photoredox catalysis and gives opportunities to design new types of photocatalytic reactions. In the same year, Stephenson and coworkers [12, 27] reported the same strategy under anerobic conditions by employing CuBr as the co-catalyst. Lately, Fu and coworkers [49] found that organic dyes such as Rose Bengal could also catalyze this reaction in the presence of 10 mol% CuI. Under

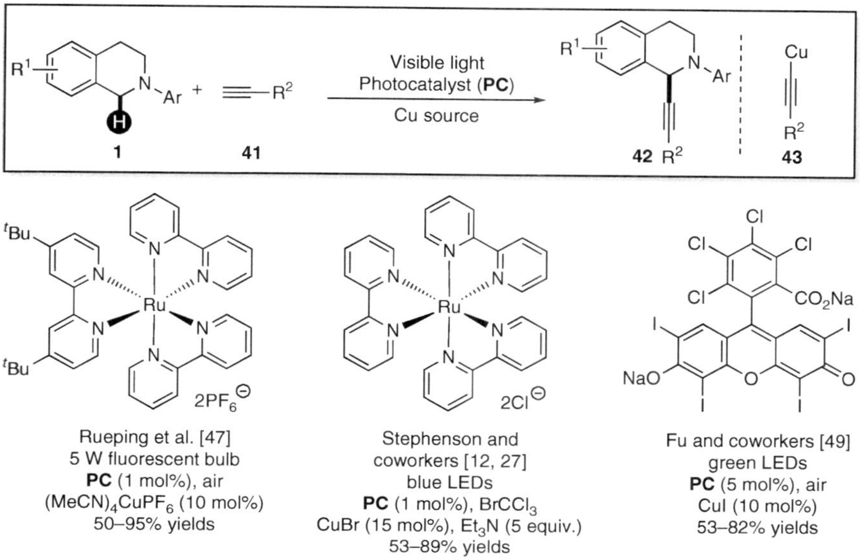

Figure 4.19 Visible light promoted α-alkynylation of tetrahydroisoquinolines.

Fu's conditions, various α-alkynylated tetrahydroisoquinolines were isolated in moderate to good yields with somewhat prolonged reaction time compared with the ruthenium complex.

4.2.6 Phosphonation Reaction

α-Amino phosphonates represent an important class of structural motifs that are widely introduced in medicinal chemistry and organic chemistry [50]. Development of new and effective methods for the synthesis of these molecules is highly desirable. In 2011, Rueping and coworkers [51] reported the first example of oxidative phosphonation reactions of tetrahydroisoquinolines by means of visible light photoredox catalysis. Under the optimal conditions, the generated iminium ion via photocatalysis was trapped by different phosphite esters to give the terminal α-amino phosphonates with good yields (Figure 4.20). Lately, the same group has carried out the reaction under heterogeneous conditions with ZnO as the catalyst [21]. In 2013, the groups of Zhu [33] and Che [13] successfully applied gold(III)-complex and palladium(II)–prophyrin complex **$PdF_{20}TPP$** into this transformation, respectively. Further, König and coworker [16] demonstrated that organic dye Eosin Y was also a good choice for this visible light photocatalytic phosphonation. Recently, the Rueping group also developed a continuous flow organocatalytic phosphonation reaction using Rose Bengal [29].

4.2.7 Addition of 1,3-Dicarbonyls

The generated iminium ions from visible light photoredox catalysis can be trapped by a wide range of nucleophiles. In this context, 1,3-dicarbonyls is a good partner for the cross-coupling reactions. The groups of Stephenson [12], Zhu [33], Che [13], and Wu [14] reported visible light promoted oxidative

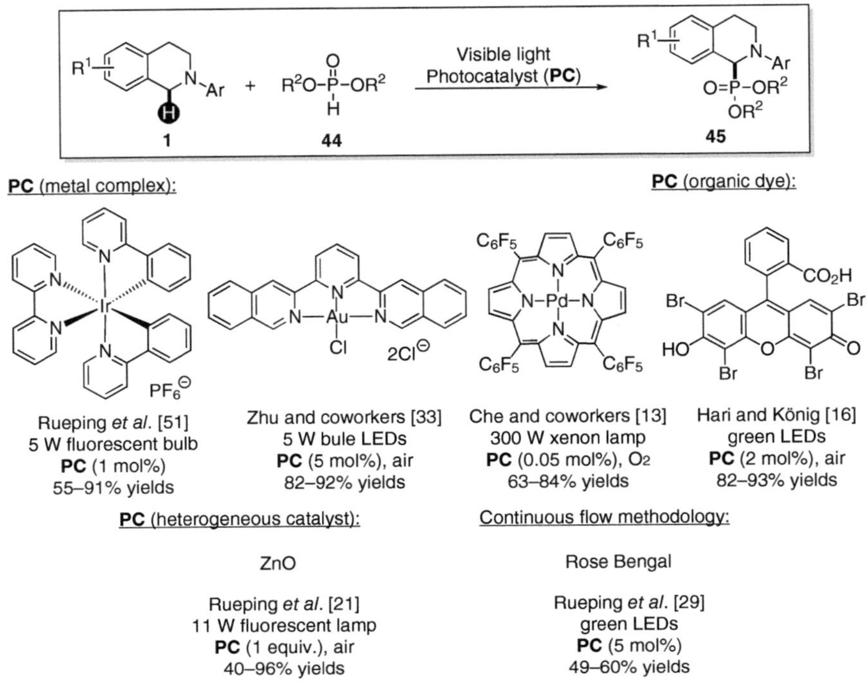

Figure 4.20 Visible light promoted α-phosphonation of tetrahydroisoquinolines.

coupling of tetrahydroisoquinolines with malonates using ruthenium, gold, and palladium and platinum complexes, respectively. Various α-diester amine derivatives could be constructed in a highly concise fashion (Figure 4.21). Organic dyes Eosin Y and TBA-eosin Y also exhibited high efficiency in the addition of malonates, which was demonstrated by König [16] and Wu [18] independently. Meanwhile, Blechert and coworkers [24] examined the feasibility of their organic heterogeneous catalyst mpg-C_3N_4. Lately, the Rueping group [29] achieved this reaction in a continuous flow reactor.

Besides *N*-aryl tetrahydroisoquinolines, secondary amines were also suitable for this transformation. In 2013, the Wu group [52] realized alkylation of secondary amines by combining visible light photoredox catalysis and transition metal catalysis. In the presence of 1 mol% $Ru(bpy)_3Cl_2$ and 10 mol% $Cu(OTf)_2$, β-keto esters could react with various *N*-substituted glycine esters efficiently, yielding the corresponding oxidative coupling products in good yields (Figure 4.22). This fine protocol provided a clean and efficient way for the functionalization of secondary amino acid esters and enriched the visible light photochemistry.

4.2.8 Formation of C—N and C—O Bonds

In 2011, the Xiao group [53] reported a visible light induced intramolecular oxidative cyclization reactions of diamines, which allowed an efficient synthesis of highly substituted tetrahydroimidazole derivatives. Under the

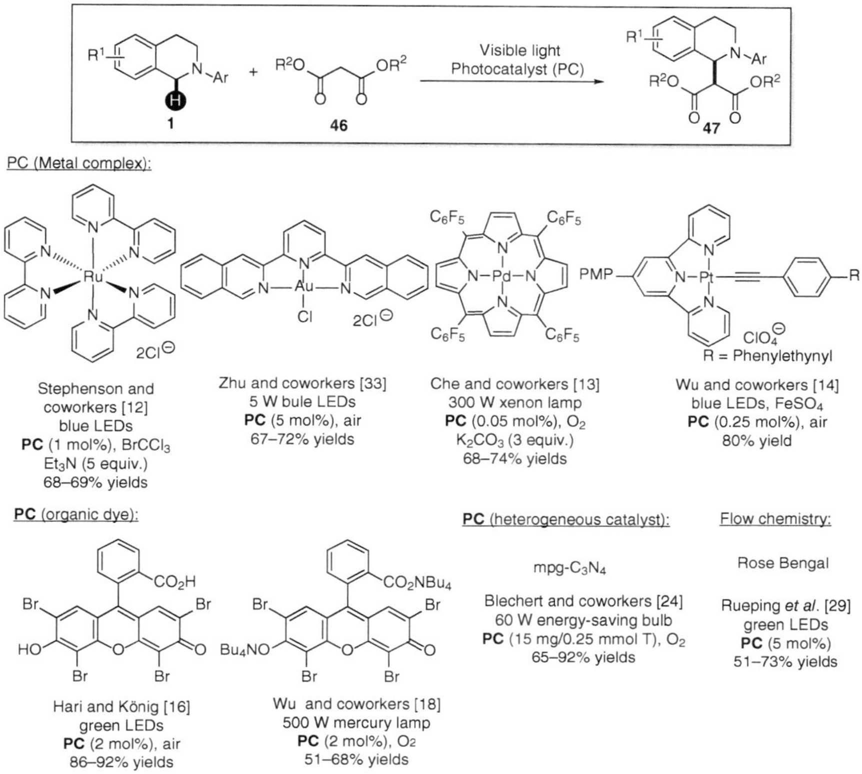

Figure 4.21 Visible light promoted oxidative coupling of tetrahydroisoquinolines with malonates.

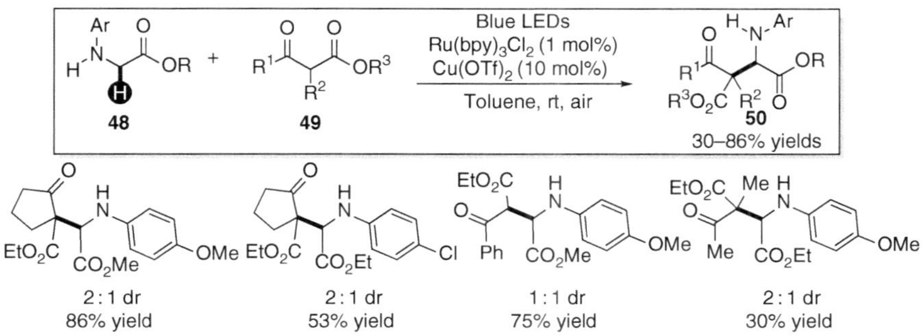

Figure 4.22 Visible light and transition metal co-catalyzed alkylation of secondary amines.

optimized reaction conditions, the benzylic C—H of diamines **51** derived from natural amino acids was first oxidized to iminium ions **53**. Subsequently, an intramolecular nucleophilic attack took place with the aid of ^{t}BuOK to afford various tetrahydroimidazoles in good yields with excellent diastereoselectivities (Figure 4.23). Based on the experimental results, they proposed that the *Re*-face attack was more favorable than the *Si*-face attack, because of steric repulsion.

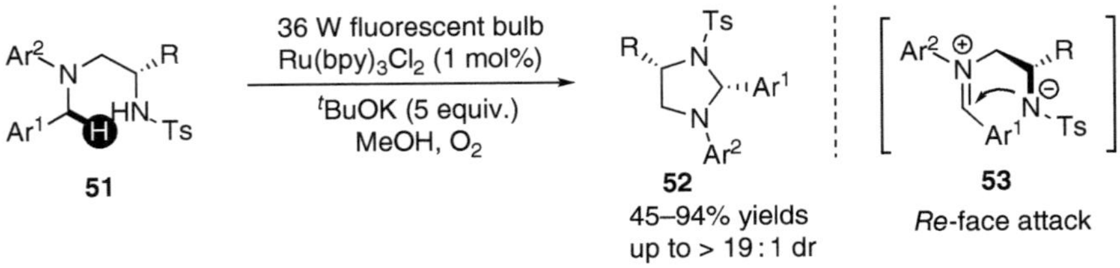

Figure 4.23 Visible light induced intramolecular oxidative cyclization of diamines.

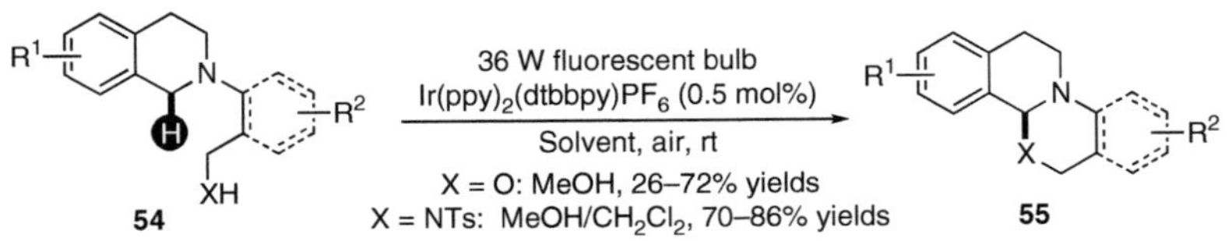

Figure 4.24 Visible light induced intramolecular oxidative cyclization reactions.

Interestingly, six-membered rings like hexahydropyrimidine derivative could also be obtained under the same conditions.

In 2012, the Xiao group [54] achieved the construction of isoquino[2,1-*a*][3,1]oxazine and isoquino[2,1-*a*]pyrimidine derivatives using the above-mentioned strategy (Figure 4.24). Using 0.5 mol% $Ir(ppy)_2(dtbbpy)PF_6$ as the photocatalyst, a variety of tetrahydroisoquinolines bearing a nucleophilic oxygen or nitrogen cite could undergo the oxidative cyclization reaction to give the corresponding products in moderate to good yields. Lately, Marvin and coworkers [55] realized the same reaction with $Ru(bpy)_3Cl_2$ as the catalyst. Notably, unactivated amines were also found suitable in Marvin's conditions. The Che group [13] found that palladium(II)–prophyrin complex $\mathbf{PdF_{20}TPP}$ was also suitable for this intramolecular reaction.

4.2.9 Miscellaneous

In 2011, Xiao and coworkers [56] disclosed a visible light induced oxidative/[3+2] cycloaddition/aromatization sequential strategy for the synthesis pyrrolo[2,1-a] isoquinolines. A variety of dipolarophiles such as maleimides, activated alkynes, and acrylates as well as maleic anhydrides reacted smoothly with dihydroisoquinoline esters to deliver the desired pyrrolo[2,1-a]isoquinolines, which formed the core structure of lamellarin alkaloids (Figure 4.25). In this regard, the Rueping group [57] reported a similar visible light photoredox catalyzed oxidative [3+2] cycloaddition reactions.

Notably, Zhao and coworkers used C_{60}-Bodipy derivative [58] and iodo-Bodipy immobilized on porous silica [59] as two kinds of novel photocatalysts to catalyze the oxidative/[3+2] cycloaddition/aromatization. A possible mechanism of the visible light induced [3+2] cycloaddition is depicted in Figure 4.26. In the irradiation of visible light, ruthenium complex was first excited to ***Ru[II]**, which was then reduced to **Ru[I]** by dihydroisoquinoline esters **56** via the reductive quenching cycle. At the same time, molecular oxygen oxidized **Ru[I]** to **Ru[II]** furnishing the catalytic cycle with the release of an oxygen radical anion. Subsequently, this highly active radical anion reacted with nitrogen radical cation **62**, yielding the

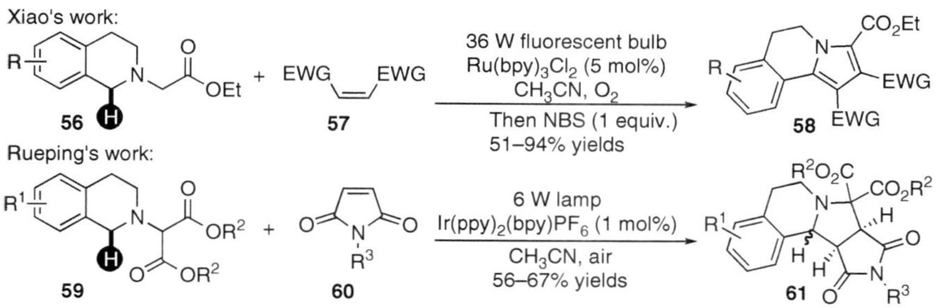

Figure 4.25 Visible light induced [3+2] cycloaddition reactions.

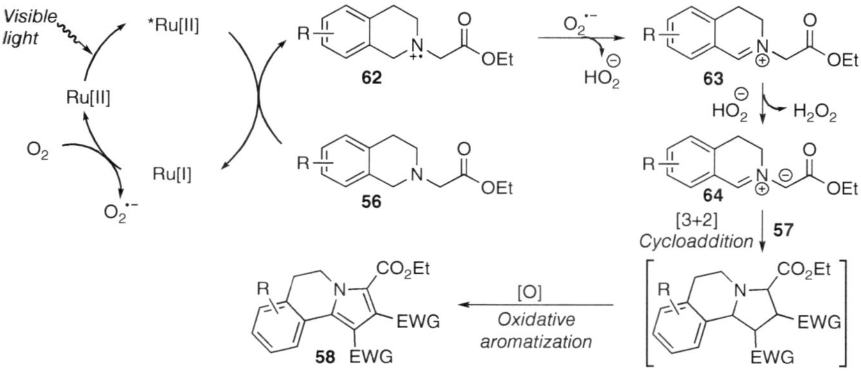

Figure 4.26 Possible mechanism of visible light induced [3+2] cycloaddition reactions.

iminium ion **63**. Next, the generated HO_2^- might serve as a base to deprotonate the α-position of ester group and afforded the 1,3-dipole azomethine **64** *in situ*. Then intermolecular [3+2] cycloaddition took place followed by an oxidative aromatization to form the pyrrolo[2,1-a]isoquinolines.

Organic molecules bearing a trifluoromethyl group play a key role in synthetic chemistry, medicinal chemistry, and functional materials. In 2012, Fu and coworkers [49] reported an elegant α-trifluoromethylation of tetrahydroisoquinolines by means of visible light photoredox catalysis. Various α-trifluoromethylated tetrahydroisoquinolines could be obtained in these mild and sustainable photocatalytic conditions (Figure 4.27).

In 2012, Rovis and coworkers [60] developed a productive dual catalysis mode for the asymmetric α-acylation of tetrahydroisoquinolines by the combination of visible light photoredox catalysis and nitrogen-heterocyclic carbene (NHC) catalysis. The authors proposed that electrophilic iminium ion was formed by photocatalysis through the oxidative quenching cycle, and NHC converted aldehydes to nucleophilic acyl anion or homoenolate equivalents. These two chemically distinct activation modes worked in conjunction under the optimized conditions, and a wide range of aliphatic aldehydes could couple with tetrahydroisoquinolines, affording the acylated products in moderate to good yields with excellent enantioselectivities (Figure 4.28). It has been found that the rational choice of the oxidant is very crucial for the high efficiency of this

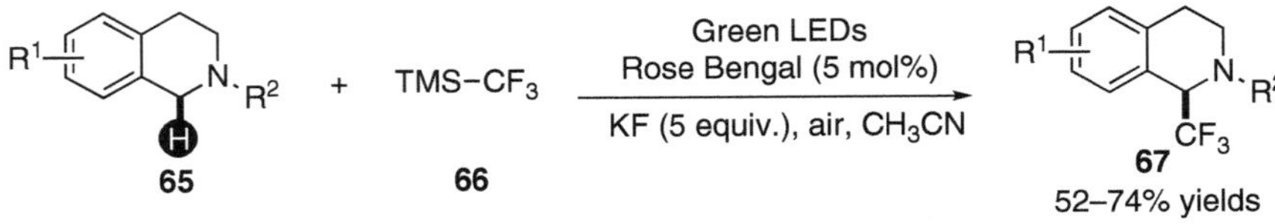

Figure 4.27 Visible light induced α-trifluoromethylation of tetrahydroisoquinolines.

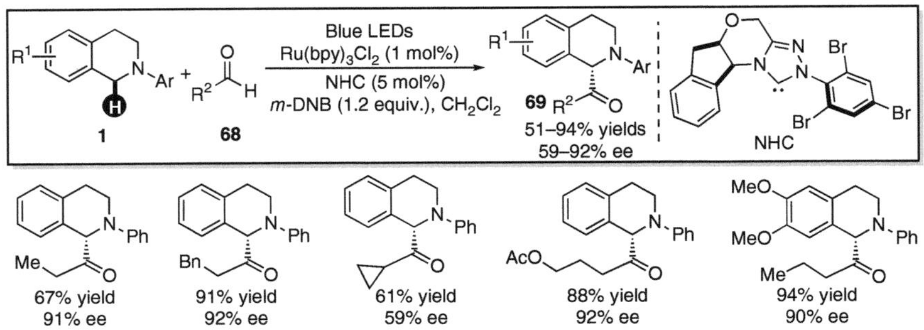

Figure 4.28 Visible light and NHC co-catalyzed α-acylation of tetrahydroisoquinolines.

transformation. *m*-Dinitrobenzene (*m*-DNB), a known oxidative quencher (OQ) of $^{*}Ru(bpy)_3^{2+}$, gave the best result.

Zhu and coworkers [61] recently uncovered a visible light induced C—H activation/retro-aza-Michael/oxidation/[3+2] cycloaddition sequence when α-ketoester was used as the coupling partner with tetrahydroisoquinolines. Under the optimized conditions, biologically important isoxazolidine derivatives could be constructed in good yields with excellent diastereoselectivities (Figure 4.29). Based on the experimental results and related literatures, the authors proposed a possible mechanism to explain the formation of the highly functionalized isoxazolidines. Firstly, iminium ion **74** was formed in the visible light photocatalytic cycle. Then α-ketoester **71a** was added to iminium ion with the aid of TfOH, delivering intermediate **75** which could be easily isomerized to enol **76**. Subsequently, retro-aza-Michael reaction occurred to give secondary amine **77**, followed by oxidation to produce imine **78**. The imine could be further oxidized to the corresponding nitrile oxide **79** that underwent a [3+2] cycloaddition to afford the desired bicyclic isoxazolidine.

Amines used in visible light photoredox catalyzed coupling reactions are mainly based on tetrahydroisoquinoline derivatives, and the direct functionalization of simple tertiary aliphatic amines is still a challenging task. In 2013, Weaver and coworkers [62] reported an interesting visible light induced C—H functionalization of various tertiary aliphatic amines, which provided a novel and effective way for the rapid access to α-azole carbinamines (Figure 4.30). It should be note that 2-chlorobenzimidazoles, -benzoxazoles, -benzothiazoles, and -thiazoles were all compatible with the optimized conditions. Reactions worked well with different tertiary aliphatic amines, yielding the desired cross-coupling products in moderate to good yields. Notably, a new analog of Nizatidine could be obtained by using this protocol.

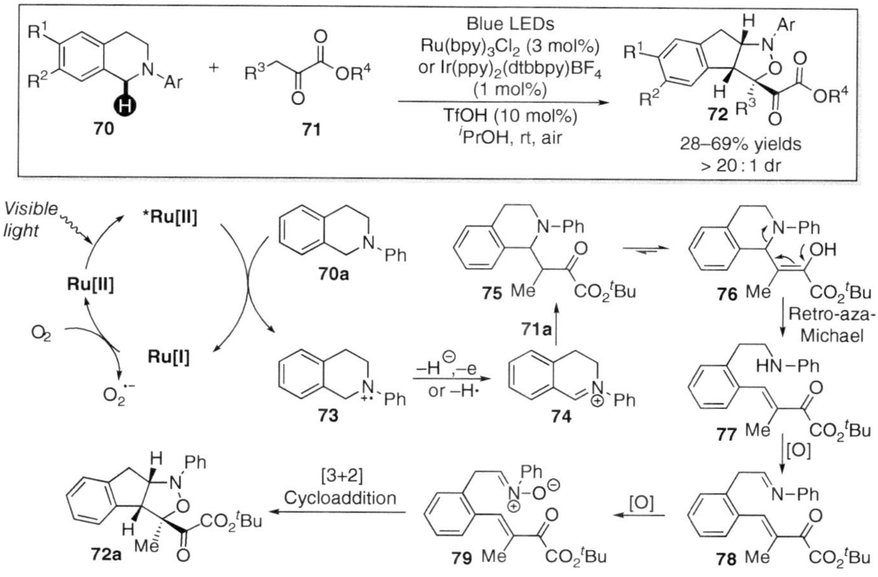

Figure 4.29 Visible light induced aerobic C—H/C—N cleavage cascade.

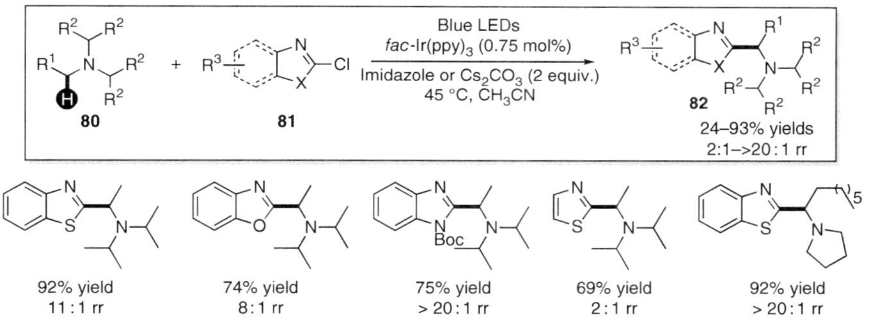

Figure 4.30 Visible light induced C—H functionalization of aliphatic amines (rr, regioselectivity).

Multicomponent reactions (MCRs), which combined three or more reactants in one reaction system to form multiple chemical bonds, represent a unique and powerful tool for the rapid construction of structurally complex and functionally diverse molecular architectures from readily available starting materials. In 2013, Rueping and Vila [63] disclosed a visible light induced three-component reaction among tertiary amines, isocyanides, and water or carboxylic acids, and various useful α-amino amides and imides were constructed under mild conditions (Figure 4.31). They proposed that the iminium ion **88** was trapped by isocyanides to form intermediate **89**. Subsequently, **89** reacted with water or carboxylic acids followed by rearrangement to give the final products. Lately, they also conducted this MCRs in a flow reactor [29].

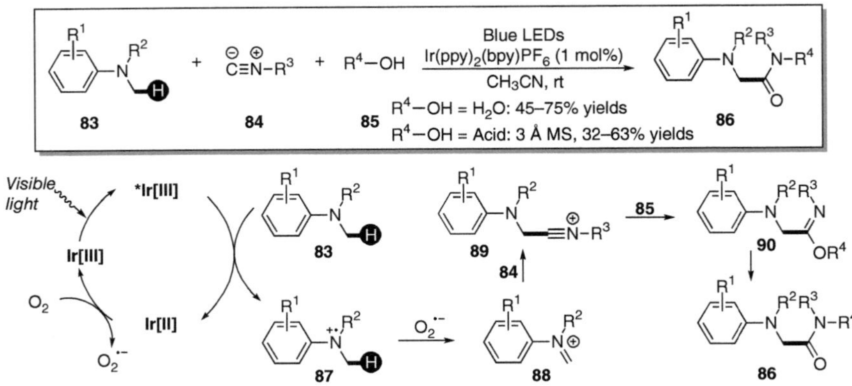

Figure 4.31 Visible light induced multicomponent reactions.

4.3 Visible Light Mediated α-Amino C—H Functionalization Via α-Amino Radicals

4.3.1 Addition to Electron-Deficient Aromatics

As highlighted in Figure 4.3, a nucleophilic α-amino radical could also be generated from amines by means of visible light photoredox catalysis. Using this highly active radical as the key intermediate to synthesize some fine chemicals is attractive. In 2011, MacMillan and coworkers [64] discovered an elegant visible light induced α-amino C—H arylation of tertiary amines using the strategy of accelerated serendipity. Cyclic and acyclic tertiary amines reacted with a wide range of electron-deficient aromatics bearing a nitrile group to give the corresponding α-arylated amines (Figure 4.32). The authors proposed a catalytic cycle to explain the described α-amino C—H arylation protocol. Presumably, the excited state ***Ir[III]** was quenched by 1,4-dicyanobenzene (DCB) **92a**, affording radical anion **94**. Subsequently, the resultant high oxidative species **Ir[IV]** oxidized the coupling partner amine **91a** to amine radical cation **95**, regenerating **Ir[III]** to complete the catalytic cycle. Next, α-amine radical **96** was formed through a deprotonation process in the presence of NaOAc. Finally, the radical–radical coupling between **96** and **94** occurred followed by releasing CN^- to give the desired α-arylation product, which could be found in many natural products and pharmaceuticals. It should be noted that the two reaction partners, amines **91** and electron-deficient aromatics **92** were activated simultaneously [65].

4.3.2 Addition to Electron-Deficient Alkenes

In 2012, Nishibayashi and coworkers [66] reported a visible light induced intermolecular addition of α-aminoalkyl radical to electron-deficient alkenes. In this reaction, solvent had an important effect on the success of this transformation, and *N*-methylpyrrolidone (NMP) was found to be the best choice. Meanwhile, the authors found that alkenes with two electron-withdrawing groups were

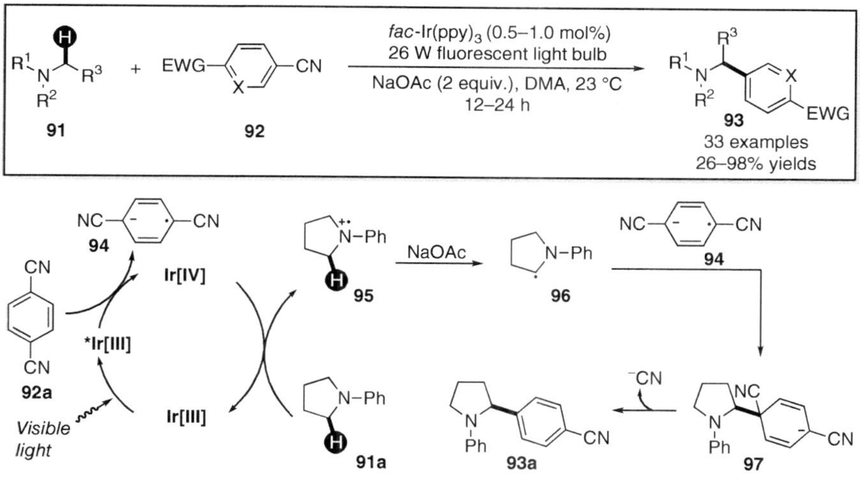

Figure 4.32 Visible light induced α-amino C—H arylation reactions.

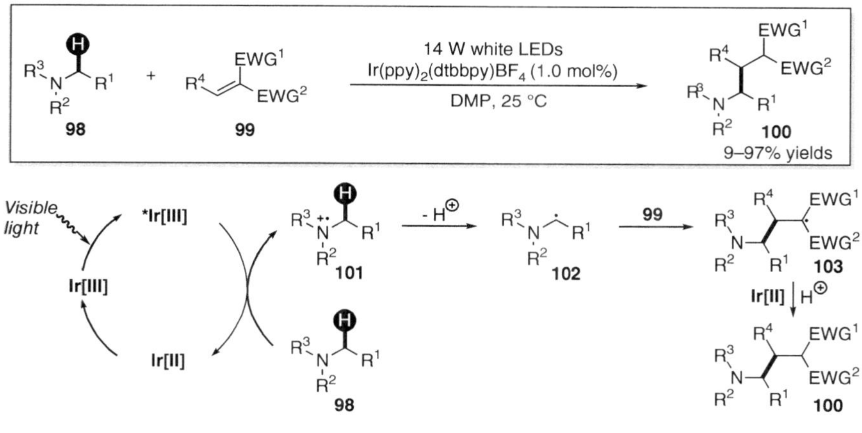

Figure 4.33 Visible light induced addition of α-aminoalkyl radical to electron-deficient alkenes.

essential for the high yields. Results of deuterium labeling experiments indicated that the activated methine proton of terminal product mainly originated from the α-position of the tertiary amine. On–off switching of the light source and control experiments showed that a radical chain process was negligible in this reaction. Based on experiment results, they proposed a sequential redox pathway to explain the formation of α-amino alkylated product (Figure 4.33). The initial step was the generation of amine radical cation **101** via the reductive quenching cycle. Subsequently, **101** was deprotonated to α-aminoalkyl radical **102** which then added to alkene **99** to yield intermediate **103**. Then **103** was reduced by reductive species **Ir[II]** followed by protonation to afford the desired product.

As mentioned above, alkenes bearing only one electron-withdrawing group such as α,β-unsaturated ketones were not suitable for visible light induced α-amino alkylation reaction. To address this issue, the Nishibayashi group [67] recently developed an alternative protocol by introducing a silyl group at the α-position of amines. Under their optimized conditions, the reactions of a variety of cyclic and acyclic ketones with α-silylamines proceeded smoothly to afford the corresponding products in moderate to good yields (Figure 4.34). Mechanistic studies indicated that silyl enol ether **107** was first formed via the visible light photoredox catalysis, which then hydrolyzed to the final product.

Another method to realize the addition of α-amino radical to α,β-unsaturated ketones is the selection of suitable tertiary amines. By using *N*-aryltetrahydroisoquinoline as the substrates, Reiser, Pandey and coworkers [68] disclosed a similar coupling reaction between tetrahydroisoquinolines and Michael acceptors. Under the irradiation of visible light, various *N*-aryltetrahydroisoquinolines reacted with α,β-unsaturated ketones in the presence of $Ru(bpy)_3Cl_2$ or $Ir(ppy)_2(dtbbpy)BF_4$, giving the desired products in moderate to excellent yields (Figure 4.35). Reactions with *exo*-methylene-γ-butyrolactone and acrylonitrile could also be accomplished, albeit with low yields. Notably, *N*-Phenylpyrrolidine was also compatible with the Iridium complex, affording the corresponding product in 28% isolated yield. At the same time, the authors investigated the intramolecular version of this reaction, and biologically important 5, 6-dihydroindolo[2,1-*a*]-tetrahydroisoquinolines could be obtained under the optimal conditions. Lately, Yoon and coworkers [69] found that the reaction

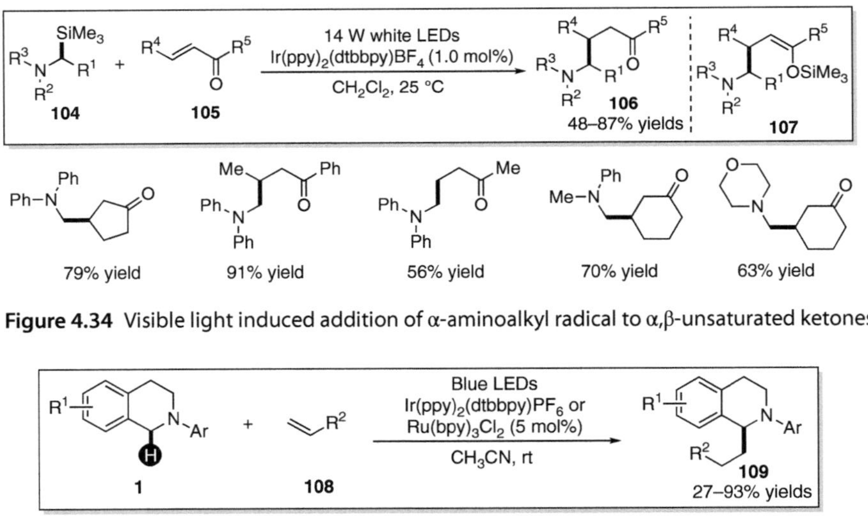

Figure 4.34 Visible light induced addition of α-aminoalkyl radical to α,β-unsaturated ketones.

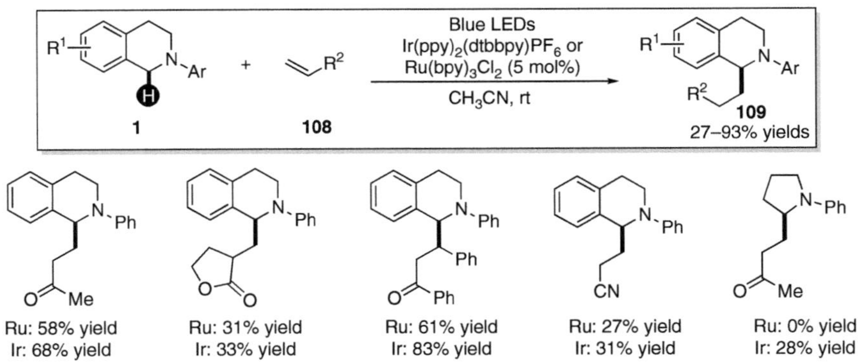

Figure 4.35 Visible light induced addition of α-aminoalkyl radical to Michael acceptors.

efficiency of the same transformation could be significantly improved when 1 equiv. of trifluoroacetic acid (TFA) was added to the reaction mixture.

Bian, Yu and coworkers [70] described a visible light promoted formal [4+2] cycloaddition reaction, providing a novel and robust method for the construction of tetrahydroquinolines from readily available amines and maleimides. Unlike the above-mentioned anerobic conditions, this transformation proceeded via an oxygen mediated pathway. α-Amino radical **114** was first formed from tertiary amine through reductive quenching in the photocatalytic cycle (Figure 4.36, path A). Then **114** was added to maleimides followed by an intramolecular radical cyclization to give intermediate **116**, which was then oxidized to the desired product by molecular oxygen. Hydrogen atom transfer between **110** and HOO$^{\bullet}$ might be another way to generate **114** (path B).

Rueping and coworkers [71] reported an interesting oxygen switchable radical additions and radical addition/cyclization reactions by means of visible light photoredox catalysis. In the presence of air and $Ir(ppy)_2(bpy)PF_6$, various anilines **117** and 2-arylidenemalononitriles **118** underwent radical addition/cyclization cascade reactions in MeCN under the irradiation of visible light, yielding tetrahydroquinoline derivatives **119** in moderate to good yields (Figure 4.37). By excluding oxygen from the reaction system, only Michael-type products were observed.

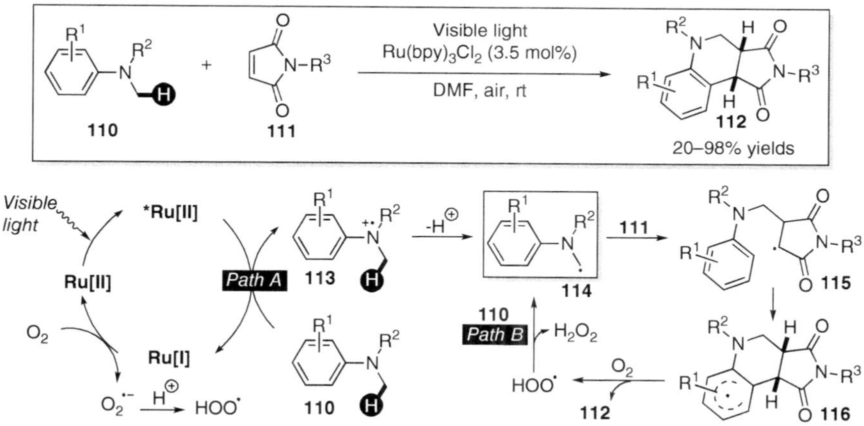

Figure 4.36 Visible light induced formal [4+2] cycloaddition.

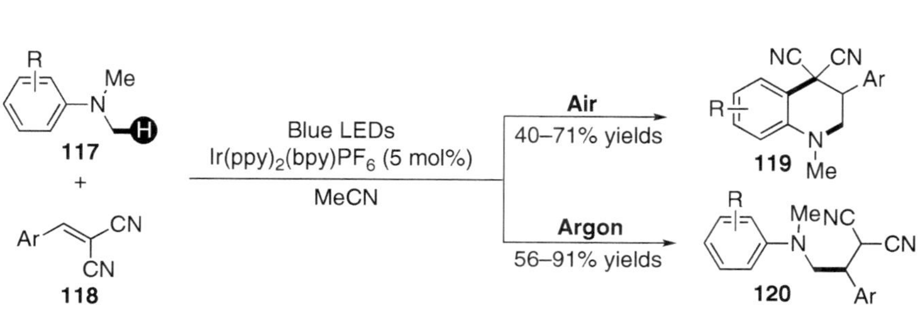

Figure 4.37 Visible light induced oxygen switchable reactions.

Recently, the Tan group [72] documented a similar photocatalytic [4+2] cycloaddition by using organic dye as the photocatalyst. In the presence of 2 mol% fluorescein, a wide range of *N*-aryl glycines underwent decarboxylation and then reacted with maleimides to afford tetrahydroquinoline derivatives with good yields (Figure 4.38). In this reaction, *N*-aryl glycines were introduced as the α-amino radical source. The authors also found that the generated α-amino radical could be further oxidized to iminium ions and trapped by various nucleophiles.

4.3.3 Miscellaneous

Besides electron-deficient aromatics and alkenes, α-amino radicals generated from nitrogen-containing benzoheterocycles could be also trapped by di-*tert*-butyl azodicarboxylate. In 2012, Nishibayashi and coworkers [73] reported a visible light induced direct sp^3 C—H amination of benzocyclic amines (Figure 4.39). In light of their oxidation or reduction potential of amines and azodicarboxylate esters, the authors proposed a possible mechanism to explain the formation of the aminated products (Figure 4.40). Tertiary amine **124** quenched the ***Ir[III]** followed by deprotonation to generate radical **128**. Meanwhile, di-*tert*-butyl azodicarboxylate served as an oxidant to oxidize **Ir[II]** to **Ir[III]**, completing the catalytic cycle and releasing radical anion **127**. Subsequently, α-amino radical **128** was added to **127** to yield anion **129**, which was then protonated to the final product.

In 2013, Li and coworkers [74] found that α-amino radicals derived from *N*,*N*-dimethylanilines can be intercepted by aryl isocyanates and aryl isothiocyanates. In this transformation, bis[2-(4,6-difluorophenyl)pyridinato-C^2,N](picolinato)-iridium(III) (FIrpic) with high $E_{0,0}$ (2.63 eV) was used in visible light

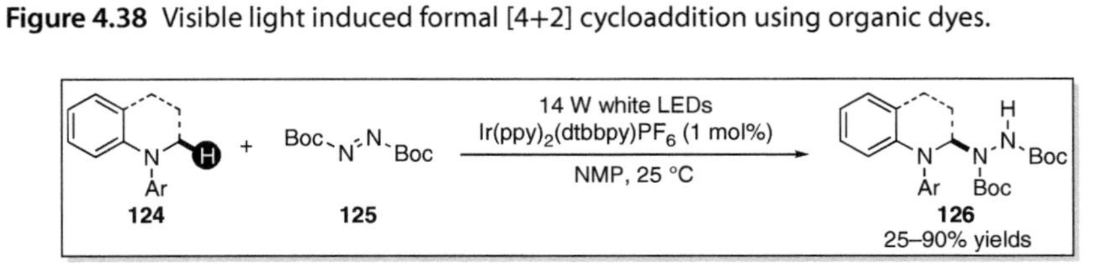

Figure 4.38 Visible light induced formal [4+2] cycloaddition using organic dyes.

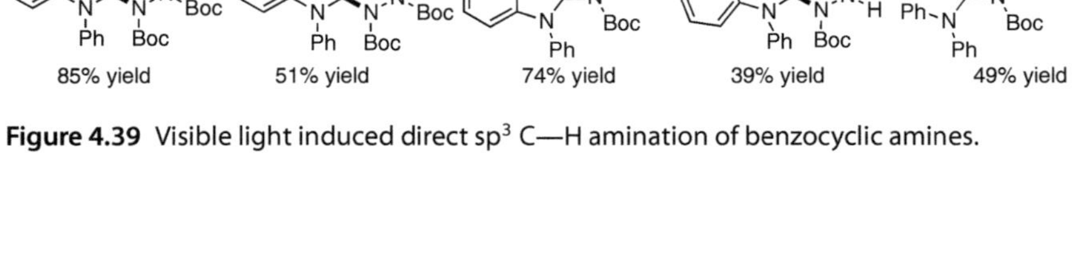

Figure 4.39 Visible light induced direct sp^3 C—H amination of benzocyclic amines.

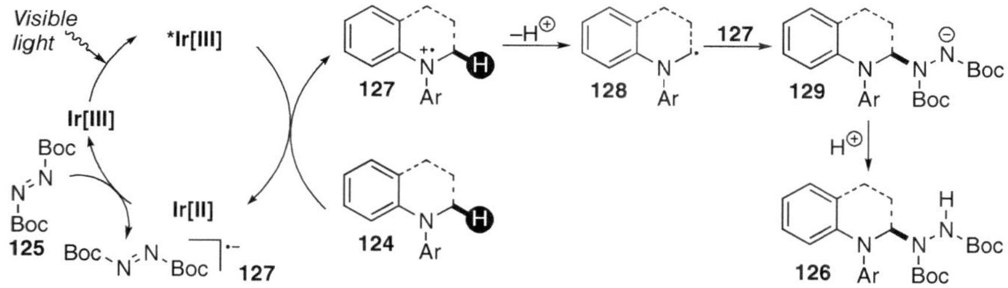

Figure 4.40 Possible mechanism for visible light induced direct sp^3 C—H amination reactions.

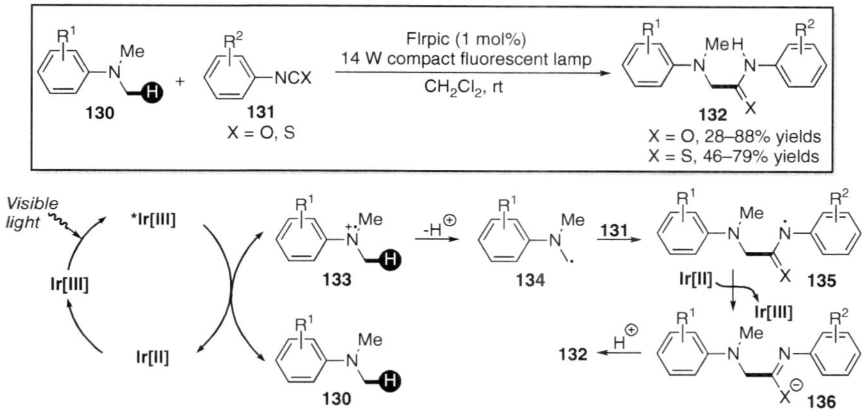

Figure 4.41 Visible light induced addition of α-amino radicals to isocyanates or isothiocyanates.

photoredox catalysis for the first time. Various *N*,*N*-dimethylaniline derivatives condensated with aryl isocyanates or aryl isothiocyanates efficiently, forming α-amino amides or α-amino thioamides in moderate to good yields (Figure 4.41). The reaction was initiated by a reductive quenching between the excited state ***Ir[III]** and amines **130**, generating radical cation **133**. Then α-amino radical **134** was generated via deprotonation of **133**. The next step was the addition of **134** to isocyanates or aryl isothiocyanates **131**, giving nitrogen radical intermediate **135**. Compound **135** was reduced by **Ir[III]** followed by protonation to form the desired product **132**.

4.4 Conclusions and Perspectives

The continuous growing interest in "green chemistry" has resulted in visible light, especially sunlight, being perceived as an ideal reagent because of its non-pollution, abundance, and sustainability. Although people's awareness of this concept can be dated back to the early twentieth century, photochemistry with visible light is largely unexplored due to the inability of most organic compounds to absorb visible light. Since 2008, visible light photoredox catalysis has been reinvestigated and has now emerged as a novel and effective method to advance these goals with the aid of the rapid development of metal polypyridyl complexes as well as organic dyes. Currently, photoredox catalysis with visible light is regarded as a popular and flourishing research field.

In this chapter, we described the recent advances in visible light mediated α-amino C—H functionalization reactions. It was observed that two active species, iminium ions and α-amino radicals derived from amines were generated by means of visible light photoredox catalysis; the former intercepted by various nucleophiles and the latter added to activated alkenes or aromatics, affording a broad range of useful and important functionalized amines. Continuous flow technology was also introduced into this field to improve the reaction efficiency, which has opened a door for the application of visible light in industrial productions. In addition, the combination of visible light photoredox catalysis with other catalytic modes such as metal catalysis and organocatalysis as well as Lewis acid catalysis will provide opportunities to design new types of reactions.

Despite these remarkable advances, several challenges and new possibilities still remain and can be summarized as follows: (1) the functionalized C—H bonds are mainly based on activated benzylic sp^3 C—H bonds adjacent to a nitrogen atom. Extending this protocol to other unactivated C—H bonds is highly desirable; (2) most of these reactions are racemic, and asymmetric synthesis using visible light photoredox catalysis is an interesting subject; (3) although some experimental data have been obtained, new progress in theory and spectroscopic techniques will help us to deeply understand the detailed mechanism of visible light photoredox catalysis; (4) using this protocol for the total synthesis of natural products and pharmaceuticals will, hopefully, become popular in the near future [75, 76].

References

1 (a) Godula, K. and Sames, D. (2006) *Science*, **312**, 67–72; (b) Brückl, T., Baxter, R.D., Ishihara, Y., and Baran, P.S. (2012) *Acc. Chem. Res.*, **45**, 826–839.

2 Blanksby, S.J. and Ellison, G.B. (2003) *Acc. Chem. Res.*, **36**, 255–263.

3 Selected reviews on transition-metal catalyzed C—H bond functionalization, see: (a) Yu, J.-Q. and Shi, Z.-J. (2010) Topics in Current Chemistry: C—H Activation, Springer, Berlin; (b) Díaz-Requejo, M.M. and Pérez, P.J. (2008) *Chem. Rev.*, **108**, 3379–3394; (c) Lyons, T.W. and Sanford, M.S. (2010) *Chem. Rev.*, **110**, 1147–1169; (d) Arockiam, P.B., Bruneau, C., and Dixneuf, P.H. (2012) *Chem. Rev.*, **112**, 5879–5918; (e) Colby, D.A., Tsai, A.S., Bergman, R.G., and Ellman, J.A. (2012) *Acc. Chem. Res.*, **45**, 814–825.

4 Selected reviews on visible light photoredox catalysis, see: (a) Yoon, T.P., Ischay, M.A., and Du, J. (2010) *Nat. Chem.*, **2**, 527–532; (b) Narayanam, J.M. and Stephenson, C.R.J. (2011) *Chem. Soc. Rev.*, **40**, 102–113; (c) Teplý, F. (2011) *Collect. Czech. Chem. Commun.*, **76**, 859–917; (d) Xuan, J. and Xiao, W.-J. (2012) *Angew. Chem. Int. Ed.*, **51**, 6828–6838; (e) Prier, C.K., Rankic, D.A., and MacMillan, D.W.C. (2013) *Chem. Rev.*, **113**, 5322–5363; (f) Zeitler, K. (2009) *Angew. Chem. Int. Ed.*, **48**, 9785–9789; (g) You, Y. and Nam, W.

(2012) *Chem. Soc. Rev.*, **41**, 7061–7084; (h) Ravelli, D., Fagnoni, M., and Albini, A. (2013) *Chem. Soc. Rev.*, **42**, 97–113; (i) Xi, Y., Yi, H., and Lei, A. (2013) *Org. Biomol. Chem.*, **11**, 2387–2403; (j) Hari, D.P. and König, B. (2013) *Angew. Chem. Int. Ed.*, **52**, 4734–4743; (k) Reckenthäler, M. and Griesbeck, A.G. (2013) *Adv. Synth. Catal.*, **355**, 2727–2744.

5 Nicewicz, D.A. and MacMillan, D.W.C. (2008) *Science*, **322**, 77–80.

6 Ischay, M.A., Anzovino, M.E., Du, J., and Yoon, T.P. (2008) *J. Am. Chem. Soc.*, **130**, 12886–12887.

7 (a) Kalyanasundaram, K. (1982) *Coord. Chem. Rev.*, **46**, 159–244; (b) Juris, A., Balzani, V., Barigelletti, F. *et al.* (1988) *Coord. Chem. Rev.*, **84**, 85–277.

8 Selected examples through energy transfer, see: (a) Zou, Y.-Q., Duan, S.-W., Meng, X.-G. *et al.* (2012) *Tetrahedron*, **68**, 6914–6919; (b) Lu, Z. and Yoon, T.P. (2012) *Angew. Chem. Int. Ed.*, **51**, 10329–10332.

9 Selected reviews on visible light induced α-amino C—H functionalization, see: (a) Shi, L. and Xia, W. (2012) *Chem. Soc. Rev.*, **41**, 7687–7697; (b) Hu, J., Wang, J., Nguyen, T.H., and Zheng, N. (2013) *Beilstein J. Org. Chem.*, **9**, 1977–2001; (c) Xie, J., Jin, H., Xu, P., and Zhu, C. (2014) *Tetrahedron Lett.*, **55**, 36–48; (d) Maity, S. and Zheng, N. (2012) *Synlett*, **23**, 1851–1856; (e) Rüping, M., Königs, R.M., and Atodiresei, I. (2013) in *Photoredox Catalysed A-Functionalization of Amines-Visible Light Mediated Carbon–Carbon and Carbon-Hetero Bond Forming Reactions* (ed. B. König), Walter de Gruyter, Berlin.

10 Li, C.-J. (2009) *Acc. Chem. Res.*, **42**, 335–344.

11 Condie, A.G., Gonzalez-Gomez, J.C., and Stephenson, C.R. (2010) *J. Am. Chem. Soc.*, **132**, 1464–1465.

12 Freeman, D.B., Furst, L., Condie, A.G., and Stephenson, C.R.J. (2012) *Org. Lett.*, **14**, 94–97.

13 To, W.P., Liu, Y., Lau, T.C., and Che, C.M. (2013) *Chem. Eur. J.*, **19**, 5654–5664.

14 Zhong, J.-J., Meng, Q.-Y., Wang, G.-X. *et al.* (2013) *Chem. Eur. J.*, **19**, 6443–6450.

15 (a) Ravelli, D., Dondi, D., Fagnoni, M., and Albini, A. (2009) *Chem. Soc. Rev.*, **38**, 1999–2011; (b) Ravelli, D. and Fagnoni, M. (2012) *ChemCatChem*, **4**, 169–171.

16 Hari, D.P. and König, B. (2011) *Org. Lett.*, **13**, 3852–3855.

17 Pan, Y., Kee, C.W., Chen, L., and Tan, C.-H. (2011) *Green Chem.*, **13**, 2682–2685.

18 Liu, Q., Li, Y.-N., Zhang, H.-H. *et al.* (2012) *Chem. Eur. J.*, **18**, 620–627.

19 Kisch, H. (2013) *Angew. Chem. Int. Ed.*, **52**, 812–847.

20 Xie, Z., Wang, C., deKrafft, K.E., and Lin, W. (2011) *J. Am. Chem. Soc.*, **133**, 2056–2059.

21 Rueping, M., Zoller, J., Fabry, D.C. *et al.* (2012) *Chem. Eur. J.*, **18**, 3478–3481.

22 Mitkina, T., Stanglmair, C., Setzer, W. *et al.* (2012) *Org. Biomol. Chem.*, **10**, 3556–3561.

23 Wang, C., Xie, Z., deKrafft, K.E., and Lin, W. (2012) *ACS Appl. Mater. Interfaces*, **4**, 2288–2294.

24 Möhlmann, L., Baar, M., Rieß, J. *et al.* (2012) *Adv. Synth. Catal.*, **354**, 1909–1913.
25 Wang, X.C., Maeda, K., Thomas, A. *et al.* (2009) *Nat. Mater.*, **8**, 76–80.
26 (a) Wirth, T. (2008) *Microreactors in Organic Synthesis and Catalysis*, Weinheim, Wiley-VCH; (b) Webb, D. and Jamison, T.F. (2010) *Chem. Sci.*, **1**, 675–680.
27 Tucker, J.W., Zhang, Y., Jamison, T.F., and Stephenson, C.R.J. (2012) *Angew. Chem. Int. Ed.*, **51**, 4144–4147.
28 Neumann, M. and Zeitler, K. (2012) *Org. Lett.*, **14**, 2658–2661.
29 Rueping, M., Vila, C., and Bootwicha, T. (2013) *ACS Catal.*, **3**, 1676–1680.
30 Cai, S., Zhao, X., Wang, X. *et al.* (2012) *Angew. Chem. Int. Ed.*, **51**, 8050–8053.
31 Selected reviews, see: (a) Córdova, A. (2004) *Acc. Chem. Res.*, **37**, 102–112; (b) J. M. M. Verkade, L. J. C. van Hemert, P. J. L. M. Quaedflieg, F. P. J. T. Rutjes, *Chem. Soc. Rev.* 2008, **37**, 29–41; (c) Kobayashi, S., Mori, Y., Fossey, J.S., and Salter, M.M. (2011) *Chem. Rev.*, **111**, 2626–2704.
32 Rueping, M., Vila, C., Koenigs, R.M. *et al.* (2011) *Chem. Commun.*, **47**, 2360–2362.
33 Xue, Q., Xie, J., Jin, H. *et al.* (2013) *Org. Biomol. Chem.*, **11**, 1606–1609.
34 Cherevatskaya, M., Neumann, M., Füldner, S. *et al.* (2012) *Angew. Chem. Int. Ed.*, **51**, 4062–4066.
35 Zhao, G., Yang, C., Guo, L. *et al.* (2012) *Chem. Commun.*, **48**, 2337–2339.
36 Bergonzini, G., Schindler, C.S., Wallentin, C.-J. *et al.* (2014) *Chem. Sci.*, **5**, 112–116.
37 For recent reviews, see: (a) Inoue, Y. and Ramamurthy, V. (2004) *Chiral Photochemsitry*, Marcel Dekker, New York; (b) Müller, C. and Bach, T. (2008) *Aust. J. Chem.*, **61**, 557–564.
38 Selected reviews, see: (a) Gröger, H. (2003) *Chem. Rev.*, **103**, 2795–2827; (b) Wang, J., Liu, X., and Feng, X. (2011) *Chem. Rev.*, **111**, 6947–6983.
39 Rueping, M., Zhu, S., and Koenigs, R.M. (2011) *Chem. Commun.*, **47**, 12709–12711.
40 To, W.-P., Tong, G.S.-M., Liu, W. *et al.* (2012) *Angew. Chem. Int. Ed.*, **51**, 2654–2657.
41 Pan, Y., Wang, S., Kee, C.W. *et al.* (2011) *Green Chem.*, **13**, 3341–3344.
42 Ushakov, D.B., Gilmore, K., Kopetzki, D. *et al.* (2013) *Angew. Chem. Int. Ed.* **53**, 557–561.
43 Meng, Q.-Y., Zhong, J.-J., Liu, Q. *et al.* (2013) *J. Am. Chem. Soc.*, **135**, 19052–19055.
44 Dai, C., Meschini, F., Narayanam, J.M.R., and Stephenson, C.R.J. (2012) *J. Org. Chem.*, **77**, 4425–4431.
45 Wang, Z.-Q., Hu, M., Huang, X.-C. *et al.* (2012) *J. Org. Chem.*, **77**, 8705–8711.
46 Zhu, S. and Rueping, M. (2012) *Chem. Commun.*, **48**, 11960–11962.
47 Rueping, M., Koenigs, R.M., Poscharny, K. *et al.* (2012) *Chem. Eur. J.*, **18**, 5170–5174.
48 Selected examples on combination of visible light photodox catalysis and metal catalysis: (a) Ye, Y. and Sanford, M.S. (2012) *J. Am. Chem. Soc.*, **134**,

9034–9037; (b) Sahoo, B., Hopkinson, M.N., and Glorious, F. (2013) *J. Am. Chem. Soc.*, **135**, 5505–5508.
49 Fu, W., Guo, W., Zou, G., and Xu, C. (2012) *J. Fluorine Chem*, **140**, 88–94.
50 (a) Allen, J.G., Atherton, F.R., Hall, M.J. *et al.* (1978) *Nature*, **272**, 56–58; (b) Hirschmann, R., Smith, A.B., Taylor, C.M. *et al.* (1994) *Science*, **265**, 234–237.
51 Rueping, M., Zhu, S., and Koenigs, R.M. (2011) *Chem. Commun.*, **47**, 8679–8681.
52 Gao, X.-W., Meng, Q.-Y., Xiang, M. *et al.* (2013) *Adv. Synth. Catal.*, **355**, 2158–2164.
53 Xuan, J., Cheng, Y., An, J. *et al.* (2011) *Chem. Commun.*, **47**, 8337–8339.
54 Xuan, J., Feng, Z.-J., Duan, S.-W., and Xiao, W.-J. (2012) *RSC Adv.*, **2**, 4065–4068.
55 Mathis, C.L., Gist, B.M., Frederickson, C.K. *et al.* (2013) *Tetrahedron Lett.*, **54**, 2101–2104.
56 Zou, Y.-Q., Lu, L.-Q., Fu, L. *et al.* (2011) *Angew. Chem. Int. Ed.*, **50**, 7171–7175.
57 Rueping, M., Leonori, D., and Poisson, T. (2011) *Chem. Commun.*, **47**, 9615–9617.
58 Huang, L. and Zhao, J. (2013) *Chem. Commun.*, **49**, 3751–3753.
59 Guo, S., Zhang, H., Huang, L. *et al.* (2013) *Chem. Commun.*, **49**, 8689–8691.
60 DiRocco, D.A. and Rovis, T. (2012) *J. Am. Chem. Soc.*, **134**, 8094–8097.
61 Xie, J., Xue, Q., Jin, H. *et al.* (2013) *Chem. Sci.*, **4**, 1281–1286.
62 Singh, A., Arora, A., and Weaver, J.D. (2013) *Org. Lett.*, **15**, 5390–5393.
63 Rueping, M. and Vila, C. (2013) *Org. Lett.*, **15**, 2092–2095.
64 A. McNally, C. K. Prier, D. W. C. MacMillan, *Science*, **334**, 1114-1117.
65 Zou, Y.-Q., Chen, J.-R., and Xiao, W.-J. (2013) *Angew. Chem. Int. Ed.*, **52**, 11701–11703.
66 Miyake, Y., Nakajima, K., and Nishibayashi, Y. (2012) *J. Am. Chem. Soc.*, **134**, 3338–3341.
67 Miyake, Y., Ashida, Y., Nakajima, K., and Nishibayashi, Y. (2012) *Chem. Commun.*, **48**, 6966–6968.
68 Kohls, P., Jadhav, D., Pandey, G., and Reiser, O. (2012) *Org. Lett.*, **14**, 672–675.
69 Espelt, L.R., Wiensch, E.M., and Yoon, T.P. (2013) *J. Org. Chem.*, **78**, 4107–4114.
70 Ju, X., Li, D., Li, W. *et al.* (2012) *Adv. Synth. Catal.*, **354**, 3561–3567.
71 Zhu, S., Das, A., Bui, L. *et al.* (2013) *J. Am. Chem. Soc.*, **135**, 1823–1829.
72 Chen, L., Chao, C.S., Pan, Y. *et al.* (2013) *Org. Biomol. Chem.*, **11**, 5922–5925.
73 Miyake, Y., Nakajima, K., and Nishibayashi, Y. (2012) *Chem. Eur. J.*, **18**, 16473–16477.
74 Zhou, H., Lu, P., Gu, X., and Li, P. (2013) *Org. Lett.*, **15**, 5646–5649.
75 For reviews pertaining to this subject that were published after the transcription of this chapter see: (a) Beatty, J.W. and Stephenson, C.R.J. (2015) Amine functionalization via oxidative photoredox catalysis: methodology development and complex molecule synthesis. *Acc. Chem. Res.*, **48** (5), 1474–1484; (b) Chen, J.-R., Hu, X.-Q., Lu, L.-Q., and Xiao, W.-J. (2016) Visible light photoredox-controlled reactions of N-radicals and radical ions. *Chem. Soc.*

Rev., **45** (8), 2044–2056; (c) Ravelli, D., Protti, S., and Fagnoni, M. (2016) Carbon–carbon bond forming reactions via photogenerated intermediates. *Chem. Rev.*, **116** (17), 9850–9913; (d) Nakajima, K., Miyake, Y., and Nishibayashi, Y. (2016) Synthetic utilization of α-aminoalkyl radicals and related species in visible light photoredox catalysis. *Acc. Chem. Res.*, **49** (9), 1946–1956; (e) Staveness, D., Bosque, I., and Stephenson, C.R.J. (2016) Free radical chemistry enabled by visible light-induced electron transfer. *Acc. Chem. Res.*, **49** (10), 2295–2306.

76 For additional reading on primary literature that has been published after the transcription of this chapter see: (a) Chandrasekhar, D., Borra, S., Kapure, J.S. *et al.* (2015) Visible-light photoredox catalysis: direct synthesis of fused β-carbolines through an oxidation/[3+2] cycloaddition/oxidative aromatization reaction cascade in batch and flow microreactors. *Org. Chem. Front.*, **2** (10), 1308–1312; (b) Chandrasekhar, D., Borra, S., Nanubolu, J.B., and Maurya, R.A. (2016) Visible light driven photocascade catalysis: Ru(bpy)3(PF6)2/TBHP-mediated synthesis of fused β-carbolines in batch and flow microreactors. *Org. Lett.*, **18** (12), 2974–2977; (c) Dai, X., Mao, R., Guan, B. *et al.* (2015) Visible light photoredox catalysis: regioselective radical addition of aminoalkyl radicals to 2,3-allenoates. *RSC Adv.*, **5** (68), 55290–55294; (d) Davies, J., Svejstrup, T.D., Fernandez Reina, D. *et al.* (2016) Visible-light-mediated synthesis of amidyl radicals: transition-metal-free hydroamination and N-arylation reactions. *J. Am. Chem. Soc.*, **138** (26), 8092–8095; (e) Ding, Y., Zhang, T., Chen, Q.-Y., and Zhu, C. (2016) Visible-light photocatalytic aerobic annulation for the green synthesis of pyrazoles. *Org. Lett.*, **18** (17), 4206–4209; (f) Fava, E., Millet, A., Nakajima, M. *et al.* (2016) Reductive umpolung of carbonyl derivatives with visible-light photoredox catalysis: direct access to vicinal diamines and amino alcohols via α-amino radicals and ketyl radicals. *Angew. Chem. Int. Ed.*, **55** (23), 6776–6779; (g) Fujiya, A., Tanaka, M., Yamaguchi, E. *et al.* (2016) Sequential photo-oxidative [3 + 2] cycloaddition/oxidative aromatization reactions for the synthesis of pyrrolo[2,1-a]isoquinolines using molecular oxygen as the terminal oxidant. *J. Org. Chem.*, **81** (16), 7262–7270; (h) Lenhart, D., Bauer, A., Pöthig, A., and Bach, T. (2016) Enantioselective visible-light-induced radical-addition reactions to 3-alkylidene indolin-2-ones. *Chem. Eur. J.*, **22** (19), 6519–6523; (i) Lin, C.-W., Hong, B.-C., Chang, W.-C., and Lee, G.-H. (2015) A new approach to nitrones through cascade reaction of nitro compounds enabled by visible light photoredox catalysis. *Org. Lett.*, **17** (10), 2314–2317; (j) Liu, X., Ye, X., Bureš, F. *et al.* (2015) Controllable chemoselectivity in visible-light photoredox catalysis: four diverse aerobic radical cascade reactions. *Angew. Chem.*, **127** (39), 11605–11609; (k) Morris, S.A., Nguyen, T.H., and Zheng, N. (2015) Diastereoselective oxidative C-N/C-O and C-N/C-N bond formation tandems initiated by visible light: synthesis of fused N-arylindolines. *Adv. Synth. Catal.*, **357** (10), 2311–2316; (l) Nicholls, T.P., Constable, G.E., Robertson, J.C. *et al.* (2016) Brønsted acid cocatalysis in copper(I)-photocatalyzed α-amino C–H bond functionalization. *ACS Catal.*, **6** (1), 451–457; (m) Rohokale, R.S., Koenig, B., and Dhavale, D.D. (2016) Synthesis of 2,4,6-trisubstituted pyridines by oxidative eosin Y photoredox catalysis. *J. Org. Chem.*, **81** (16), 7121–7126; (n) Ruiz Espelt, L., McPherson, I.S., Wiensch, E.M., and Yoon, T.P. (2015) Enantioselective conjugate additions of α-amino radicals via cooperative photoredox and lewis acid catalysis. *J. Am. Chem. Soc.*, **137** (7), 2452–2455; (o) Shen, Z., Yang, P., and Tang, Y. (2016) Transition metal-free visible light-driven photoredox oxidative annulation of arylamidines. *J. Org. Chem.*, **81** (1), 309–317; (p) Tang, J., Grampp, G., Liu, Y. *et al.* (2015) Visible light mediated cyclization of tertiary anilines with maleimides using nickel(II) oxide surface-modified titanium dioxide catalyst. *J. Org. Chem.*, **80** (5), 2724–2732; (q) Uraguchi, D., Kinoshita, N., Kizu, T., and Ooi, T. (2015) Synergistic catalysis of ionic Brønsted acid and photosensitizer for a redox neutral asymmetric α-coupling of N-arylaminomethanes with aldimines. *J. Am. Chem. Soc.*, **137** (43), 13768–13771; (r) Wang, C., Ren, X., Xie, H., and Lu, Z. (2015) [3+2] Redox-neutral cycloaddition of nitrocyclopropanes with styrenes by visible-light photocatalysis. *Chem. Eur. J.*, **21** (27), 9676–9680; (s) Wang, C., Qin, J., Shen, X. *et al.* (2016) Asymmetric radical–radical cross-coupling through visible-light-activated iridium catalysis. *Angew. Chem. Int. Ed.*, **55** (2), 685–688; (t) Wang, J. and Zheng, N. (2015) The cleavage of a C—C bond in cyclobutylanilines by visible-light photoredox catalysis: development of a [4+2] annulation method. *Angew. Chem. Int. Ed.*, **54** (39), 11424–11427; (u) Wei, G., Zhang, C., Bureš, F. *et al.* (2016) Enantioselective aerobic oxidative C(sp3)–H olefination of amines via cooperative photoredox and asymmetric catalysis. *ACS Catal.*, **6** (6), 3708–3712.

5

Visible Light Mediated Cycloaddition Reactions

Scott Morris, Theresa Nguyen, and Nan Zheng

University of Arkansas, Department of Chemistry and Biochemistry, Fayetteville, AR, USA

5.1 Introduction

Cycloaddition reactions are a class of highly useful and versatile reactions that are particularly effective for the construction of carbocycles and heterocycles. Two notable examples in this class of reactions are the Diels–Alder reaction and the 1,3-dipolar cycloaddition, also known as the Huisgen cycloaddition. Cycloaddition reactions generally make multiple bonds and stereocenters in one single step with predictable stereochemistry. They also often exhibit excellent atom economy. These appealing features make them a preferred reaction in organic synthesis. Therefore, continuous efforts have been made by organic chemists to uncover new ways to promote cycloaddition reactions.

Light typically activates cycloaddition reactions by electron transfer or energy transfer. A photocatalyst containing a visible chromophore is often employed in order to use visible light, which cannot be absorbed by most organic molecules. The excited state of the photocatalyst can serve as a one-electron acceptor via a reductive quenching cycle, as a one-electron donor via an oxidative quenching cycle, or as a photosensitizer via energy transfer. The photocatalyst's versatile mode of activation allows a variety of transformations to be explored, including cycloaddition reactions. During the past 6 years, a number of new cycloaddition reactions promoted by visible light have been reported and have significantly expanded the scope of chemistry that can be accomplished. Although great work has been performed on cycloaddition reactions using UV light, the focus of this chapter is aimed at reviewing newly reported cycloaddition methods that use visible light. The cycloaddition reactions discussed in this chapter are loosely defined, and some of them are actually annulation reactions rather than cycloaddition reactions.

Visible Light Photocatalysis in Organic Chemistry, First Edition.
Edited by Corey R. J. Stephenson, Tehshik P. Yoon and David W. C. MacMillan.

5.2 [2+2] Cycloadditions: Formation of Four-Membered Rings

5.2.1 Introduction to [2+2] Cycloadditions

The [2+2] cycloaddition of α,β-unsaturated ketones or esters to alkenes, alkynes, or allenes generating cyclobutanes [1] is one of the most applied reactions in organic synthesis [2–6]. Cyclobutanes are important, as they are basic structural elements in many biologically active natural products, and they also produce strain-releasing fragmentations, which are useful in synthesizing medium-sized rings of higher complexity. The ability to generate up to four new stereocenters and two new σ bonds in a high-energy cyclobutane ring is no easy task. This is why [2+2] cycloaddition is arguably the most straightforward way to make such structures. In fact, about 100 years ago, Ciamician first saw the potential in using visible light to promote [2+2] cycloaddition [7]. As outlined in his 1912 report, carvone was converted to camphor carvone after 1-year exposure to intense sunlight (Figure 5.1). Expansion of this work has been seen in many UV-mediated processes. However, the use of visible light possesses a number of advantages over UV, as it can mimic sunlight's photon density, tolerate functional groups, and be compatible with regular glassware. Yoon and others have had the most impactful contributions to the development of these visible light mediated processes in recent years.

5.2.2 Utilization of the Reductive Quenching Cycle

In the early 2000s, Krische and coworkers revealed the first metal-catalyzed intramolecular [2+2] cycloaddition of tethered aryl enones using cobalt and copper to produce bicyclic [3.2.0] rings, which, in a subsequent work, was confirmed to be initialized by a one-electron reduction of enone [8]. The radical anion intermediate was authenticated by the formation of identical [2+2] cycloadducts using cathodic and homogeneous one-electron reduction. In analyzing these results, Yoon and coworkers hypothesized that visible light photoredox catalysis could be used in place of cobalt and copper to generate the same radical anion intermediate [9]. Figure 5.2 details their findings.

Indeed, a variety of aromatic enones efficiently undergo the desired transformation with high diastereoselectivity. Upon exposure to a 275 W floodlight, the reactions were completed within 2 h in degassed acetonitrile. At least one aromatic enone must be present for the transformation because only aromatic enones are reducible under photoredox conditions. Reactivity trends that are similar to Krische's work are also observed in Yoon's, as symmetrical and unsymmetrical aryl bis enones with a number of substituents react smoothly. Figure 5.3 surveys the scope of this chemistry. Various aryl groups were incorporated into

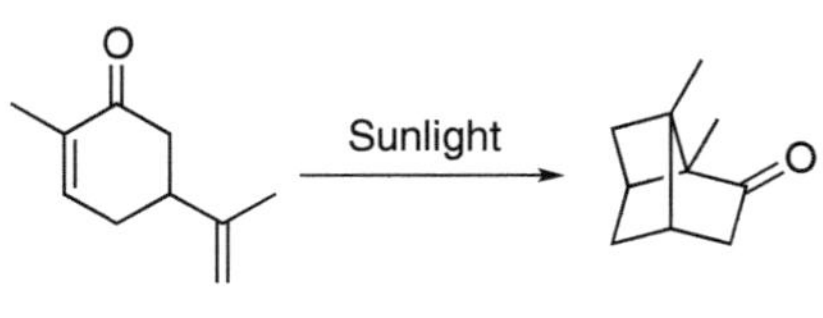

Figure 5.1 Conversion of carvone to camphor carvone using sunlight, first reported by Giacomo Ciamician.

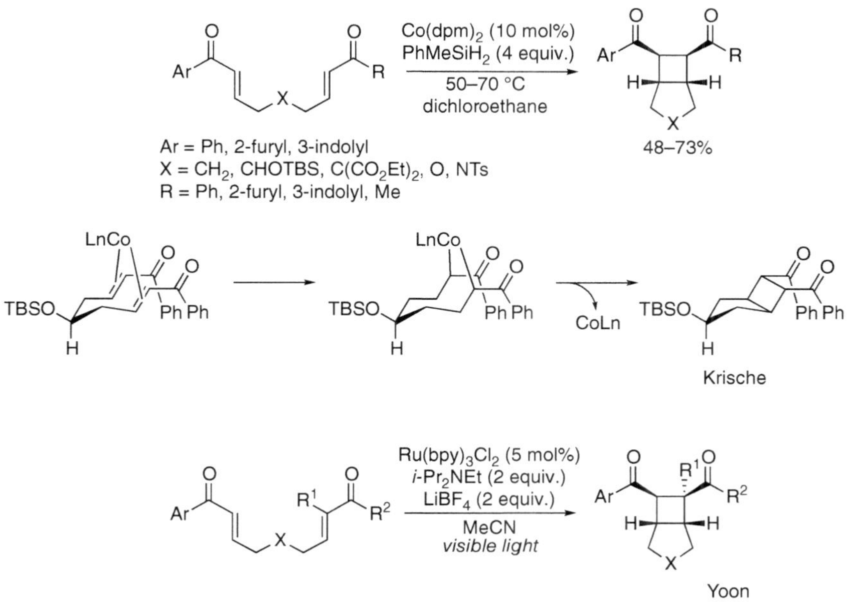

Figure 5.2 Initial [2+2] cycloadditions of enones, reported by Krische and Yoon.

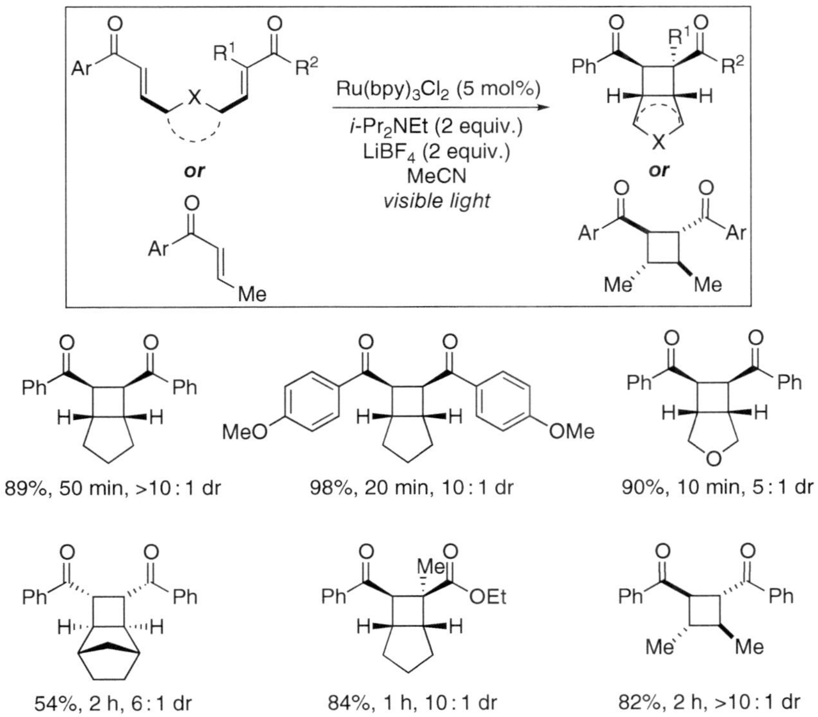

Figure 5.3 Reaction scope of the [2+2] cycloaddition of aryl enones.

the desired product, as well as into the oxygen tether. Trisubstituted alkenes cleanly produced the desired bicycles containing the quaternary stereocenter. Interestingly, homodimerization reactions proceeded, forming the corresponding cyclobutanes in high yields and diastereoselectivities, although racemic isomers were obtained, a phenomenon not seen with the intramolecular cases.

While investigating the mechanism, it was found that no reaction occurred in the absence of the amine or $LiBF_4$. When conducted in the absence of light or a catalyst, the reaction also did not proceed. These observations are in agreement with the involvement of the radical anion intermediate. The amine is considered to serve as a sacrificial electron donor to generate reduced $Ru(bpy)_3{}^+$, which subsequently reduces the activated enone to the radical anion intermediate, while regenerating $Ru(bpy)_3{}^{2+}$ (Figure 5.4). $LiBF_4$ is proposed to serve two roles in the catalysis. The BF_4 anion exchanges the chloride anion of $Ru(bpy)_3Cl_2$, thus increasing its solubility in CH_3CN. The Li cation binds to the oxygen atom in order to activate the enone toward the one-electron reduction.

One year later, in 2009, Yoon showcased an intermolecular crossed [2+2] cycloaddition, branching from the previously reported intramolecular cycloadditions of aryl enones [10]. The biggest challenge to developing this reaction was to circumvent the problem of forming the homodimerization product. They reasoned that a reducible aryl enone is needed for the initial formation of the radical anion intermediate, with the other coupling partner resulting in a nonreducible but still a better Michael acceptor than the aryl enone. Following this thought, they chose aryl enone **1** and methyl vinyl ketone **2** to test the proposed reaction which, upon exposure to a simple 23 W compact fluorescent bulb, gratifyingly gave the desired cycloadduct **3** in 84% yield with excellent chemo- and diastereoselectivity. Figure 5.5 illustrates their initial finding using the optimal reaction conditions.

It was critical that aryl enone was β-substituted, ensuring that it was less reactive than the other coupling partner, the Michael acceptor. Within this constraint,

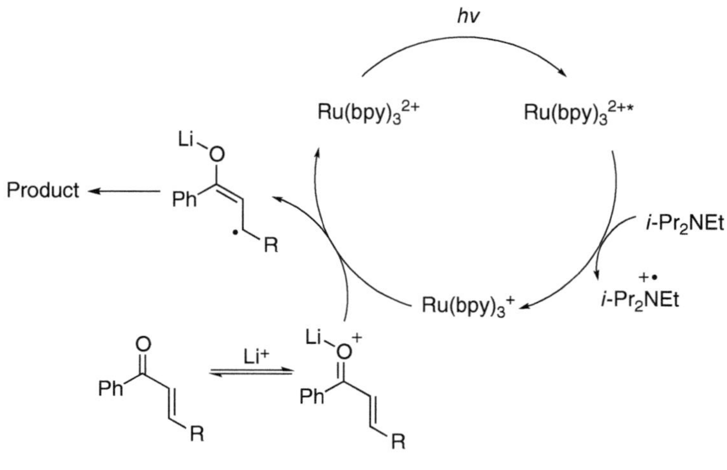

Figure 5.4 Proposed catalytic cycle for the [2+2] cycloaddition of aryl enones.

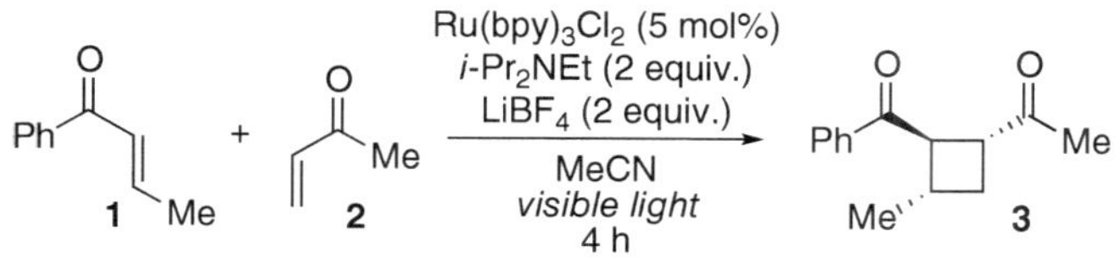

Figure 5.5 Crossed intermolecular [2+2] cycloaddition.

it was found that various aryl enones are effective in the transformation, as well as at different β-substituents, although it was observed that the steric bulk of the β-substituent retarded the rate of the reaction. Selected examples from this reaction scope are shown in Figure 5.6.

Furthermore, variation of the Michael acceptor is tolerated. In addition to enones, acrylic esters and acrylic thioesters also participated in the cycloaddition. The α-substitution at the Michael acceptor is compatible, allowing the formation of cyclobutanes with an all-carbon quaternary stereocenter in decent yield and selectivity.

In a continued study, Yoon and coworkers addressed the issue regarding the necessity of an aryl enone. In turn, they designed what they identified as a "redox auxiliary," which modulates the reduction potential of an otherwise redox-inactive enoate substrate [11]. They tested this hypothesis in an intermolecular [2+2] homodimerization reaction using a variety of α,β-unsaturated carbonyl compounds as surrogates for carboxylate esters. It was found that α,β-unsaturated acylimidazoles were superior to other surrogates, such as unsaturated phosphonate esters, *N*-acyl pyrroles, and *N*-acyl pyrazoles (Figure 5.7).

Once the ideal redox auxiliary was found, it was applied to the crossed [2+2] intermolecular cycloaddition. Interestingly, it was found that a high Lewis

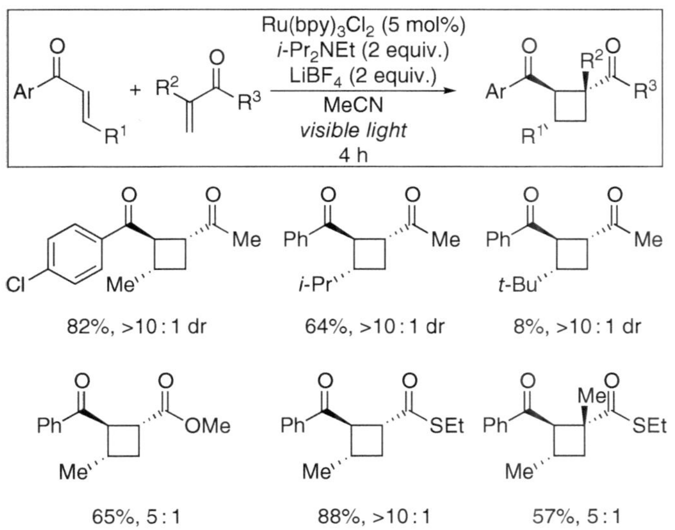

Figure 5.6 Reaction scope of the crossed intermolecular [2+2] cycloaddition.

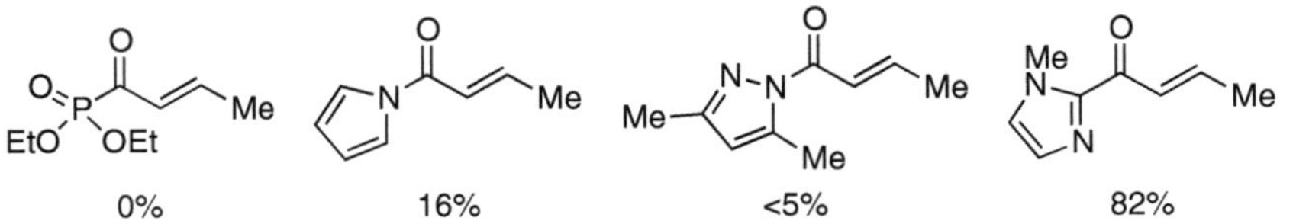

Figure 5.7 Redox auxiliary substrates examined as carboxylate ester surrogates.

acid-to-substrate ratio was necessary to avoid low yields, increase diastereoselectivity, and minimize the homodimerization. In addition, lowering the catalyst loading from 5 to 2.5 mol% modestly increased the formation of the heterodimer over the undesired homodimer. Ultimately, adding the acylimidazole via a syringe pump significantly improved the formation of the heterodimer product, presumably by minimizing the concentration of substrate to acrylate while maintaining a high Lewis acid-to-substrate ratio. Under the optimal conditions, they were able to use 2-acylimidazole-containing substrates to generate various crossed intermolecular [2+2] cycloaddition products. A survey of the photogenerated cycloadducts is shown in Figure 5.8.

The authors were gratified that a number of Michael acceptor, such as esters, ketones, and amides, efficiently yielded the desired products under the optimal reaction conditions. Interestingly, intramolecular reactions of enones were applicable, and lowering the loading of $LiBF_4$ to 0.5 equiv. from 2 equiv. somewhat increased the product yields (Figure 5.9).

Finally, they developed a two-step protocol to transform the 2-acylimidazole moiety into a carboxylic acid derivative. Upon *N*-alkylation with MeOTf, the resulting imidazolium salt can be successfully displaced by oxygen-, sulfur-,

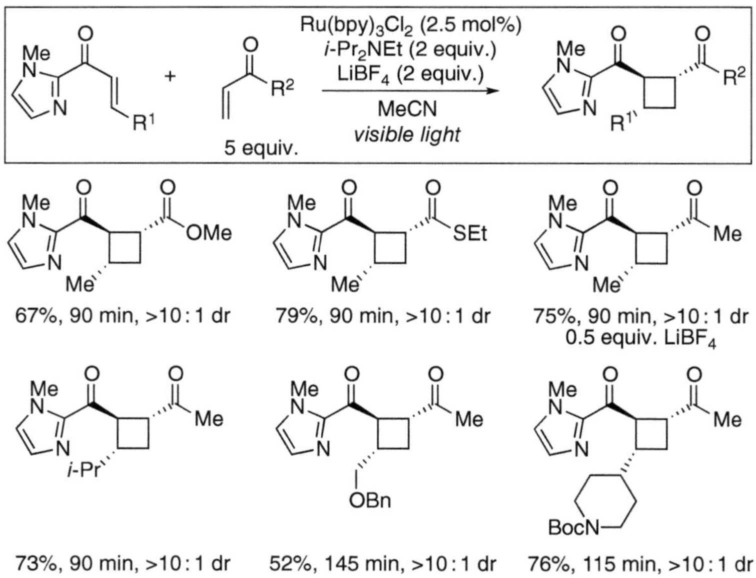

Figure 5.8 Crossed intermolecular redox auxiliary cycloadducts generated using visible light.

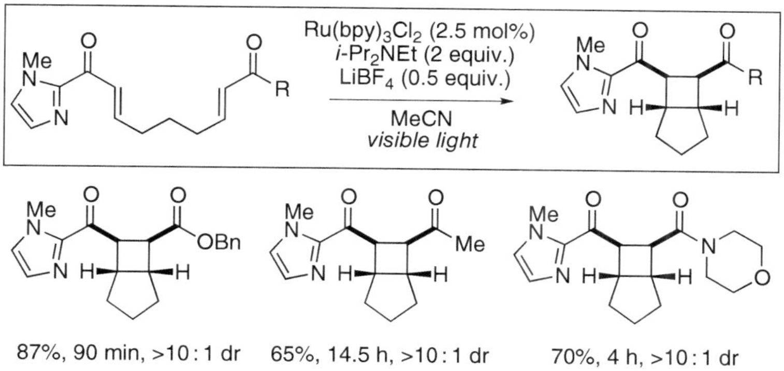

Figure 5.9 Intramolecular redox auxiliary cycloadducts generated using visible light.

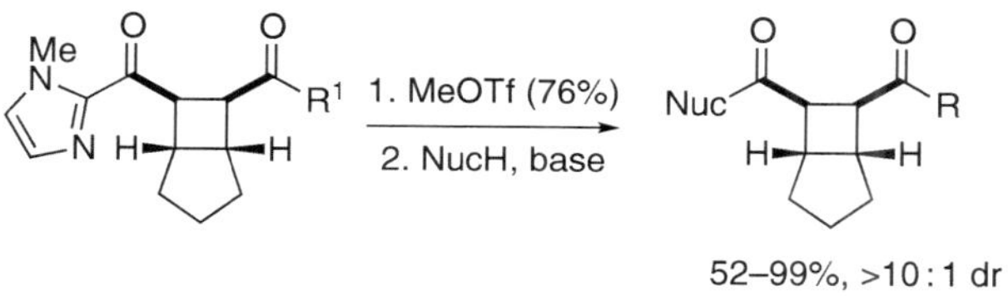

Figure 5.10 Transformation of the redox auxiliary to the carboxylic acid derivative.

or nitrogen-containing nucleophiles to afford the corresponding carboxylic esters/acids, thioesters, or amides while retaining stereochemistry (Figure 5.10).

As discussed above, pairing visible light photoredox catalysis with the [2+2] cycloaddition has significantly expanded the methodology's scope, which is evident by Yoon's pioneered works. These works utilizing the reductive quenching cycle suffer some limitations, such as the need for Lewis acids and reducible substrates, which hinder the substrate generality of this chemistry. To fully exploit the [2+2] cycloaddition mediated by visible light, not only continuous efforts on discovering new chemistries using this cycle but also efforts on uncovering new pathways to merge visible light photoredox catalysis with the [2+2] cycloaddition will be needed. In the next section, we will discuss the [2+2] cycloaddition utilizing the oxidative quenching cycle.

5.2.3 Utilization of the Oxidative Quenching Cycle

The [2+2] cycloaddition based on the reductive quenching cycle requires the use of at least one alkene moiety that can be reduced by $Ru(bpy)_3^+$ (i.e., electron-poor alkenes). To overcome this constraint in the substrate to include electron-rich alkenes, Yoon and coworkers developed a complementary method utilizing the oxidative quenching cycle. Electron-rich alkenes have been previously demonstrated to undergo the [2+2] cycloaddition upon one-electron oxidation by chemical oxidants or photoinduced electron transfer with organic sensitizers. $Ru(bpy)_3^{3+}$, produced by oxidatively quenching $Ru(bpy)_3^{2+*}$ with an electron acceptor (e.g., methyl viologen, MV^{2+}), is a powerful oxidant capable of oxidizing electron-rich olefins to their corresponding radical cations. As expected, with 5 mol% $Ru(bpy)_3(PF_6)_2$, 15 mol% $MV(PF_6)_2$, and a regular household light bulb, bis(styrene) **4** undergoes the intramolecular [2+2] cycloaddition to afford cyclobutane **5** in excellent yield and diastereoselectivity, as depicted in Figure 5.11 [12]. Control studies revealed that each of the components in the catalyst system is essential. MV was found to be much more effective than other known electron acceptors, such as nitroarenes and quinones. Nitromethane was the solvent of choice over more polar solvents, such as dimethyl sulfoxide (DMSO) and dimethylformamide (DMF). The reaction was modestly sensitive to water, as the addition of magnesium sulfate increased the yield and aided reproducibility.

The proposed mechanism, shown in Figure 5.12, follows the pattern described for an oxidative quenching cycle. The photoexcited $Ru(bpy)_3^{2+*}$ is oxidized by MV^{2+}, affording the reduced analog MV^{+} and the highly oxidizing $Ru(bpy)_3^{3+}$. The ruthenium species then oxidizes the electron-rich styrene **4** to generate the radical cation $\mathbf{4}^{\bullet+}$, regenerating the ground-state catalyst $Ru(bpy)_3^{2+}$. Radical cation $4^{\bullet+}$ then undergoes cyclization to afford cycloadduct **5**.

While studying the substrate scope, it was found that an electron-donating substituent must be present either in the *ortho* or *para* position of at least one styrene to achieve the desired transformation (Figure 5.13). This is in agreement with the mechanistic hypothesis in which the radical cation must be initially formed. With respect to the other alkene moiety, aliphatic olefins did not react

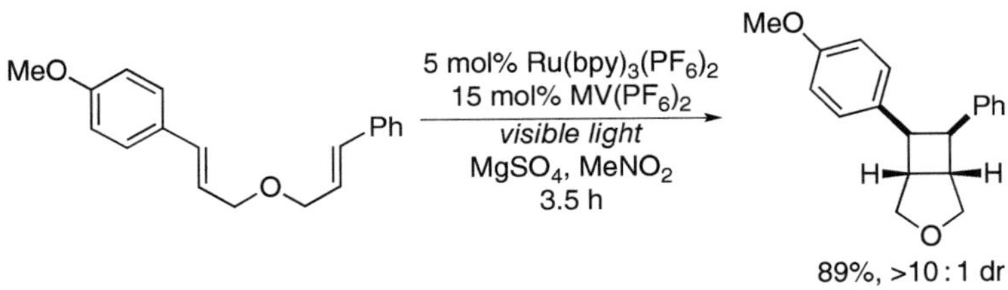

Figure 5.11 Intramolecular [2+2] cycloaddition of bis(styrenes).

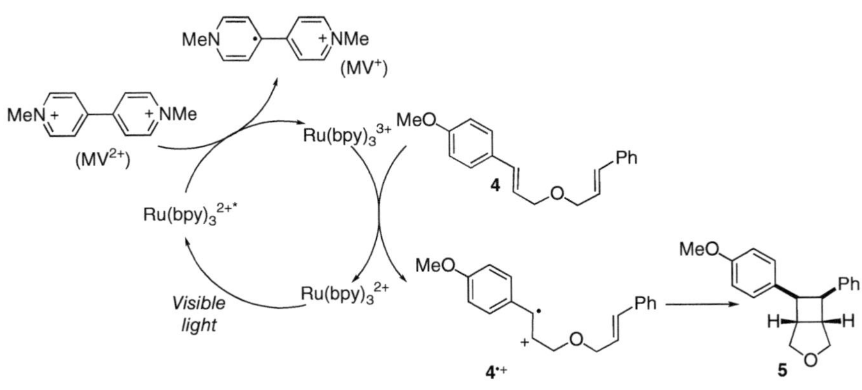

Figure 5.12 Proposed mechanism for the intramolecular [2+2] cycloaddition of bis(styrenes).

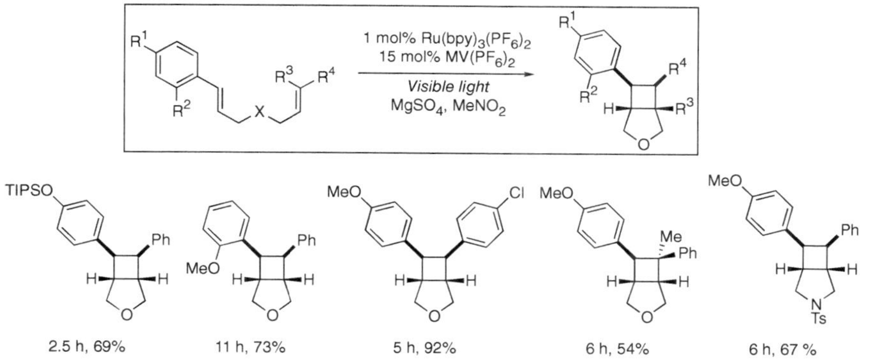

Figure 5.13 Selected cycloadducts for the intramolecular [2+2] cycloaddition of bis(styrenes).

under the designed conditions. On the other hand, styrenes were viable coupling partners. Electron-rich and electron-poor substituents on the arene, as well as substituents at the α-position of the styrene were tolerated. Surprisingly, the cycloaddition was found to be sensitive to the tether's structure, with only oxygen- and nitrogen-containing tethers providing the products in useful yields.

The Yoon group then expanded this method to the crossed [2+2] heterodimerization reaction of styrenes to generate asymmetrical cyclobutanes [13]. Using the intermolecular [2+2] homodimerization of styrenes as a model reaction, initial studies showed that the conditions used in the intramolecular case were ineffective for the intermolecular homodimerization. The authors were then able to obtain the homodimerization product in a modest yield (57% after 2 h) using more oxidizing conditions ($Ru(bpz)_3^{2+}$ and air). It was discovered during the optimization, however, that the electron-rich cyclobutane product is likely to be oxidized by $Ru(bpz)_3^{2+*}$, initiating a cycloreversion process to regenerate the starting material (Figure 5.14).

In order to promote the product conversion while inhibiting the cycloreversion process, an ideal photocatalyst should oxidize the styrene without oxidizing the dimerization product. In other words, its oxidation potential ($E^{2+*/+}$) should be between 1.1 and 1.27 V. To test this hypothesis, the authors examined less-oxidizing catalysts than $Ru(bpz)_3^{2+}$ in the homodimerization of the electron-rich styrene and found that $Ru(bpm)_3^{2+}$ ($E^{2+*/+} = 1.20$ V) was the

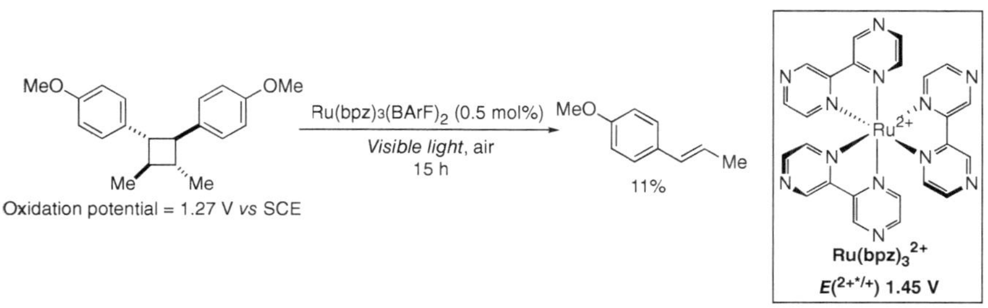

Figure 5.14 Observed cycloreversion of electron-rich cyclobutanes.

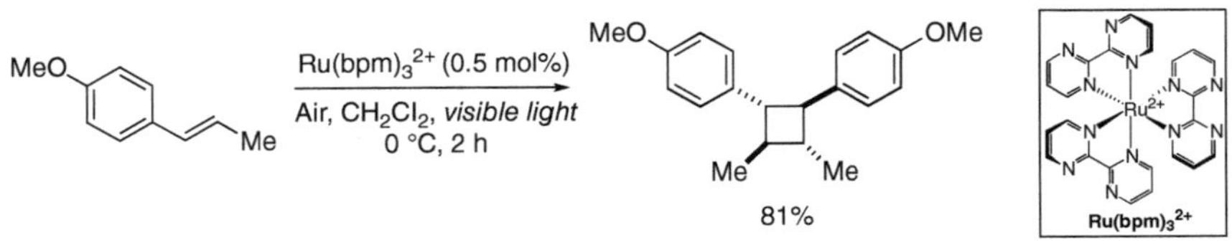

Figure 5.15 Initial discovery of the [2+2] dimerization of electron-rich styrenes.

optimal catalyst, providing 81% isolated yield at 0 °C in only 2 h. Figure 5.15 illustrates this observation.

In developing the substrate scope of the heterodimerization, the authors made several strategic considerations. The coupling partner must be less electron-rich than the electron-rich styrene, ensuring that only the styrene is oxidized to the key radical cation. Additionally, the coupling partner needs to be β-unsubstituted, in order that steric hindrance may favor heterodimerization over the undesired homodimerization. The rate of the cycloreversion process induced by one-electron oxidation of cyclobutanes is known to be temperature-dependent, with higher temperature favoring the process. Consistent with this phenomenon, the authors noticed marked increase in rate and yield when the reaction temperature was lowered to −15 °C. Lastly, when the homodimerization of the electron-rich styrene became a significant problem, a syringe pump was used to deliver the styrene to the reaction mixture. After these adjustments, a variety of oxidizable electron-rich styrenes participated in the crossed heterodimerization with a number of terminal alkenes, including styrenes, vinyl ethers, and allyl silanes (Figure 5.16). The yields and diastereoselectivities were generally high, providing an extra benefit to this methodology.

The authors proposed a mechanism based on a chain-reaction process, as shown in Figure 5.17. Upon irradiation with visible light, the ruthenium catalyst absorbs a photon, generating the photoexcited catalyst, which then

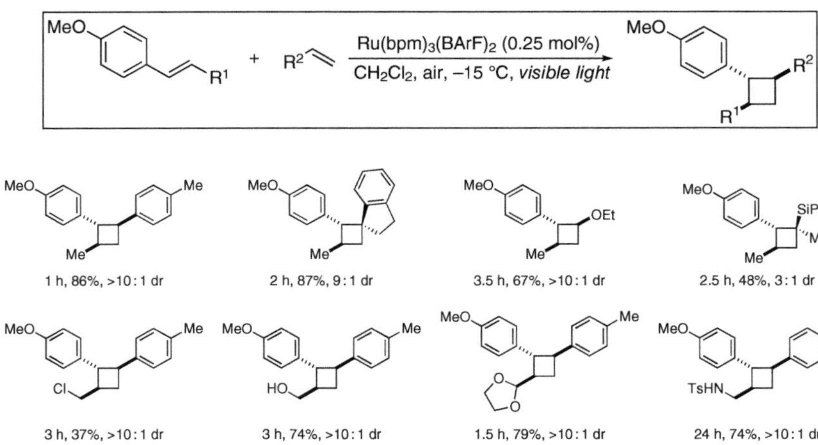

Figure 5.16 Selected cycloadducts of the [2+2] heterodimerization of styrenes.

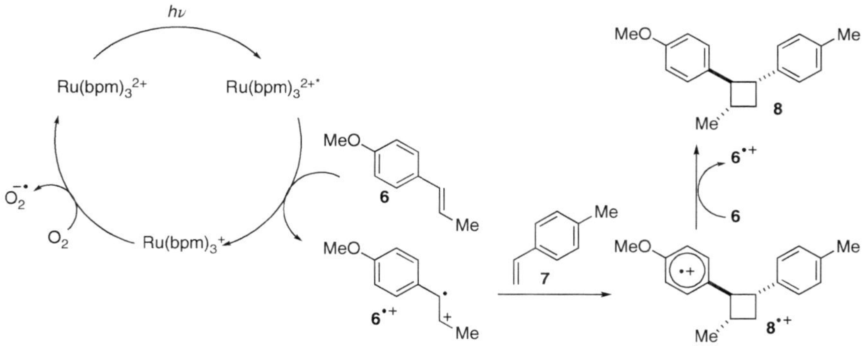

Figure 5.17 Proposed mechanism for the [2+2] heterodimerization of styrenes.

oxidizes styrene **6**. The radical cation styrene $\mathbf{6}^{\bullet+}$ then reacts with the coupling partner **7**, while the reduced catalyst is turned over by molecular oxygen. One electron reduction of cyclobutane intermediate $\mathbf{8}^{\bullet+}$ from **6** provides the desired product **8**, regenerating another molecule of $\mathbf{6}^{\bullet+}$, which then continues the chain reaction. The authors favored the reduction of $\mathbf{8}^{\bullet+}$ by unreacted **6** over the photoreduced $Ru(bpm)_3{}^+$ complex. This was based on the notion that since the concentrations of the catalyst and $\mathbf{8}^{\bullet+}$ were both very low, quenching the catalyst by $\mathbf{8}^{\bullet+}$ will be significantly slower than quenching by oxygen.

The [2+2] cycloaddition reaction that focuses on the oxidative quenching cycle has been briefly addressed. The cycloaddition via this cycle allows electron-rich substrates to participate in these important reactions and thus broadens the previously hindered scope of this methodology via the reductive quenching cycle. Both inter- and intramolecular reactions are applied to visible light photoredox catalysis using various oxidants, such as methyl viologen and oxygen. In the upcoming section, we will examine [2+2] cycloaddition methodology that undergoes the energy transfer mechanism.

5.2.4 Utilization of Energy Transfer

The [2+2] cycloaddition utilizing the reductive quenching cycle or oxidative quenching cycle relies on an efficient electronic transfer between the catalyst and the substrate, which is governed by their redox properties. In the case of the reductive quenching cycle, the alkene substrate's peak reduction potential needs to be less negative than the reduction potential of M/M^-, while in the case of the oxidative quenching cycle, the alkene substrate's peak oxidation potential has to be less positive than the reduction potential of M^+/M. This constraint on the substrate's redox properties has fundamentally limited the cycloaddition's substrate scope. In 2012, an alternative pathway via energy transfer was reported which involves triplet sensitization of the alkene substrate by the photoexcited catalyst to promote the [2+2] cycloaddition [14, 15]. The outcome of the energy transfer is dictated by the excited state triplet energies (E_T) of the substrate and photocatalyst, which, when the two are properly matched, can generate a substrate scope different from that of electron transfer, as energy transfer reactions can proceed without the structural limitations oftentimes needed for electron transfer reactions. Although the triplet sensitization by organic chromophores (sensitizers) has been extensively exploited in UV-promoted photochemistry, examples citing the use of visible light active organometallic compounds to sensitize organic substrates remain scarce.

In the first report, the Yoon group carefully chose styrene **9** to examine the feasibility of conducting the [2+2] cycloaddition by energy transfer [14]. Styrene **9** has a rather high oxidation potential (+1.42 V vs SCE), which prevents it from participating in the [2+2] cycloaddition by electron transfer. Since E_T of styrene **9** is estimated to be about 60 kcal mol^{-1}, it becomes apparent that Ru-based photocatalysts will not be effective, as their E_T ranges from 45 to 48 kcal mol^{-1}. The authors thus shifted their focus to Ir(dF(CF_3)ppy)$_2$(dtbbpy)$^+$, a fluorinated iridium catalyst that possesses an E_T of 61 kcal mol^{-1}. Upon exposure to a 23 W compact fluorescent light bulb, styrene **9** smoothly underwent the cycloaddition with just 1 mol% iridium catalyst to provide cyclobutane **10** in 83% isolated yield after 15 h (Figure 5.18).

Through other control studies and the substrate scope study, the authors noticed some significant differences between the energy transfer process and the electron transfer process. In the [2+2] cycloaddition via energy transfer, solvent polarity had little-to-no effect on reaction efficiency, whereas polar solvents proved to be beneficial for the cycloaddition by electron transfer. The [2+2] cycloaddition is generally favored in the energy transfer processes. On the other hand, the [4+2] cycloaddition is the major reaction pathway in the electron transfer processes. Alkene isomerization is often encountered in the energy transfer mediated cycloaddition, resulting in different reactivity and/or stereoisomers. This phenomenon is seldom observed in the electron transfer process.

The [2+2] cycloaddition by energy transfer enjoys a broader substrate scope. Substituted styrenes, regardless of their electronic character, electron-rich and -poor heterostyrenes, are all suitable for initializing the cycloaddition. For the other alkene moiety, there appears no constraint on the structure of the alkene and a wide variety of alkenes participate in the cycloaddition. The cycloaddition also tolerates diverse functional groups, including some sensitive ones (e.g., I) that cannot survive under traditional UV conditions. Figure 5.19 outlines a variety of cycloadducts the authors obtained using the visible light promoted energy transfer.

Xiao and coworkers have applied the [2+2] cycloaddition mediated by energy transfer to homodimerization of 3-ylideneoxindoles, producing densely

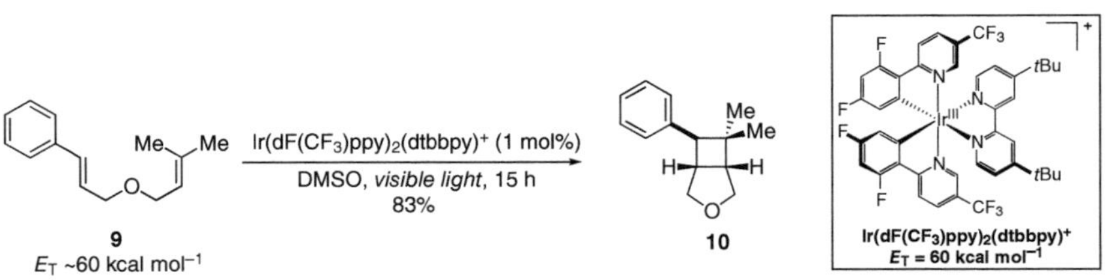

Figure 5.18 Intramolecular [2+2] cycloaddition of styrenes using energy transfer.

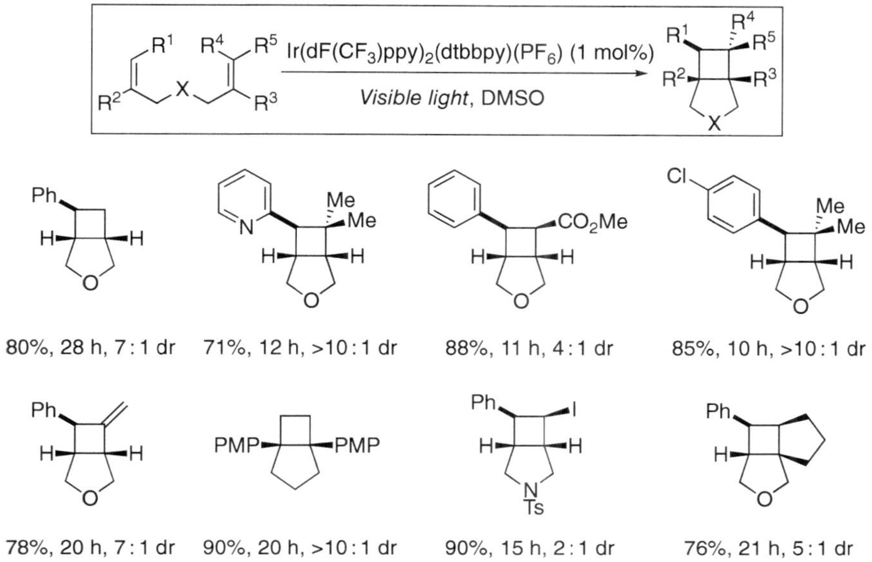

Figure 5.19 Selected cycloadducts obtained using energy transfer.

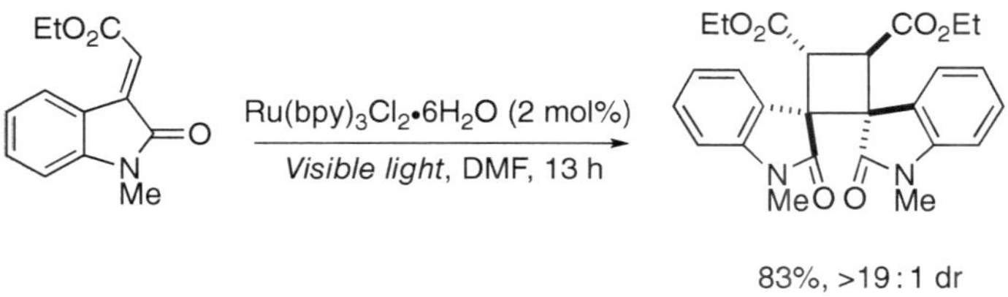

Figure 5.20 [2+2] Homodimerization of 3-ylideneoxindoles using energy transfer.

functionalized oxindoles bearing two spirocycles and four stereocenters in excellent diastereo- and regioselectivities (Figure 5.20) [15].

The exceptional reactivity exhibited by 3-ylideneoxindoles in the [2+2] cycloaddition is probably attributed to the constraint on the alkene, which limits E/Z isomerization and thus extends the lifetime of the triplet. The Yoon group noticed a similar reactivity exhibited by indene that also has the same constraint on the alkene. The homodimerization was conducted in DMF with $Ru(bpy)_3Cl_2$ as the energy transfer catalyst ($E_T = 49\,kcal\,mol^{-1}$) and a 36 W fluorescent light bulb as the light source. The process appears to be quite general, and a variety of substituents on the benzene ring regardless of their electronic nature are tolerated. Several examples from their report are illustrated in Figure 5.21.

The authors proposed a mechanism involving energy transfer, shown in Figure 5.22. Upon visible light irradiation, the triplet excited state $Ru(bpy)_3{}^{2+*}$ sensitizes the ground state of 3-ylideneoxindole **11** (**S_0**), resulting in the triplet excited state **11** (**T**). The resulting triplet state then reacts with another ground state starting material **11** (**S_0**), forming the biradical intermediate **12**, which cyclizes to give the cyclobutane product **13**.

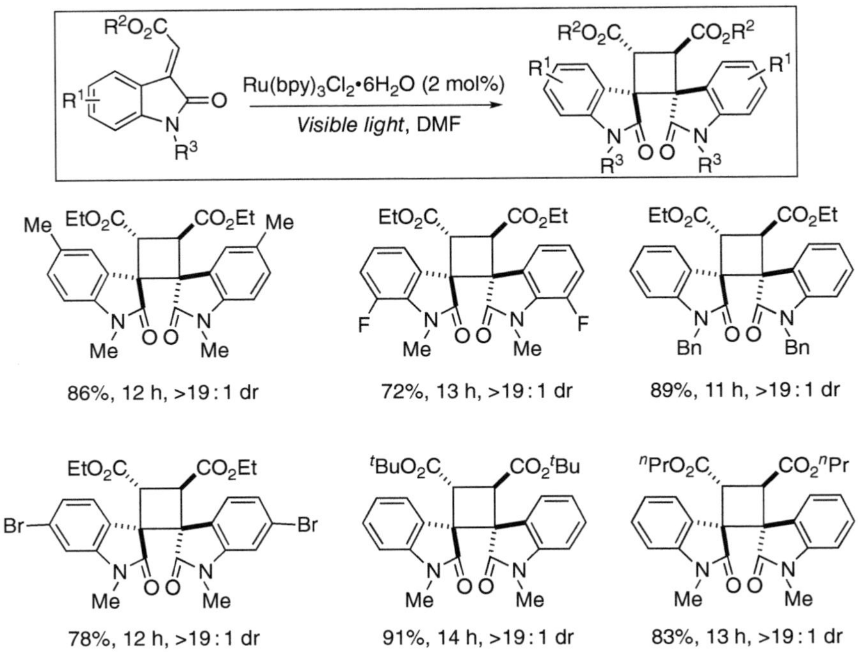

Figure 5.21 Various cycloadducts obtained via homodimerization of 3-ylideneoxindoles.

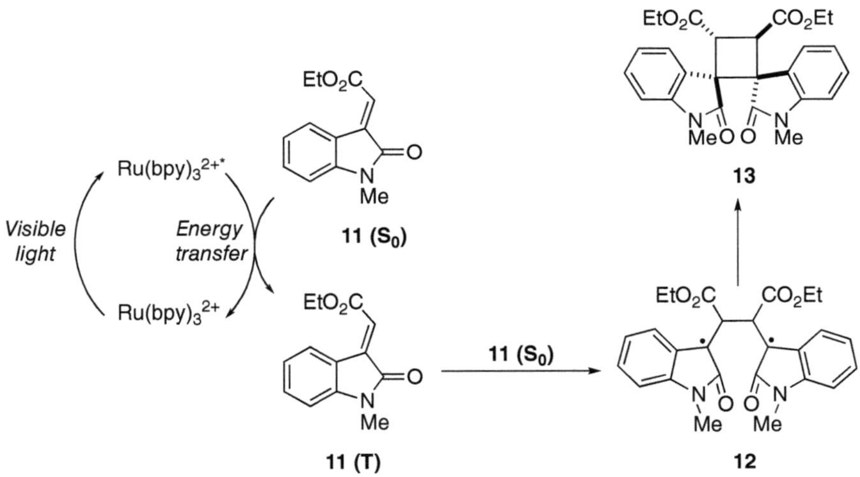

Figure 5.22 Proposed mechanism of the [2+2] homodimerization of 3-ylideneoxindoles.

5.2.5 [2+2] Conclusion

In summary, the [2+2] cycloaddition is arguably the most powerful method in producing densely functionalized cyclobutanes. Within the realm of visible light catalyzed cycloadditions, the [2+2] cycloaddition is the most developed, and advanced largely by the work of Yoon and coworkers. In the next section, the focus is shifted to advances made on [3+2] cycloadditions/annulations using visible light photoredox catalysis.

5.3 [3+2] Cycloadditions: Formation of Five-Membered Rings

5.3.1 Introduction to [3+2] Cycloadditions

Five-membered rings, such as cyclopentanes and the heterocyclic analogs, are another important class of building blocks in organic synthesis. One of the main approaches to access them is through [3+2] cycloadditions. Similar to [2+2] cycloadditions, [3+2] cycloaddition is a powerful method for constructing cyclic compounds, allowing the simultaneous formation of two new σ bonds and up to four stereocenters. The merging of visible light photocatalysis with [3+2] cycloadditions is primarily achieved by using photocatalysts to manipulate substrates' oxidation states in order to generate the established reactive intermediates in these reactions. Reductive quenching of the photoexcited catalysts has been observed to be the dominant pathway employed in these reported examples.

5.3.2 [3+2] Cycloaddition of Cyclopropylamines

Cyclopropanes' propensity to undergo an irreversible ring opening is largely owing to their inherent ring strain. Cyclopropylamines participate in this ring opening by oxidation of the amine to the amine radical cation that triggers the ring opening to produce a distonic ion. The distonic ion possesses a dual functionality: a nucleophilic carbon radical and an electrophilic iminium ion, which allows sequential addition to olefins and hence completes a formal [3+2] cycloaddition. In 2012, Zheng and coworkers applied visible light photoredox catalysis to this ring opening process and developed an intermolecular [3+2] annulation of olefins with mono- and bicyclic cyclopropylanilines [16]. Zheng's method features good to excellent yields, excellent regiocontrol with respect to the alkene, and tolerance of various functional groups, although an aryl group on the amine is a structural requirement. Cycloadditions with monocyclic cyclopropylanilines gave poor diastereoselectivity, and modest diastereoselectivity was achieved with bicyclic cyclopropylanilines (Figure 5.23). It is worth noting that conjugated dienes underwent the annulation successfully; the annulation reaction only occurred at the terminal double bond, providing the desired product with the favored *trans* isomer.

The Zheng group proposed a catalytic cycle involving a $Ru(bpz)_3(PF_6)_2$-catalyzed photo-oxidation of *N*-cyclopropylanilines to initiate the reaction (Figure 5.24). Initially, the ruthenium catalyst absorbs a photon, generating the photoexcited $Ru(bpz)_3^{2+*}$, which oxidizes the cyclopropylaniline **14** to the amine radical cation **14**$^{•+}$. The amine radical cation **14**$^{•+}$ subsequently undergoes ring opening, forming the β-carbon radical iminium ion (distonic ion) **15**. The distonic ion then adds intermolecularly to the olefin (via Giese radical addition),

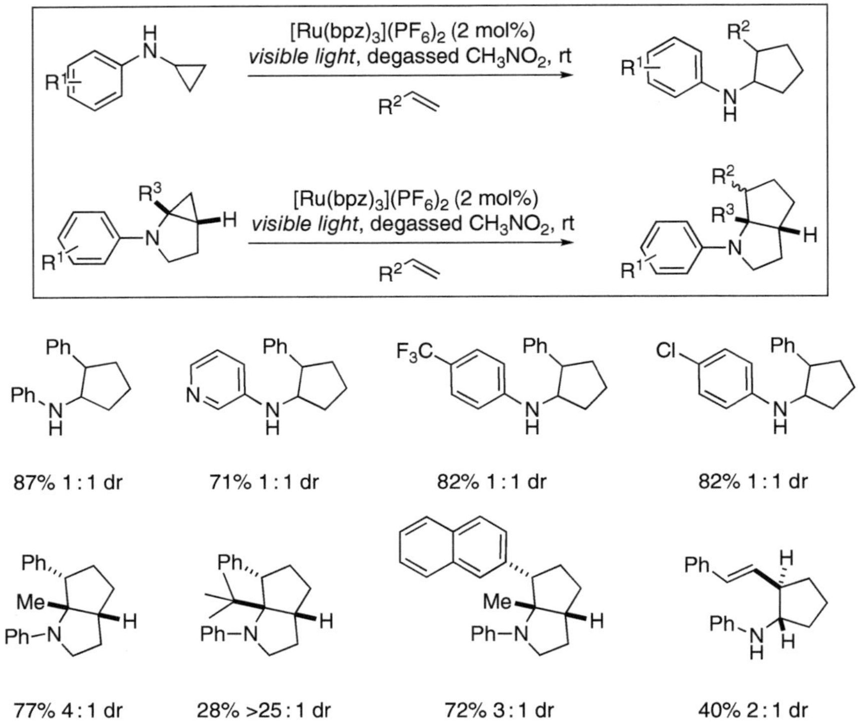

Figure 5.23 Scope of the redox-neutral [3+2] annulation developed by Zheng.

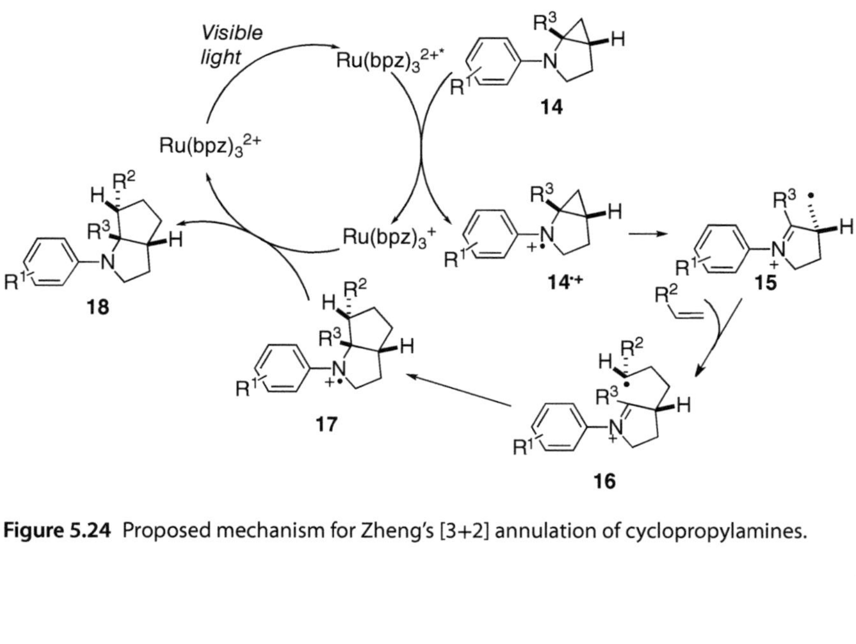

Figure 5.24 Proposed mechanism for Zheng's [3+2] annulation of cyclopropylamines.

furnishing the more stable radical **16**. Intramolecular addition of the secondary radical **16** to the iminium ion produces amine radical cation **17**. To complete the catalytic cycle, the reduced $Ru(bpz)_3^+$ reduces the amine radical cation **17** providing the annulation product **18** and regenerating the ground-state catalyst.

5.3.3 1,3-Dipolar Cycloaddition of Azomethine Ylides

Stabilized azomethine ylides are an important class of ylides in the [3+2] dipolar cycloaddition. Taking advantage of facile photo-oxidation of tetrahydroisoquinolines to the iminium ions, Xiao [17] and Rueping [18] independently recognized that when tetrahydroisoquinolines are substituted with an active methylene group, the resulting iminium ion is readily converted to a stabilized azomethine ylide. Both groups showed that the azomethine ylides generated *in situ* by this method are intercepted by a range of dipolarophiles to produce fused pyrrolidines and pyrroles (upon oxidation) in good to excellent yields (Figure 5.25).

The authors proposed a plausible mechanism of the [3+2] dipolar cycloaddition, illustrated in Figure 5.26. Upon irradiation by visible light, the photoexcited state of Ru^{2+} oxidizes tetrahydroisoquinoline **19** to amine radical cation **20**. The catalyst is turned over by oxidation with molecular oxygen, providing superoxide radical anion, which is anticipated to abstract a hydrogen atom from amine radical cation **20**. The newly formed iminium ion **21** is then deprotonated by a molecule of hydroperoxide anion, generating azomethine ylide **22** and hydrogen

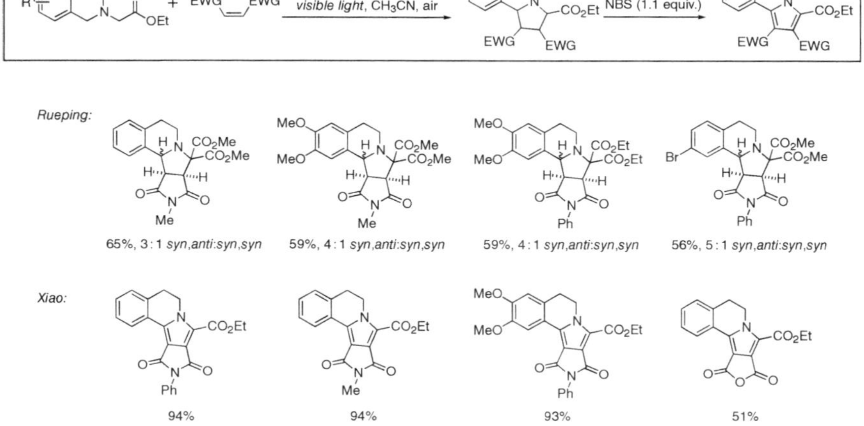

Figure 5.25 General overview and scope of Rueping and Xiao's independent [3+2] cycloaddition of azomethine ylides.

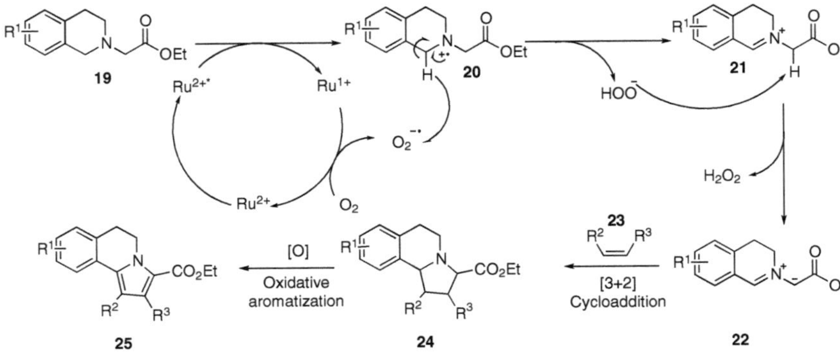

Figure 5.26 Proposed mechanism for the [3+2] cycloaddition of azomethine ylides.

peroxide. A subsequent [3+2] dipolar cycloaddition with a dipolarophile **23** generates fused pyrrolidine **24**. In the work of Xiao, further oxidation aromatizes the pyrrolidine moiety, providing pyrrole **25**.

5.3.4 [3+2] Cycloaddition of Aryl Cyclopropyl Ketones

Ring opening of cyclopropyl ketones has been previously established by one-electron reduction of the ketone to the radical anion via chemical reductants, which subsequently undergoes the ring opening to produce the distonic radical anion [19]. Yoon and coworkers surmised that photoreduced $[Ru(bpy)_3]^{1+}$ not only reduces the ketone to initialize the ring opening, but also catalyzes two C–C bond formations to complete a net [3+2] cycloaddition. In addition to $Ru(bpy)_3Cl_2$, a strong Lewis acid $La(OTf)_3$ and the stoichiometric reductant TMEDA played essential roles to afford the intramolecular cycloadduct in acceptable to excellent yields (55–86%). Various esters, ketones, and thioesters (in particular α-substituted) were well tolerated to provide excellent diastereoselectivity of the quaternary carbon stereocenters generated in the product (Figure 5.27). Similar to the [2+2] cycloaddition developed by the same group (vide supra), an aryl group on the ketone is a required constraint.

5.3.5 [3+2] Cycloaddition via ATRA/ATRC

Atom transfer radical addition or cyclization (ATRA/ATRC) has become a successful tool in constructing highly functionalized cyclic frameworks by forming two new C—C bonds. The use of strongly reducing photoexcited $Ir(ppy)_3$ provides a convenient way to generate carbon radicals from alkyl, alkenyl, or aryl halides without resorting to the use of stoichiometric hazardous radical initiators, which are usually employed. Following this approach, Yao reported a nonreductive, intermolecular visible light photoredox [3+2] ATRC reaction of unactivated alkyl iodides as radical precursors [20]. Initially optimizing the reaction with dimethyl-2-(iodomethyl) cyclopropane-1,1-dicarboxylate **26** and an electron-rich terminal alkene, the ideal conditions required $[Ir(ppy)_3]$ as the photocatalyst using a 14 W CFL, while utilizing *N*,*N*-diisopropylethylamine

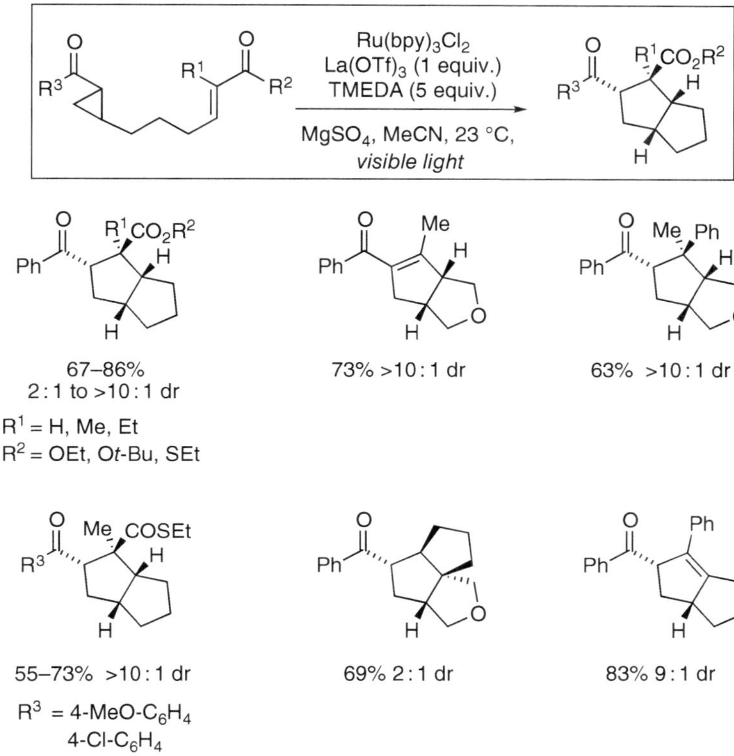

Figure 5.27 Visible light mediated [3+2] cycloaddition of aryl cyclopropyl ketones.

(DIEA) as a sacrificial electron donor to produce cycloadduct **27**. Polar solvents (e.g., CH_2Cl_2/H_2O (4 : 1)) displayed a significant effect in promoting the reaction. The aid of a Lewis acid, which is considered to coordinate to the carbonyl groups of the dicarboxylate, enhanced the reactivity toward less electron-rich alkenes. $Zn(OAc)_2{\bullet}2H_2O$ revealed to be the best Lewis acid among those screened. The reaction conditions were applied to other substrates, proceeding smoothly to furnish the desired cyclopentane/cyclopentene derivatives (Figure 5.28). Terminal alkynes and terminal alkenes with various functional groups were tolerated to afford the desired products in moderate to excellent yields (23–92%). Adding to the substrate scope, 1,1- and 1,2-disubstituted alkenes were subjected to ATRC successfully.

A radical chain reaction mechanism was proposed by Yao, as depicted in Figure 5.29. Upon irradiation, the photoexcited Ir complex is oxidized by malonic ester **26**. As the halogen atom in **26** accepts the electron, it is anticipated to leave as a halide anion while generating homoallylic radical **28**. Sequentially, the [3+2] cyclization occurs with an olefin to furnish cyclopentyl carbinyl radical **29**. From here, two pathways occur for conversion to the desired product **31**. Along *pathway a*, radical intermediate **29** is oxidized by the Ir^+ complex to give the cyclopentyl carbinyl cation **30**, which is then intercepted by the iodide to afford

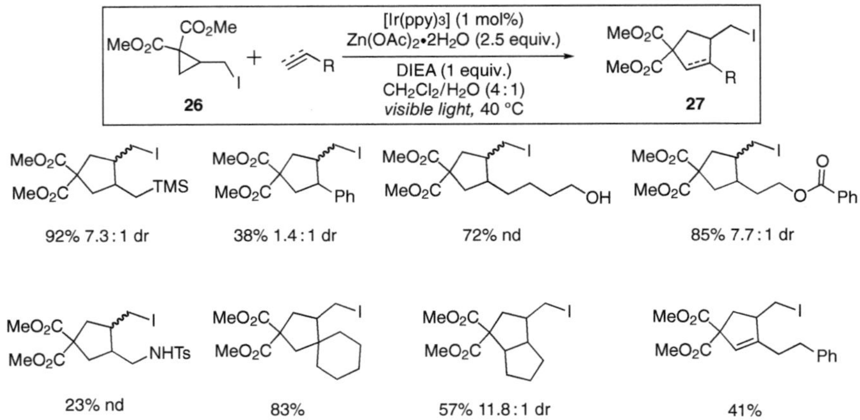

Figure 5.28 General reaction and scope of Yao's ATRC [3+2] cycloaddition.

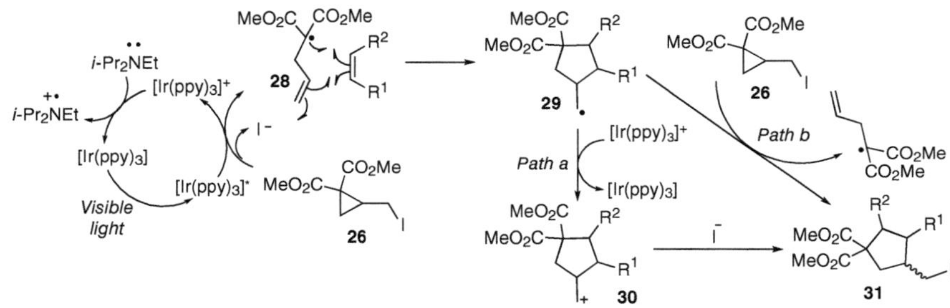

Figure 5.29 Proposed mechanisms for the [3+2] ATRC cycloaddition.

cycloadduct **31**. Along *pathway b*, abstraction of the iodine from **26** by **29** gives cycloadduct **31**, thus regenerating radical **28**. In studying the radical propagation step, Yao and coworkers employed an on/off light-switching experiment. Their observations support a radical/polar crossover pathway (*pathway a*); however, *pathway b* cannot be excluded.

5.3.6 [3+2] Conclusion

Once again, visible light photoredox catalysis has proven to be a powerful tool to promote cycloaddition reactions. The formation of five-membered rings via light mediated [3+2] cycloadditions has been revealed, including the opening of cyclopropylamines to produce reactive distonic ions, dipolar cycloaddition of stabilized azomethine ylides formed via photo-oxidation of tetrahydroisoquinolines, opening of cyclopropyl aryl ketones, and ATRA/ATRC of (iodomethyl)cyclopropanes. These examples have clearly demonstrated the utility of visible light photoredox catalysis in atom-economical reactions, and expansion of the scope of this type of catalysis to the formation of other reactive intermediates for [3+2] cycloadditions is currently underway.

5.4 [4+2] Cycloadditions: Formation of Six-Membered Rings

5.4.1 Introduction to [4+2] Cycloadditions

The Diels–Alder reaction has proven to be an extremely important and synthetically useful reaction in the realm of organic synthesis. The utility of this reaction is showcased in that it forms two new C–C bonds and potentially multiple stereocenters in a single step. However, it is well known that synthetically useful Diels–Alder reactions require the use of electronically matched systems, usually being an electron-rich diene reacting with an electron-poor dienophile or vice versa. Mismatched systems need to be forced (usually by high temperature or Lewis acid activation) and hence are often not synthetically useful. Alternatively, radical cations of electron-rich alkenes (dienophiles) undergo facile Diels–Alder cycloaddition with electron-rich dienes. These radicals have been shown to possess unique reactivity, often allowing a Diels–Alder cycloaddition to occur that thermal cases cannot suffice. Often, the radicals are generated using one-electron chemical oxidants or organic photosensitizers in high loadings, with the organic photosensitizer usually requiring high-intensity xenon lamps. In the next section, radical ion Diels–Alder cycloadditions using visible light and a ruthenium photocatalyst, to provide unique structures that Diels–Alder cycloadditions of the corresponding neutral species cannot generate will be examined.

5.4.2 [4+2] Cycloadditions Using Radical Anions

In 2011, Yoon endeavored to investigate hetero-Diels–Alder cycloadditions of radical anions [21]. The study was promoted by the formation of the Diels–Alder product while studying the intramolecular [2+2] cycloaddition (see Section 1.2.2.) of diene **32**, which contained one extra methylene group within its tether. The authors were surprised that no [2+2] product was observed, and that the Diels–Alder adduct **33** was formed instead along with a reductive monocyclization product **34** (Figure 5.30). Prompted by the data, the authors opted to develop conditions that had the potential to maximize the Diels–Alder adduct **33**.

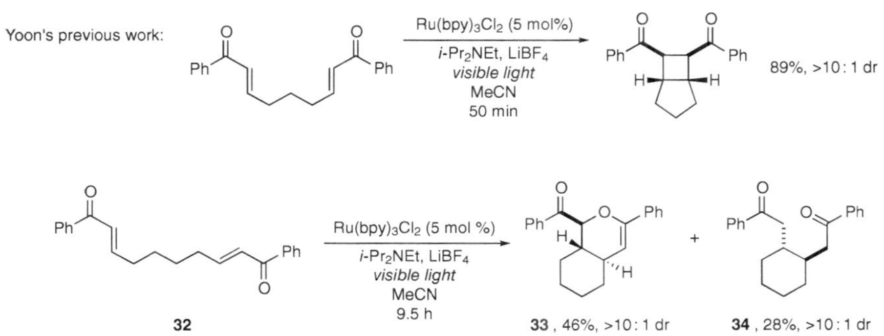

Figure 5.30 Yoon's initial discovery of a visible light induced radical anion Diels–Alder cycloaddition.

Initial investigations led the authors to believe that the reductive monocyclization product **34** was produced from a decomposition of **33** from over-reduction and reductive cleavage. This hypothesis was then corroborated by subjecting pure **33** to the reaction conditions for 24 h, leading to 33% product yield of **34**. Encouraged by the results, the authors sought to study the rate of product conversion while minimizing the undesired side product. As with many Diels–Alder cycloadditions, it was found that water dramatically increased the rate of product conversion as well as decreased the formation of the undesired reductive monocyclization product **34** (<5%). It was also discovered that $LiBF_4$ was a necessary additive, indicating that the Lewis acidity of the lithium cation is crucial for this cycloaddition, an observation also made in the [2+2] intramolecular cycloaddition presented by the Yoon group [9]. The optimized conditions are shown in Figure 5.31.

The group then shifted their focus to the substrate scope, which is shown in Figure 5.32. Symmetrical aryl bis(enones) reacted smoothly under the reaction conditions, which included both electron-rich and -deficient aryl enones. Similarly, unsymmetrical bis(enones) containing only one aryl group reacted favorably in isolated cases, while other α-alkoxy enones required further optimization to provide moderate yields. It was noted that as steric bulk increases, product yield decreases. The authors were also delighted to discover that electronically mismatched pairs, namely two electron-deficient components, worked effectively in this transformation.

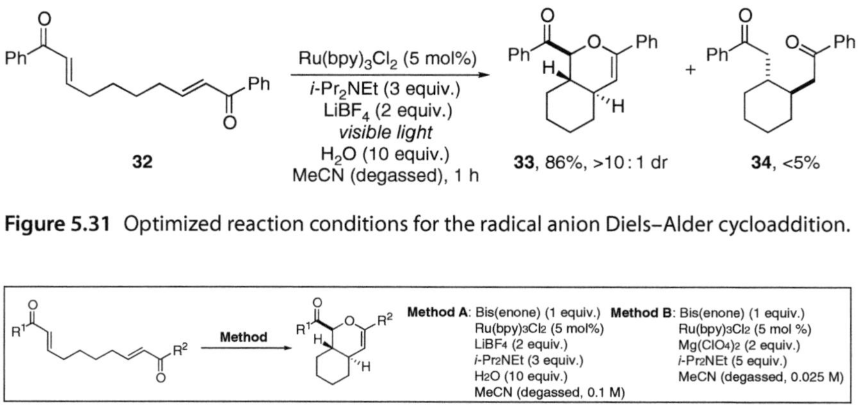

Figure 5.31 Optimized reaction conditions for the radical anion Diels–Alder cycloaddition.

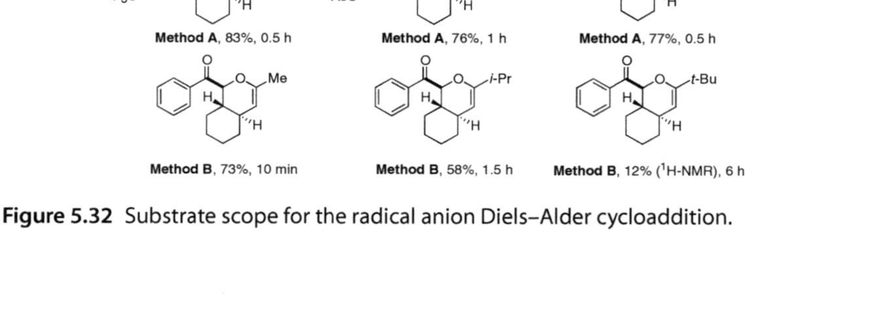

Figure 5.32 Substrate scope for the radical anion Diels–Alder cycloaddition.

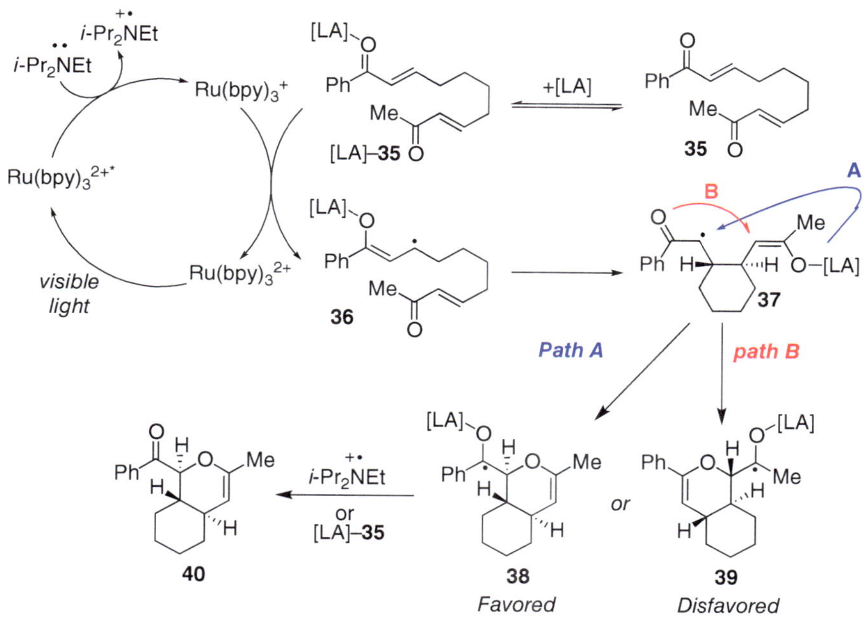

Figure 5.33 Yoon's proposed mechanism for the radical anion Diels–Alder cycloaddition.

In all cases, one regioisomer was formed. The authors rationalized this observation using a step-wise process that was inspired by Krische and Bauld's proposal for radical anion cycloadditions (Figure 5.33) [22]. Mechanistically, the photoexcited $Ru(bpy)_3^{2+}$ (generated via visible light) undergoes reductive quenching by *i*-Pr_2NEt. The reduced $Ru(bpy)_3^{+}$ then reduces the Lewis acid-activated enone complex [LA]–**35**, generating the activated radical anion **36** while turning over the catalyst. The radical anion **36** efficiently undergoes β,β coupling, producing a monocyclized distonic radical anion intermediate **37**, which forms the carbon–oxygen bond via path A or B. The authors speculated that the aryl ketyl radical **38** is more stable than its regioisomer **39**, thereby forming the sole isomer. A one-electron loss of radical anion **38** to either the amine radical cation or the Lewis acid enone complex results in the neutral Diels–Alder cycloadduct **40**.

Although the authors showed that further modifications can be made to the tetrahydropyran products without losing selectivity (such as acetylation of the enol ether, dihydroxylation of the olefin, and catalytic hydrogenation), the chemoselectivity favoring the [4+2] cycloaddition over the [2+2] cycloaddition remains unanswered thus far.

5.4.3 [4+2] Cycloadditions Using Radical Cations

Soon after their discovery of the intramolecular radical anion hetero-Diels–Alder cycloaddition, the Yoon group published a facile intermolecular Diels–Alder cycloaddition between a photogenerated radical cation of an electron-rich olefin and an electron-rich diene [23]. The radical cation was planned to be produced by a visible light excited ruthenium complex, and hence circumvented the need for high loadings of chemical oxidants (such as aminium salts), photoinitiated electron transfer (PET) with an organic photosensitizer, and the use of high-intensity xenon lamps. Having gained information from the [2+2] photocycloaddition of electron-rich olefins using $Ru(bpy)_3^{2+}$ and methyl viologen (MV^{2+}), Yoon and coworkers applied the same catalyst system to a Diels–Alder cycloaddition (Figure 5.34). Indeed, it was found that in just 1 h, cycloadduct **43** is formed from electronically mismatched anethole **41** and isoprene **42** (Eq. (1)). Not surprisingly, when **41** and **42** were subjected to thermal Diels–Alder conditions, no reaction occurred (Eq. (2)).

More studies have shown that the catalyst system was unfortunately not robust enough to generate a useful substrate scope. The authors speculated that the reason might be the reduced MV^{+}, which might be participating in back electron transfer to the dienophile radical cation, thus reducing overall efficiency. The use of air as an irreversible co-oxidant was unsuccessful, presumably because of its inefficiency in oxidatively quenching $Ru(bpy)_3^{2+*}$. The authors decided to utilize the tris-bipyrazine analog, $Ru(bpz)_3^{2+}$, which maintains a sufficient excited state oxidation potential (1.4 V vs. SCE) to directly oxidize **41**. The authors further fine-tuned the catalyst system based on $Ru(bpz)_3^{2+}$. $MgSO_4$ was removed because it had detrimental effects on the yield. The diene loading was lowered without decreasing the yield. The BArF counterion gave slightly better yields and allowed the reaction to be carried out in less polar solvents (CH_2Cl_2) with lower amounts of catalyst (0.5 mol%). Figure 5.35 depicts the broad substrate scope that the designed system tolerates. Various dienes are reactive, including bulky, aryl substituted, and cyclic cases, which produce *endo* selectivity when applicable. Additionally, a variety of substituted dienophiles are tolerated, including allylic substitutions, cyclic styrenes, and enamines (instead of styrenes). It is of note, however, that certain dienes and dienophiles, including the electron-rich Danishefsky diene and electron-poor diene esters, are not reactive using this chemistry. It is noteworthy that β,β-disubstituted dienophiles, as well as stilbenes and unprotected phenols are unreactive. A final difficulty arises from the necessity of an electron-donating group in either the *ortho* or *para* position of the styrene.

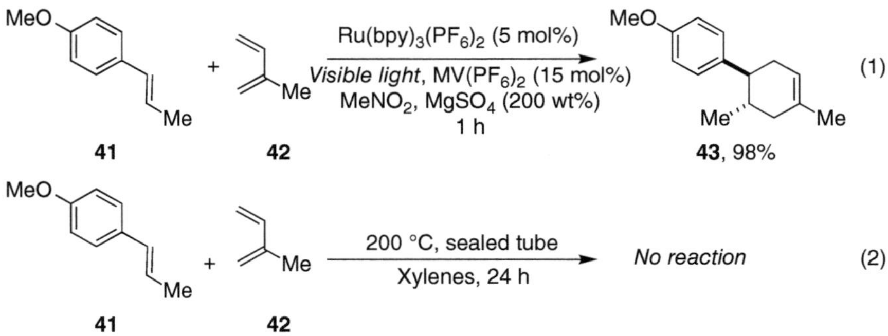

Figure 5.34 Radical cation Diels–Alder cycloaddition compared to thermal Diels–Alder cycloaddition.

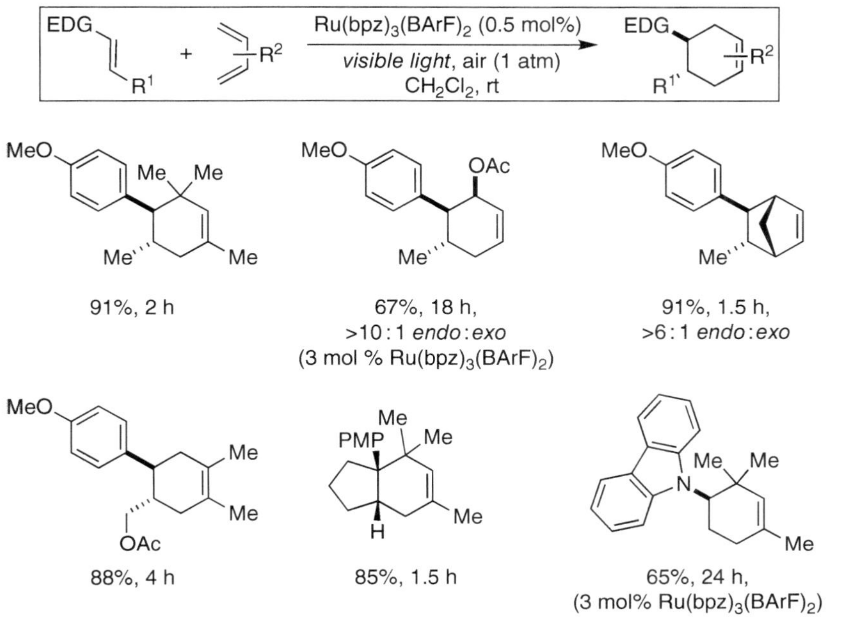

Figure 5.35 Substrate scope of the radical cation Diels–Alder cycloaddition.

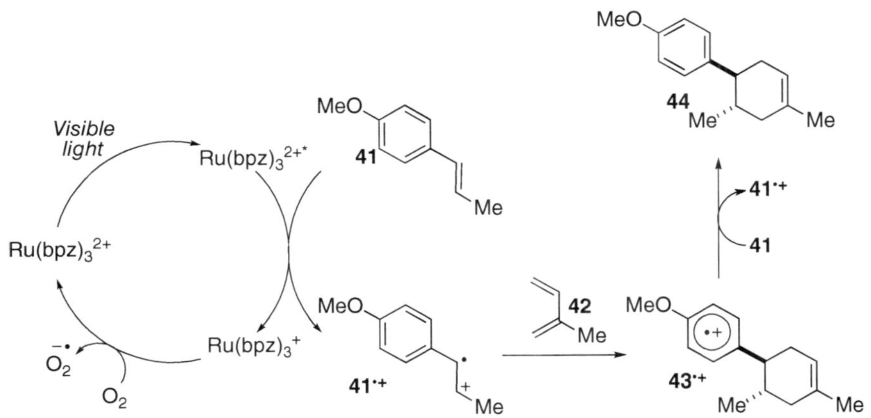

Figure 5.36 Proposed mechanism for the radical cation Diels–Alder cycloaddition.

The authors proposed a mechanism capable of describing the overall process, which is outlined in Figure 5.36. Photoexcited $Ru(bpz)_3^{2+*}$ was considered to directly oxidize anethole **41**, generating its radical cation $\mathbf{41}^{\bullet+}$ and the reduced catalyst $Ru(bpz)_3^+$. The catalyst was turned over by the oxygen present in air, while the radical cation $\mathbf{41}^{\bullet+}$ reacted with the electron-rich isoprene **42**. The newly formed aromatic radical cation $\mathbf{43}^{\bullet+}$ was then expected to form the neutral Diels–Alder cycloadduct **44** via one electron reduction by anethole **41**, a chain-propagation step determined from a control experiment where the authors found that in degassed solvents in the absence of air, cycloadduct **44** was formed but not to completion.

Following this report, the Yoon group published an intramolecular version of this Diels–Alder reaction to complement the previous studies [24]. The authors were surprised that the conditions used for the intermolecular Diels–Alder reactions did not translate well to the intramolecular case, as the cycloadduct **46** was obtained in only 35% combined yields as a 3 : 1 *endo* : *exo* mixture after 2 h from Diels–Alder substrate **45** (Eq. (3), Figure 5.37). However, when subjected to the less oxidizing $Ru(bpy)_3(PF_6)_2$ with methyl viologen as the co-oxidant, **46** was obtained in both a higher yield and a better *endo*:*exo* ratio (Eq. (4), Figure 5.37).

A substrate scope was then developed using the optimized reaction conditions, as shown in Figure 5.38. Systems with a three-carbon tether react smoothly to give the desired product, while the four-carbon tether does not react in a 24-h time period. As expected, the styrene component requires an electron-donating group in either the ortho or para position; however, additional substitutions are tolerated. Electron-rich olefin components also work under the optimized conditions, although the yield decreases in the case of the 2-furanyl group. When analyzing the diene, various substitutions are amenable, including internal substitution (R^4), providing a quaternary stereocenter at the ring junction.

5.4.4 [4+2] Conclusion

Visible light has been successfully utilized in generating radical cations or radical anions that have reactivity toward the Diels–Alder cycloaddition reaction. Yoon and coworkers successfully revealed an intramolecular hetero-Diels–Alder cycloaddition, which was elaborated to a robust intermolecular Diels–Alder cycloaddition between electron-rich olefins and electron rich-dienes. In the final report, Yoon and coworkers developed a similar methodology that extended its scope to the intramolecular system, generating complex polycycles in a single step. In all the cases discussed, moderate to high yields and selectivities were

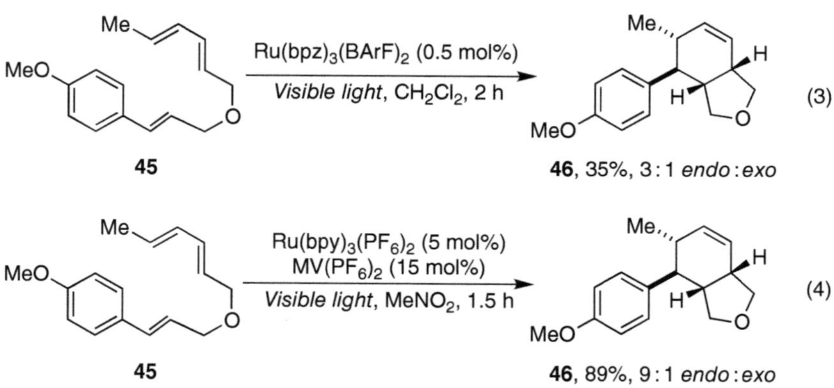

Figure 5.37 Development of the intramolecular radical cation Diels–Alder cycloaddition.

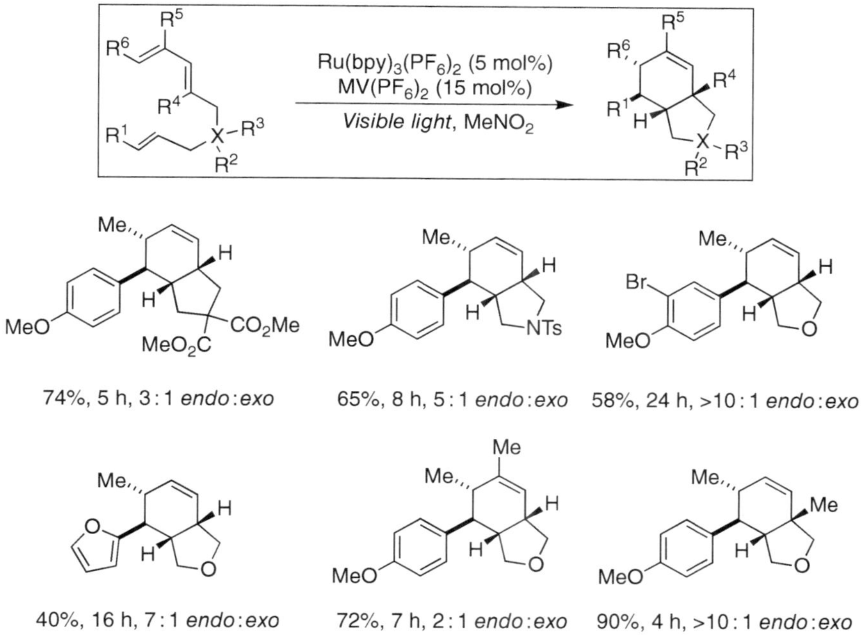

Figure 5.38 Substrate scope of the intramolecular radical cation Diels–Alder cycloaddition.

obtained for these powerful C—C bond-forming reactions, all of which can be applied to the synthesis of complex natural products and can be performed on the gram scale. The successful applications of photogenerating radical anions or cations to enable electronically mismatched Diels–Alder cycloadditions can be potentially parlayed into broader electronically mismatched reactions.

5.5 Conclusion

Selected recent examples of visible light mediated cycloadditions and annulations are discussed in this chapter for the construction of four-, five-, and six-membered rings. Although not discussed, it is worth mentioning the recent reports by Yoon and coworkers [25], Zheng and coworkers [26], and Konig and coworkers [27] that provide more examples of using visible light for accessing carbo- and hetero-cycles. The discussed examples that utilize reductive quenching cycle, oxidative quenching cycle, or energy transfer pathway, amply demonstrate the prowess of visible light mediated cycloadditions and annulations in generating multiple bonds and stereocenters in a single step. The impact of utilizing visible light to promote cycloadditions and annulations has been realized throughout this chapter, and it seems that the discoveries reported have merely scratched the surface of what can be accomplished using visible light photoredox catalysis [28].

References

1 Series, P. (2005) *The Chemistry of Functional Groups*, Wiley, Chichester.

2 (a) Crimmins, M.T. and Reinhold, T.L. (1993) Enone olefin [2+2] photochemical cycloadditions. *Org. React. (N.Y.)*, **44**, 297–588; (b) Winkler, J.D., Bowen, C.M., and Liotta, F. (1995) [2+2] Photocycloaddition/fragmentation strategies for the synthesis of natural and unnatural products. *Chem. Rev. (Washington, D.C.)*, **95** (6), 2003–2020; (c) Magereta, P. (2005) *Molecular and Supramolecular Photochemistry*, vol. **12**, Marcel Dekker, New York; (d) Namyslo, J.C. and Kaufmann, D.E. (2003) The application of cyclobutane derivatives in organic synthesis. *Chem. Rev.*, **103** (4), 1485–1537; (e) Fu, N.-Y., Chan, S.-H., and Wong, H.N.C. (2005) *Patai Series: The Chemistry of Functional Groups*, Wiley, Chichester; (f) Horspool, W. (2005) *Patai Series: The Chemistry of Functional Groups*, Wiley, Chichester.

3 Schuster, D.I., Lem, G., and Kaprinidis, N.A. (1993) New insights into an old mechanism: [2+2] photocycloaddition of enones to alkenes. *Chem. Rev.*, **93** (1), 3–22.

4 Lee-Ruff, E. and Mladenova, G. (2003) Enantiomerically pure cyclobutane derivatives and their use in organic synthesis. *Chem. Rev.*, **103** (4), 1449–1483.

5 Bach, T. (1998) Stereoselective intermolecular [2+2] photocycloaddition reactions and their application in synthesis. *Synthesis*, **5**, 683–703.

6 Mattay, J., Conrads, R., and Hoffman, R. (1996) *Methoden der Organischen Chemie (Houben-Weyl)*, 4th edn, vol. **5**, Thieme Verlag, Stuttgart.

7 Ciamician, G. (1912) Photochemistry of the future. *Science (Washington, DC, USA)*, **36**, 385–394.

8 Baik, T.-G., Luis, A.L., Wang, L.-C., and Krische, M.J. (2001) A diastereoselective metal-catalyzed [2+2] cycloaddition of bis-enones. *J. Am. Chem. Soc.*, **123** (27), 6716–6717.

9 Ischay, M.A., Anzovino, M.E., Du, J., and Yoon, T.P. (2008) Efficient visible light photocatalysis of [2+2] enone cycloadditions. *J. Am. Chem. Soc.*, **130** (39), 12886–12887.

10 Du, J. and Yoon, T.P. (2009) Crossed intermolecular [2+2] cycloadditions of acyclic enones via visible light photocatalysis. *J. Am. Chem. Soc.*, **131** (41), 14604–14605.

11 Tyson, E.L., Farney, E.P., and Yoon, T.P. (2012) Photocatalytic [2+2] cycloadditions of enones with cleavable redox auxiliaries. *Org. Lett.*, **14** (4), 1110–1113.

12 Ischay, M.A., Lu, Z., and Yoon, T.P. (2010) [2+2] Cycloadditions by oxidative visible light photocatalysis. *J. Am. Chem. Soc.*, **132** (25), 8572–8574.

13 Ischay, M.A., Ament, M.S., and Yoon, T.P. (2012) Crossed intermolecular [2+2] cycloaddition of styrenes by visible light photocatalysis. *Chem. Sci.*, **3** (9), 2807–2811.

14 Lu, Z. and Yoon, T.P. (2012) Visible light photocatalysis of [2+2] styrene cycloadditions by energy transfer. *Angew. Chem. Int. Ed.*, **51** (41), 10329–10332, S10329/1-S10329/128.

15 Zou, Y.-Q., Duan, S.-W., Meng, X.-G. *et al.* (2012) Visible light induced intermolecular [2+2]-cycloaddition reactions of 3-ylideneoxindoles through energy transfer pathway. *Tetrahedron*, **68** (34), 6914–6919.

16 Maity, S., Zhu, M., Shinabery, R.S., and Zheng, N. (2012) Intermolecular [3+2] cycloaddition of cyclopropylamines with olefins by visible-light photocatalysis. *Angew. Chem. Int. Ed.*, **51** (1), 222–226.
17 Zou, Y.-Q., Lu, L.-Q., Fu, L. *et al.* (2011) Visible-light-induced oxidation/[3+2] cycloaddition/oxidative aromatization sequence: a photocatalytic strategy to construct pyrrolo[2,1-a]isoquinolines. *Angew. Chem. Int. Ed.*, **50** (31), 7171–7175, S7171/1-S7171/66.
18 Rueping, M., Leonori, D., and Poisson, T. (2011) Visible light mediated azomethine ylide formation-photoredox catalyzed [3+2] cycloadditions. *Chem. Commun. (Cambridge, UK)*, **47** (34), 9615–9617.
19 Lu, Z., Shen, M., and Yoon, T.P. (2011) [3+2] Cycloadditions of aryl cyclopropyl ketones by visible light photocatalysis. *J. Am. Chem. Soc.*, **133** (5), 1162–1164.
20 Gu, X., Li, X., Qu, Y. *et al.* (2013) Intermolecular visible-light photoredox atom-transfer radical [3+2]-cyclization of 2-(iodomethyl)cyclopropane-1,1-dicarboxylate with alkenes and alkynes. *Chem. Eur. J.*, **19** (36), 11878–11882.
21 Hurtley, A.E., Cismesia, M.A., Ischay, M.A., and Yoon, T.P. (2011) Visible light photocatalysis of radical anion hetero-Diels–Alder cycloadditions. *Tetrahedron*, **67** (24), 4442–4448.
22 (a) Roh, Y., Jang, H.-Y., Lynch, V. *et al.* (2002) Anion radical chain cycloaddition of tethered enones: intramolecular cyclobutanation and Diels–Alder cycloaddition. *Org. Lett.*, **4** (4), 611–613; (b) Yang, J., Felton, G.A.N., Bauld, N.L., and Krische, M.J. (2004) Chemically induced anion radical cycloaditions: intramolecular cyclobutanation of bis(enones) via homogeneous electron transfer. *J. Am. Chem. Soc.*, **126** (6), 1634–1635.
23 Lin, S., Ischay, M.A., Fry, C.G., and Yoon, T.P. (2011) Radical cation Diels–Alder cycloadditions by visible light photocatalysis. *J. Am. Chem. Soc.*, **133** (48), 19350–19353.
24 Lin, S., Padilla, C.E., Ischay, M.A., and Yoon, T.P. (2012) Visible light photocatalysis of intramolecular radical cation Diels–Alder cycloadditions. *Tetrahedron Lett.*, **53** (24), 3073–3076.
25 Parrish, J.D., Ischay, M.A., Lu, Z. *et al.* (2012) Endoperoxide synthesis by photocatalytic aerobic [2+2+2] cycloadditions. *Org. Lett.*, **14** (6), 1640–1643.
26 (a) Nguyen, T.H., Maity, S., and Zheng, N. (2014) Visible light mediated intermolecular [3+2] annulation of cyclopropylanilines with alkynes. *Beilstein J. Org. Chem.*, **10**, 975–980, 6 pp; (b) Nguyen, T.H., Morris, S.A., and Zheng, N. (2014) Intermolecular [3+2] annulation of cyclopropylanilines with alkynes, enynes, and diynes via visible light photocatalysis. *Adv. Synth. Catal.*, **356** (13), 2831–2837.
27 Hari Durga, P., Hering, T., and Konig, B. (2012) Visible light photocatalytic synthesis of benzothiophenes. *Org. Lett.*, **14** (20), 5334–5337.
28 For additional reading of primary literature that was published after the transcription of this chapter see: (a) Chandrasekhar, D., Borra, S., Nanubolu, J.B., and Maurya, R.A. (2016) Visible light driven photocascade catalysis: Ru(bpy)3(PF6)2/TBHP-mediated synthesis of fused β-carbolines in batch and flow microreactors. *Org. Lett.*, **18** (12), 2974–2977; (b) Chen, J.-R., Hu, X.-Q., Lu, L.-Q., and Xiao, W.-J. (2016) Exploration of visible-light photocatalysis in

heterocycle synthesis and functionalization: reaction design and beyond. *Acc. Chem. Res.*, **49** (9), 1911–1923; (c) Fujiya, A., Tanaka, M., Yamaguchi, E. *et al.* (2016) Sequential photo-oxidative [3+2] cycloaddition/oxidative aromatization reactions for the synthesis of pyrrolo[2,1-a]isoquinolines using molecular oxygen as the terminal oxidant. *J. Org. Chem.*, **81** (16), 7262–7270; (d) Liu, Q., Zhu, F.-P., Jin, X.-L. *et al.* (2015) Visible-light-driven intermolecular [2+2] cycloadditions between coumarin-3-carboxylates and acrylamide analogs. *Chem. Eur. J.*, **21** (29), 10326–10329; (e) Ruhl, K.E. and Rovis, T. (2016) Visible light-gated cobalt catalysis for a spatially and temporally resolved [2+2+2] cycloaddition. *J. Am. Chem. Soc.*, **138** (48), 15527–15530; (f) Southgate, E.H., Pospech, J., Fu, J. *et al.* (2016) Dearomative dihydroxylation with arenophiles. *Nat. Chem.*, **8** (10), 922–928; (g) Wang, C., Ren, X., Xie, H., and Lu, Z. (2015) [3+2] Redox-neutral cycloaddition of nitrocyclopropanes with styrenes by visible-light photocatalysis. *Chem. Eur. J.*, **21** (27), 9676–9680; (h) Yang, B. and Lu, Z. (2016) Visible-light-promoted oxidative [4+2] cycloadditions of aryl silyl enol ethers. *J. Org. Chem.*, **81** (16), 7288–7300; (i) Yuan, Z.-G., Wang, Q., Zheng, A. *et al.* (2016) Visible light-photocatalysed carbazole synthesis via a formal (4+2) cycloaddition of indole-derived bromides and alkynes. *Chem. Commun.*, **52** (29), 5128–5131; (j) Zeng, T.-T., Xuan, J., Ding, W. *et al.* (2015) [3+2] Cycloaddition/oxidative aromatization sequence via photoredox catalysis: one-pot synthesis of oxazoles from 2H-azirines and aldehydes. *Org. Lett.*, **17** (16), 4070–4073.

6

Metal-Free Photo(redox) Catalysis

Kirsten Zeitler

Institute of Organic Chemistry, Faculty of Chemistry and Mineralogy, University of Leipzig, Johannisallee 29, Leipzig, D-04103, Germany

6.1 Introduction

In recent years, visible light-mediated photocatalysis has gained increasing attention as a powerful synthetic technique that enables formerly challenging or even impossible chemical transformations [1, 2]. Apart from its promises for efficient and sustainable processes by employing visible light as a waste-free, non-toxic reagent from abundant renewable resources, for example, for the production of solar fuels [3], and hence the transfer of light energy into (storable) chemical energy [3a], photocatalysis offers a unique platform for the development of novel bond constructions that are currently elusive, but are of strategic interest for the synthetic organic chemists' toolbox [4].

Particularly, photocatalysis with visible light does not require specialized equipment such as UV irradiation sources used in classical photochemistry [5] and hence offers advantageous, mild conditions for reactions that mostly proceed at room temperature. Additionally, the lack of absorption of most organic molecules in the visible light region minimizes the potential for deleterious side reactions that might occur upon direct photoexcitation of the corresponding substrates.

In general, several diverse mechanisms are available for the interaction of photocatalysts (also often referred to as photosensitizers or photosensibilization agents) with organic compounds. Based on the specific properties of the catalyst's photoexcited state, which is reached upon irradiation, this interaction may generate highly reactive intermediates that are activated for further transformations. Three common modes of activation are briefly outlined here, namely, energy transfer, electron transfer, and atom transfer: so-called direct *energy transfer* to suitable compounds (to be estimated by the relative triplet state energies E_T of catalyst and substrate) leads to their electronic excitation, representing a rise in energy and hence allowing for reactivity, which is considerably different to that known for their ground states (such as [2 + 2]-cycloaddition, double bond isomerization, etc.). Furthermore, as a result of the photoexcited state's remarkable property to function both as a stronger oxidant and as a stronger

Visible Light Photocatalysis in Organic Chemistry, First Edition.
Edited by Corey R. J. Stephenson, Tehshik P. Yoon and David W. C. MacMillan.

reductant than its corresponding ground state, radical intermediates (i.e., radicals/radical ions) may most commonly be accessed either by single *electron transfer* (photoinduced electron transfer (PET) [6], photoredox catalysis) or by *atom transfer*, most commonly via *hydrogen atom transfer* (HAT). The latter can be especially useful if electrochemical constraints that usually apply for such single-electron redox processes (single electron transfer, SET) could curtail the accessibility of these versatile radical intermediates (thermodynamically unfavorable electron transfers indicated by the standard reduction potentials $E_{1/2}$). However, only few photocatalysts allow for *direct* HAT processes (e.g., polyoxometalates, aromatic aldehydes) [7]. Overcoming this shortage is possible by combining photocatalysts with suitable HAT catalysts, such as thiyl radicals, as powerful hydrogen acceptors [8] that are reversibly accessed by oxidation of their corresponding thiolate anions and thereby also allow for selective radical C—H functionalization (predictable by comparison of bond dissociation energies (BDEs) [9]).

As mentioned in preceding text, by far the most common pathway of photocatalysts' "productive deactivation," as compared with radiative or non-radiative energy loss of their long-lived excited states, is their behavior as highly potent SET reagents offering reductive or oxidative power (indicated by the corresponding reduction potentials $E_{1/2}$) either from these excited states or via the transiently generated oxidized or reduced catalyst intermediates (for details on such secondary processes, see Sections 6.2.2 and 6.2.3). This transient generation of oxidant and reductant within the same flask offers additional mechanistic possibilities for overall redox-neutral reactions "shuttling" the electrons to distinct points in the mechanism; [10] despite the corresponding reducing or oxidizing "ground state" intermediates may also be accessed by using sacrificial reagents (which may not end up as the desired product) (Scheme 6.1).

As part of the increasing interest in using visible light photocatalysis as a novel platform to design new transformations during the last 5–10 years, although its potential usefulness had been proposed by Stobbe and Ciamician more than 100 years ago [11], a great variety of organic and inorganic compounds have been examined as photocatalysts for synthetic applications. Besides the employment of inorganic [12] and organic [13] semiconductors [14], most activity in this field has centered on transition-metal complexes [2a,b,d,e,15]. The majority of these organometallic photocatalysts are ruthenium and iridium polypyridyl complexes, although recent investigations have also focused on gold [16], chromium [17], cerium [18], and copper complexes [19]. While these photocatalysts are broadly applicable and may be tuned in their properties by ligand manipulations [20], especially the high costs and limited future availability have triggered greater interest in metal-free alternatives. Additional challenges, for example, being critical in connection with the synthesis of biologically active materials, such as potential toxicities and separation problems for such metal-based catalysts from final products to meet high levels of purity, [21, 22], have further increased the demand for alternative photocatalysts. Purely organic dyes as photocatalysts are typically less expensive, widely available, and easy to handle and modify and hence may have properties that can even out-perform organometallic and inorganic photocatalysts. Despite this attractiveness, there are far fewer reports on the application of these metal-free catalyst counterparts [23].

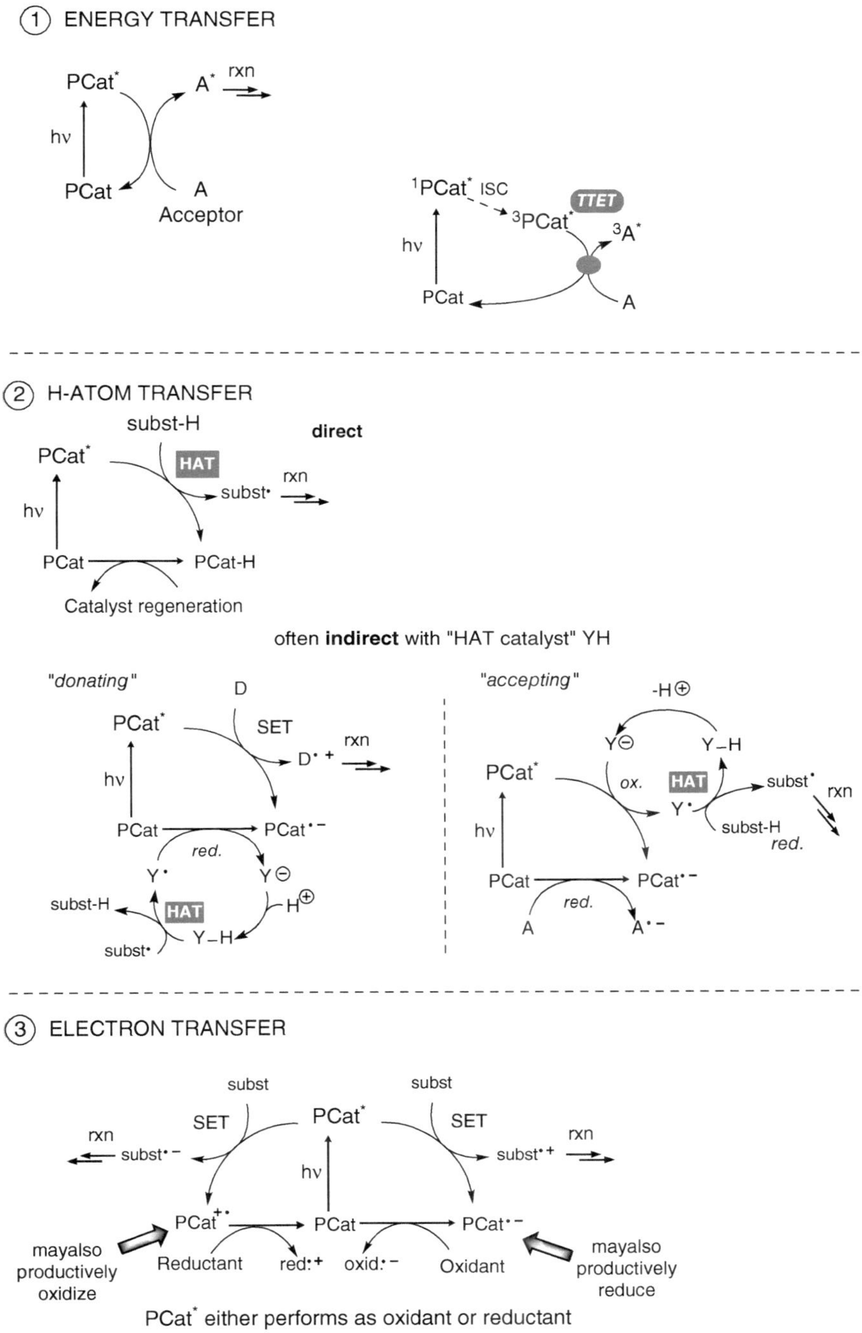

Scheme 6.1 Different mechanisms in photocatalysis.

This chapter will provide an overview on recent selected advances for the application of organic photocatalysts in synthesis.

6.1.1 Background

Taking nature as a blueprint for the design of suitable photocatalysts one will realize that nature does not simply rely on metal complexes as photoactive pigments, but also employs organic dyes for interaction with light. Besides ubiquitous Mg-containing substituted tetrapyrroles including chlorophylls and bacteriochlorophylls [24] the purely organic structures range from quinone scaffolds (e.g., stentorin and fringelites [25]) to flavin and pterin derivatives such as found in photolyases and cryptochromes [26].

Similar scaffolds are reflected in the great diversity of currently used *organic* photocatalysts (see Section 6.1.2).

While organic dyes are widely used for energy transfer processes to generate singlet oxygen, which subject is covered in great detail by several recent reviews [27, 28] and is beyond the scope of this chapter, their applications for other synthetic transformations, especially for new bond-forming reactions are rather scarce. In material sciences, metal-free approaches in photoinitiated polymerization reactions have gained recent interest as an enabling technology to novel tailor-made macromolecular architectures and in avoiding residual metal trace, which is supposed to be beneficial to the polymers' properties. Current efforts and advances in this field have been compiled recently [29, 30] and will not be covered within this chapter.

From a historical point of view, the use of organic dyes has been long-standing. However, early applications of cyanoarenes, (anthra)quinones, and electron-deficient heterocycles such as methylquinolinium or pyrylium salts [31] as well as aryl ketones [32] and xanthones often relied on the irradiation with UV or broadband, UV-encompassing light sources (Xenon lamps, etc.,) despite the existence of (minor) absorptions for some of these compounds in the visible region ($\lambda \geq 400$ nm). Hence, only recently, chemists have started to take advantage of the full potential of these compounds for *visible light photocatalysis* based on the outstanding seminal work on *PET* (PET, a term that covers both classical UV- and visible light-photoinitiated electron transfers (photoredox transformations)) [1c,d,6,31,33].

6.1.2 Classes of Organic Photocatalysts

Structures of organic photocatalysts are dominated by aromatic scaffolds providing extended conjugated systems as a prerequisite for their light absorption in the visible range. In connection with the surge of interest for alternative photo(redox) catalysts, typical organic dyes like methylenblue, riboflavin, and related aza-heterocycles [34] or xanthene-type dyes including fluorescein, eosin Y [35], eosin B, erythrosin, and rose bengal (RB), as well as related rhodamines have been successfully examined for their photocatalytic properties. Advancements in the development of electron-poor heterocycles as visible light photocatalysts have led to aryl *N*-methyl acridinium dyads, known as Fukuzumi-type catalysts, which are among the most powerful and broadly applicable photo-oxidants [21c,36]. In addition, however less commonly used, phenoxazine (such as nile red) and phenothiazine scaffolds [37], larger aromatic systems such as pyrenes or perylenes as well as boron dipyrromethene (BODIPY) derivatives [38] and donor–acceptor dyad systems [39] based on dicyanoarene or dicyanopyrazine chromophores have become promising, versatile catalyst scaffolds for synthetic transformations. The compiled list of organophotocatalysts as shown in Figures 6.1 and 6.2 demonstrates the immense diversity of catalysts used.

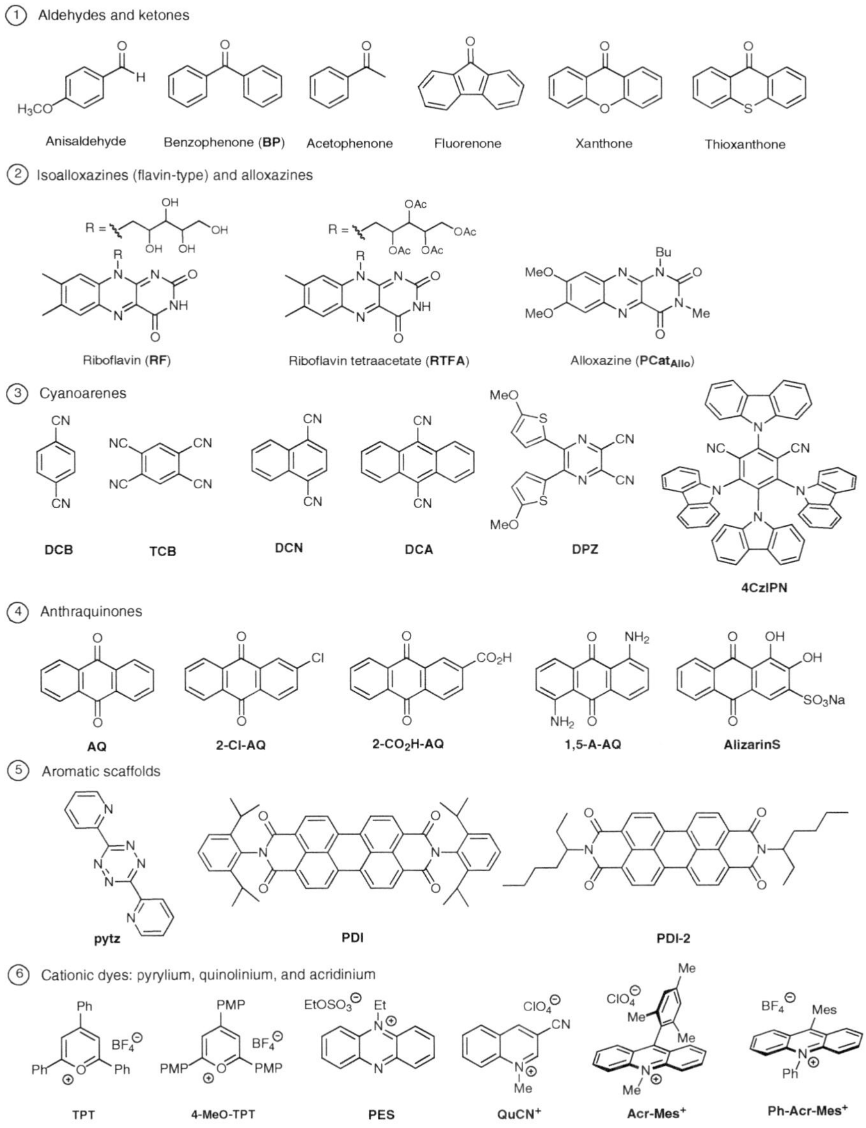

Figure 6.1 Chemical structures of common organic photocatalysts (part 1).

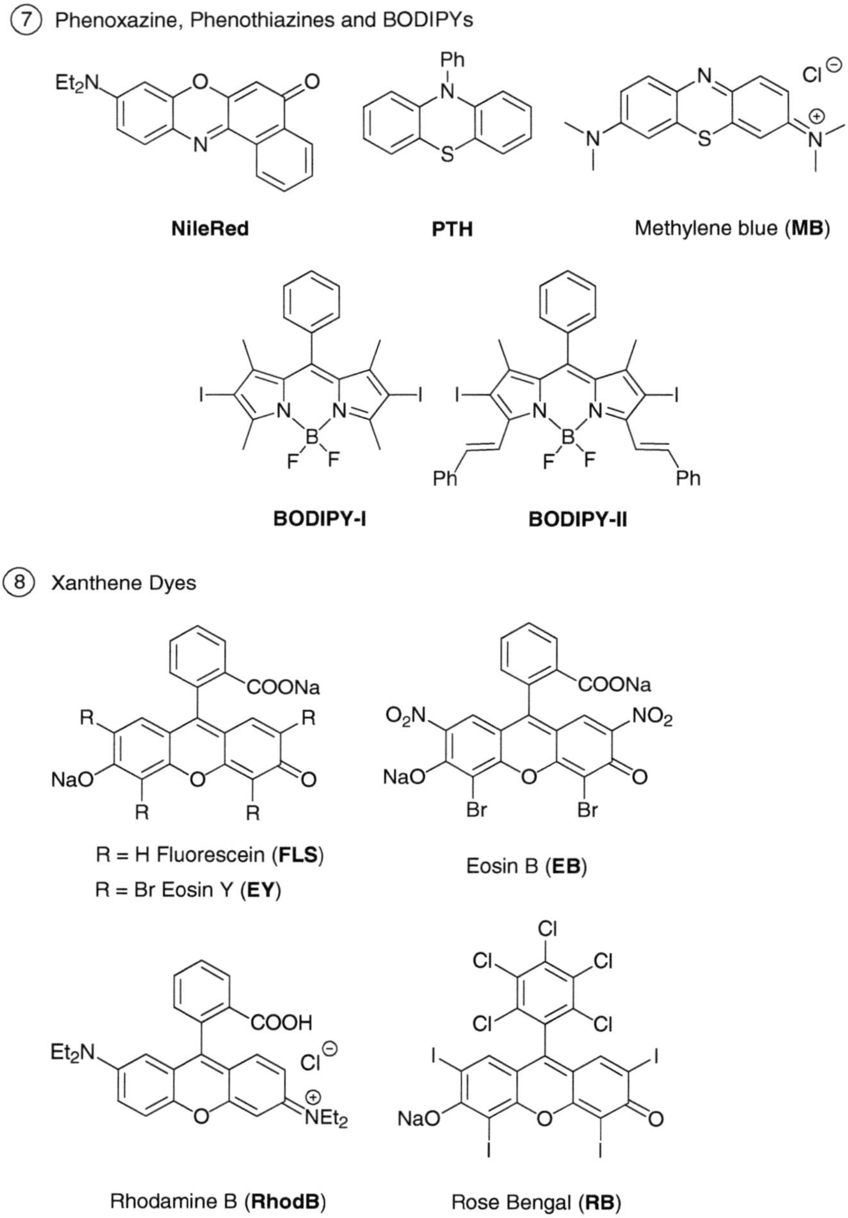

Figure 6.2 Chemical structures of common organic photocatalysts (part 2).

Purely organic, but distinct to the above mentioned photocatalytic processes substrates may get amenable to visible light irradiation and hence reaching excited states of altered reactivity by an initial (ground state) formation of so-called electron donor–acceptor complexes (EDA) which influence the absorption of the substrate and often will facilitate a direct electron transfer within the complex to yield reactive radical or radical ionic species [40]. This catalyst-free methodology is beyond the scope of this short chapter; readers are directed to some recent illustrative examples [41].

Apart from its visible light absorption together with appropriate redox properties and/or excited state energies (depending on its activation mechanism) a suitable organic photocatalysts should also display sufficient photostabilitity and excited state life-time. While these requirements are often associated with organometallic complexes [2b,e,18] due to their rapid *intersystem crossing* (ISC) to give a long-lived, lowest-energy triplet *metal-to-ligand-charge-transfer* (MLCT) state, organic dyes are often supposed to lack both stability and enhanced excited state life-time.

As an illustrating example, moderating the putative weaknesses, Scheme 6.2 provides a direct comparison of the photophysical and redox properties of $Ru(bpy)_3{}^{2+}$ with those of eosin Y to demonstrate their similarities that may account for the fact that eosin Y can often be employed as a metal-free surrogate for $Ru(bpy)_3{}^{2+}$, especially for reactions following a reductive quenching mechanism (right-hand side of the mechanism; for selected examples *vide infra*). Enhanced spin-orbit coupling as facilitated by the heavy atom substituents (here: Br) promotes *intersystem crossing (ISC)* to the longer-lived triplet state of eosin Y.

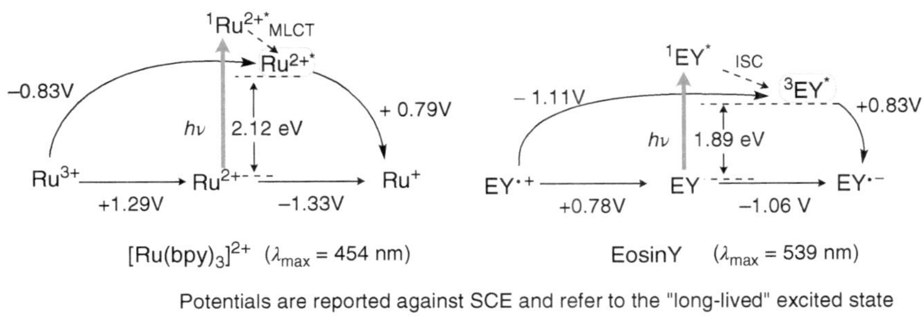

Scheme 6.2 Comparison of photophysical and.redox properties of $Ru(bpy)_3{}^{2+}$ with eosin Y (EY).

Since the early beginnings in using organic photocatalysts some decades ago the synthetic community's recent surge of interest for sustainable methods has also triggered considerable progress in the field of metal-free visible light photocatalysis. While a clear-cut categorization for the great diversity of photocatalysts and their variable modes of action might be difficult to be achieved, this chapter aims at highlighting selected advances in the field grouped according to different mechanistic pathways and the redox function taken by the catalysts. In a final part, newer developments of this catalytic manifold are discussed, including multicatalytic approaches [42] to point out the great promises of this methodology: providing access to a great variety of intermediates with unique reactivity under mild reaction conditions.

6.2 Applications of Organic Photocatalysts

Apart from the remarkable increase in both reductive and oxidative power of photoexcited species and hence their ability to engage in electron transfer with organic substrates (so-called photoredox transformations), a second fundamental decay pathway for photoexcited states is *energy transfer* to promote other organic molecules in an excited state. This convenient approach allows access to electronically excited states without the requirement of (short-wavelength) UV photons as compared to a direct, non-sensitized photoexcitation via a singlet state.

6.2.1 Energy Transfer Reactions

Despite the great success of UV-absorbing organic chromophores as triplet photosensitizers in classical photochemistry applications [43], only few examples have been reported for the synthetic use of triplet sensitization with organic dyes that absorb in the visible range. This is more remarkable as the energy of a $\lambda = 400$ nm photon is as high as around 300 kJ mol^{-1} ($\approx$71 kcal mol^{-1}) and thereby exceeds the triplet energy of many potential substrates for photocatalytic synthetic applications [44]. Most commonly, chromophores for efficient triplet–triplet energy transfer (TTET) do not only show favorably situated low lowest singlet and relatively high, long-lived lowest triplet states, but also exhibit high *intersystem crossing (ISC)* rates; typical scaffolds are found among carbonyl compounds, enones, heavy-atom substituted dyes, and so on.

Only recently, Melchiorre and coworkers could show that simple aromatic aldehydes are valuable photo-organocatalysts to promote ATRA reactions (atom transfer radical additions) between haloalkanes and alkenes upon irradiation with a household compact fluorescent lamp (23 W Compact fluorescent lamp (CFL) bulb) [45]. Their mechanistic investigations proved the near UV part of the CFL emission spectrum (band pass filter with $\lambda \geq 360$ nm, no reaction with $\lambda \geq 385$ nm) to be responsible for the population of the *p*-methoxy benzaldehyde's triplet state (circa 300 kJ mol^{-1}) allowing for an efficient energy transfer to the alkyl halide to effect the homolytic cleavage of the C—Hal bond (BDEs: 260–300 kJ mol^{-1} [8]). While the *para* methoxy group led to red-shifted absorption and increased triplet state life-time as compared to benzaldehyde, and no reaction was observed in the presence of typical triplet quenchers (such as oxygen, 2,5-dimethylhexa-2,4-diene or pyridazine); other organic sensitizers could similarly catalyze these reactions if provided with adequate excitation and triplet state energies (Scheme 6.3).

With respect to the construction of cyclobutane scaffolds photochemical and photocatalyzed [2 + 2]-cycloaddition reactions [46] are among the most efficient approaches to access these structures, which are also widespread in natural products [47]. While both reductive and oxidative methods have been used in photoredox radical ion cyclization reactions to yield the corresponding four-membered ring derivatives (for variants using organic photocatalysts, *vide infra*), metal-free photocatalysts relying on an energy transfer mode of action have been applied to circumvent typical substrate and cycloreversion

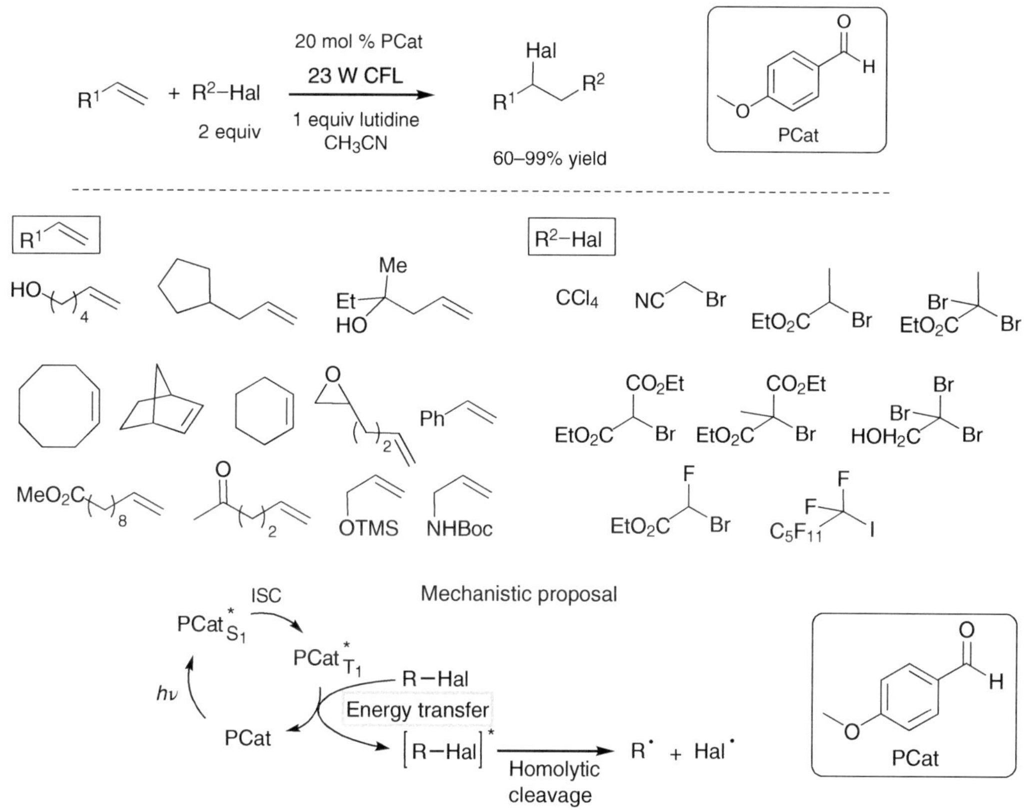

Scheme 6.3 ATRA reactions using *p*-methoxy benzaldehyde as energy transfer photocatalyst.

restrictions (according to their redox properties) that occur for photocatalytic electron transfer processes as well as to develop a highly enantioselective transformation. Based on their successful work with chiral xanthones [48] as sensitizer Bach and coworkers demonstrated that thioxanthones, which show a bathochromic shift ($\lambda_{max} \approx 390$ nm) in absorption, could be correspondingly integrated in a related hydrogen-bonding scaffold [49]. The newly designed chiral thioxanthone catalyst with a triplet energy of about 264 kJ mol^{-1} proved to be efficient for enantioselective intramolecular [2 + 2] photocycloaddition reactions of 4-(pent-4-enyl)quinolones and their heteroatom analogs (Scheme 6.4).

Thioxanthone was also successfully applied for the cyclization of atropisomeric maleimides transferring the axial chirality to products with central chirality [50].

A broadly applicable, metal-free [2+2] photocycloaddition approach for the intramolecular cyclization of both electron-rich and electron-poor dienes was recently disclosed by the Cibulka group [51]. As blue light absorbing riboflavin tetraacetate (**RFTA**, $\lambda_{max} = 450$ nm; $E_T = 209$ kJ mol^{-1}) could not match the triplet energies necessary to excite different styrene derivatives (e.g., β-methylstyrene: $E_T = 249$ kJ mol^{-1} or 1-phenylbut-2-en-1-one: $E_T = 259$ kJ mol^{-1}) altering the scaffold from the flavin-type isoalloxazines to the isomeric alloxazines ($E_T = 245$ kJ mol^{-1} for 1,3,7,8-tetramethyl-alloxazine), which are also known for their superior ISC yields, was key to success (Scheme 6.5).

E/Z-photoisomerization of the styrene double bond was detected to be a competitive process to yield a single major product regardless of the configuration of the starting styrene; however, this triplet excited state isomerization outcompetes productive *inter*molecular cyclizations.

Gilmour and coworkers recently reported on a bioinspired, energy transfer-mediated *Z*-selective photoisomerization of activated olefins [52], i.e.,

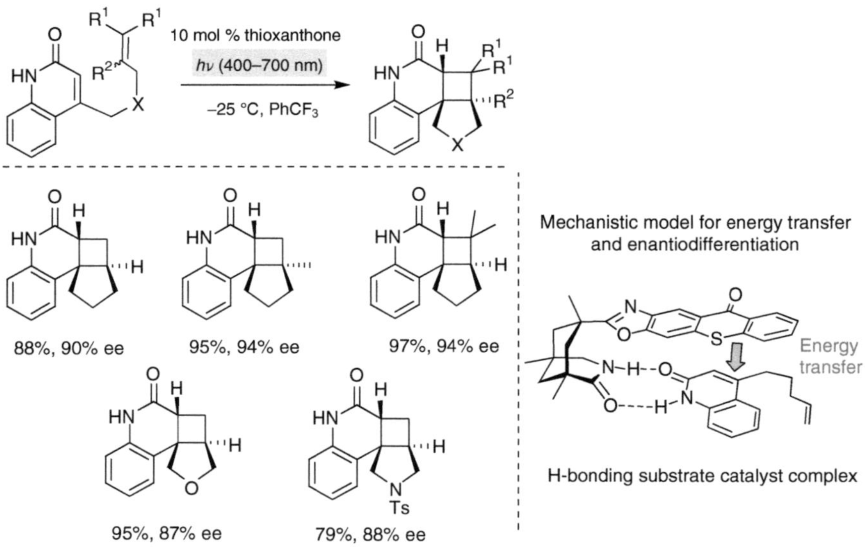

Scheme 6.4 Enantioselective intramolecular [2 + 2] photocycloaddition reactions with a chiral thioxanthone.

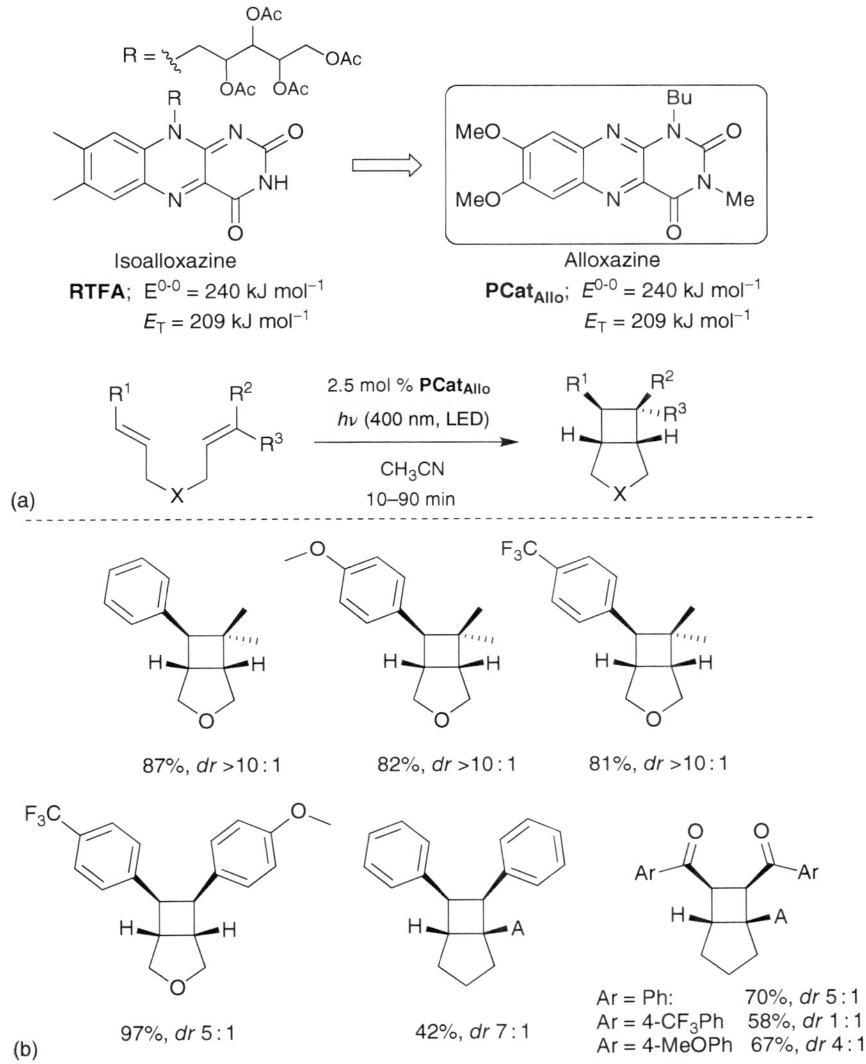

Scheme 6.5 Alloxazine-photocatalyst for the visible light, intramolecular [2 + 2]-photocycloaddition of dienes.

α,β-unsaturated carbonyl compounds, using riboflavin **RF** @ 402 nm irradiation as the triplet sensitizer (cf. E_T (*E*-cinnamoic ester) = 235 kJ mol^{-1}, E_T (*Z*-cinnamoic ester) 239 kJ mol^{-1}) [53]. Selective discrimination between substrate and product is supposed to be based on an exploitation of breaking the conjugation of the *Z*-isomer's π-system, which is enhanced by introducing a third olefin substituent in β-position to augment the A1,3-strain upon $E \rightarrow Z$-isomerization. This mechanistic hypothesis is supported by a comparative analysis of the isomerization using different organic and transition-metal-based photosensitizers of varying triplet energies. Accumulation of the *Z*-isomer could only be effected using sensitizers with E_T ranging between 180 and 255 kJ mol^{-1}, while both lower and higher values led to a significant decrease in selectivity. Interestingly, again, riboflavin here allows for an energy transfer-type reaction instead of the common photo-oxygenation reactions of sulfides, amines, and benzylic positions [32] (Scheme 6.6).

By far, more visible light promoted transformations follow an *electron transfer* pathway. As mentioned earlier, photoexcited species are both stronger oxidants and reductants as compared to the corresponding ground state catalyst. However, not only these electronically excited catalyst species may productively interact with a substrate but could also be quenched by a sacrificial electron donor or acceptor to generate ground state species of high reducing or oxidative power for a "dark reaction"-type interaction with substrates in net reductive or net oxidative reactions; depending on the substrates employed this may lead to the generation of radicals or radical ions. Redox-neutral reactions, which are highly desirable with respect to the development of sustainable atom-economic methods, may occur if both the excited and the quenched catalyst species are involved in productive steps.

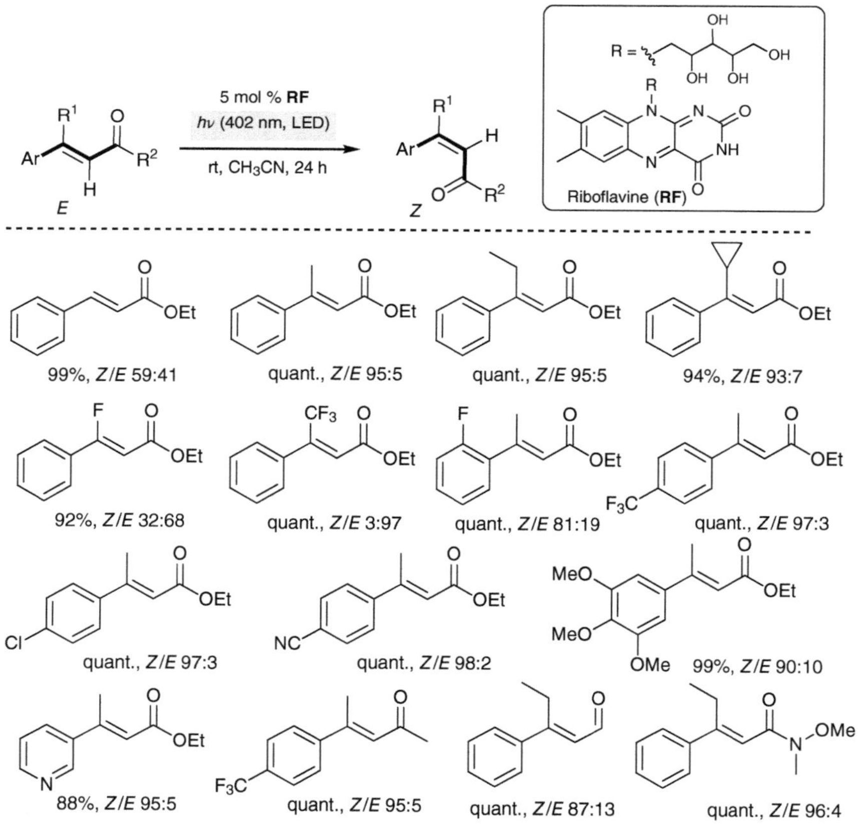

Scheme 6.6 Catalytic $E \rightarrow Z$ isomerization of activated olefins (for reported E/Z ratios IUPAC rules apply!).

6.2.2 Reductive Quenching of the Catalyst

Reductive quenching of the photocatalyst's excited state is a common pathway in many aerobic oxidations (presence of O_2 or air) where catalyst regeneration from its reduced species is achieved by reoxidation with oxygen upon formation of superoxide radical anions. These may accumulate or could also be involved in follow-up reactions (e.g., as shown in Scheme 6.45), including further oxidations, H-abstractions, or protonations. Alternatively, net reductive reactions can be achieved in the presence of sacrificial electron donors that range from simple tertiary amines to ascorbic acid or dihydropyridine derivatives, and so on, which often are supposed to play a dual role both as electron and hydrogen atom donors (Scheme 6.7).

Further, different photocatalyst classes (sorted by structure) that follow such a reductive quenching mechanism will be discussed with a selection of known applications.

6.2.2.1 Cyanoarenes

Several cyanoarenes (such as **DCB**, **DCN**, and **DCA**) have been widely used as exited state oxidants for a great variety of transformations, including amine α-functionalization [54], and have been reviewed extensively [33a,55]. Catalyst re-oxidation is commonly achieved by O_2 under aerobic conditions; the use of "electron relay reagents (ER)" such as methyviologen MV^{2+} or dimethoxynaphthalene (DMN) is known to accelerate the reaction [1d]. Recent applications have been disclosed by Pandey and coworkers to affect regioselective, direct benzylic CH activation for intramolecular cycloetherifications as well as for

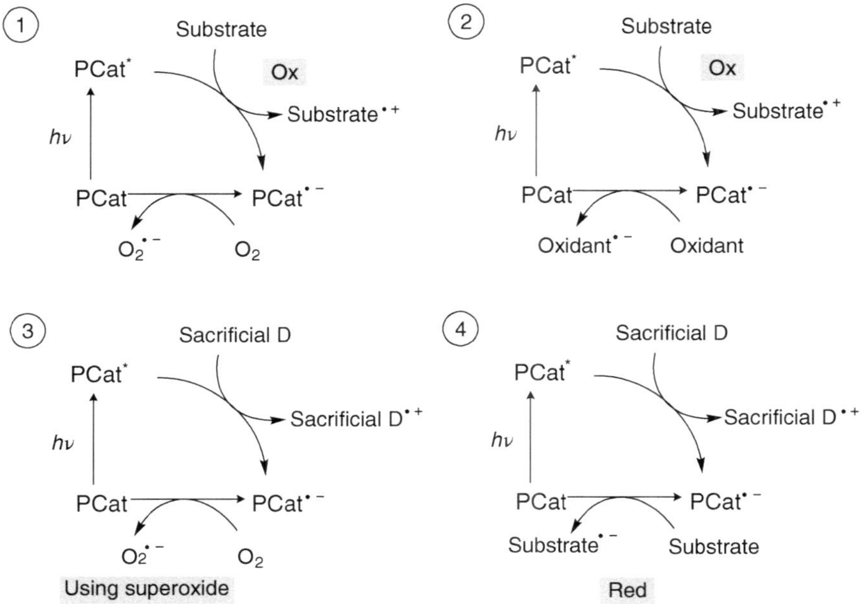

Scheme 6.7 Overview on possible pathways for reductive quenching catalysts.

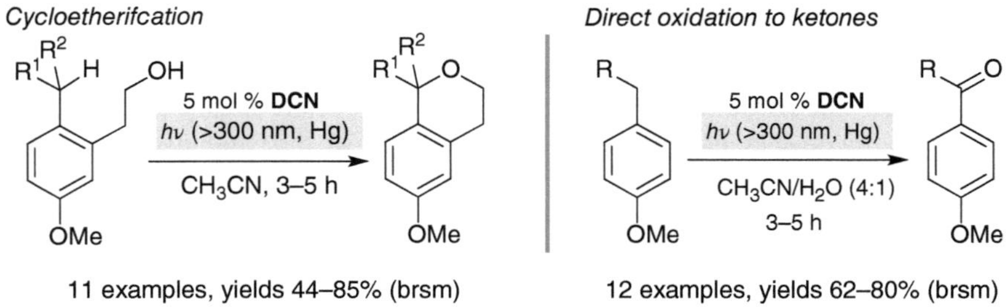

Scheme 6.8 Cyanoarene DCN-mediated photoredox cycloetherification and benzylic oxidation.

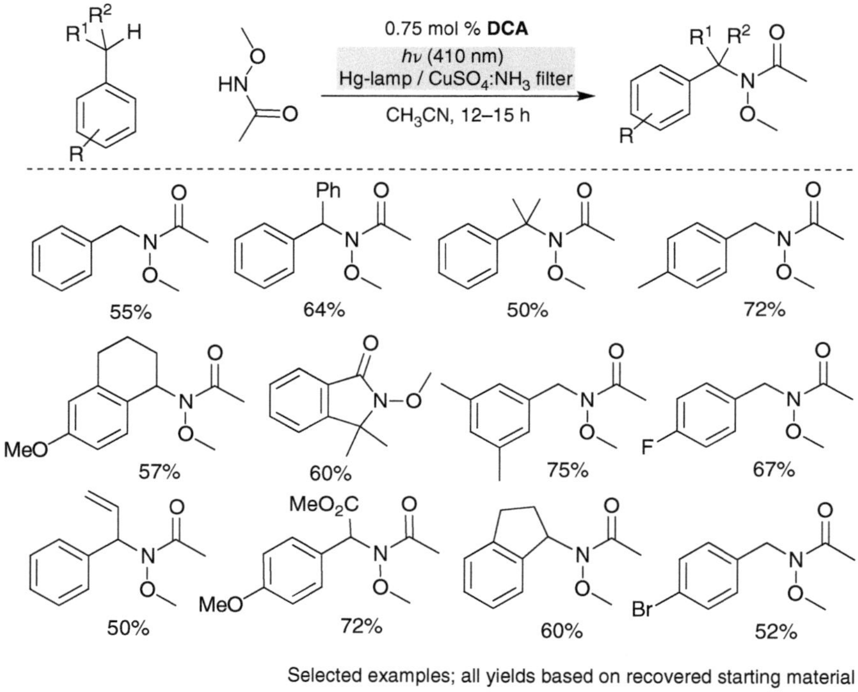

Scheme 6.9 DCA-mediated benzylic amination by cross-dehydrogenative coupling.

the synthesis of aryl ketones using **DCN** as organic photoredox catalyst with irradiation at $\lambda > 300$ nm; these transformations hence still require UV light [56] (Scheme 6.8).

With **DCA** as photo-oxidant and using visible light the same group could realize mild benzylic amination reactions in atom-economic cross-dehydrogenative coupling (CDC) using Weinreb amides as amination reagent [57] (Scheme 6.9).

6.2.2.2 Quinones

Quinones are not only well-suited ground state oxidants, but their excited (triplet) states even show strongly increased oxidative power. While excitation of DDQ and 9,10-anthraquinone (**AQ**) requires UV irradiation, substituted anthraquinones are useful visible light photocatalysts. Importantly, although photo-oxygenations both by production of superoxide radical anions or singlet oxygen seem to be predominant transformations, typical reactions of their excited triplet states are twofold, allowing either for H-atom transfer and/or for oxidative photoelectron transfer (PET) depending on the pH of the reaction mixture [58].

Itoh and coworkers built on the strong oxidative power to afford arene trifluoromethylation as well as pefluoralkylation; radicals were accessed by photooxidation of the corresponding sodium sulfinates in the presence of **2-CO$_2$H-AQ** and TFA as additive. Here, catalyst regeneration might involve the reduction of SO_2 to its radical anion [59] (Scheme 6.10).

Using molecular oxygen as the terminal oxidant, **2-Cl-AQ** was a competent catalyst for the Henry reaction of diverse nitroalkanes with *N*-aryl-tetrahydroisoquinolines (THIQs) [60] (Scheme 6.11).

An extension of this reaction was recently examined in the Brasholz lab, where, in the presence of 1,4-diamino anthraquinone (**1,5-A-AQ**) as well as K_3PO_4 as base tetracyclic 12-nitroindolo-[2,1-*a*] isoquinolines were formed upon subsequent 6π-electrocyclization and oxidation [61] (Scheme 6.12).

6.2.2.3 Cationic Dyes: Pyrylium, Quinolinium, and Acridinium Scaffolds

Even stronger oxidative power can be observed for cationic photocatalysts than for neutral sensitizers allowing for higher PET efficiency due to simple charge exchange when the radical cation of the donor is formed.

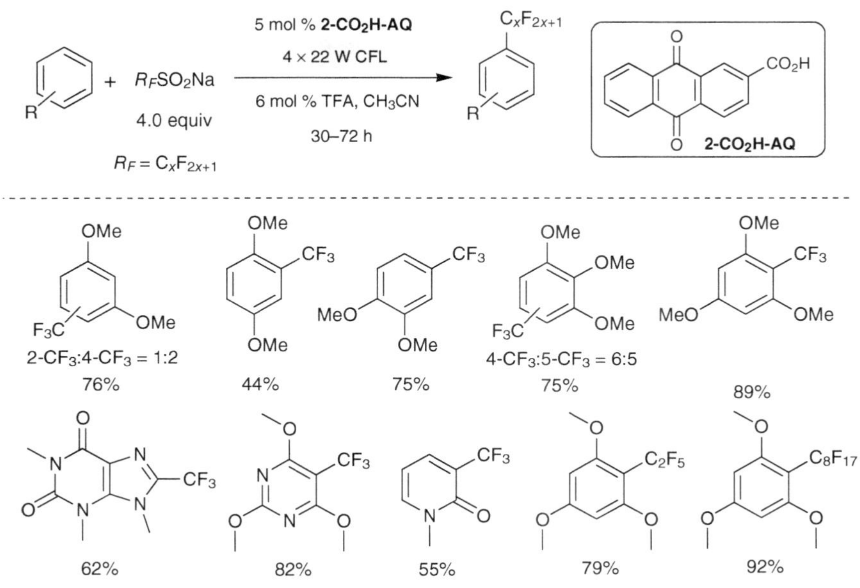

Scheme 6.10 Arene perfluoroalkylation by photoredox oxidation of corresponding Na sulfinates.

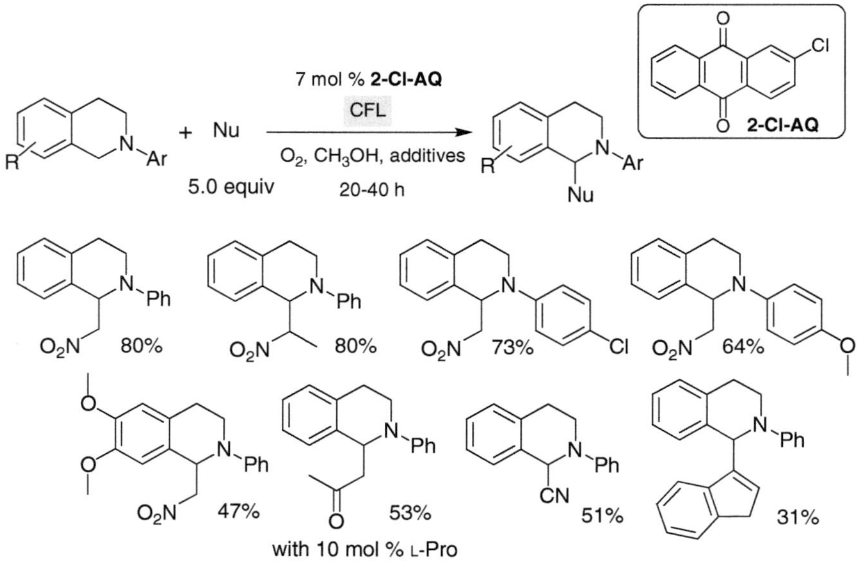

Scheme 6.11 Aerobic photo-oxidative C-functionalization of THIQs.

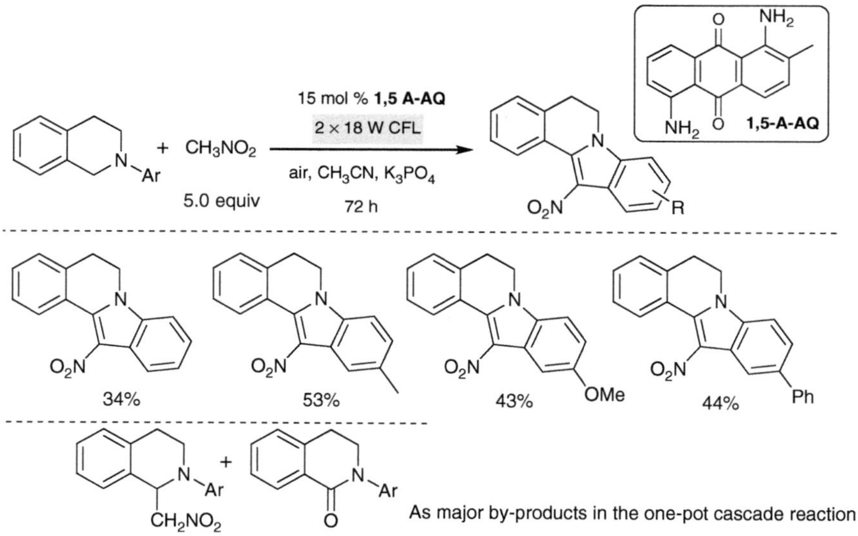

Scheme 6.12 Photocatalytic CDC coupling/dehydrogenation/6π-electrocyclization/oxidation cascade.

Due to their cationic nature, pyrylium salt photocatalysts are powerful excited state oxidants with high PET efficiency, which exhibit easily tunable redox potentials depending on the aryl substituents [31]. Since the 1980s, their advantages over common cyanoarenes with respect to longer wavelength irradiation (visible light) in lower catalyst concentrations have been noted. A seminal application example using 2,4,6-triphenylpyrylium tetrafluoroborate

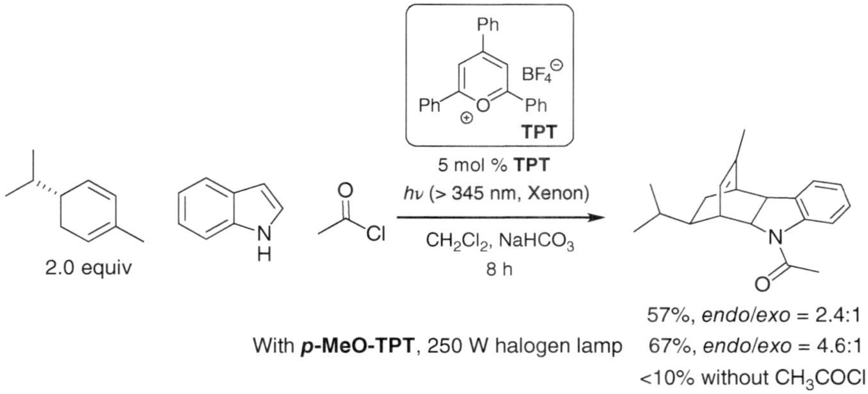

Scheme 6.13 Photocatalytic TPT-induced radical cation Diels–Alder reaction of indoles and dienes.

(**TPT**) is the photochemically initiated electron transfer, radical chain-type [62] Diels–Alder reaction [63] where electron-rich alkenes as their oxidized radical cations are good electron-poor dienophiles offering a means to redox umpolung-type reactions.

As an example, 1,3-cyclohexadienes undergo efficient radical cation Diels–Alder reaction with indoles as dienophiles, albeit only in the presence of an acylation reagent to prevent the carbazole products' unwanted oxidation by increasing their redox potential via *N*-acylation as compared to the parent indole substrate [33c] (Scheme 6.13).

Another way to circumvent potential problems caused by the (too) strong oxidative power of TPT-type catalysts was presented by the Nicewicz laboratory. To reduce undesired, photo-oxidatively induced cycloreversion of cyclobutane adducts generated by [2 + 2]-cycloadditive dimerization of electron-rich styrenes, they introduced polyarenes as "electron relay" (ER); thereby the excited state potential of the *p*-MeO-TPT photocatalyst can be "attenuated" in the presence of these co-sensitizers (naphthalene (1.61 V), anthracene (1.21 V)) [30a] (Scheme 6.14).

Due to a redox potential being a little higher than the starting alkene but lower than the cycloadduct, the radical cation of this electron relay is capable of oxidizing the alkene starting material, but minimizes the unproductive cycloreversion. The practical utility of this methodology was demonstrated by the direct synthesis of natural lignan magnosalin from *E*-asarone as well as by the fast access to an advanced intermediate within the synthesis of endiandrin.

Newman–Kwart rearrangement of *O*-aryl carbamothioates to the corresponding *S*-aryl carbamothioates can be achieved by photoinduced SET offering mild conditions as alternative to typical harsh thermal requirements for this transformation (200–300 °C) [64]. Starting with the oxidation of the thiocarbonyl moiety a broad range of thiophenol derivatives can be easily accessed, although larger *ortho*-substituents as well as *ortho*-alkenyl or aryl substitution inhibited the rearrangement (Scheme 6.15).

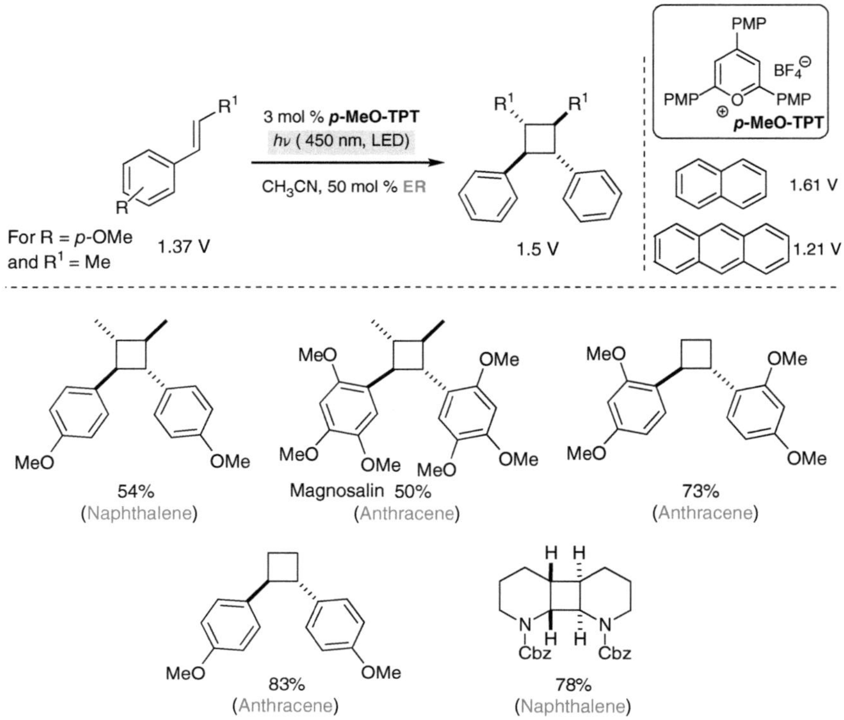

Scheme 6.14 Photodimerization to cyclobutanes by an electron relay attenuated photo-oxidation (vs. SCE).

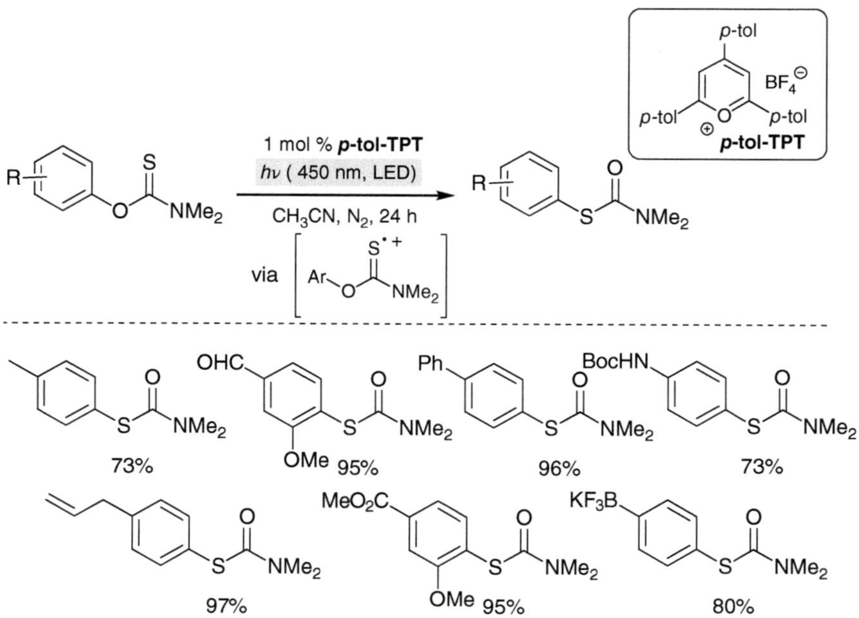

Scheme 6.15 Mild conditions for Newman–Kwart rearrangement by organic photoredox catalysis.

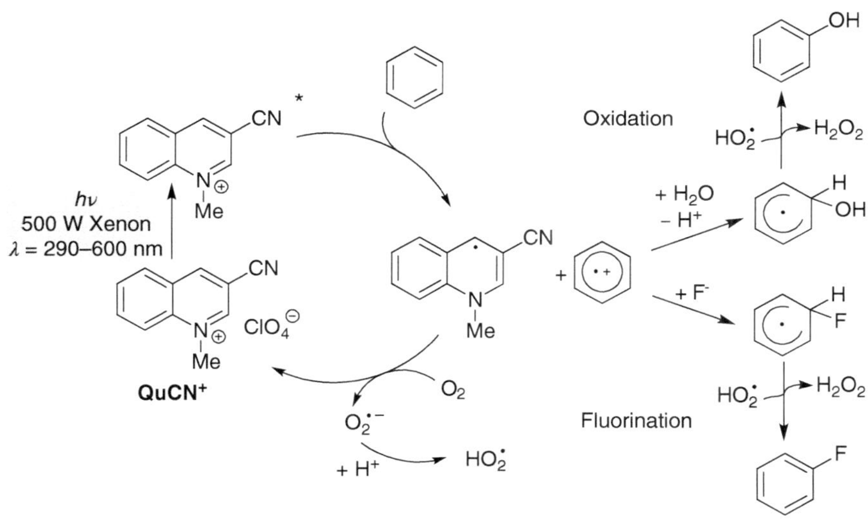

Scheme 6.16 Oxidative photocatalytic arene functionalization using $QuCN^+$ as catalyst.

The Fukuzumi group has recently developed a quinolinium catalyst (3-cyano-1-methylquinolinium perchlorate salt: **QuCN$^+$** ClO_4^-), which does not only allow for selective phenol formation upon benzene oxidation, but could also be used for monofluorination of benzenes when fluoride rather than water traps the intermediate radical arene cation [23b,c] (Scheme 6.16).

This approach is a valuable alternative to the conventional Balz–Schiemann protocol for converting anilines to fluorides. While these reactions require broad band irradiation with 500 W Xe-lamps ($\lambda > 290$ nm), a related phenanzinium catalyst (phenazine ethosulfate: **PES**) mediates the oxidative amidation of aromatic aldehydes aided by light derived from a household CFL lamp [65]. *In situ*-formed H_2O_2 during catalyst regeneration under the O_2 atmosphere is believed to oxidize the hemiaminal intermediate to the target amide (Scheme 6.17).

Intrigued by the remarkable redox capabilities of cationic aromatic scaffolds as visible light photocatalysts, Fukuzumi and coworkers also developed a mesityl acridinium catalyst (**Acr-Mes$^+$**), which, as a strong electron acceptor, displays great oxidative power assisted by the long-lived charge-separated excited state where the two aromatic planes are in an almost perpendicular orientation. As known for the pyrylium catalysts acridinium dyes are not prone to produce singlet oxygen and hence can be used for other diverse photo-oxidation reactions. A full review of all these methods is beyond the scope of this chapter, but has recently been summarized [30a]. The formation of dioxetanes from alkenes or anthracenes hence follows a SET-promoted cycloaddition pathway and is complementary to $^1O_2$ methods. The essential formation of arene radical cations also plays a key role for xylene oxidation as well as for the selective, oxidative monobromination of electron-rich arenes. Using O_2 as a terminal oxidant and HBr as the bromine source aryl bromides are proposed to be generated by interception

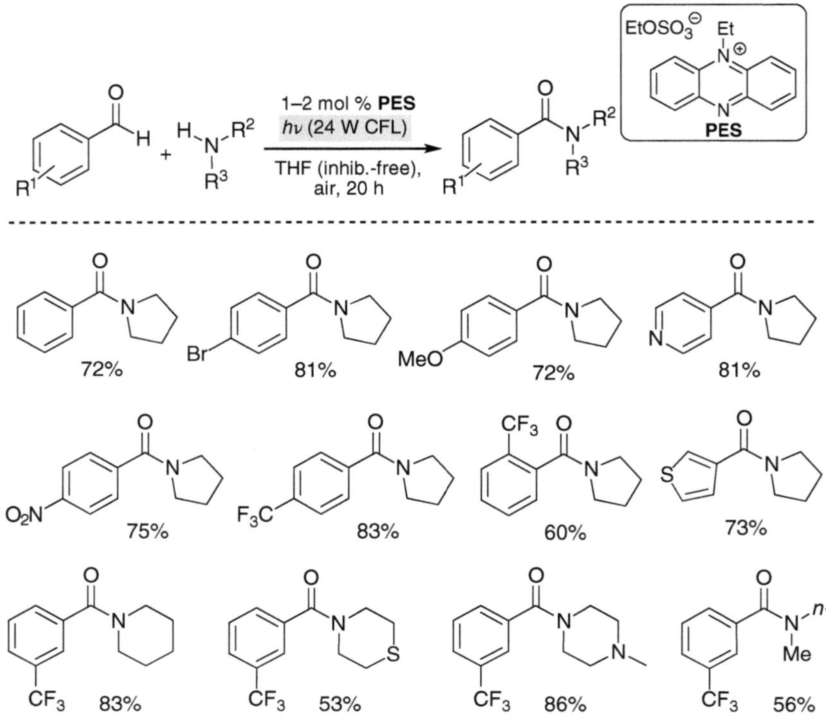

Scheme 6.17 Oxidative amidation of aromatic aldehydes.

of the arene radical cation intermediate with the bromide followed by a hydrogen atom transfer (HAT) (Scheme 6.18).

Using a modified acridinium catalyst, the Nicewicz group was able to greatly extend this aromatic C—H-functionalization to a direct arene amination protocol under aerobic conditions with H_2O_2 as the only by-product [66]. Both azoles and ammonium salts can be used as the nitrogen source in the presence of TEMPO as a key additive. Its beneficial role is proposed to be multifaceted: it may scavenge detrimental oxygen radical species and also could facilitate the final aromatization of the products (Scheme 6.19).

As an alternative approach to the previously described aerobic catalyst regeneration (and hence the concurrent formation of $O_2^{\bullet -}$: see Scheme 6.7, example 1) and the utilization of the quite unique oxidative power of Fukuzumi's catalyst **Acr-Mes**$^+$ was presented by Nicewicz and coworkers in a series of alkene hydrofunctionalization reactions [23a] [67]. For both intra- and intermolecular anti-Markovnikov alkene additions the regeneration of the reduced catalyst is coupled to the redox cycle of a redox-active HAT mediator (HAT catalyst) to achieve not only hydroalkoxylation and hydroamination reactions, but also the direct anti-Markovnikov addition of mineral acids, such as HCl, HF, and also phosphoric and sulfonic acids. Following the oxidation of the

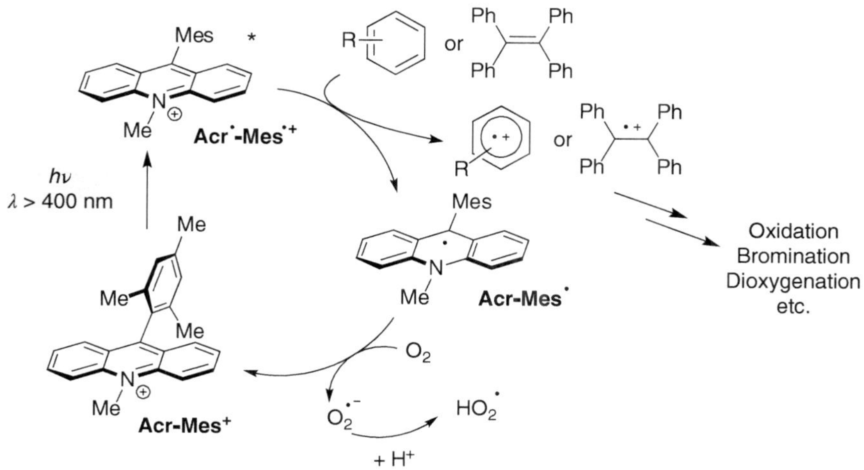

Scheme 6.18 Proposed general mechanism for oxidative transformations with 9-mesityl-10-methylacridinium photocatalysts.

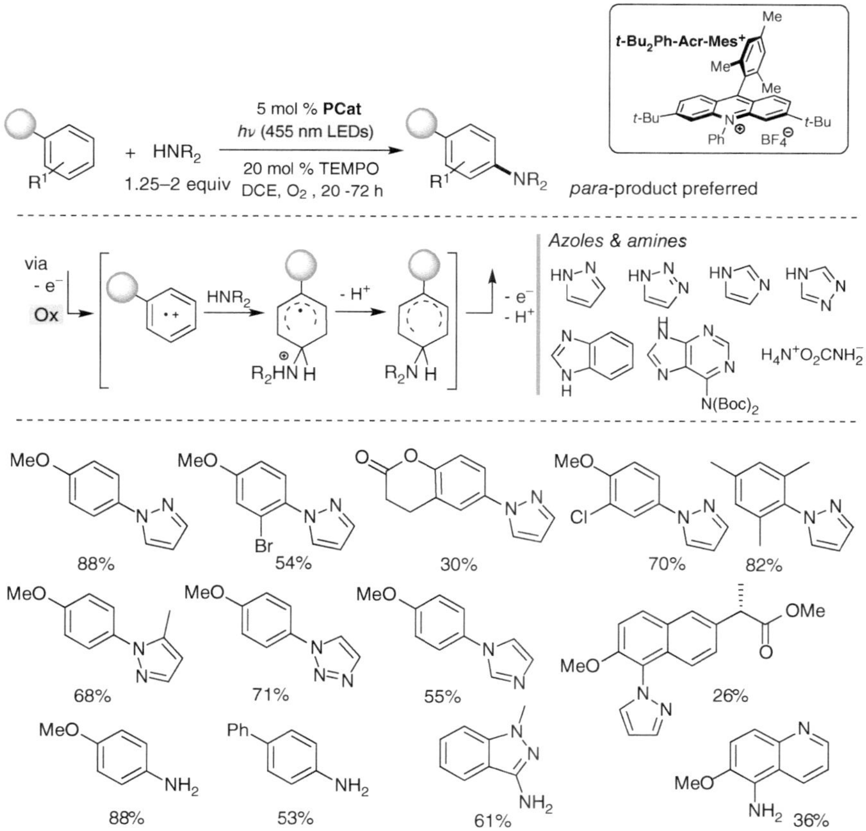

Scheme 6.19 Site-selective C—H amination of arenes with modified Fukuzumi photocatalyst.

styrenes to their radical cation the dual function of the HAT catalyst includes accepting an electron to reoxidize the reduced photocatalyst, but also – upon protonation – to transfer a hydrogen atom to the radical intermediate which is generated upon interception of a nucleophile and thereby to complete the formation of the desired product. Apart from α-C—H acidic compounds (e.g. phenylmalononitrile, **PMN**) sulfinic acids, thiols, or disulfides (homolysis of the weak S—S-bond can also yield the active catalytic thiyl species) as S—H donors have proven to be versatile redox-active HAT catalysts (Scheme 6.20).

Along these lines the Nicewicz group has demonstrated the usefulness of this concept for a broad set of heterocycle syntheses [68], including 5- to 7-membered cyclic ethers [69] and lactones [70] and their nitrogen counterparts, such as pyrrolidines [71] and γ-lactams [72] were furnished for *N*-tosyl and *N*-Boc derivatives avoiding a detrimental oxidation of the amine substrates (Scheme 6.21).

A direct intermolecular amination strategy takes advantage of the common catalyst system with $(PhS)_2$ as HAT catalyst and hence provides a simple access to phenylethylamines using triflylamides $TfNH_2$ as the nitrogen source in the presence of a base [8d] (Scheme 6.22).

The anti-Markovnikov addition of mineralic acids could be performed more conveniently using the corresponding and easy to handle hydrohalide salts, even allowing for hydrofluorination with triethylamine trihydofluoride $NEt_3 \times 3HF$ as commercially available, suitable fluoride source; here, the optimal HAT donor was *p*-nitrodisulfide [73]. Sulfonates and phosphates could also be introduced by applying this methodology (Scheme 6.23).

The strong oxidative power of the acridinium type catalysts is also well suited to generate radicals from other electron-rich precursors, including sulfinates and carboxylates as well as potassium trifluoroborates.

Hydrotrifluoromethylation of alkenes can be achieved with Langlois reagent CF_3SO_2Na (under extrusion of SO_2) in the presence of trifluoroethanol TFE and a thiophenol HAT catalyst, securing the reoxidation of the catalyst and product formation by HAT to the intermediary radical [74] (Scheme 6.24).

Based on seminal reports of Yoshimi *et al.* [75] using a DCN/phenanthrene catalyst system the irradiation of carboxylates (often prepared *in situ*) in the presence of Fukuzumi's catalyst allows for a comparable oxidative generation of radicals upon decarboxylation (expelling CO_2) [76].

While Wallentin used this approach for the selective decarboxylation of *N*-protected amino acids and phenyl acetic acids in the presence of a base and $(4\text{-ClPhS})_2$ as HAT catalyst in the presence of a *N*-Me acridinium photocatalyst [77], the Nicewicz group found more general conditions for the decarboxylation applying the more stable *N*-Phe acridinium catalyst (**Phe-Acr-Mes$^+$**) [78]. They also revealed the strong dependence of this transformation on the reaction conditions with TFE present as solvent being key to the success of this reaction. Decarboxylation under mild conditions was achieved for both mono- and dicarboxylic acids, such as malonates (Schemes 6.25 and 6.26).

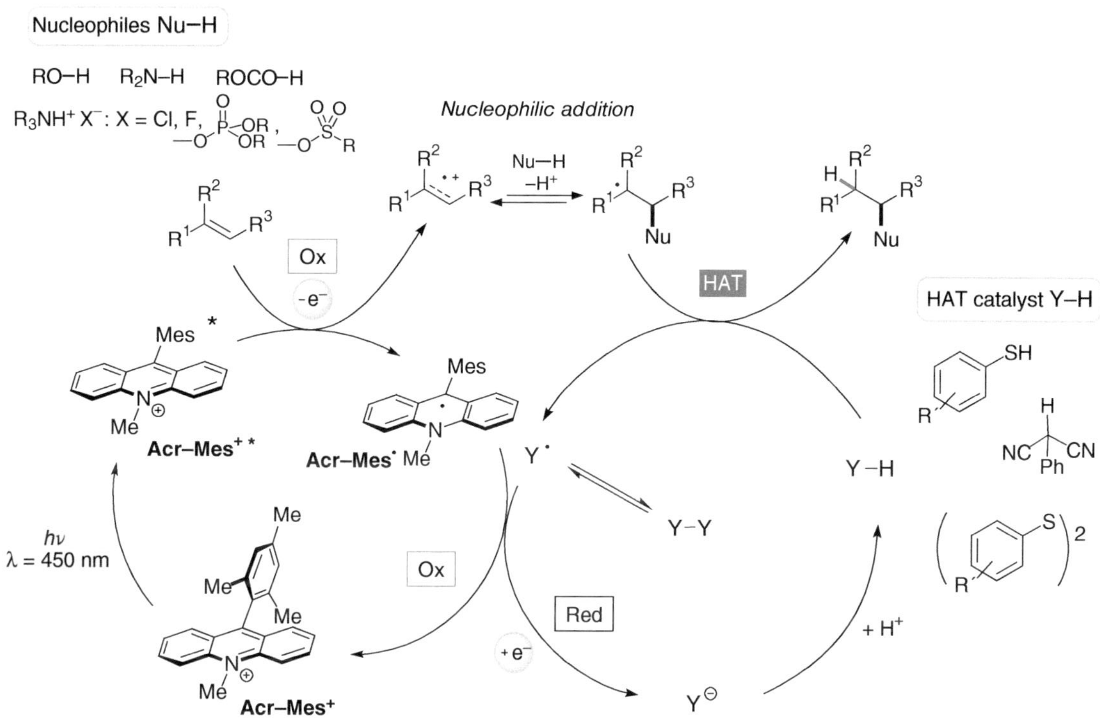

Scheme 6.20 General mechanism for anti-Markovnikov functionalization of alkenes.

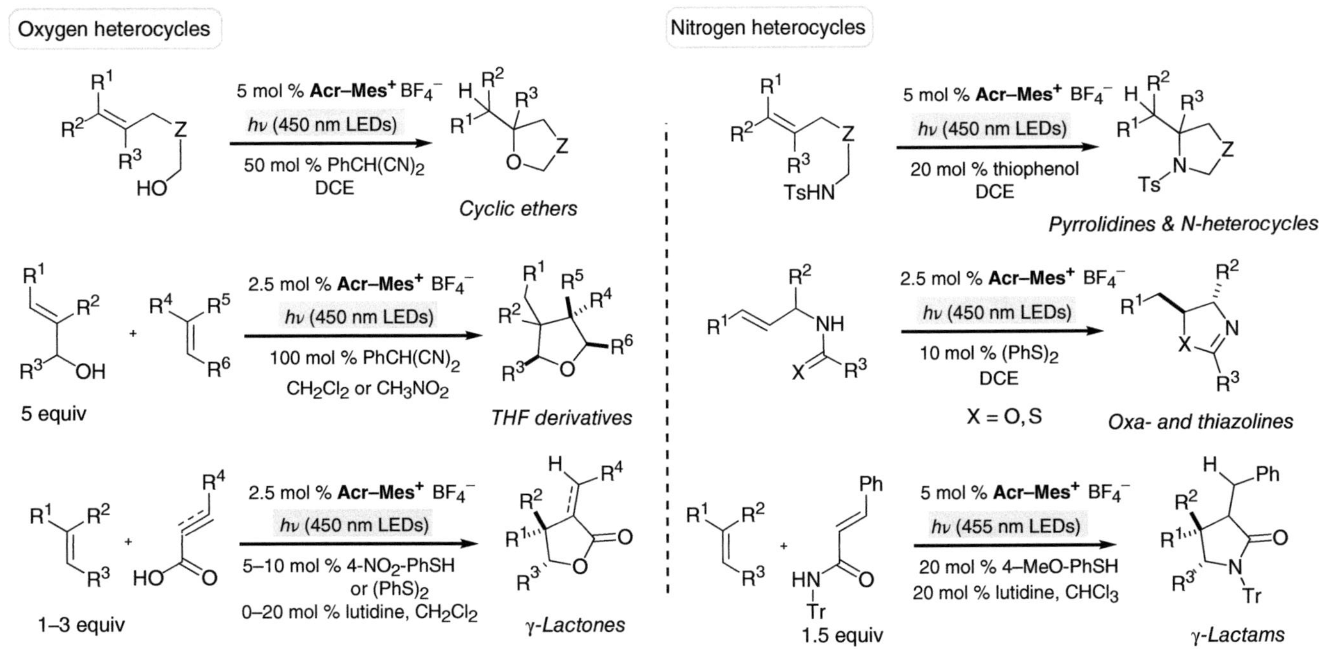

Scheme 6.21 Inter- and intramolecular Markovnikov hydrofunctionalization for heterocycle synthesis.

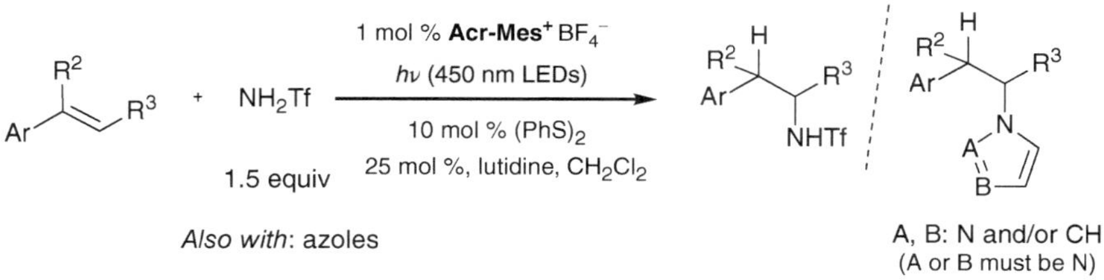

Scheme 6.22 Access to phenylethylamines via anti-Markovnikov hydroamination of styrene derivatives.

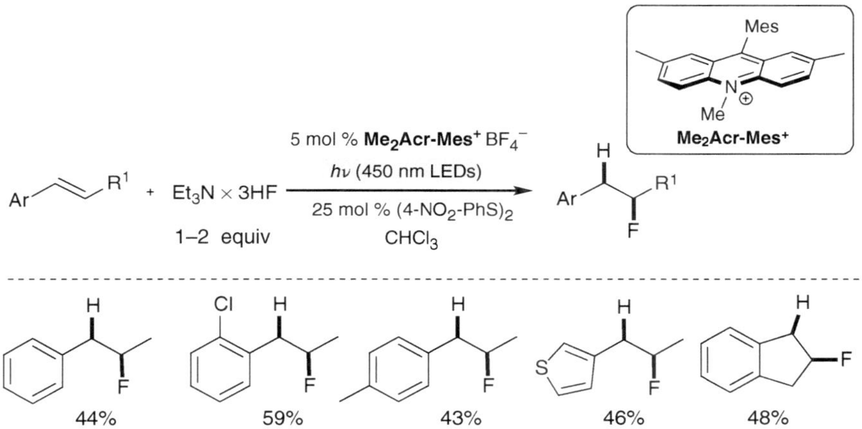

Scheme 6.23 Anti-Markovnikov hydrofluorination of styrenes.

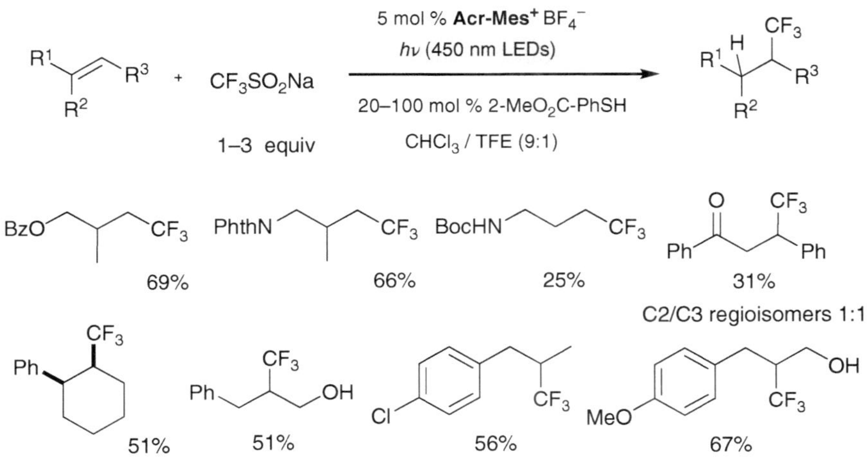

Scheme 6.24 Catalytic sodium hydrotrifluoromethylation of unactivated alkenes and styrenes with trifluoromethanesulfinate salt (Langlois reagent).

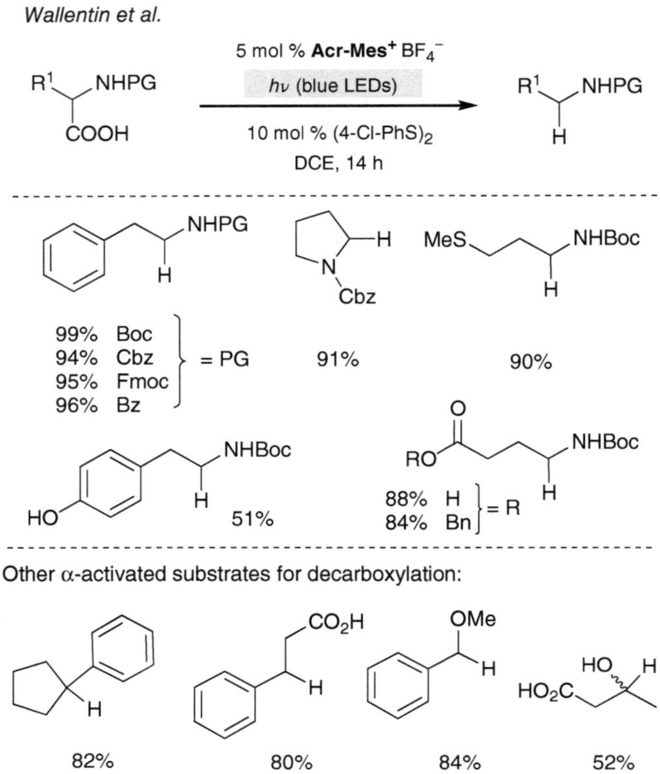

Scheme 6.25 Photocatalytic reductive decarboxylation of α-activated carboxylic acids.

A protocol for decarboxylative fluorination uses selectfluor to trap the stabilized radicals by fluor atom transfer. The remaining radical cation of the fluorinating reagent is proposed to regenerate the catalyst [79] (Scheme 6.27).

Akita utilized the photodecarboxylation to generate stabilized radicals (*sec.* and *tert.*) for the addition to strongly activated (electron-poor) olefins [80]. They could also extend the concept of oxidative radical generation with **Acr-Mes**$^+$ as catalyst to potassium trifluoroborates to be trapped with electron-deficient alkenes. The intermediate radical is supposed to be reduced by the reduced ground state species of the catalyst to render the whole transformation redox-neutral (Scheme 6.28).

Expulsion of CO_2 was inhibited for arylcarboxylic acids in favor of a dehydrogenative lactonization to benzo-3,4-coumarins [81]. The initially generated benzoyloxy radical forms the aryl radical intermediate in a 6-*endo*-trig cyclization. The excess of ammonium persulfate realizes catalyst turnover and may also facilitate the final aromatization (Scheme 6.29).

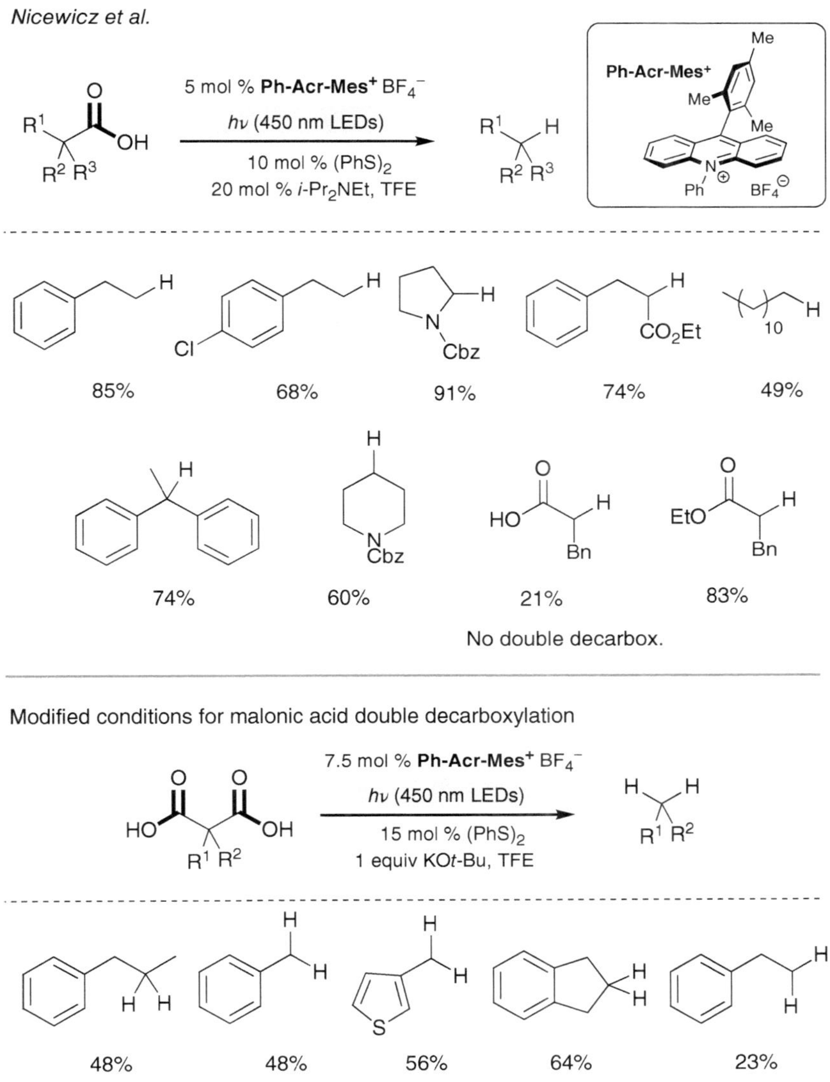

Scheme 6.26 Hydrodecarboxylation of carboxylic and malonic acid derivatives.

Starting with 2*H*-azirines their photo-oxidation with **Acr-Mes**$^+$ induces ring opening to a distonic radical cation capable of performing a formal [3 + 2] cycloaddition with activated alkynes and generates tetra-substituted pyrrols [82] (Scheme 6.30).

In the presence of aldehydes instead of alkynes the distonic azaallenylradical cation is intercepted by the aldehyde's nucleophilic attack; subsequent radical cyclization and reduction allowing for catalyst regeneration leads to 2,5-dihydrooxazoles [83].

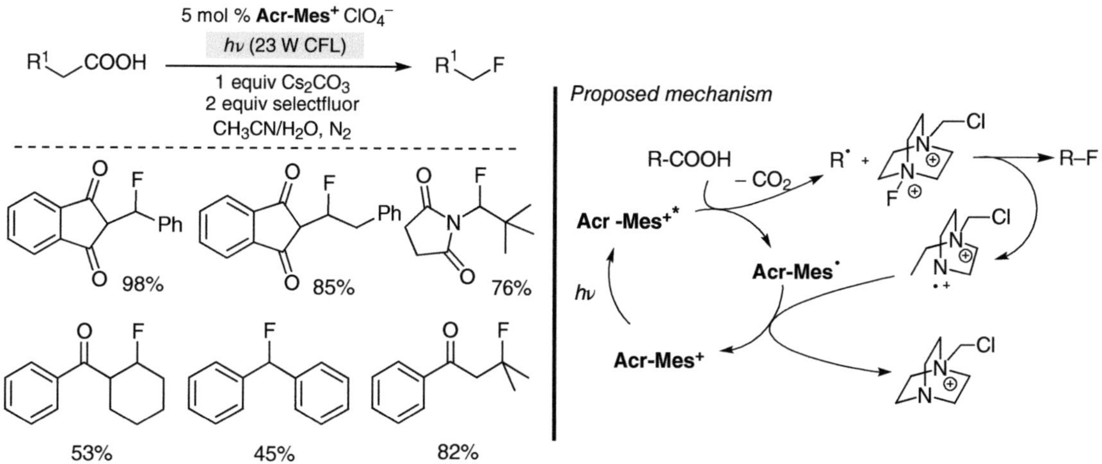

Scheme 6.27 Selected examples for the decarboxylative fluorination of aliphatic carboxylic acids and mechanistic proposal.

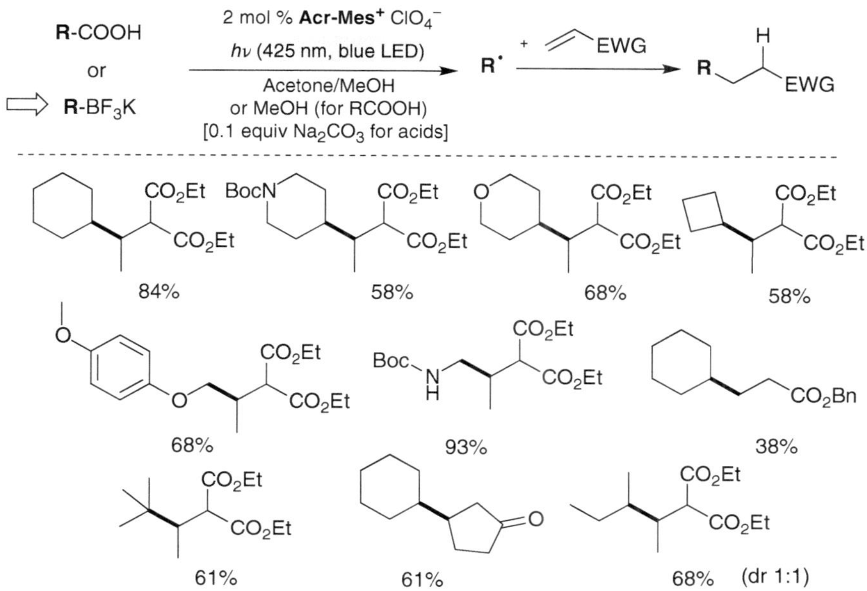

Scheme 6.28 Oxidative radical generation from trifluoroborates and carboxylates for C—C-coupling reactions.

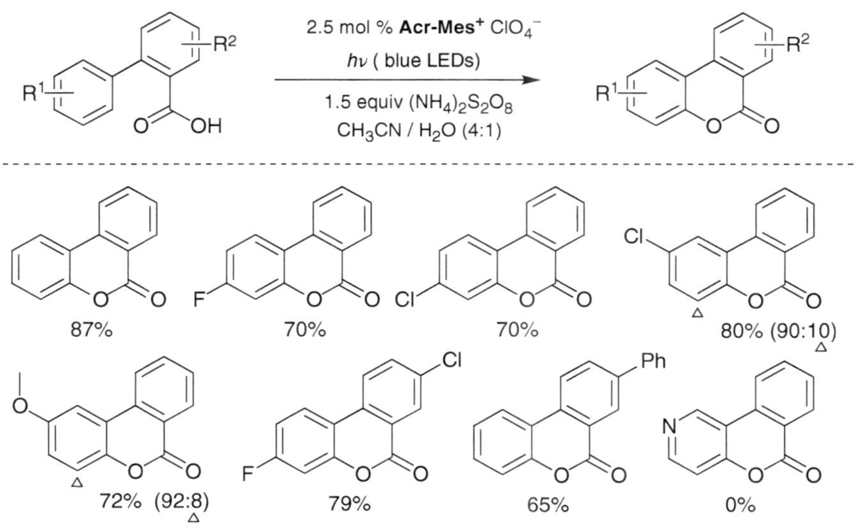

Scheme 6.29 Oxidative dehydrogenative lactonization of 2-aryl benzoic acids (Δ flags alternative regioisomers).

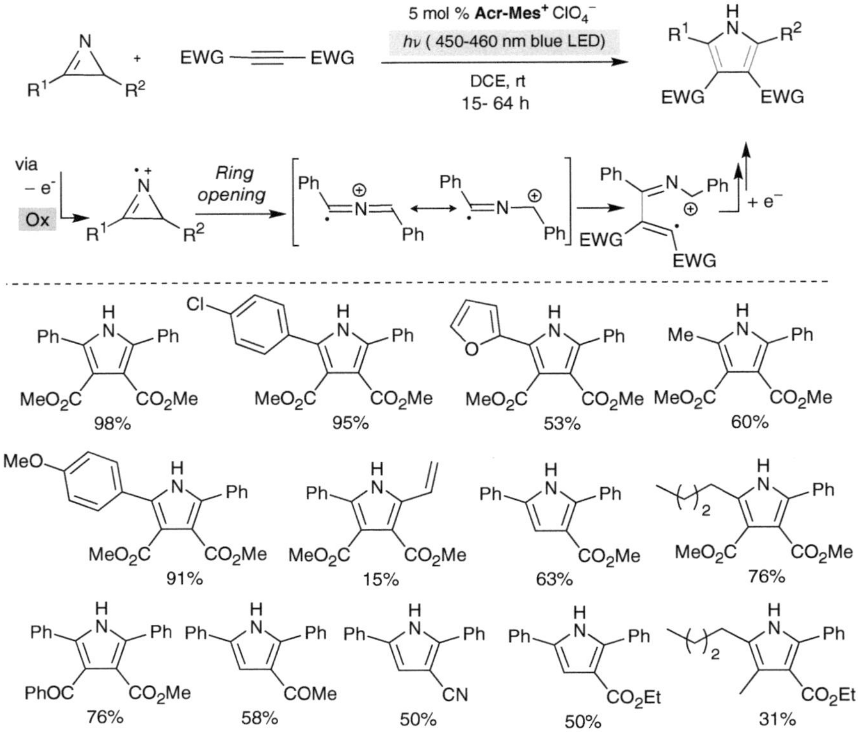

Scheme 6.30 Photocatalytically induced formal [3 + 2]-cycloaddition for the synthesis of pyrroles.

6.2.2.4 Xanthene Dyes and Further Aromatic Scaffolds

Broadly applicable organic dyes, such as xanthene-type and related dyes, and also phenothiazinium (methyleneblue) and perylene dyes, are also well known to function as versatile photo-oxidants, followed by their *in situ* regeneration with molecular O_2.

The oxidative transformation of tetrahydroisoquinolines (THIQs) [84] in the presence of various nucleophiles has evolved to a kind of a benchmark reaction for testing the performance of photocatalysts, mostly employing aerobic conditions. The generation of iminium ions from the initially formed α-amino radical via a deprotonation/oxidation sequence or HAT can be affected by a great variety of organic (but also metal-organic) photocatalysts. Common *C*-nucleophiles are nitroalkanes, malonates [85], or isonitriles, but also electron-rich alkenes and heteroarenes, such as silyl enol ethers and indoles; C—C bond connection with iminium ions could also be achieved with the silylated reagents TMSCN or $TMSCF_3$ [86]. Other tertiary aryl amines can undergo similar reactions. Reactions of the nucleophilic α-amino radical are mainly described for metal–organic photocatalysts [84, 87] (Scheme 6.31).

As part of a multicomponent approach using rose bengal (**RB**) as photocatalyst the reaction of dimethylanilines and isocyanides affords the corresponding α-amino acids in an oxidative Ugi reaction [88] (Scheme 6.32).

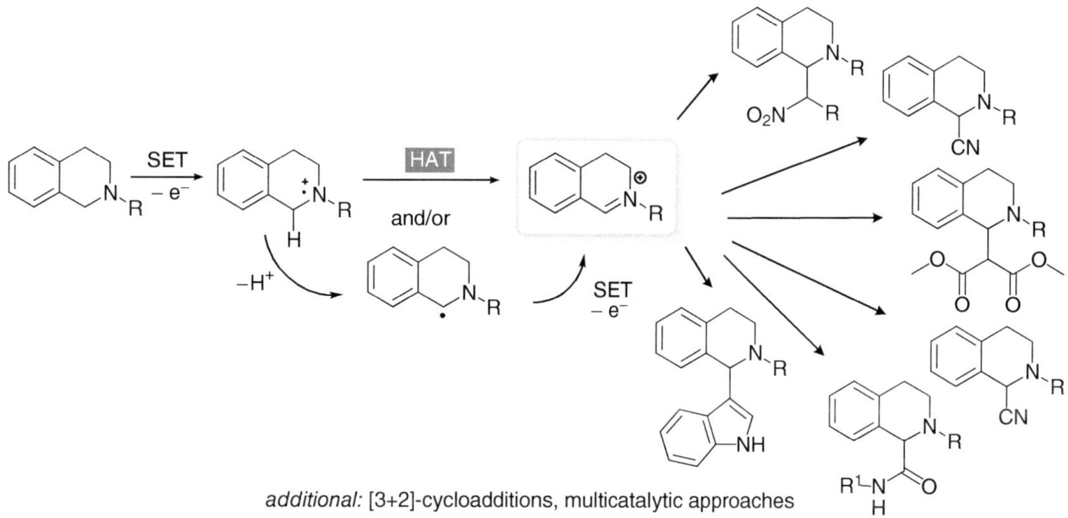

Scheme 6.31 Overview C—C-bond forming reactions involving THIQs (tetrahydroisoquinolines).

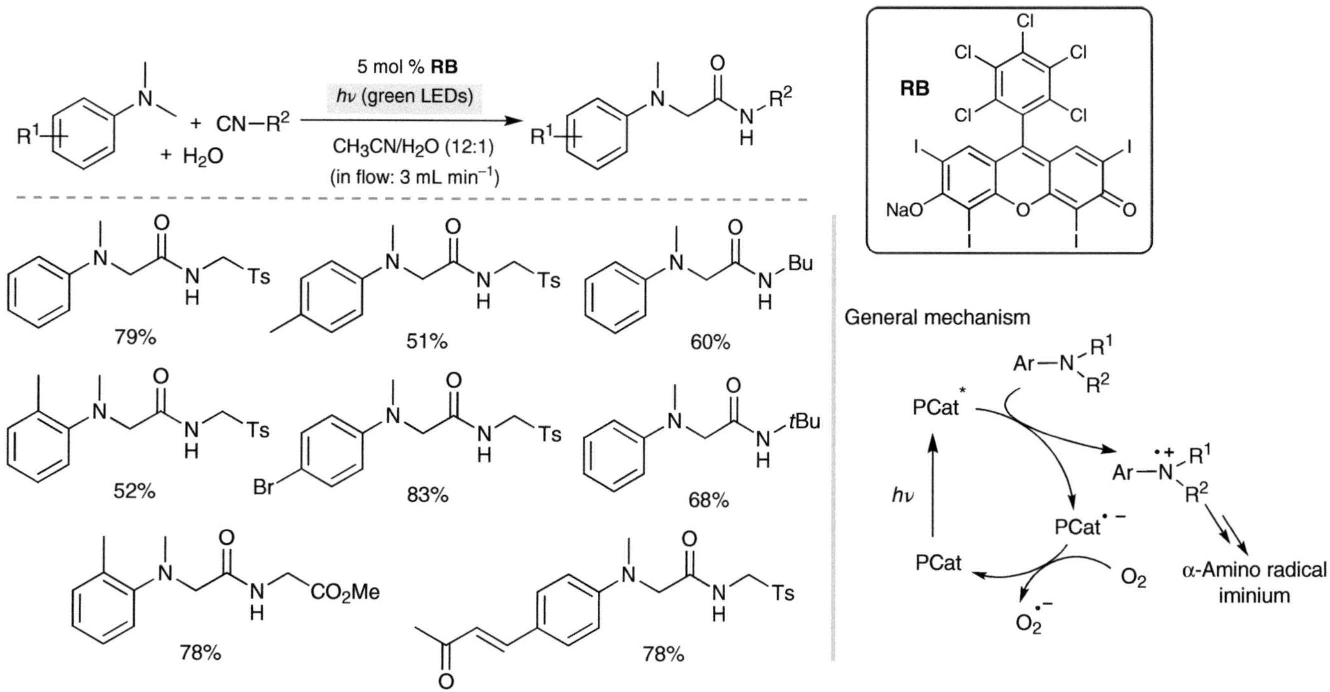

Scheme 6.32 Photocatalytic Ugi reaction with dimethyl anilines.

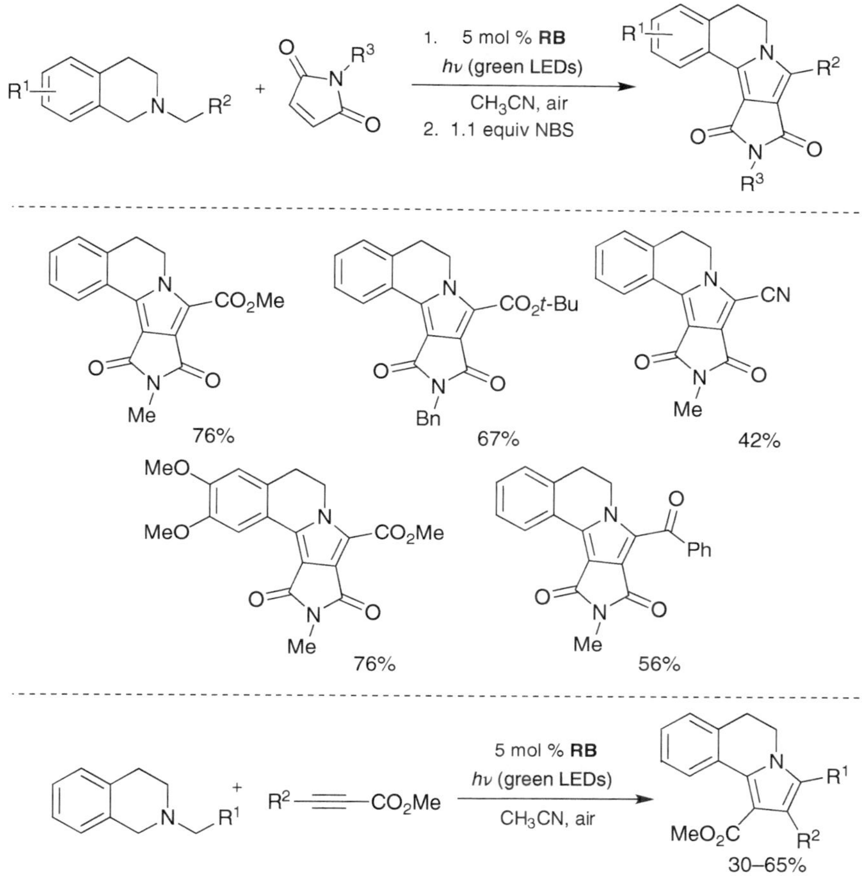

Scheme 6.33 Synthesis of pyrroloisoquinolines via an oxidation/[3 + 2] cycloaddition/aromatization cascade.

Additionally, heteroatom nucleophiles such as phosphites [85] or amines [89] are also competent substrates for transformation of tertiary amine-derived iminium ion species. Approaches to include this photocatalytic oxidative, iminium generation into multicatalytic processes are discussed in Section 6.2.4.3.

The *in situ* generation of 1,3-dipoles from *N*-alkylated THIQs is a highly attractive approach to yield pyrrolo-[2,1-*a*]-isoquinoline scaffolds, which are common structural elements in natural products, in a [3 + 2]-cycloaddition/aromatization sequence triggered by the later-stage addition of NBS [90]. Electron-poor alkenes such as *N*-aryl maleimides and electron-deficient alkynes are competent substrates for this transformation (Scheme 6.33).

Subjecting *N*-dialkyl anilines to related photoredox conditions in the presence of maleimides affords tricyclic tetrahydroquinolines in a radical addition/cyclization sequence; [91] final aromatization may occur upon either a HAT or an oxidation/deprotonation step.

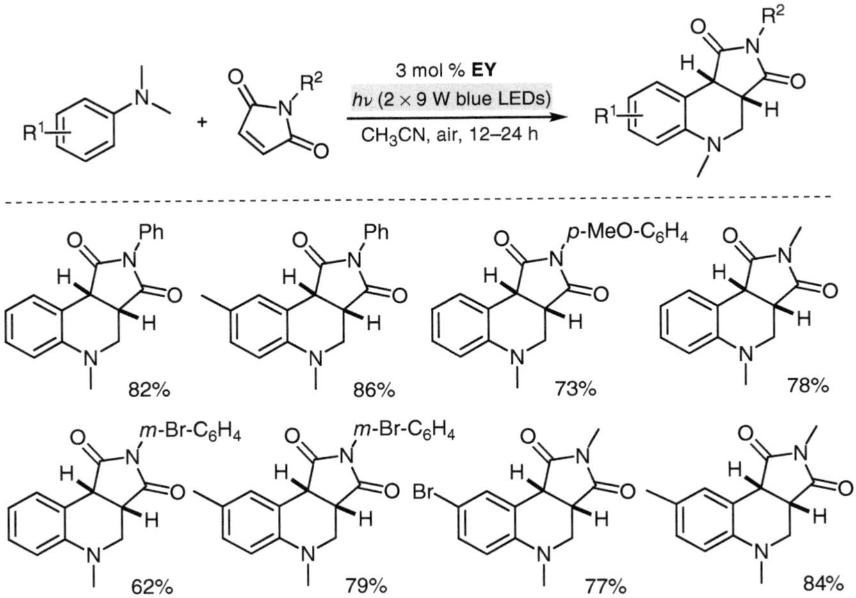

Scheme 6.34 Eosin Y catalyzed, aerobic oxidative cyclization of *N,N*-dimethylanilines with maleimides.

A variant for the iminium generation and a number of subsequent reactions has been contributed by Tan and coworkers, who employed *N*-aryl-α-amino acids that can also yield α-amino radicals (and hence also iminium ions by further oxidation) from the oxidatively generated amino radical cations and their subsequent decarboxylation in the presence of fluorescein (**FLS**) as organic photoredox catalyst [92]. As described earlier (see Scheme 6.34), in the presence of maleimides as reaction partner an annulation reaction is observed yielding tricyclic quinoline-type products.

Improved results for α-amino functionalization triggered by photocatalytic oxidation can be obtained applying so-called flow conditions [88, 93]. While not only light penetration and hence the excitation of the photocatalyst is improved as compared to batch reactors, potential light-mediated product degradation can also be minimized by precise reaction control in flow systems. Moreover, possible automatization [94] and high reproducibility as well as a facilitated and predictable scale-up to increase productivity rank among the advantages of continuous flow-methodologies.

Photoredox-promoted generation of symmetrical disulfides from thiols in the presence of base using either a batch or a flow set-up under aerobic conditions was recently reported by Noël and co workers [95]. Here, eosin Y (**EY**) clearly proved to be superior as compared to other organic and organometallic photocatalysts (Scheme 6.35).

Rhodamine B (**RhodB**) outperformed related organic dyes as photocatalyst for the oxidative generation of *P*-centered radicals from secondary phosphine oxides (in equilibrium with their phosphinic acid tautomer). Addition to terminal

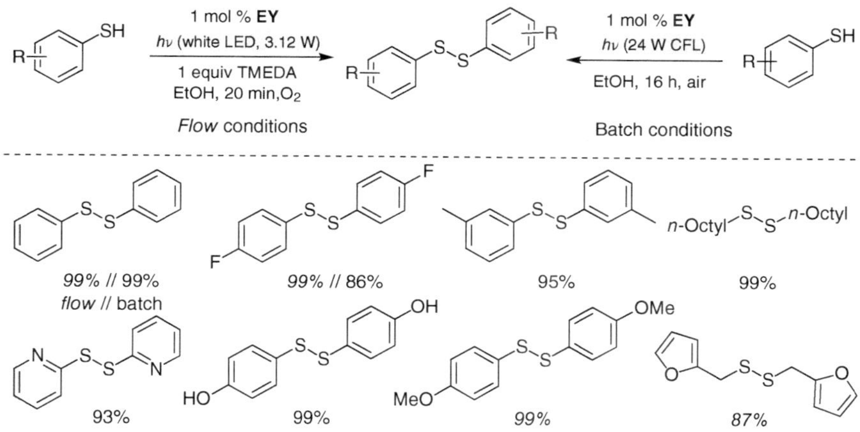

Scheme 6.35 Eosin Y as photocatalyst for batch and flow formation of disulfides from corresponding thiols.

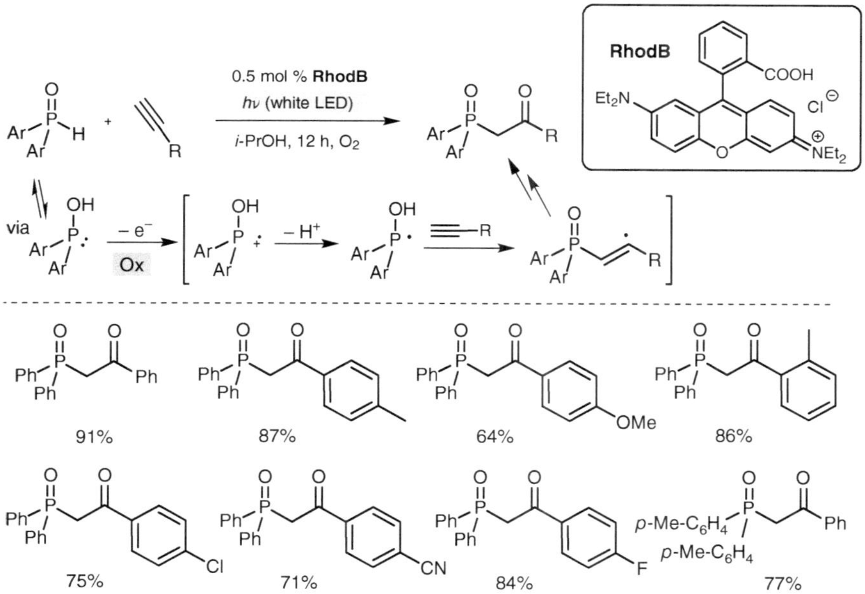

Scheme 6.36 Rhodamine B (**RhodB**) promoted oxidative phosphinylation of aromatic alkynes.

alkynes under aerobic conditions gives rise to a wide range of α-keto-phosphine oxides [96] (Scheme 6.36).

Rose bengal (**RB**) has proven to be the catalyst of choice for the photo-oxidative generation of acylnitroso intermediates from benzylhydroxy carbamate precursors, which then undergo an acylnitroso ene reaction with a wide range of different alkenes [97].

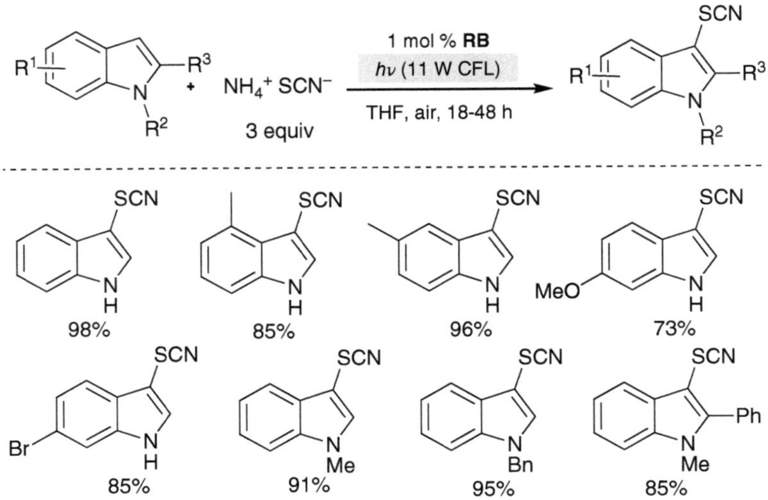

Scheme 6.37 Aerobic C-3 thiocyanation of indoles with rose bengal (**RB**) as photoredox catalyst.

The oxidation of ammonium thiocyanates was also achieved by excited state rose bengal (**RB**) (as proven by Stern–Volmer type quenching studies) and allows for the convenient aerobic 3-thiocyanation of indoles [98] (Scheme 6.37).

In a related process, the generation of radical cations from thioamides upon the reductive quenching of Eosin was used by Yadav *et al.* for the desulfurization to amides [99]. If primary amides are employed, these compounds undergo a cyclodesulfurization to 1,2,4-thiadiazoles [100].

Oxidative radical generation from aryl and alkyl hydrazines has recently been demonstrated to be possible in the context of phenanthridine syntheses starting from 2-isocyanobiphenyls with eosin B **EB** as photoredox catalyst. Three consecutive eosin-mediated photo-oxidative steps are assumed to gain access to the attacking radical upon N_2 extrusion [101] (Scheme 6.38).

Polyene cyclizations with alcohols or 1,3-dicarbonyl compounds as terminating group have been realized in HFIP upon initial radical cation formation [102]. As photocatalyst recycling in this process is proposed to take place during the reduction of the cyclized intermediate, the full transformation is net redox-neutral (Scheme 6.39).

Tetraethylenediamine (TMEDA) can be used as a C1-formylation reagent [103]. Initial photo-oxidative iminium generation (catalyzed by a reductive quench of excited **RB**) from TMEDA sets the stage for an electrophilic aromatic substitution with electron-rich (hetero)arenes, such as indoles. A second oxidation of the aryl-attached TMEDA fragment, which preferably takes place in the benzylic position, followed by hydrolysis of this iminium part yields the formylated product and an equivalent of monodemethylated TMEDA (Scheme 6.40).

While current literature seems to suggest that xanthene dyes and structurally related organic chromophores play a superior role for the oxidative, visible light

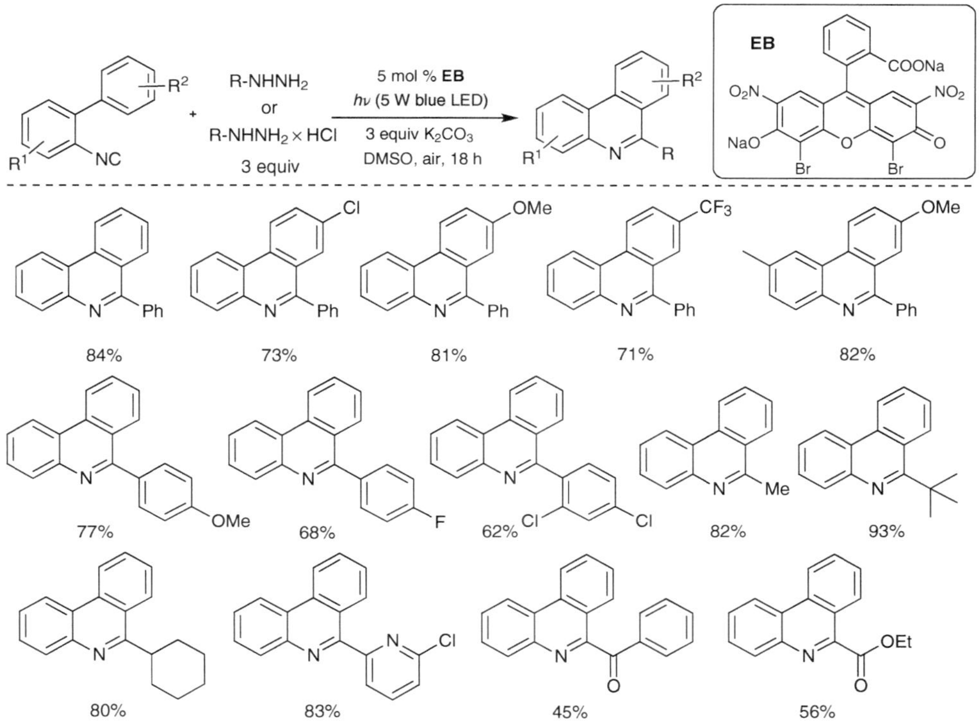

Scheme 6.38 Eosin B **EB**-mediated cyclization of 2-isocyanobiphenyls with hydrazines as radical source to 6-substituted phenanthridines.

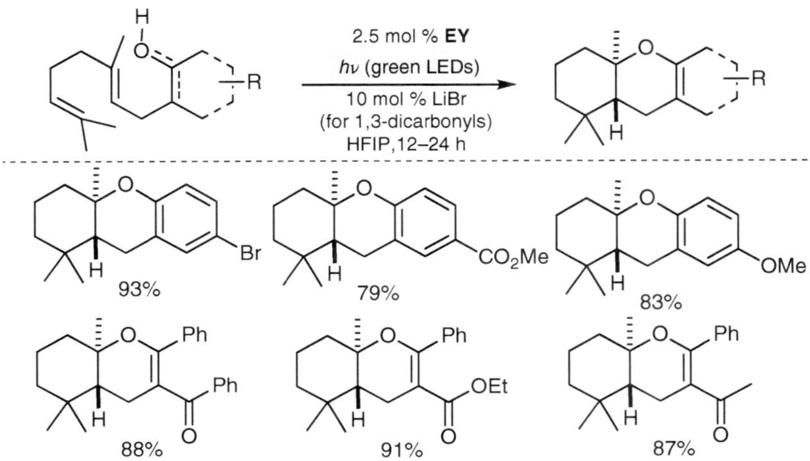

Scheme 6.39 Polyene cascade cyclization with nucleophilic termination.

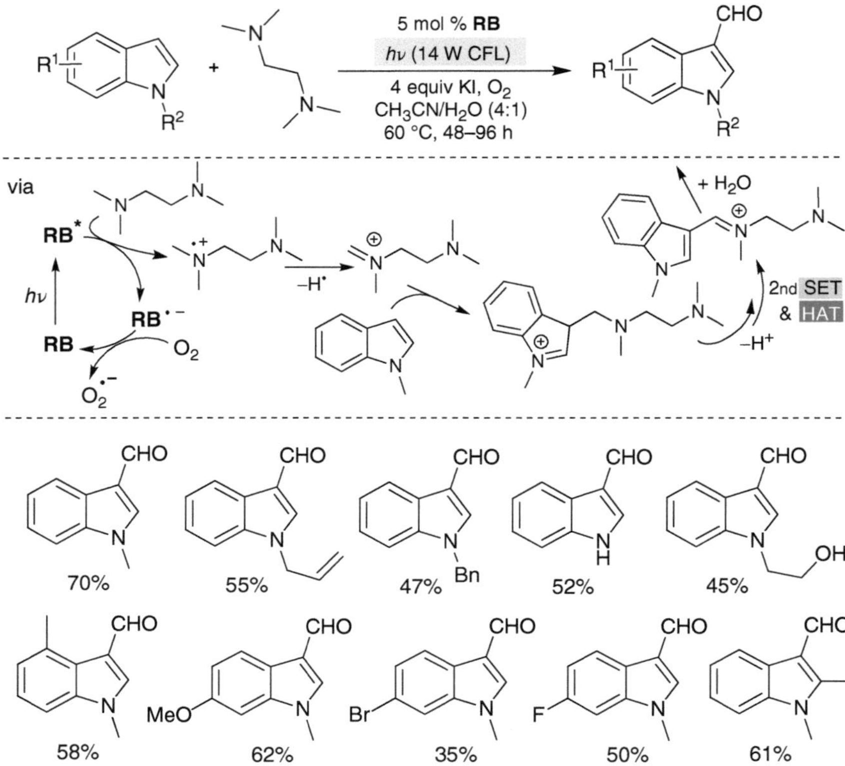

Scheme 6.40 Indole C-3 formylation using TMEDA as one-carbon-source and O_2 as terminal oxidant.

photocatalysis under aerobic conditions, in recent years a number of alternative organic scaffolds have found successful applications in a number of photoredox catalytic processes.

Using donor–acceptor substituted dicyanopyrazine-derived chromophores (**DPZ**) as photoresponsive catalysts, THIQs and other activated tertiary amines can undergo a wide scope of CDC-coupling reactions with different nucleophiles (*vide supra*) [104] (Scheme 6.41).

A reaction condition dependent switch in terms of product formation was observed for the **DPZ**-catalyzed transformation of THIQs with itaconimides as electrophiles [105]. Here, deprotonation of the initially generated amino radical cation leads to a nucleophilic α-amino radical, which is prone to attack electron-deficient alkenes. In the presence of $LiPF_6$ as Lewis-acidic additive the transformation of THIQs with *N*-phenyl itaconimides yields spirocyclic products in an addition/cyclization process. Unlike simple addition/protonation products are majorly observed if the reaction is conducted in the presence of an excess of K_3PO_4 as the inorganic base at low temperatures (at higher temperatures with Li_3PO_4 as base an addition/elimination becomes prevalent) (Scheme 6.42).

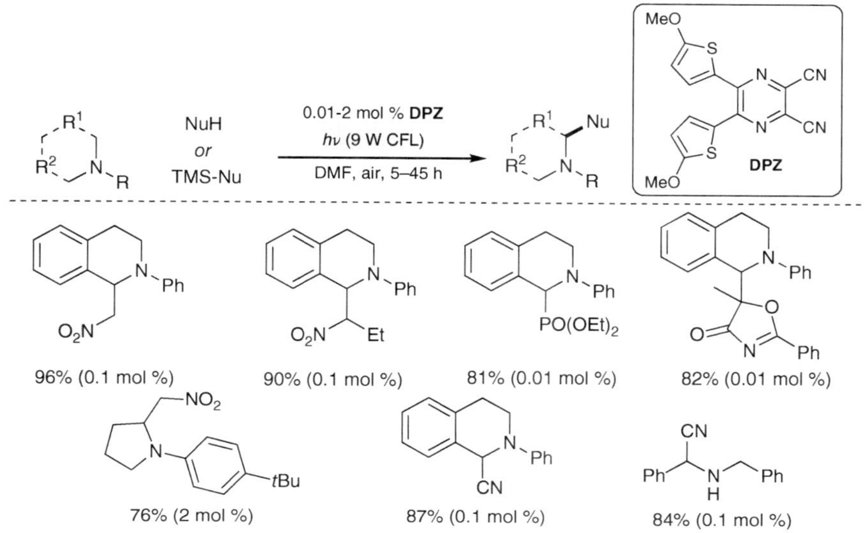

Scheme 6.41 2-Methoxythienyl dicyanopyrazine (**DPZ**) push–pull chromophor for photoredox CDC reactions.

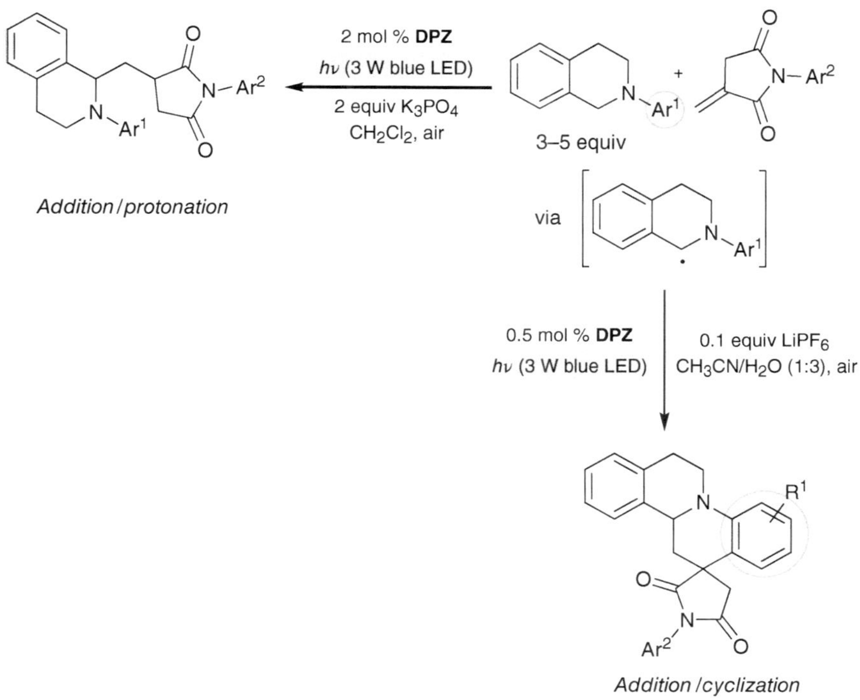

Scheme 6.42 Condition-dependent chemoselectivity in aerobic photoredox radical cascade reactions.

Dipyridyltetrazines as related push-pull chromophores were shown to be competent photo-oxidants for the synthesis of 2-substituted benzothiazoles and benzimidazoles upon conversion of aldehydes with the corresponding *ortho*-amino thiophenol or *ortho*-phenylendiamine [106]. Initially generated thio- or amino radical cations can cyclize onto the imine; protonation and rearomatizing oxidation yields the heterocyclic products.

BODIPY dyes which are widely used fluorescent chromophores featuring high photostability have mainly been investigated as robust and tunable fluorophores [107], and lately also as photosensitizers for photodynamic therapy [108] or within dye-sensitized solar cells and photovoltaics [109]. Only recently their first applications as photoredox catalysts have been reported. The group of Zhao employed iodo-substituted BODIPYs that are supposed to gain their photocatalytic activity from long-lived triplet states to catalyze a full range of the typical aforementioned CDC reactions of THIQs [110], including aza-Henry reactions as well as [3 + 2]-cycloadditions with *N*-aryl maleimides (Scheme 6.43).

Fullerene C_{60}-BODIPY dyads have also been applied for similar reactions [111]. Interestingly, very recently, the groups of Meng and Wu demonstrated that simple BODIPYs (without heavy atoms) can also catalyze aerobic CDC reactions

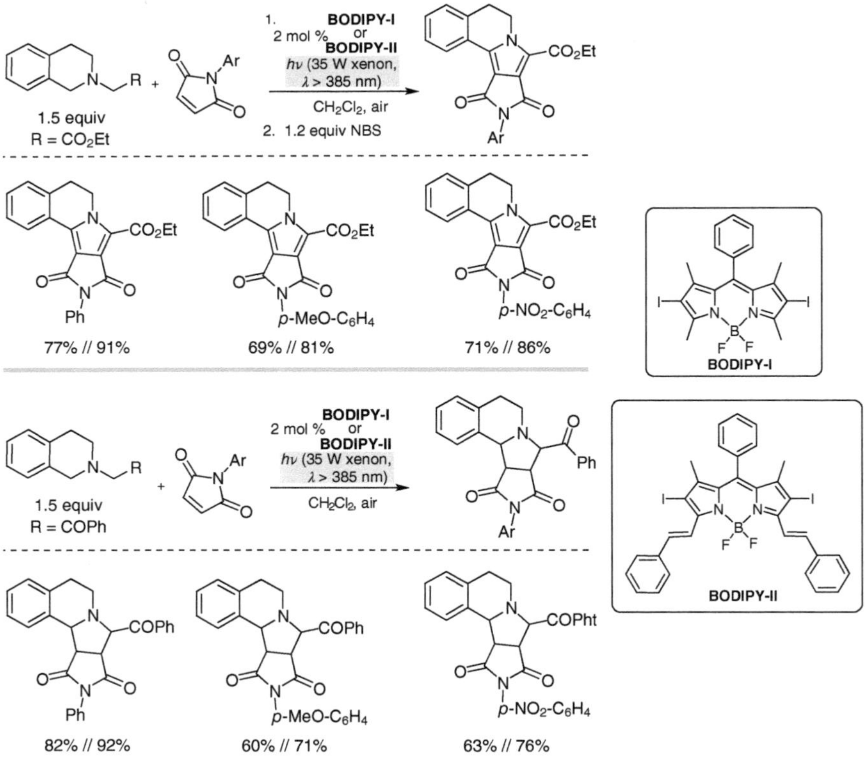

Scheme 6.43 Exemplary CDC reactions using iodo-BODIPYs as photoredox catalyst (reductive quenching).

for C—C and C—P coupling [112]. Their mechanistic studies indicate a photoredox activity from the dye's singlet state and its regeneration with molecular oxygen.

Most of the aerobic photoredox catalytic reactions discussed in this chapter rely on the regeneration of the photocatalyst's reduced form upon reoxidation with molecular oxygen (either supplied directly or from "open air"), which concurrently leads to the formation of superoxide radical anions [113] and in turn may partially also produce H_2O_2 by protonation and subsequent H-abstraction (see Scheme 6.7/③). Reductively quenched catalysts, however, may also be regenerated directly with other sacrificial oxidants (see Scheme 6.7/②). Some examples for the latter approach are provided here.

A chemoselective removal of *p*-methoxybenzyl ether protecting groups (PMB-ether) can be effected in the presence of excess H_2O_2 as oxidant and eosin Y **EY** as photocatalyst [114] (Scheme 6.44).

In a facile synthesis of multisubstituted quinazolines from arylamidines both $CBrCl_3$ and CBr_4 can reoxidize the photocatalyst's radical anion, which results from an initial reductive quenching of rose bengal (**RB** as superior photocatalyst for this transformation with *N*-benzyl-*N*′-arylamides [115]. Deprotonation/oxidation of the amino radical cation to the iminium ion (or its formation via HAT) followed by an S_EAr-type cyclization and rearomatization represents one possible mechanistic pathway.

On the other hand, the productive use of the superoxide species arising from the catalyst's reoxidation under aerobic conditions (see Scheme 6.7/③) has been reported for the hydroxylation of boronic acids to the corresponding phenols [116] using methylene blue (**MB**) (and for a single example also eosin Y [117]) as photocatalyst. The reductive quenching of the catalyst is enabled by using *N*,*N*-diisopropylethylamine (DIPEA) (or other tertiary amines) as potent sacrificial electron donor (Scheme 6.45).

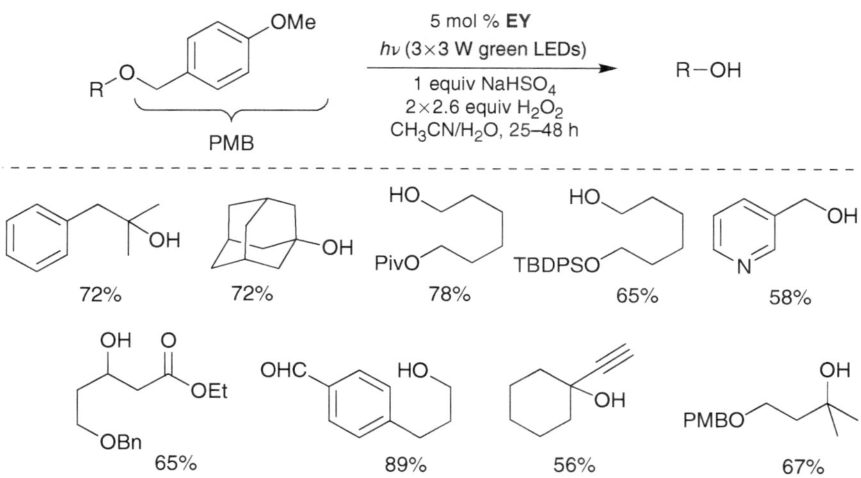

Scheme 6.44 PMP-ether deprotection using H_2O_2 to reoxidize reductively quenched eosin Y **EY** photocatalyst.

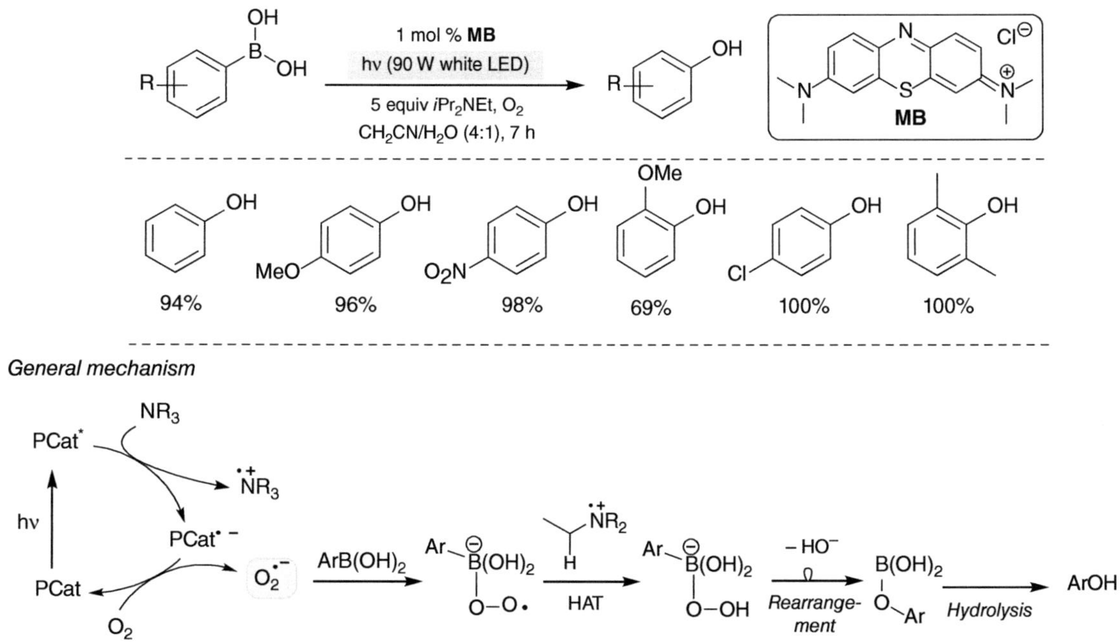

Scheme 6.45 Oxidative hydroxylation of arylboronic acids by visible light photoredox catalysis.

Under optimized aerobic conditions this hydroxylation of boronic acids can be performed with **RB** in an EtOH/H_2O mixture as solvent and DIPEA as electron-donor [118].

Following a reductive quenching pathway of the catalyst does not necessarily lead to an oxidative synthetic method as the desired transformation can also build on the hereby obtained, stronger reductive properties of the catalyst's reduced ground state. This albeit requires the employment of sacrificial electron donors such as tertiary amines (*vide supra*) or other reductants. Following these lines an early seminal example of Kellogg and coworkers describes the eosin Y-mediated reduction of phenacyl sulfonium salts with dihydropyridines [119]. These Hantzsch-ester type reductants may play a dual role in this reaction both as electron and hydride donor (hydrogen atom donor, respectively) (Scheme 6.46).

The dehalogenation of activated halogenides (α-carbonyl derivatives) is also possible with a wide range of yellow to red organic dyes of sufficient reductive power ($E_0/E_0^{\bullet-}$) and tertiary amines as sacrificial electron donors [120] (Scheme 6.47). For the integration of this generation of electrophilic radicals in a synergistic catalytic method for the enantioselective α-alkylation of aldehydes, see Section 6.2.4.3 (Scheme 6.72).

A methylene blue (**MB**)-catalyzed, reductive generation of trifluoromethyl radicals has been developed by the Scaiano group using Togni's hypervalent iodine CF_3-reagent resulting in both the release of the desired $CF_3^\bullet$ radical and the formation of 2-iodo-benzoate [121]. Tertiary amines like NEt_3 and DIPEA were competent reductive quenchers while the group observed a rate enhancement for DBU and TMEDA, possibly due to the formation of EDA-complexes between these amines and **MB** aiding the charge-separating

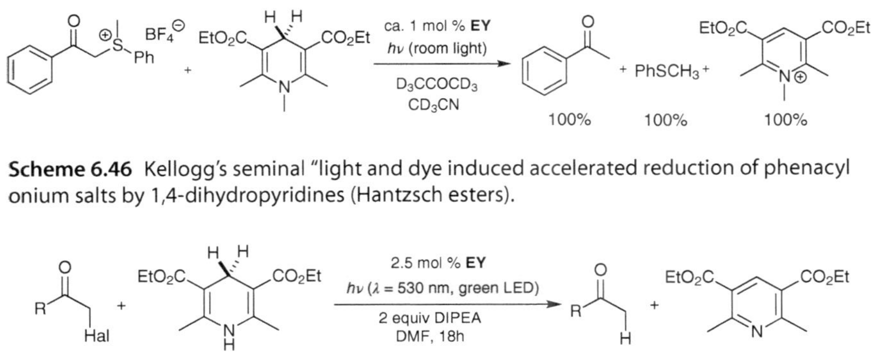

Scheme 6.46 Kellogg's seminal "light and dye induced accelerated reduction of phenacyl onium salts by 1,4-dihydropyridines (Hantzsch esters).

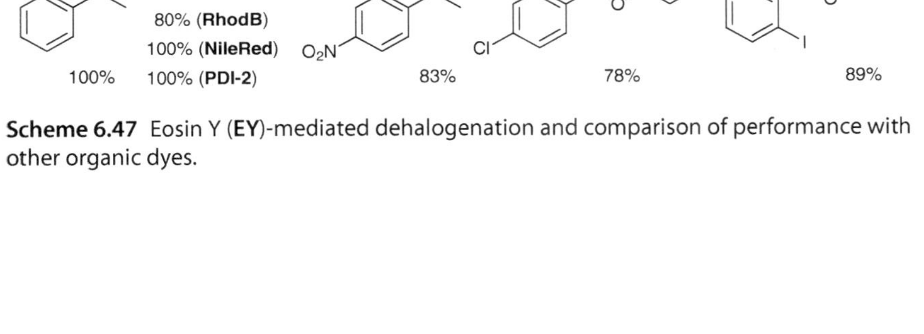

Scheme 6.47 Eosin Y (**EY**)-mediated dehalogenation and comparison of performance with other organic dyes.

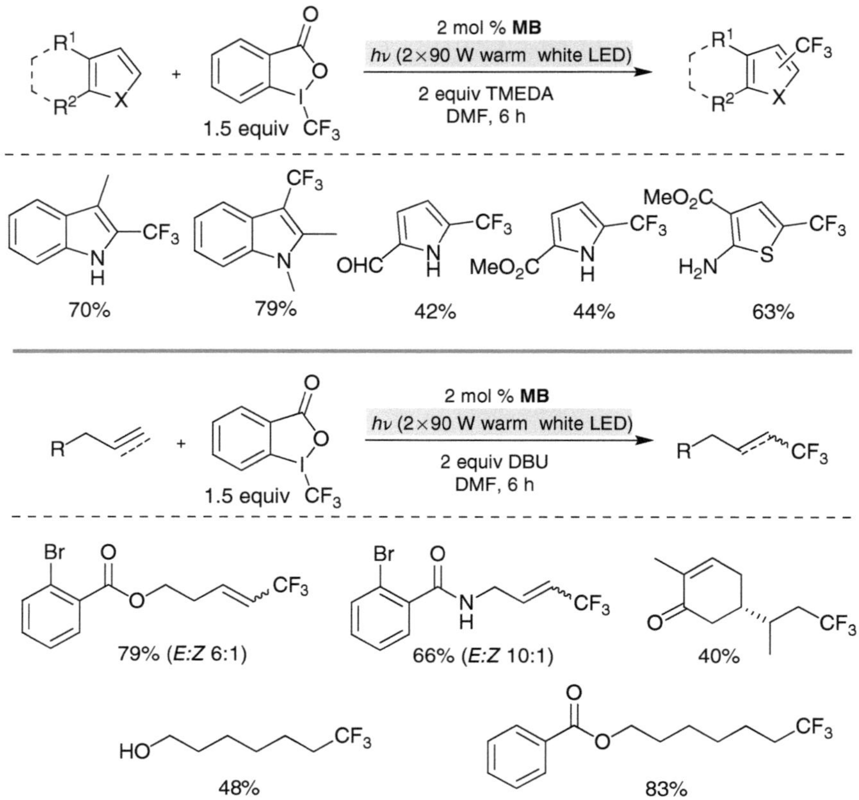

Scheme 6.48 Methylenblue **MB** as photocatalyst for trifluoromethylation of e-rich hetarenes and hydrotrifluoromethylation of terminal alkenes and alkynes.

electron transfer. Electron-rich heteroarenes were trifluoromethylated under optimized conditions while unactivated, terminal alkenes and alkynes undergo hydrotrifluoromethylation. The presence of multiple H-atom donors is assumed to favor this hydrotrifluoromethylation over an elimination pathway yielding trifluoromethylated products (Scheme 6.48).

Symmetrical and unsymmetrical diarylacetates are available from the eosin Y-catalyzed activation of ethyl bromofluoroacetate and electron-abundant indoles and/or protected anilines in a proposed radical addition/elimination/nucleophilic addition sequence [122] and allows fast access to relevant bisindolylmethane building blocks. Interestingly, the same reaction failed with ethyl dibromoacetate as reagent (Scheme 6.49).

Taking advantage of the reductive power of the eosin radical anion König and coworkers have developed a versatile, metal-free perfluoroarylation of arenes using bromopentafluorobenzene as radical precursor and triethylamine as sacrificial electron donor [123] (Scheme 6.50).

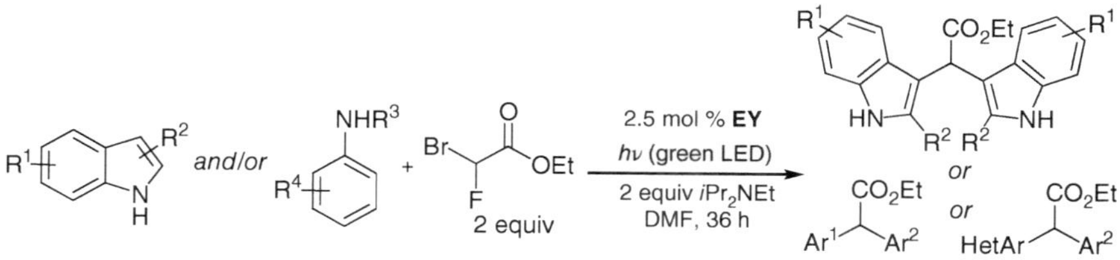

Scheme 6.49 Coupling of e-rich arenes and hetarenes with bromofluoroacetate for symmetrical and unsymmetrical products.

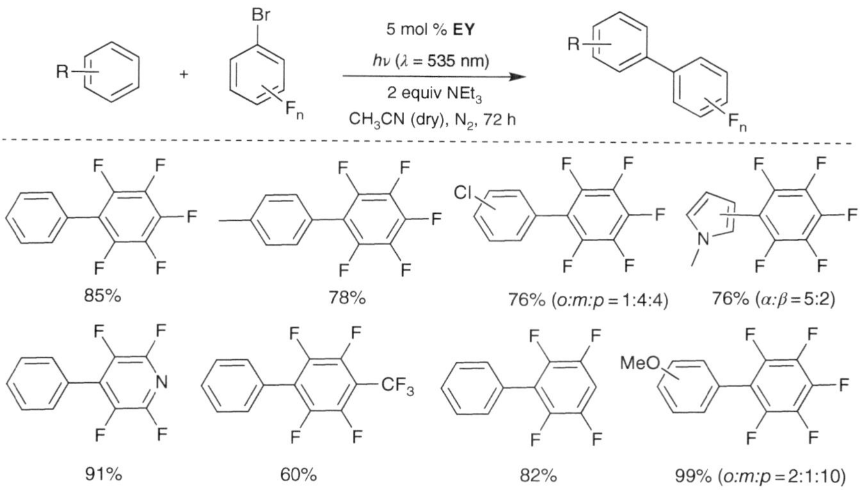

Scheme 6.50 Fluorinated aryl bromides for **EY**-catalyzed perfluoroarylation of simple arenes.

6.2.3 Oxidative Quenching of the Catalyst

The direct interaction of the photocatalyst's excited state with oxidants (to be reduced themselves) leads to oxidative quenching and the formation of an oxidized ground state species of the photocatalyst which itself may then serve as an oxidant. Both steps can be used productively to either reduce substrates and sacrificial acceptors respectively, and/or to use the oxidative power of the quenched catalyst for further activation. As this oxidative quenching pathway is not favorable for organic photocatalysts with cationic nature (pyrylium and acridinium type of catalysts) xanthene type dyes, most commonly eosin Y **EY** and **RB**, are prevalent showing oxidative quenching behavior in a number of diverse reactions. Typical oxidative quenchers for excited states quenching range from polyhalomethanes and sulfonyl compounds to aryl diazonium salts and thereby produce radicals for further transformations (Scheme 6.51).

Using CBr_4 as oxidative quencher for eosin Y generates tribromomethyl radicals $CBr_3^{\bullet}$ that in the presence of DMF can form Vilsmeyer-type iminium ions. These in turn can react with (oxygen) nucelophiles such as the hydroxy group of ketoximes to create an appropriate leaving group and hence lead to

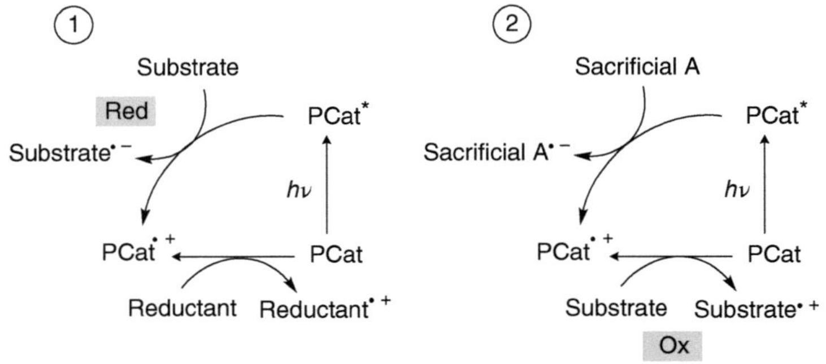

Scheme 6.51 Overview on possible pathways for oxidative quenching catalysts.

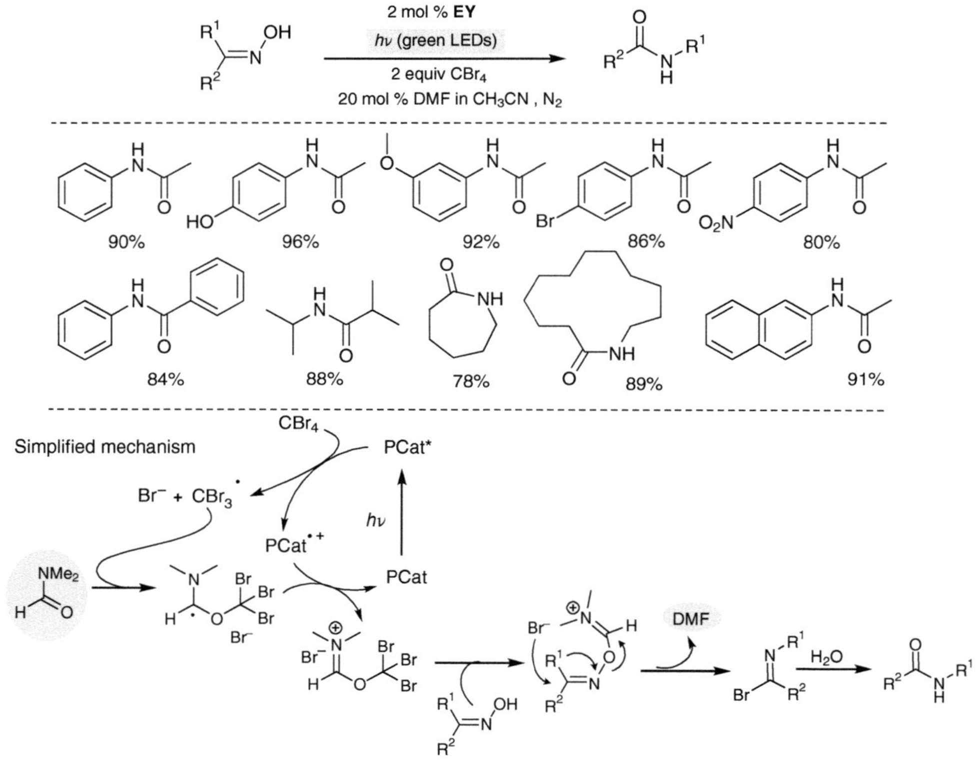

Scheme 6.52 Photoredox-mediated Beckmann rearrangement of oximes via Vilsmeyer-type intermediate.

a Beckmann-type rearrangement generating the corresponding amines [124]. Aldoximes and primary amides are transformed in a formal dehydration to the corresponding nitriles [125] (Scheme 6.52).

Zwitterionic rhodamine B is shown to be oxidatively quenched by *i*-C_3F_7I or an excess of trichlorobromomethane in water [126]. The concurrently generated electrophilic (perfluoroalkyl) radicals may react with bisalkenes to yield formal

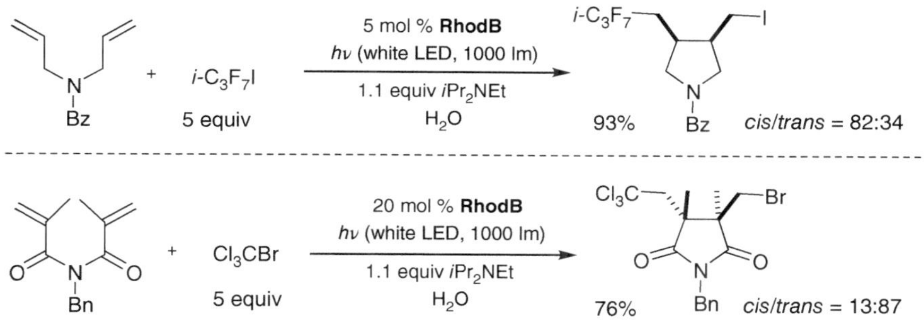

Scheme 6.53 Rhodamine B for atom transfer radical cyclization reactions.

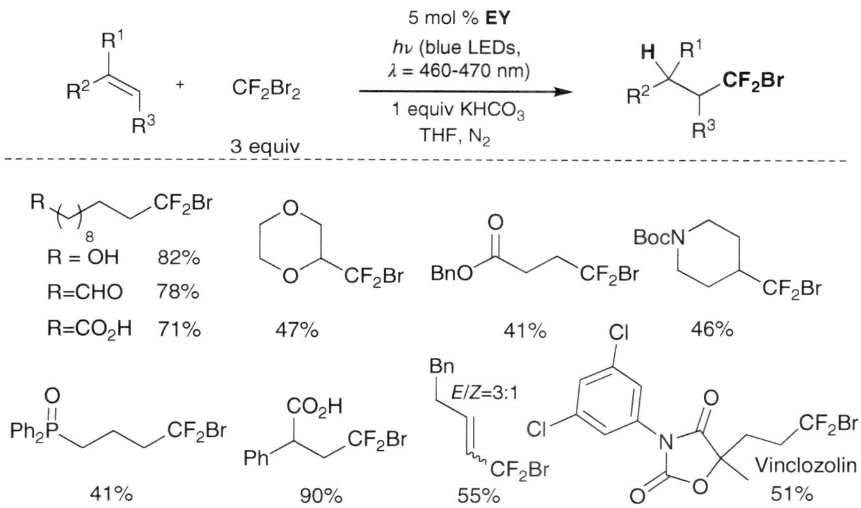

Scheme 6.54 Eosin Y-catalyzed hydrobromodifluoromethylation of alkenes with dibromodifluoromethane.

cyclization-ATRA products. DIPEA is needed as sacrificial reductant to regenerate the rhodamine catalyst (Scheme 6.53).

Terminal alkenes and alkynes can undergo selective hydrobromodifluoromethylation with CF_2Br_2 in the presence of $KHCO_3$ as additive in THF [127]. Stern–Volmer studies hint to an oxidative quenching mechanism. Upon addition of the $CF_2Br^•$ radical the product forms by H-abstraction from the solvent THF. Photocatalyst regeneration may then occur also with the aid of the solvent, while ATRA- and elimination by-products can be suppressed with the optimized reaction conditions. This protocol also allows for "late-stage-functionalization" of natural products and drugs in modest yields (Scheme 6.54).

tert-Butyl hydroperoxide (TBHP) is also been used as oxidative quencher and leads to *t*BuO$^•$ radicals and eosin Y **EY**$^{•+}$ [128]. The alkoxy radical is supposed to generate a sulfonyl radical from the sulfinic acid precursor which then may add on the alkyne which induces cyclization to the coumarin-type radical species.

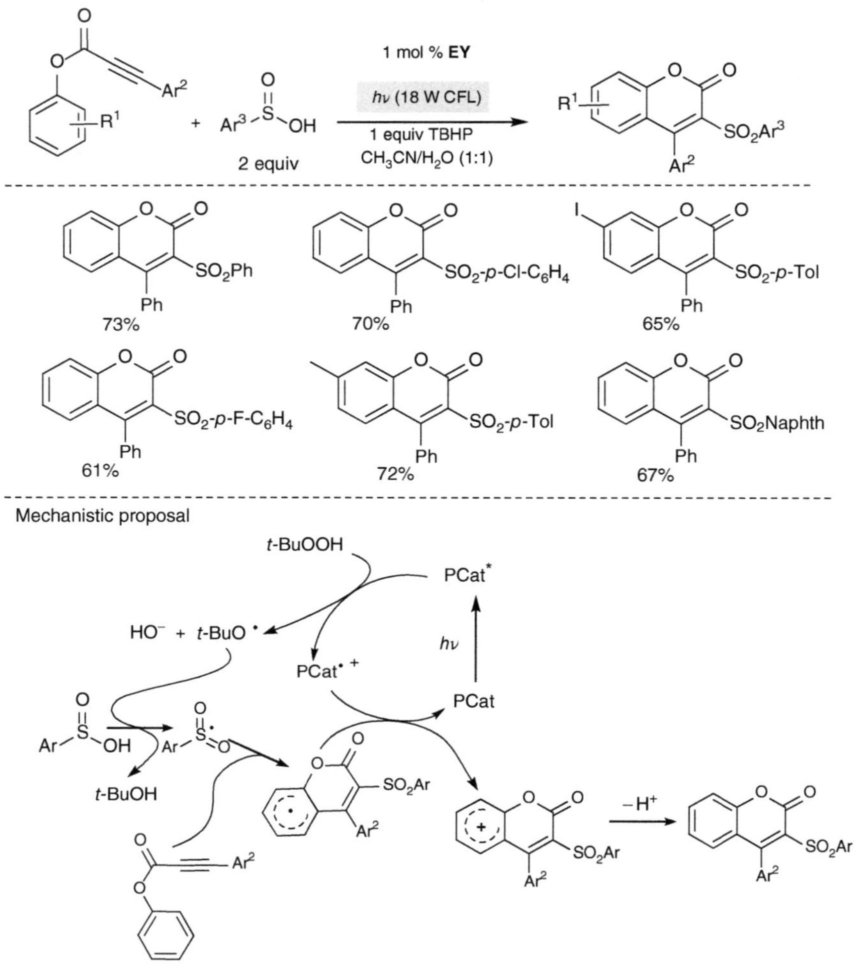

Scheme 6.55 Oxidative cyclization of phenyl propriolates to coumarins induced by sulfinic acid derived radicals.

Final oxidation to the carbocation followed by deprotonative rearomatization is promoted by **EY**$^{\bullet+}$ and regenerates the eosin photocatalyst (Scheme 6.55).

A single report of Tan deals with the oxidative quenching of **RB** by β-keto esters [129], which in turn are transformed to their radical anions and upon HAT to the enolates, respectively. SET oxidation of the enolate then allows for trapping with persistent TEMPO radical and affords α-oxygenated products.

Sulfones and sulfonyl halides can also be employed as oxidative quenchers, most commonly for eosin Y. The formation of the substrates' radical anions causes a weakening of the adjacent C—S-bond and hence leads to the release of the corresponding C-radical together with a sulfonate anion or halide and SO_2, respectively. Along these lines β-aryl ketosulfones can efficiently be desulfonated

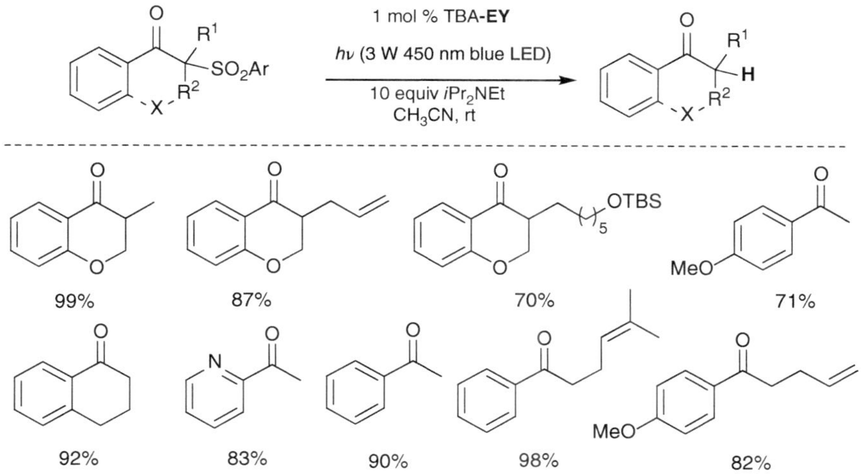

Scheme 6.56 Eosin Y-mediated desulfonylation of β-arylketosulfones.

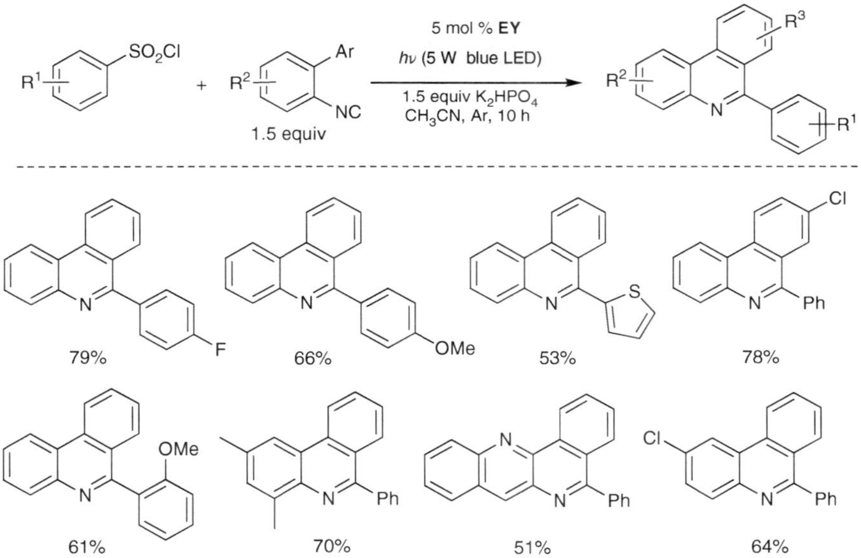

Scheme 6.57 Access to phenanthridines with aryl sulfonyl chloride-derived aryl radicals.

using DIPEA as the terminal reductant to regenerate the eosin Y photocatalyst [130]. Mechanistic studies prove the oxidative quenching cycle (Scheme 6.56).

Both the generation of aryl and $CF_3^{\bullet}$ radicals have been reported for the corresponding sulfonyl chlorides. The aryl radicals were used for the syntheses of 6-aryl phenanthridine in a desulfonation/addition/cyclization radical cascade of 2-isocyanobiphenyls [131] (Scheme 6.57).

Trifluoromethyl radicals generated from CF_3SO_2Cl were employed by Kappe *et al.* for the trifluoromethylation of ketones in a flow system [132]. First

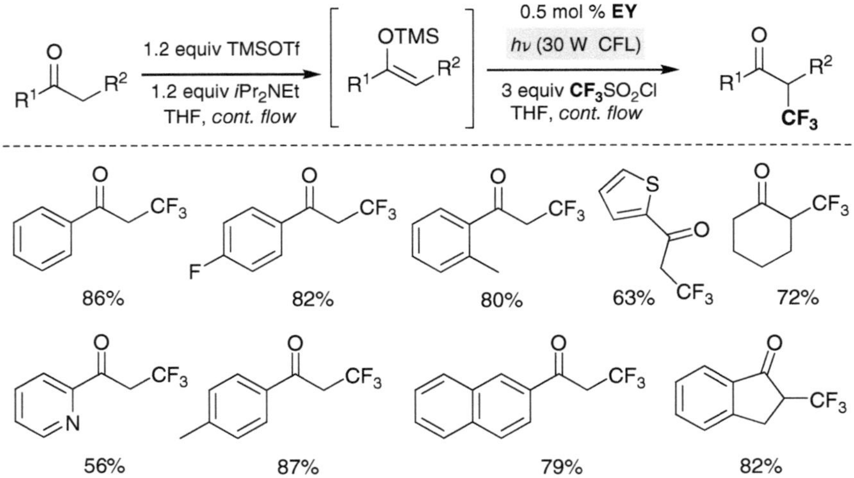

Scheme 6.58 Continuous flow α-trifluoromethylation of ketones.

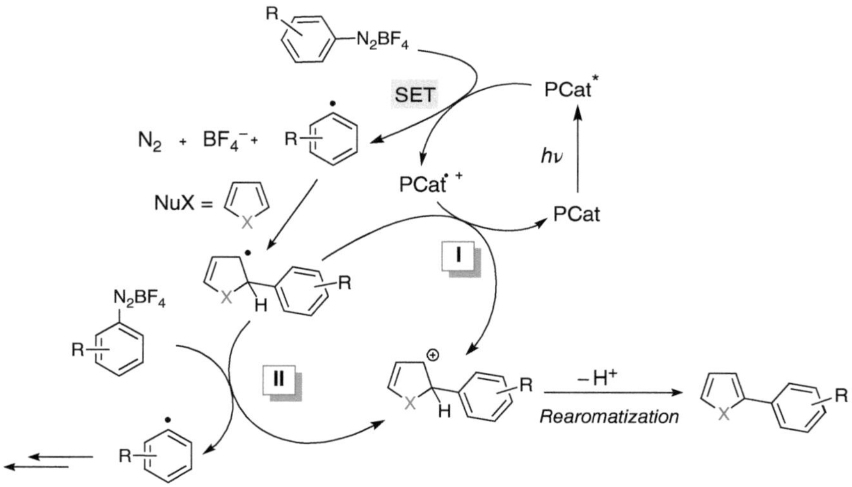

Scheme 6.59 Proposed general mechanism for eosin Y-mediated arylation reactions using arene diazonium salts.

manipulation in this flow reactor involves the silyl enolether formation followed by the eosin Y photocatalyzed trifluoromethylation (Scheme 6.58).

The reductive *in situ* generation of aryl radicals from diazonium salts is most commonly used for (metal-free) arylation reactions [35]. These radicals add to electron-rich heteroarenes (furans, pyrrols, thiophenes) yielding an intermediate radical which then may be oxidized to its cationic counterpart either by the radical cation of the oxidized dye (pathway **I**) or by a radical chain transfer pathway (pathway **II**). Rearomatizing deprotonation of the cation affords the desired cross-coupling product [133] (Scheme 6.59).

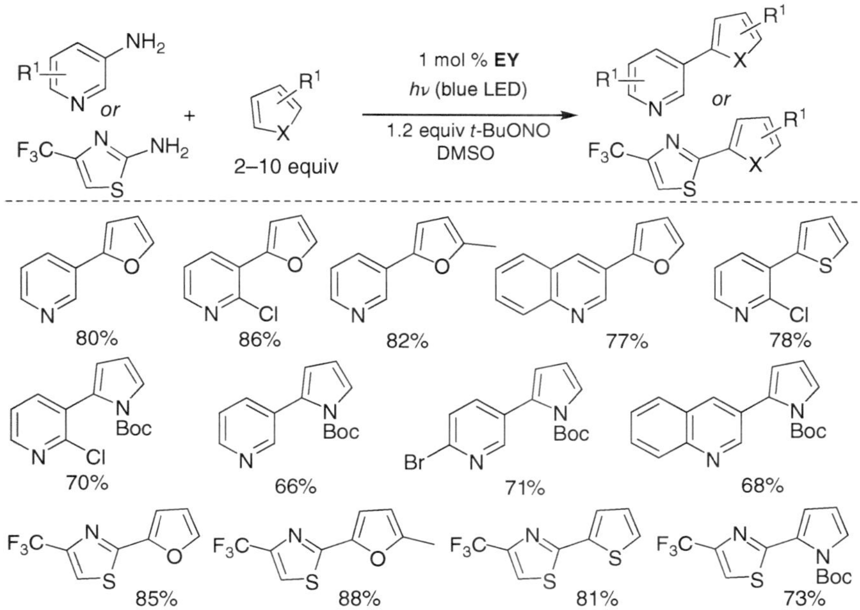

Scheme 6.60 Access to biheteroaryls via eosin Y catalyzed C—H heteroarylation of heteroarenes.

An extension of the scope of this reaction was reported for heteroarene-derived radical precursors; here, the *in situ* generation of the diazonium salts from the corresponding heteroaromatic amines seems to be favorable [134] (Scheme 6.60).

For *ortho*-amido aryl diazonium salts bearing an α-hydrogen next to the nitrogen of the cyclic amide, the initially formed aryl radical is transposed to this position by a 1,5-hydrogen transfer to generate an α-amido radical [135]. This is then oxidized to the corresponding iminium ion that subsequently can be trapped by external nucleophiles (such as e-rich heteroarenes: indoles, etc.). Acid-promoted ring opening and addition of a second nucleophile generates ω-functionalized benzamides (Scheme 6.61).

Phenanthrenes can be accessed with eosin Y as photocatalyst if biaryl diazonium salts are trapped with alkynes to furnish an intermediate vinyl radical, which then forms the phenanthrene scaffold upon radical cyclization [136] (Scheme 6.62).

Using *ortho*-methylthio arene diazonium salts a similar intermolecular process allows to access benzothiophenes by initial capture of the aryl radical with alkynes and subsequent cyclization to generate a sulfur-centered radical [137]. Following its SET oxidation by $\mathbf{EY}^{\bullet+}$ or a radical chain transfer, rearomatization is proposed to occur via a methyl group transfer to the solvent DMSO as nucleophile (Scheme 6.63).

In more recent studies, diazonium salts are also a starting point for photoredox catalytic, metal-free carbonylation reactions. In the presence of eosin Y [138] or fluorescein [139] these salts can be smoothly transformed into the corresponding

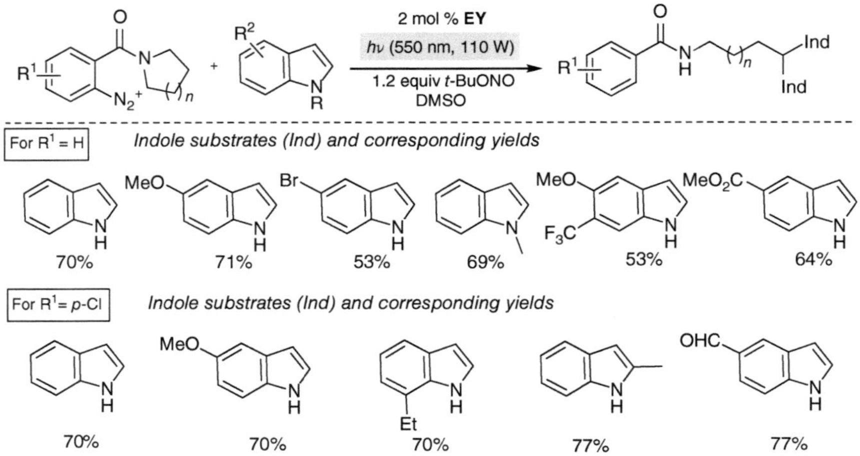

Scheme 6.61 Preparation of bis(indolyl)alkanes via photoredox induced α-amino C(sp³)—H activation through 1,5-hydrogen migration.

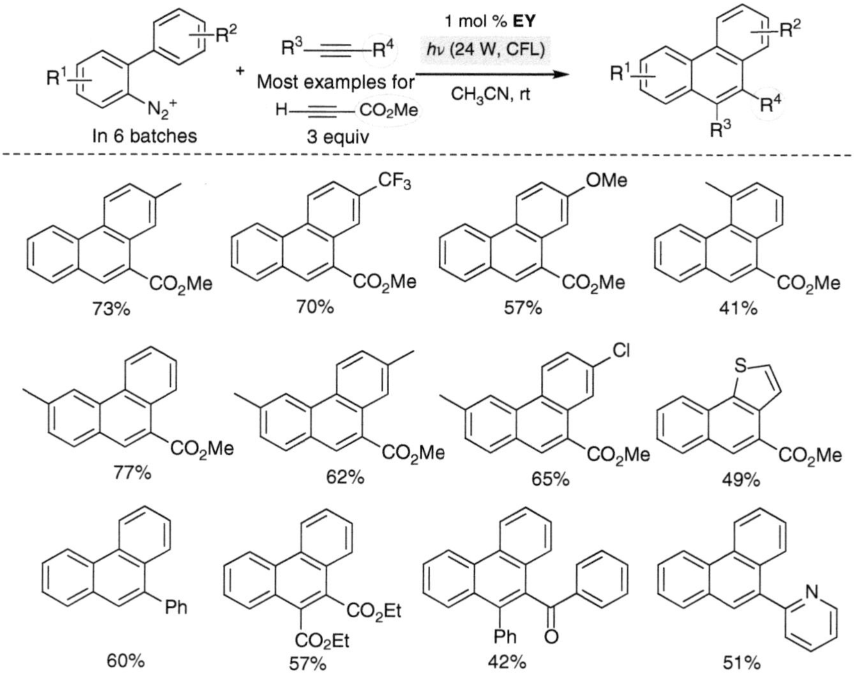

Scheme 6.62 Preparation of phenanthrenes by benzannulation of biaryldiazonium salts with alkynes.

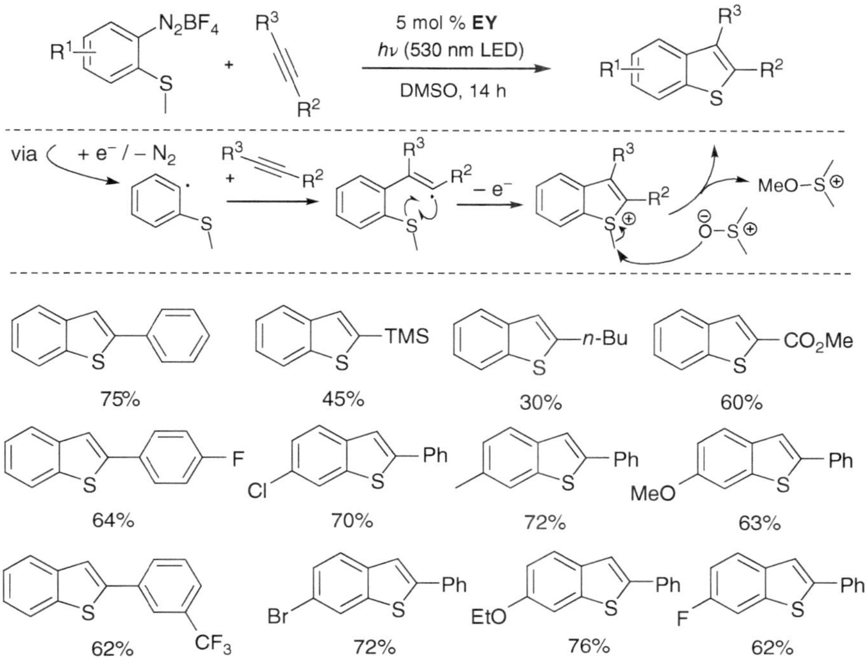

Scheme 6.63 Benzothiophenes by radical annulation of *o*-methylthio-arene diazonium salts with alkynes.

benzoate esters in a pressurized mixture of CO (50 and 81 bar, respectively) with MeOH as solvent upon irradiation. The mechanism of this alkoxycarbonylation is proposed to start with the dye's oxidative quenching and concurrent formation of an aryl radical, which in turn rapidly reacts with CO to the acyl radical. Its further oxidation to the highly electrophilic acylium ion is promoted by the radical cation of the dye and final reaction with the nucleophilic alcohol solvent yields the benzoate ester (Scheme 6.64).

Ketones can also be accessed if the intermediate acyl cation is trapped by other (e-rich) arenes and heteroarenes [140], including indoles for the selective formation of indol-3-yl-aryl ketones [141].

Due to the instability of diazonium salts the choice of suitable nucleophilic reaction partners in direct or cascade arylation–addition reactions (such as for Meerwein type transformations) is limited. In an aminoarylation of styrenes with diazonium salts and nitriles as nucleophiles in a Ritter-type final trapping step organic photocatalysts showed only low activity [142]. However, for a related intermolecular formyloxyarylation with DMF as the final nucleophilic trapping agent eosin Y **EY** and other xanthene dyes such as **RB** and different rhodamines could affect the test reaction in moderate yields [143] (Scheme 6.65).

Further functionalizations as well as defunctionalizing reactions of aryl diazonium salts are also well known. Apart from C—C-bond formations (*vide supra*) this also includes C-heteroatom bond-forming transformations. Eosin Y proved

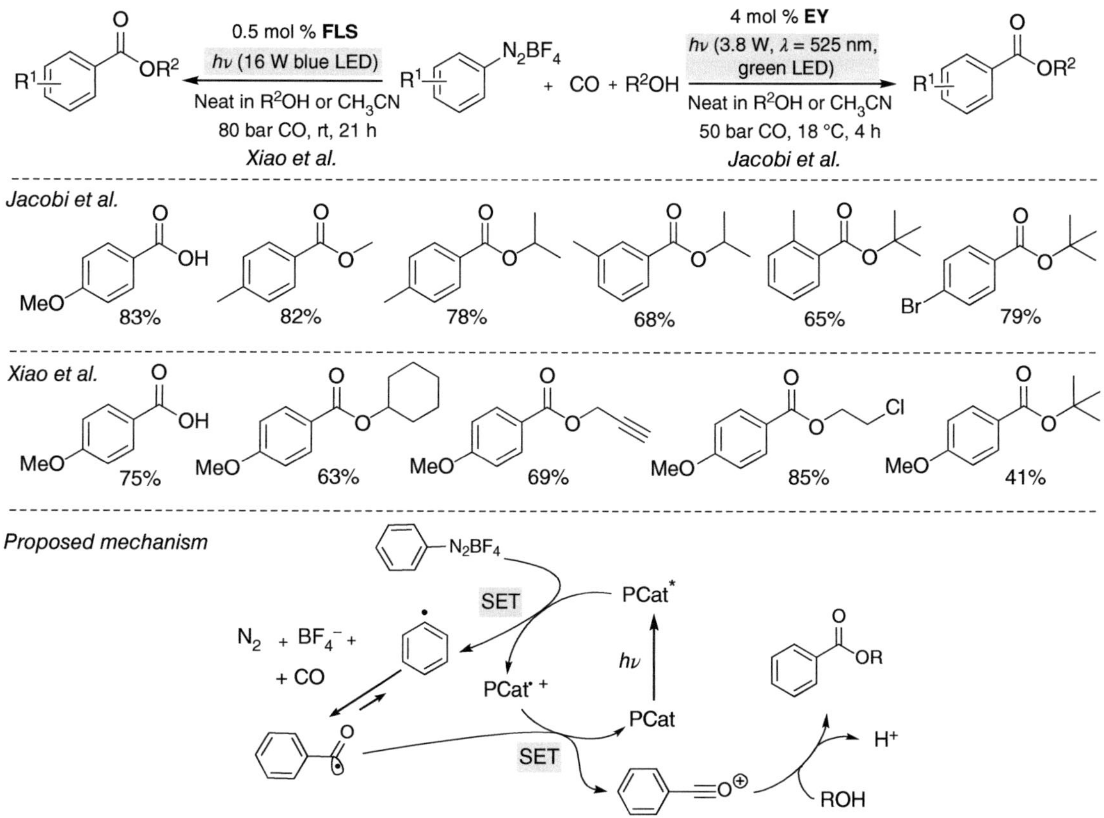

Scheme 6.64 Metal-free carbonylations of arene diazonium salts mediated by xanthene type dyes.

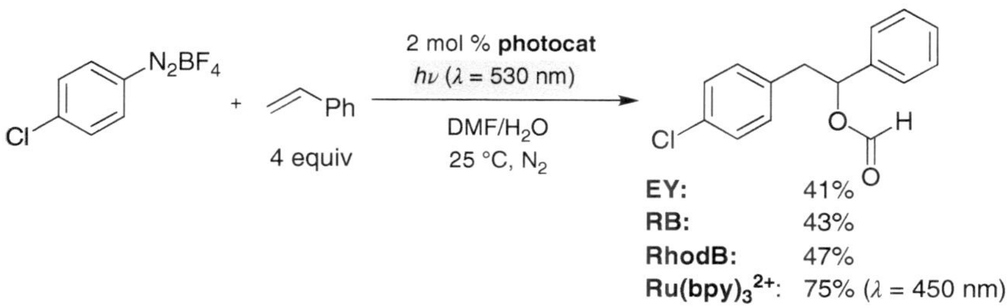

Scheme 6.65 Comparison of performance of different photocatalysts for the formyloxyarylation of alkenes.

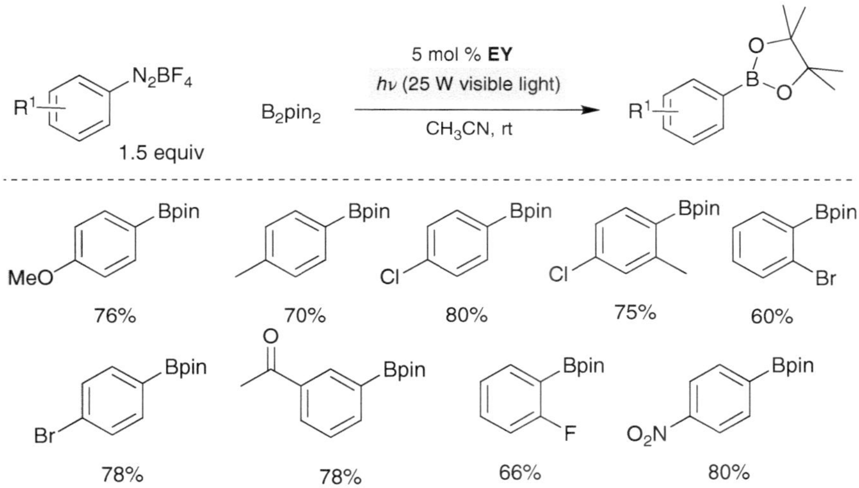

Scheme 6.66 Access to arylboronates via borylation of arene diazonium salts.

to be a suitable catalyst for the photocatalytic borylation of aryl diazonium salts using B_2Pin_2 as reagent, which may be activated by the BF_4^- counterion [144] (Scheme 6.66).

A similar cationic transfer of a methyl group to the solvent DMSO (acting as nucleophile) as suggested for the benzothiophene synthesis (see Scheme 6.63) is proposed for the eosin Y-mediated synthesis of aryl sulfides from the corresponding diazonium salts and disulfides [145]; aryl selenides are also accessible if diselenides are employed. This method avoids the intermediate formation of explosive diazosulfides, which occurs in base-induced methods (Scheme 6.67).

Hydro- and deuterodefunctionalization of aryl diazonium salts is best effected by catalytic eosin B (**EB**) in DMF (and d^7-DMF respectively), which is proposed to act as HAT donor to furnish the defunctionalized (or deuterated) arene [146] (Scheme 6.68).

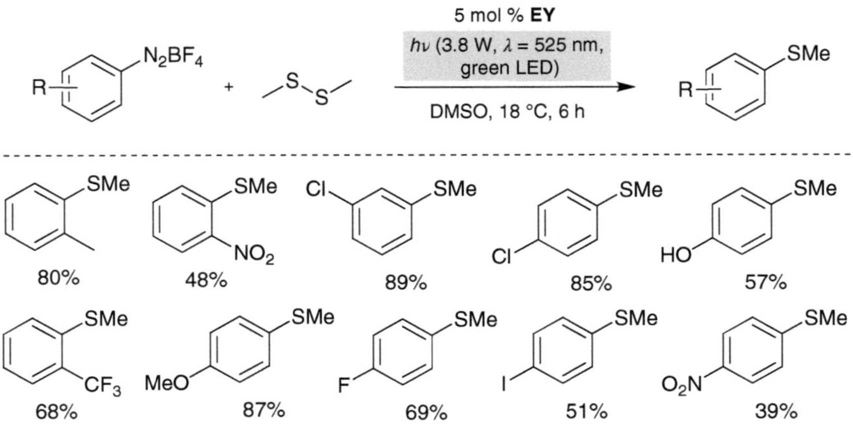

Scheme 6.67 Eosin Y photocatalyzed synthesis of arylsulfides from arene diazonium salts.

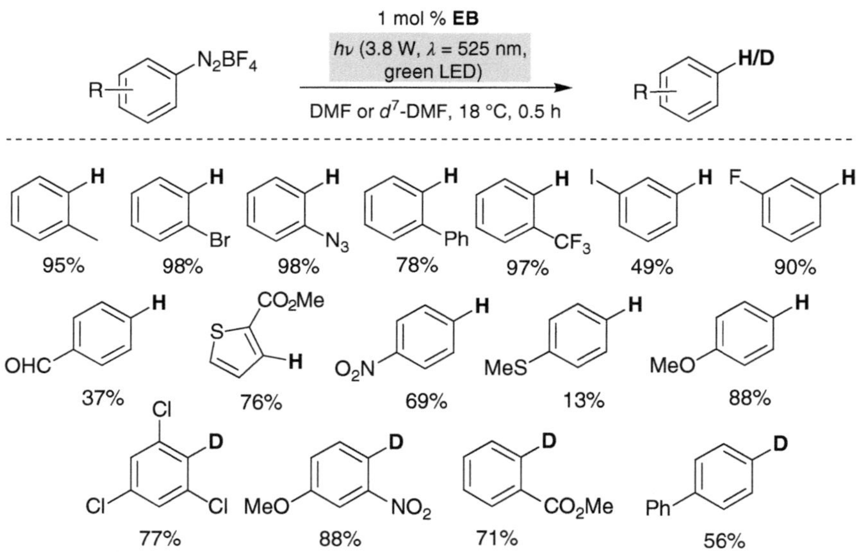

Scheme 6.68 Eosin B-mediated hydro-/deuterodefunctionalization of aniline- derived aryl diazonium salts.

6.2.4 New Developments

This chapter summarizes a few recent developments in visible light photocatalysis with organic dyes, which have recently contributed to substantially broaden the scope of possible applications.

As the additional energy that can be added using photocatalysts, which are excited with visible light ($\lambda \geq 400\,\text{nm}$; $E \approx 300\,\text{kJ mol}^{-1}$), is inherently limited, bond activation by such corresponding excited states is restricted to this energy

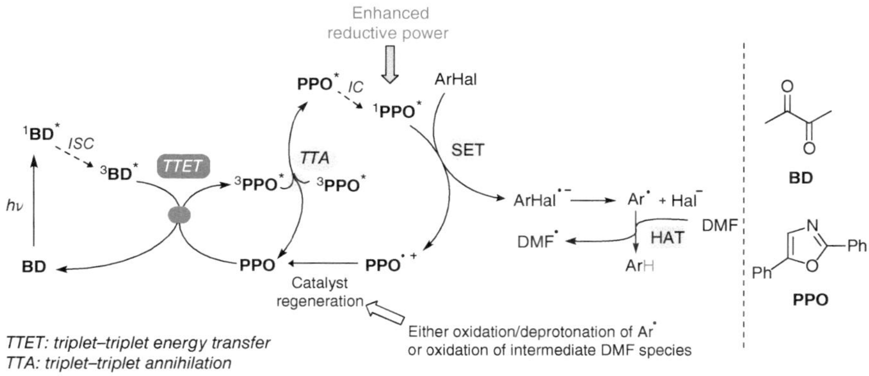

Scheme 6.69 Triplet–triplet annihilation upconversion (TTA-UC) allowing for SET activation of aryl bromides.

level, or the resulting excited state redox potential, respectively. Two approaches to work around these limitations and hence to allow for broader applicability of (organic) photo(redox) catalysis have been recently introduced.

6.2.4.1 Upconversion

The first approach relies on energy upconversion (UC) by non-linear two-quantum processes, such as two-photon-absorption (TPA) or triplet–triplet annihilation (TTA) [147]. A first proof of principle using TTA-UC for photocatalytic reactions has only recently been brought forward by Jacobi von Wangelin and co workers [148]. They employed Castellano's record annihilation donor/acceptor system (anti-Stokes shift of 0.64 eV; $\lambda_{em} \approx 370$ nm) 2,3-butadione (**BD**) and 2,5-diphenyloxazole (**PPO**) for the reductive dehalogenation of aryl bromides in DMF, which additionally may also act as the H-atom transfer agent (see Scheme 6.69).

Upon irradiation with a pulsed blue light laser (430 nm) **BD** as triplet photosensitizer harvests the excitation energy and transfers it after *intersystem crossing* (*ISC*) via triplet–triplet energy transfer (TTET) to **PPO** as acceptor. Subsequently, two nearby ^{3}**PPO*** (excited triplet acceptor) can collide under TTA and a sufficient population of the **PPO**'s excited singlet state ^{1}**PPO*** (characteristic delayed fluorescence at 370 nm) could be achieved. Thus, this allowed for the initial formation of the aryl bromide's radical anion $ArBr^{\bullet-}$ to rapidly dissociate into bromide and the aryl radical. According to the authors' calculations this initial SET is only in reach for the excited singlet state ^{1}**PPO***, while an ET mechanism from ^{3}**PPO*** seems thermodynamically unfeasible.

6.2.4.2 Consecutive Photoelectron Transfer

A second elegant approach of König and coworkers rather builds on consecutive electron transfers to create a species of increased reductive powers as compared to the aforementioned energy transfer process [149]. Mimicking the *Z*-scheme of natural photosynthesis using a two-step photoexcitation with NEt_3 as a sacrificial electron donor, the gained reductive power allows to activate stronger C_{aryl}—Hal bonds, that is, aryl chlorides instead of aryl iodides or aryl bromides. Apart from photoreduction with either NEt_3 or DMF acting as the hydrogen atom donor the highly reactive aryl radicals can also be trapped with electron-rich heteroarenes for C—C-bond forming aryl-cross coupling reaction (also see Schemes 6.59 and 6.60) if the reaction is conducted in DMSO instead of in DMF (Scheme 6.70).

Under oxygen-free conditions, perylenbisimides (**PDI**) known for their high thermo- and photostability can generate colored, stable radical anions $\mathbf{PDI}^{\bullet-}$ upon irradiation in the presence of electron donors (e.g., NEt_3, etc.,). This $\mathbf{PDI}^{\bullet-}$ intermediate is stable enough to undergo a second excitation with light to populate the radical anion's excited state $\mathbf{PDI}^{*\bullet-}$ which thereby has gained increased reductive power. This now allows the reduction of less-activated aryl halides and hence provides an alternative access to aryl radicals under mild and metal-free conditions, even employing stable aryl chlorides as inexpensive bulk material.

6.2.4.3 Multicatalysis

Organic photocatalysts have also been involved in quite a number of multicatalytic processes [42, 150]. The possibility to access otherwise unattainable transformations has stimulated increasing interest in the last few years. Based on

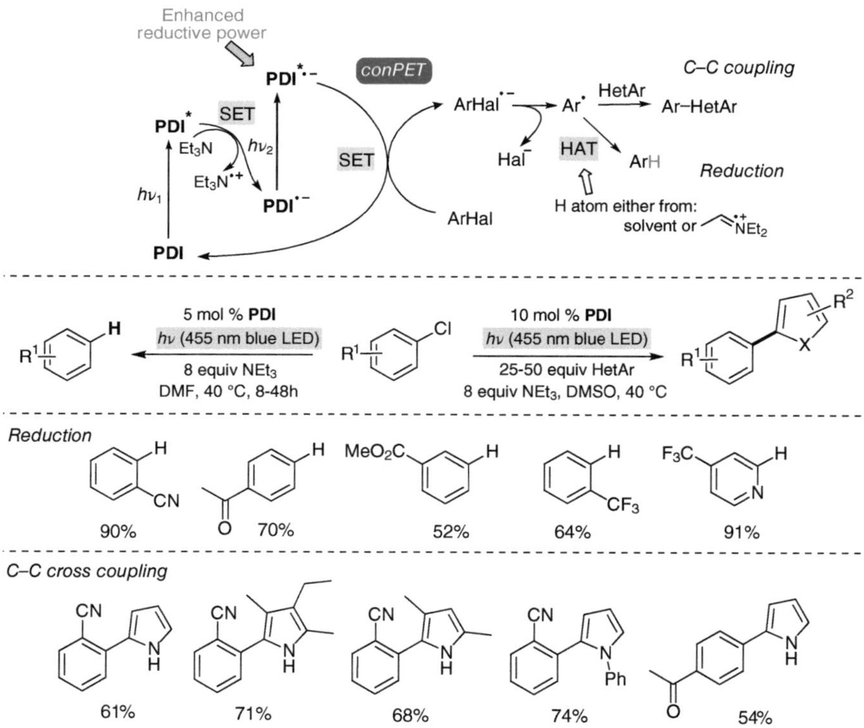

Scheme 6.70 Consecutive photoelectron transfer (conPET) using perylene bisimides as photocatalyst to increase reductive power for transformations of aryl chlorides.

the aforementioned activation modes, visible light photo(redox) catalysis offers a versatile "interface" for connections with other catalytic cycles. This includes not only the parallel generation of reactive intermediates as demonstrated for the combination of photoredox catalytic iminium generation by **RB** together with proline-derived catalytic enamines in Mannich type reactions [113, 151], but also includes synergistic and cooperative catalytic processes with metal or organocatalysts.

Eosin Y **EY** as a surrogate for $Ru(bpy)_3{}^{2+}$ can be used for a metal-free, cooperative catalytic reductive cyclization of (aryl) bisenones in the presence of thioureas or other H-bonding catalysts (TADDOLs, etc.) [152], which promote the initial electron transfer by LUMO-lowering (lowest unoccupied molecular orbital) and proton-coupled electron transfer (PCET) (Scheme 6.71).

In a more complex interlocking of catalytic cycles eosin Y plays a double role in the asymmetric alkylation of aldehydes with activated halogenides (α-carbonyl halogenides or perfluoroalkylhalides) [120a,153]. Similar to the $Ru(bpy)_3{}^{2+}$-mediated reaction [154], the oxidative power of the excited state is proposed to promote the aminocatalytic cycle by oxidation of the α-amino radical to the corresponding iminium ion while the photocatalyst's radical anion ground state species reduces the halide partner for radical formation upon mesolytic C—Hal bond fission (Scheme 6.72).

Only recently organic dyes as photocatalysts have also been combined with metal catalysis. The combination of a reductive quenching of eosin Y (**EY** or other related xanthene dyes) for iminium ion generation and its subsequent Friedel–Crafts-type addition to indoles can be productively coupled with a Co(III) catalyst ($Co^{III}(dmgH)_2Cl_2$; dmgH = dimethylglyoximate) to productively use the generated reduction equivalents for the formation of molecular hydrogen H_2 [155]. This approach satisfyingly avoids aerobic conditions and the need for sacrificial oxidants (*vide supra*). In the presence of nitrobenzenes their reduction to the corresponding anilines can also be effected in a cascade-type reaction [156] (Scheme 6.73).

A comparable procedure of combining visible light photocatalysis with concurrent Co-co-catalyzed H_2-evolution uses the stronger oxidizing Fukuzumi catalyst **Mes-Acr⁺** for the functionalization of isochromans with 1,3-dicarbonyl compounds which require an additional activation with a Cu(II) co-catalyst [157] (Scheme 6.74).

Highly versatile eosin Y (**EY**) also allows for the generation of acyl radicals from α-keto acids upon decarboxylation. In the presence of $Pd(OAc)_2$ these acyl radical can be trapped by *ortho*-C—H-activated acetanilides for their *ortho*-acylation [158] (Scheme 6.75).

Although transition-metal complex-based photocatalysts seem to be more versatile partners in the combined Au/photoredox catalysis for the synthesis of α-arylated ethers, both eosin Y (**EY**) and fluorescein (**FLS**) showed good performance for the dual catalytic oxyarylation of alkenes using different diazonium salts as reaction partners [159] (Scheme 6.76).

Carbazolyl dicyanobenzene- based donor–acceptor fluorophores are versatile photoredox catalysts for both decarboxylative as well as deborylative dual photoredox/Ni-catalytic SET cross-coupling reactions with aryl halides and

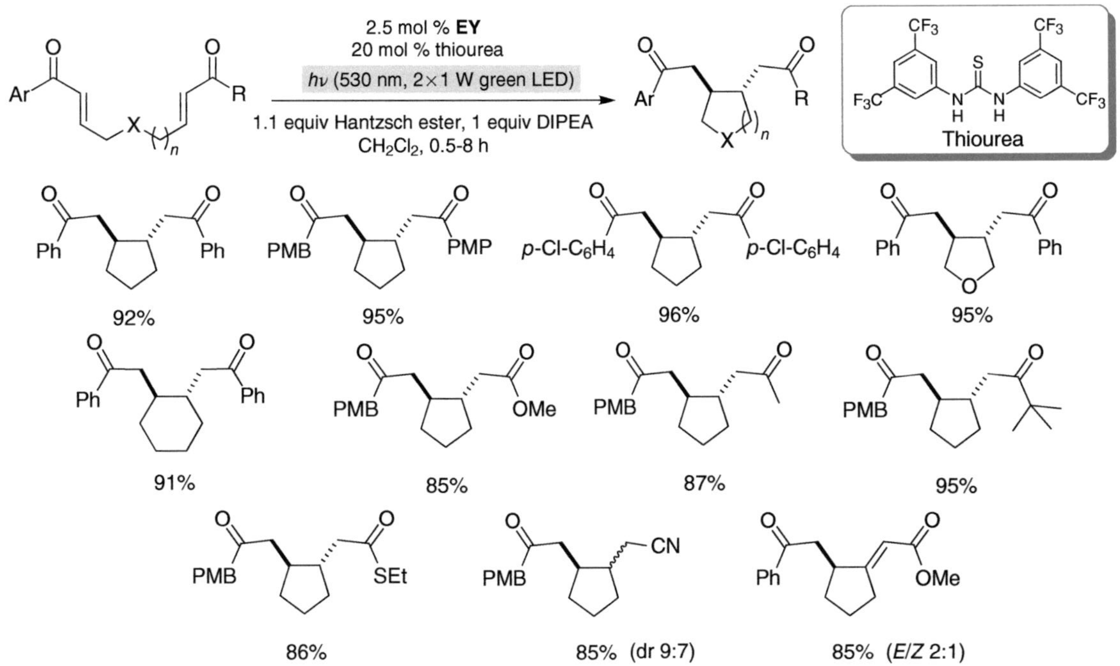

Scheme 6.71 Cooperative catalysis with eosin Y and H-bonding catalysts to effect reductive cyclization by PCET (PMP = *p*-methoxyphenyl).

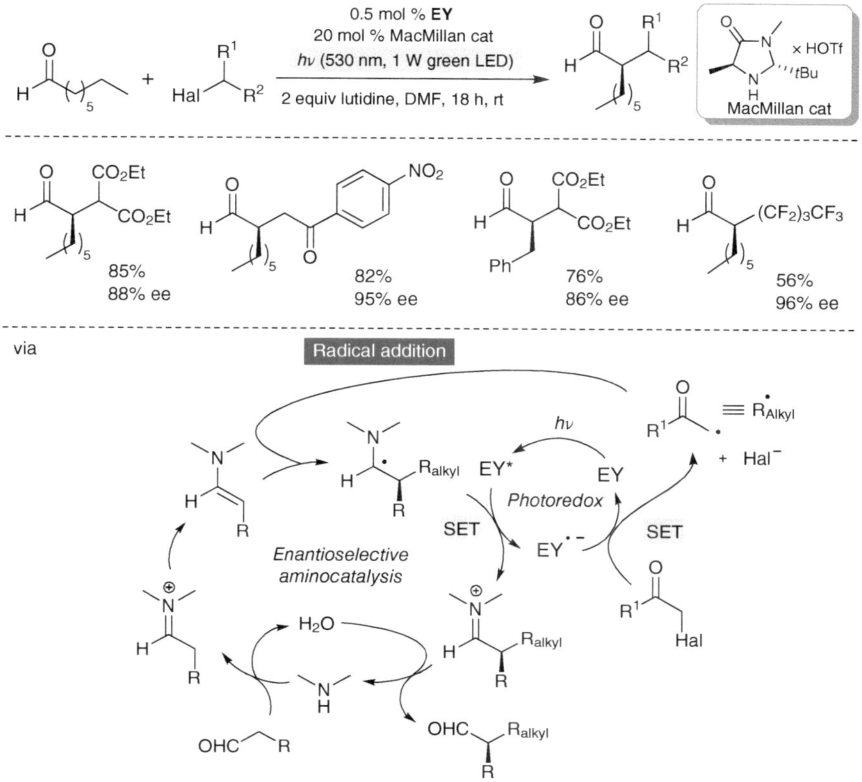

Scheme 6.72 Metal-free, enantioselective α-alkylation and -perfluoroalkylation via synergistic combination of aminocatalysis and photoredox activation.

C_{sp^3}-hybridized nucleophiles which are *in situ* generated as radicals from the corresponding carboxylic acid or the potassium trifluoroborate precursor, respectively, [160] (Scheme 6.77).

Obtained yields using **4CzIPN** as photoredox catalyst are comparable to the seminal examples employing Ir-based photocatalysts [161] (Scheme 6.78).

A synergistic combination of riboflavin (**RF**) and a non-heme iron catalyst allows for the efficient C—H-bond photo-oxygenation of alkyl benzenes to the corresponding carbonyl compounds [162]. The Fe-catalyst significantly increases the catalytic activity of **RF** and is proposed to play a dual role as H_2O_2 disproportionation catalyst (important avoidance of **RF** bleaching) and as oxygenation catalyst (Scheme 6.79).

A special case of multicatalysis where the photocatalyst riboflavin **RF** promotes two different activation modes in a cascade fashion was recently disclosed by Gilmour and coworkers [163]. In their biomimetic coumarin synthesis initial energy transfer photoisomerization of *ortho*-hydroxy-β-substituted cinnamoic acids (see Section 6.2.1, Scheme 6.6) is supposed to be followed by a

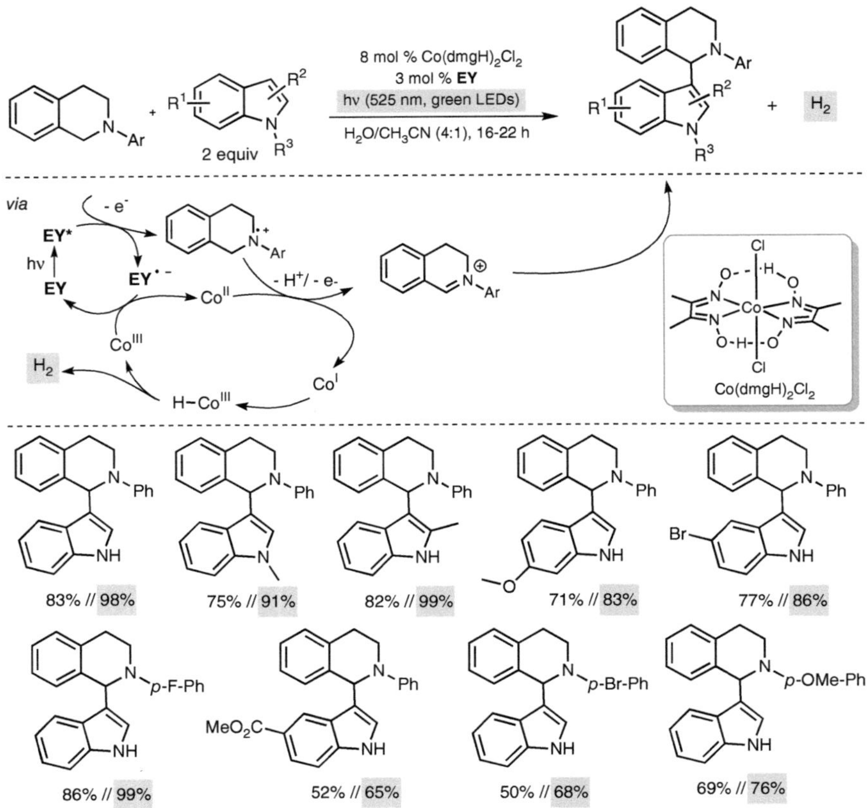

Scheme 6.73 Cross-coupling reactions combined with Co-co-catalyzed hydrogen evolution.

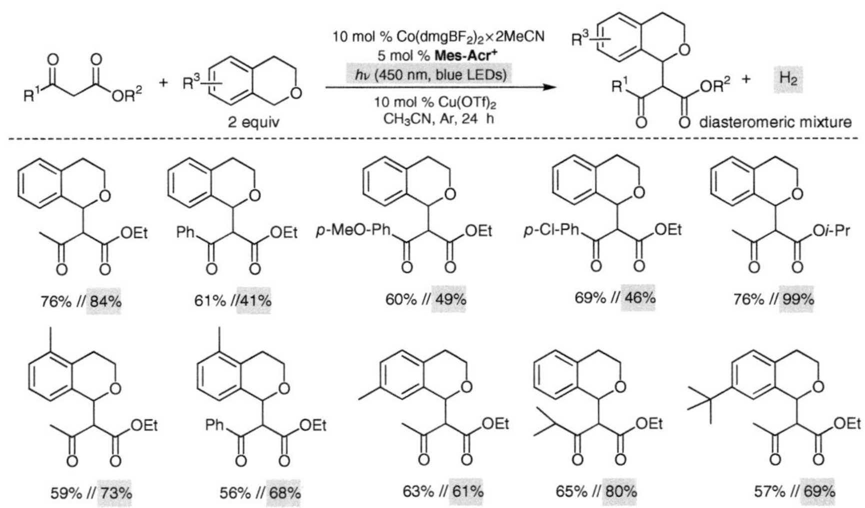

Scheme 6.74 Cross-coupling transformation of isochromans and β-keto esters combined with hydrogen-evolution.

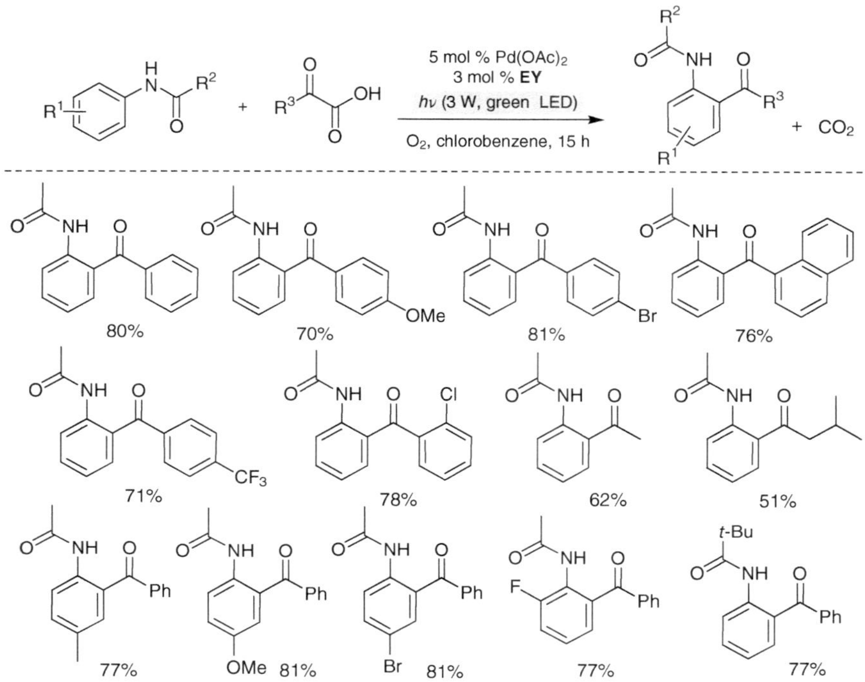

Scheme 6.75 Decarboxylative *ortho*-acylation of acetanilides with α-oxocarboxylic acids by merging Pd and photoredox catalysis.

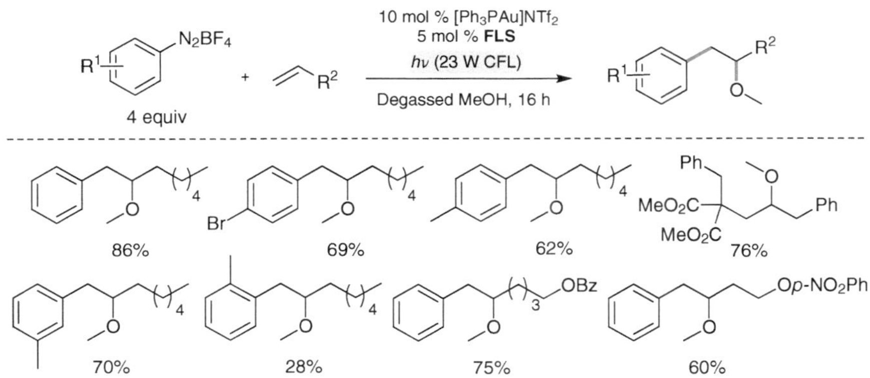

Scheme 6.76 Intermolecular oxyarylation of alkenes by dual photoredox and gold catalysis.

photocatalytic SET-oxidation/deprotonation sequence to generate a carboxylic radical. This then cyclizes and yields the target coumarin upon a formal HAT transfer (or SET/deprotonation) to riboflavin which is released in its reduced form $\mathbf{RFH_2}$. Oxygen was found to be essential as terminal oxidant and for the reoxidation of the reduced riboflavin photocatalyst (Scheme 6.80).

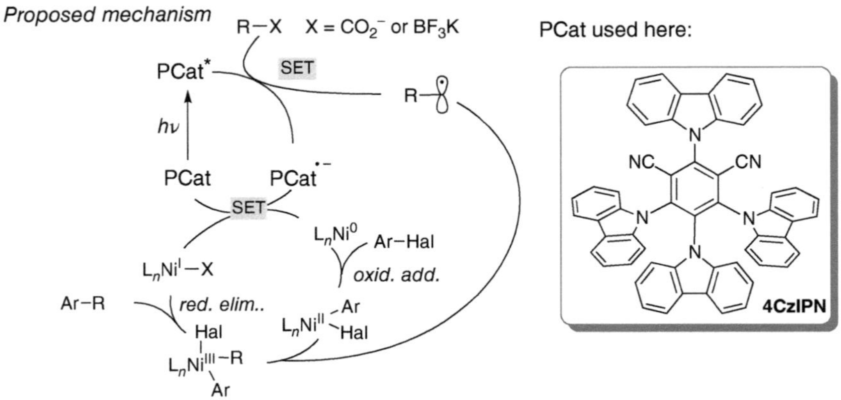

Scheme 6.77 Organic photoredox/Ni dual catalytic C(sp^3)—C(sp^2) cross-coupling – mechanistic proposal.

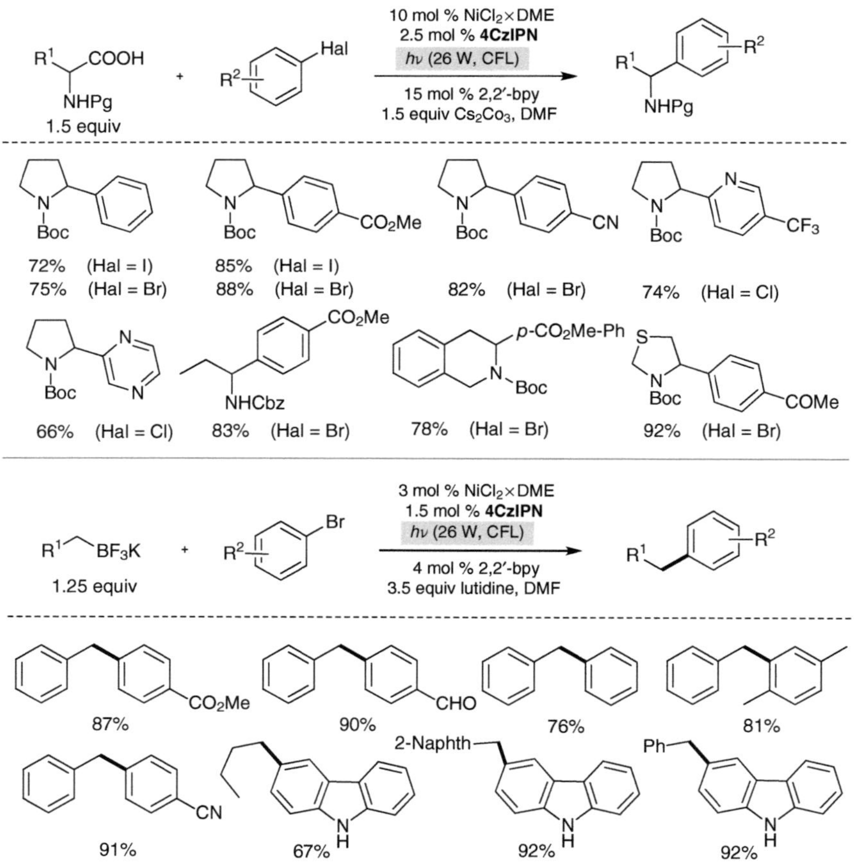

Scheme 6.78 Organic photoredox/Ni dual catalytic C(sp^3)—C(sp^2) cross-coupling employing either carboxylic acids or potassium trifluoroborates as alkyl radical precursor.

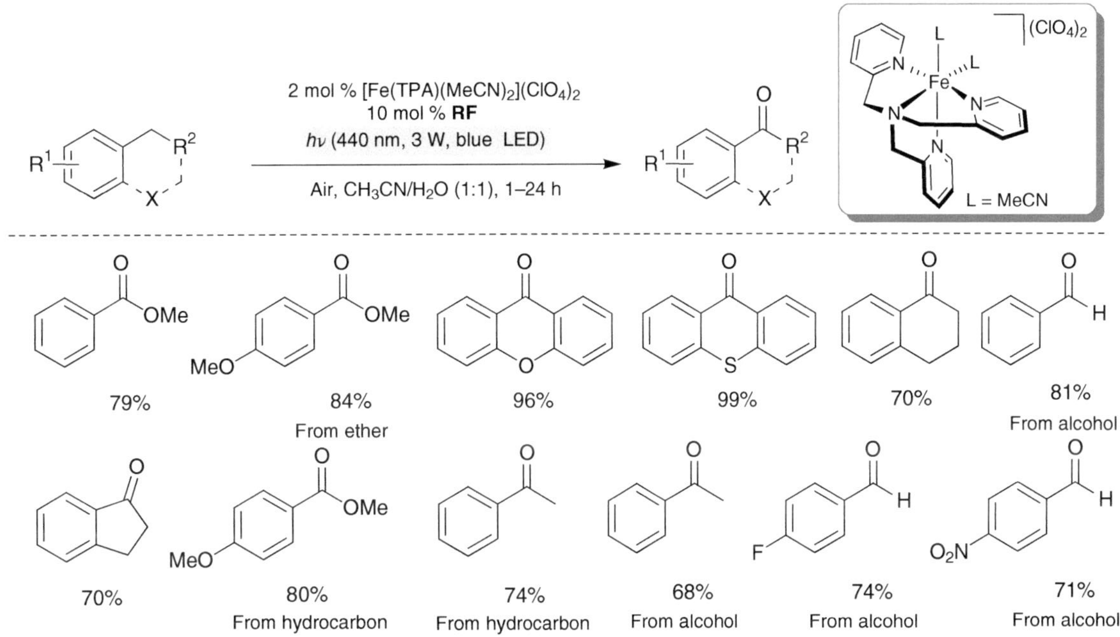

Scheme 6.79 C—H photo-oxygenation of alkyl benzenes catalyzed by Riboflavin and a non-heme Fe catalyst.

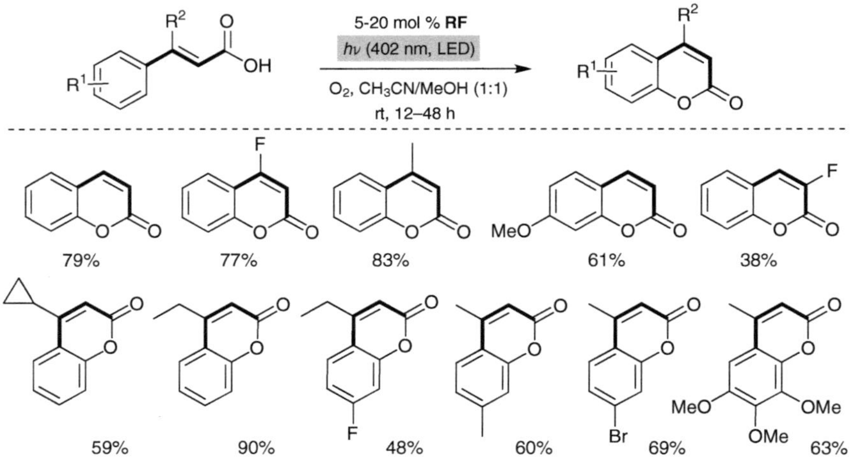

Scheme 6.80 Cascade energy transfer/SET catalysis with ribolflavin for the synthesis of coumarins.

6.3 Conclusion and Outlook

As outlined in this chapter early seminal work has set the stage for a huge range of possible applications of purely organic, visible light photocatalysis. Following the significant improvements over the last few years the way is now cleared for new opportunities and developments in synthetic methodology to exploit the benefits of metal-free photocatalysis for sustainable and formerly unattainable transformations.

References

1 For general synthetic photochemistry: (a) Albini, A. (2016) *Photochemistry Past, Present and Future*, Springer, Berlin, Heidelberg; (b) Bach, T. and Hehn, J.P. (2011) *Angew. Chem., Int. Ed.*, **50**, 1000–1045; (c) Hoffmann, N. (2008) *Chem. Rev.*, **108**, 1052–1103; (d) Fagnoni, M., Dondi, D., Ravelli, D. and Albini, A. (2007) *Chem. Rev.*, **107**, 2725–2756.

2 For recent reviews on visible light photocatalysis, see: (a) Schultz, D.M. and Yoon, T.P. (2014) *Science*, **343**, 1239176; (b) Prier, C.K., Rankic, D.A., and MacMillan, D.W.C. (2013) *Chem. Rev.*, **113**, 5322–5363; (c) König, B. (ed.) (2013) *Chemical Photocatalysis*, de Gruyter, Berlin; (d) Narayanam, J.M.R. and Stephenson, C.R.J. (2011) *Chem. Soc. Rev.*, **40**, 102–113; (e) Telpý, F. (2011) *Collect. Czech. Chem. Commun.*, **76**, 859–917; (f) Zeitler, K. (2009) *Angew. Chem., Int. Ed.*, **48**, 9785–9789.

3 (a) Balzani, V., Bergamini, G., and Ceroni, P. (2015) *Angew. Chem., Int. Ed.*, **54**, 11320–11337; (b) Joya, K.S., Joya, Y.F., Ocakoglu, K., and van de Krol, R. (2013) *Angew. Chem., Int. Ed.*, **52**, 10426–10437; (c) Cogdell, R.J., Brotosudarmo, T.H.P., Gardiner, A.T. *et al.* (2010) *Biofuels*, **1**, 861–876.

4 (a) Bach, T. (2015) *Angew. Chem., Int. Ed.*, **54**, 11294–11295; (b) Brimioulle, R., Lenhart, D., Maturi, M.M., and Bach, T. (2015) *Angew. Chem., Int. Ed.*, **54**, 3872–3890; (c) Griesbeck, A.G., Hoffmann, N., and Warzecha, K.-D. (2007) *Acc. Chem. Res.*, **40**, 128–140.

5 Albini, A. and Fagnoni, M. (eds) (2010) *Handbook of Synthetic Photochemistry*, Wiley-VCH, Weinheim.

6 Kavarnos, G.J. (1993) *Fundamentals of Photoinduced Electron Transfer*, VCH, New York.

7 For selected recent example of radical generation by HAT with an organic photocatalyst, albeit using UV light, see:Amaoka, Y., Nagatomo, M., Watanabe, M. *et al.* (2014) *Chem. Sci.*, **5**, 4339–4345.

8 For some selected recent examples, see: (a) Protti, S., Fagnoni, M., and Ravelli, D. (2015) *ChemCatChem*, **7**, 1516–1523; (b) Romero, N. and Nicewicz, D.A. (2015) *J. Am. Chem. Soc.*, **136**, 17024–17035; (c) Cuthbertson, J.D. and MacMillan, D.W.C. (2015) *Nature.*, **519**, 74–77; (d) Nguyen, T.M., Manohar, N., and Nicewicz, D.A. (2014) *Angew. Chem., Int. Ed.*, **53**, 6198–6201; (e) Hager, D. and MacMillan, D.W.C. (2014) *J. Am. Chem. Soc.*, **136**, 16986–16989; (f) Wilger, D.J., Gesmundo, N.J., and Nicewicz, D.A. (2013) *Chem. Sci.*, **4**, 2625–2629.

9 Luo, Y.-R. (2003) *Handbook of Bond Dissociation Energies in Organic Compounds*, CRC Press LLC, Boca Raton.

10 Studer, A. and Curran, D.P. (2014) *Nature Chem.*, **6**, 765–773.

11 (a) Ciamician, G. (1912) *Science*, **36**, 385–394; (b) Stobbe, H. (1908) *Z. Elektrochem. Angew. Phys. Chem.*, **33**, 473–483.

12 (a) Kisch, H. (2013) *Angew. Chem., Int. Ed.*, **52**, 812–847; (b) Maldotti, A. and Molinari, A. (2011) *Top. Curr. Chem.*, **303**, 185–216.

13 (a) Zheng, Y., Lin, L., Wang, B., and Wang, X. (2015) *Angew. Chem., Int. Ed.*, **54**, 12868–12884; (b) Cao, S., Low, J., Yu, J., and Jaroniec, M. (2015) *Adv. Mater.*, **27**, 2150–2176; (c) Wang, Y., Wang, X., and Antonietti, M. (2012) *Angew. Chem., Int. Ed.*, **51**, 68–89; (d) Rouch, W.D., Zhang, M., and McCulla, R.D. (2012) *Tetrahedron Lett.*, **53**, 4942–4945.

14 (a) Kisch, H. (2015) *Semiconductor Photocatalysis: Principles and Applications*, Wiley-VCH, Weinheim; (b) Cherevatskaya, M. and König, B. (2014) *Russ. Chem. Rev.*, **83**, 183–195.

15 Juris, A., Balzani, V., Barigelletti, F. *et al.* (1988) *Coord. Chem. Rev.*, **84**, 85–277.

16 For selected recent examples, see: (a) McTiernan, C.D., Morin, M., McCallum, T. *et al.* (2016) *Catal. Sci. Technol.*, **6**, 201–207; (b) Larsen, M.H., Houk, K.N., and Hashmi, A.S.K. (2015) *J. Am. Chem. Soc.*, **137**, 10668–10676; (c) Revol, G., McCallum, T., Morin, M. *et al.* (2013) *Angew. Chem., Int. Ed.*, **52**, 13342–13345.

17 Stevenson, S.M., Shores, M.P., and Ferreira, E.M. (2015) *Angew. Chem., Int. Ed.*, **54**, 6506–6510.

18 Yin, H., Carroll, P.J., Manor, B.C. *et al.* (2016) *J. Am. Chem. Soc.*, **138**. doi: 10.1021/jacs.6b02248

19 For a recent review, see: Paria, S. and Reiser, O. (2014) *ChemCatChem*, **6**, 2477–2483.

20 (a) Singh, A., Teegardin, K., Kelly, M. *et al.* (2015) *J. Organomet. Chem.*, **776**, 51–59; (b) Koike, T. and Akita, M. (2014) *Inorg. Chem. Front.*, **1**, 562–576; (c) Ischay, M.A., Ament, M.S., and Yoon, T.P. (2012) *Chem. Sci.*, **3**, 2807–2811.

21 For illustrating examples of the removal of Ru residues in the context of olefin metathesis, see: (a) Szczepaniak, G., Urbaniak, K., Wierzbicka, C. *et al.* (2015) *ChemSusChem,*, **8**, 4139–4148; (b) French, J.M., Caras, C.A., and Diver, S.T. (2013) *Org. Lett.*, **15**, 5416–5419; (c) Vougioukalakis, G.C. (2012) *Chem. Eur. J.*, **18**, 8868–8880.

22 For a review on removal of transition metals in a ATRA reaction, see: Ding, M., Jiang, X., Zhang, L. *et al.* (2015) *Macromol. Rapid Commun.*, **36**, 1702–1721.

23 For recent reviews, see: (a) Nicewicz, D.A. and Nguyen, T.M. (2014) *ACS Catal.*, **4**, 355–360; (b) Fukuzumi, S. and Ohkubo, K. (2014) *Org. Biomol. Chem.*, **12**, 6059–6071; (c) Fukuzumi, S. and Ohkubo, K. (2013) *Chem. Sci.*, **4**, 561–574; (d) Ravelli, D., Fagnoni, M. and Albini, A. (2013) *Chem. Soc. Rev.*, **42**, 97–113; (e) Marin, M. L., Santos-Juanes, L., Arques, A., Amat, A. M. and Miranda, M. A. (2012) *Chem. Rev.*, **112**, 1710–1750; (f) Ravelli, D. and Fagnoni, M. (2012) *ChemCatChem*, **4**, 169–171.

24 Croce, R. and van Amerongen, H. (2014) *Nature Chem. Biol.*, **10**, 492–501.

25 Falk, H. (1999) *Angew. Chem., Int. Ed.*, **38**, 3116–3136.

26 (a) Chaves, I., Pokorny, R., Byrdin, M. *et al.* (2011) *Annu. Rev. Plant Biol.*, **62**, 335–364; (b) Sancar, A. (2000) *Annu. Rev. Biochem.*, **69**, 31–67.

27 (a) Zamadar, M. and Greer, A. (2010) in *Handbook of Synthetic Photochemistry* (eds A. Albini and M. Fagnoni), Wiley-VCH, Weinheim, pp. 353–386; (b) Montagnon, T., Tofi, M., and Vassilikogiannakis, G. (2008) *Acc. Chem. Res.*, **41**, 1001–1011; (c) Clennan, E.L. and Pace, A. (2005) *Tetrahedron*, **61**, 6665–6691; (d) DeRosa, M.C. and Crutchley, R.J. (2002) *Coord. Chem. Rev.*, **233–234**, 351–371.

28 For an extended compliation on (photo)physical data of common singlet oxygen photosensitizers, see:Redmond, R.W. and Gamlin, J.N. (1999) *Photochem. Photobiol.*, **70**, 391–475.

29 For recent reviews, see: (a) Dadashi-Silab, S., Doran, S., and Yagci, Y. (2016) *Chem. Rev.*, 116. doi: 10.1021/acs.chemrev.5b00586 (b) Trotta, J.T. and Fors, B.P. (2016) *Synlett*, **27**, 702–713.

30 For selected recent examples using visible light photooxidants, see: (a) Perkowski, A.J., You, W., and Nicewicz, D.A. (2015) *J. Am. Chem. Soc.*, **137**, 7580–7583; (b) Ogawa, K.A., Goetz, A.E., and Boydston, A.J. (2015) *J. Am. Chem. Soc.*, **137**, 1400–1403; (c) Aguirre-Soto, A., Lim, C.-H., Hwang, A.T. *et al.* (2014) *J. Am. Chem. Soc.*, **136**, 7418–7427; (d) Treat, N.J., Sprafke, H., Kramer, J.W. *et al.* (2014) *J. Am. Chem. Soc.*, **136**, 16096–16101.

31 Miranda, M. A. and García, H. (1994) *Chem. Rev.*, **94**, 1063–1089.

32 Pérez-Prieto, J., Galian, R. E. and Miranda, M. A. (2006) *Mini-Rev. Org. Chem.*, **3**, 117–135.

33 (a) Pandey, G., Ghorai, M.K., and Hajra, S. (1996) *Pure & Appl. Chem.*, **68**, 653–658; (b) Pandey, G. and Hajra, S. (1994) *Angew. Chem., Int. Ed.*, **33**, 1169–1171; (c) Gieseler, A., Steckhan, E., Wiest, O., and Knoch, F.

(1991) *J. Org. Chem.*, **56**, 1405–1411; (d) Fox, M.A. and Chanon, M. (eds) (1988) *Photoinduced Electron Transfer, Parts A & C*, Elsevier, Amsterdam; (e) Mattay, J., Gersdorf, J., and Mertes, J. (1985) *J. Chem. Soc., Chem. Commun.*, 1088–1090; (f) van Bergen, T. J., Hedstrand, D. M., Kruizinga, W. M. and Kellogg, R. M. (1979) *J. Org. Chem. A*, **44**, 4953–4962; (g) Hedstrand, D.M., Kruizinga, W.M., and Kellogg, R.M. (1978) *Tetrahedron Lett.*, **14**, 1255–1258.

34 For recent reviews, see: (a) Iida, H., Imada, Y., and Murahashi, S.-I. (2015) *Org. Biomol. Chem.*, **13**, 7599–7613; (b) Insińska-Rak, M. and Sikorski, M. (2014) *Chem. Eur. J.*, **20**, 15280–15291; (c) de Gonzalo, G. and Fraaije, M.W. (2013) *ChemCatChem*, **5**, 403–415; (d) König, B., Kümmel, S., and Cibulka, R. (2013) in *Chemical Photocatalysis* (ed. B. König), de Gruyter, Berlin, pp. 45–66.

35 (a) Hari, D.P. and König, B. (2014) *Chem. Commun.*, **50**, 6688–6699; (b) Majek, M., Filace, F., and Jacobi von Wangelin, A. (2014) *Beilstein J. Org. Chem.*, **10**, 981–989.

36 (a) Benniston, A.C., Harriman, A., Li, P. *et al.* (2005) *J. Am. Chem. Soc.*, **127**, 6054–16064; (b) Fukuzumi, S., Kotani, H., Ohkubo, K. *et al.* (2004) *J. Am. Chem. Soc.*, **126**, 1600–1601.

37 Jose, J. and Burgess, K. (2006) *Tetrahedron*, **62**, 11021–11037.

38 (a) Zhao, J., Xu, K., Yang, W., Wang Z. and Zhong, F. (2015) *Chem. Soc. Rev.*, **44**, 8904–8939; (b) Lincoln, R., Greene, L. E., Krumova, K., Ding, Z. and Cosa, G. (2014) *J. Phys. Chem. A*, **118**, 10622–10630.

39 For a recent review, see: Bureš, F. (2014) *RSC Adv.*, **4**, 58826–58851.

40 For a recent overview on theory and applications: (a) Lima, C. G. S., Lima, T. d. M., Duarte, M., Jurberg, I. D. and Paixão, M. W. (2016), *ACS Catal.*, **6**, 1389–1407; For general information:(b) Fukuzumi, S. (1997) *Bull. Chem. Soc. Jpn.*, **70**, 1–28; (c) Kochi, J.K. (1988) *Angew. Chem., Int. Ed.*, **27**, 1227–1266; (d) Foster, R. (1969) *Organic Charge-Transfer Complexes*, Academic Press, New York.

41 (a) Kandukuri, S.R., Bahamonde, A., Chatterjee, I. *et al.* (2014) *Angew. Chem., Int. Ed.*, **53**, 1485–1489; (b) Nappi, M., Bergonzini, G., and Melchiorre, P. (2015) *Angew. Chem., Int. Ed.*, **54**, 4921–4925; (c) Arceo, E., Bahamonde, A., Bergonzini, G., and Melchiorre, P. (2014) *Chem. Sci.*, **5**, 2438–2442; (d) Arceo, E., Jurberg, I.D., Álvarez-Fernández, A., and Melchiorre, P. (2013) *Nature Chem.*, **5**, 750–756.

42 For recent reviews, see: (a) Wende, R.C. and Schreiner, P.R. (2012) *Green Chem.*, **14**, 1821–1849; (b) Allen, A.E. and MacMillan, D.W.C. (2012) *Chem. Sci.*, **3**, 633–658.

43 (a) Zhao, J., Wu, W., Sun, J., and Guo, S. (2013) *Chem. Soc. Rev.*, **42**, 5323–5351; (b) Albini, A. (1981) *Synthesis*, 249–262.

44 For recent synthetic applications of triplet sensitization with Ru and Ir transition metal complexes in the range of 190 to 250 kJ mol^{-1}, see: (a) Farney, E.P. and Yoon, T.P. (2014) *Angew. Chem., Int. Ed.*, **53**, 793–797; (b) Lu, Z. and Yoon, T.P. (2012) *Angew. Chem., Int. Ed.*, **51**, 10329–10332.

45 Arceo, E., Montroni, E., and Melchiorre, P. (2014) *Angew. Chem., Int. Ed.*, **53**, 12064–12068.

46 For reviews, see: (a) Poplata, S., Tröster, A., Zhou, Y.-Q., and Bach, T. (2016) *Chem. Rev.*, **116**. doi: 10.1021/acs.chemrev.5b00723 (b) Hehn, J. P., Müller, C. and Bach T. (2010) in *Handbook of Synthetic Photochemistry*, (eds Albini, A. and Fagnoni, M.), Wiley-VCH, Weinheim, pp. 171–215.

47 Dembitsky, V.M. (2008) *J. Nat. Med.*, **62**, 1–33.

48 (a) Maturi, M.M., Wenninger, M., Alonso, R. *et al.* (2013) *Chem. Eur. J.*, **19**, 7461–7472; (b) Müller, C., Bauer, A., and Bach, T. (2009) *Angew. Chem., Int. Ed.*, **48**, 6640–6642.

49 Alonso, R. and Bach, T. (2014) *Angew. Chem., Int. Ed.*, **53**, 4368–4371.

50 Kumarasamy, E., Raghunathan, R., Jockusch, S., Ugrinov, A. and Sivaguru, J. (2014) *J. Am. Chem. Soc.*, **136**, 8729–8737.

51 Mojr, V., Svobodová, E., Straková, K., Neveselý, T.,Chudova, J., Dvořáková, H. and Cibulka, R. (2015) *Chem. Commun.*, **51**, 12036–12039.

52 Metternich, J.B. and Gilmour, R. (2015) *J. Am. Chem. Soc.*, **137**, 11254–11257.

53 Wang, H.-B., Zhai, B.-C., Tang, W.-J. *et al.* (2007) *Chem. Phys.*, **333**, 229–235.

54 Pandey, G., Rani, K.S., and Lakshmaiah, G. (1992) *Tetrahedron Lett.*, **35**, 5107–5110.

55 Pandey, G. (1992) *Synlett*, 546–552.

56 Pandey, G., Pal, S., and Laha, R. (2013) *Angew. Chem., Int. Ed.*, **52**, 5146–5149.

57 Pandey, G. and Laha, R. (2015) *Angew. Chem., Int. Ed.*, **54**, 14875–14879.

58 Lerch, S., Unkel, L.N., Wienefeld, P., and Brasholz, M. (2014) *Synlett*, **25**, 2673–2680.

59 Cui, L., Matusaki, Y., Tada, N. *et al.* (2013) *Adv. Synth. Catal.*, **355**, 2203–2207.

60 Yamaguchi, T., Nobuta, T., Tada, N. *et al.* (2014) *Synlett*, **25**, 1453–1457.

61 Rusch, F., Unkel, L.N., Alpers, D. *et al.* (2015) *Chem. Eur. J.*, **21**, 8336–8340.

62 Mloch, J. and Steckhan, E. (1985) *Angew. Chem., Int. Ed.*, **24**, 412–414.

63 Cismesia, M.A. and Yoon, T.P. (2015) *Chem. Sci.*, **6**, 5426–5434.

64 Perkowski, A. J., Cruz, C. L.and Nicewicz, D. A. (2015) *J. Am. Chem. Soc.*, **137**, 5684–15687.

65 Leow, D. (2014) *Org. Lett.*, **16**, 5812–5815.

66 Romero, N.A., Margrey, K.A., Tay, N.E., and Nicewicz, D.A. (2015) *Science*, **349**, 1326–1330.

67 Margrey, K.A. and Nicewicz, D.A. (2016) *Acc. Chem. Res.*, **49**, 1997–2006.

68 Nicewicz, D. A. and Hamilton, D. S. (2014) *Synlett*, **25**, 1191–1196.

69 Grandjean, J.-M.M. and Nicewicz, D.A. (2013) *Angew. Chem., Int. Ed.*, **52**, 3967–3971.

70 Zeller, M.A., Riener, M., and Nicewicz, D.A. (2014) *Org. Lett.*, **16**, 4810–4813.

71 Nguyen, T.M. and Nicewicz, D.A. (2013) *J. Am. Chem. Soc.*, **135**, 9588–9591.

72 Gesmundo, N.J., Grandjean, J.-M.M., and Nicewicz, D.A. (2015) *Org. Lett.*, **17**, 1316–1319.

73 Wilger, D. J., Grandjean, J.-M. M., Lammert, T. R. and Nicewicz, D. A. (2014) *Nature Chem.*, **6**, 720–726.

74 Wilger, D.J., Gesmundo, N.J., and Nicewicz, D.A. (2013) *Chem. Sci.*, **4**, 3160–3165.

75 (a) Yoshimi, Y., Itou, T., and Hatanaka, M. (2007) *Chem. Commun.*, 5244–5246; (b) Yoshimi, Y., Masuda, M., Mizunashi, T. *et al.* (2009) *Org. Lett.*, **11**, 4652–4655.

76 For a recent review on visible light photocatalytic decarboxylations, see: Xuan, J., Zhang, Z.-G., and Xiao, W.-J. (2015) *Angew. Chem., Int. Ed.*, **54**, 15632–15641.

77 Cassani, C., Bergonzini, G., and Wallentin, C.-J. (2014) *Org. Lett.*, **16**, 4228–4231.

78 (a) Griffin, J.D., Zeller, M.A., and Nicewicz, D.A. (2015) *J. Am. Chem. Soc.*, **137**, 11340–11348.

79 Wu, X., Meng, C., Yuan, X. *et al.* (2015) *Chem. Commun.*, **51**, 11864–11867.

80 Chinzei, T., Miyazawa, K., Yasu, Y. *et al.* (2015) *RSC Adv.*, **5**, 21297–21300.

81 Ramirez, N.P., Bosque, I., and Gonzalez-Gomez, J.C. (2015) *Org. Lett.*, **17**, 4550–4553.

82 Xuan, J., Xia, X.-D., Zeng, T.-T. *et al.* (2014) *Angew. Chem., Int. Ed.*, **53**, 5653–5656.

83 Zeng, T.-T., Xuan, J., Ding, W. *et al.* (2015) *Org. Lett.*, **17**, 4070–4073.

84 For recent reviews regarding the photoredox catalytic transformation of amino compounds, see: (a) Hu, J., Wang, J., Nguyen, T.H., and Zheng, N. (2013) *Beilstein J. Org. Chem.*, **9**, 1977–2001; (b) Shi, L. and Xia, W. (2012) *Chem. Soc. Rev.*, **41**, 7687–7697.

85 Hari, D.P. and König, B. (2011) *Org. Lett.*, **13**, 3852–3855.

86 Pan, Y., Wang, S., Kee, C.W. *et al.* (2011) *Green Chem.*, **13**, 3341–3344.

87 (a) Ruiz Espelt, L., Wiensch, E.M., and Yoon, T.P. (2013) *J. Org. Chem.*, **78**, 4107–4114; (b) Miyake, Y., Nakajima, K., and Nishibayashi, A. (2012) *J. Am. Chem. Soc.*, **134**, 3338–3341; (c) Kohls, P., Jadhav, D., Pandey, G., and Reiser, O. (2012) *Org. Lett.*, **14**, 672–675.

88 Rueping, M., Vila, C., and Bootwicha, T. (2013) *ACS Catal.*, **3**, 1676–1680.

89 Xuan, J., Feng, Z.-J., Duan, S.-W., and Xiao, W.-J. (2012) *RSC Adv.*, **2**, 4065–4068.

90 Vila, C., Lau, J., and Rueping, M. (2014) *Beilstein J. Org. Chem.*, **10**, 1233–1238.

91 Liang, Z., Xu, S., Tian, W., and Zhang, R. (2015) *Beilstein J. Org. Chem.*, **11**, 425–430.

92 Chen, L., Chao, C.S., Pan, Y. *et al.* (2013) *Org. Biomol. Chem.*, **11**, 5922–5925.

93 For recent reviews on the photo(redox) catalysis in flow systems, see: (a) Cambié, D., Bottecchia, C., Straathof, N.J.W. *et al.* (2016) *Chem. Rev.*, **116**. doi: 10.1021/acs.chemrev.5b00707 (b) Plutschack, M.B., Correia, C.A., Seeberger, P.H., and Gilmore, K. (2016) *Top. Organomet. Chem.*, **57**, 43–76; (c) Garlets, Z.J., Nguyen, J.D., and Stephenson, C.R.J. (2014) *Isr. J. Chem.*, **54**, 351–360.

94 Fabry, D.C., Sugiono, E., and Rueping, M. (2016) *React. Chem. Eng.*, **1**, 129–133.

95 Talla, A., Driessen, B., Straathof, N.J.W. *et al.* (2015) *Adv. Synth. Catal.*, **357**, 2180–2186.

96 Bu, M.-J., Lu, G.-P., and Cai, C. (2016) *Catal. Sci. Technol.*, **6**, 413–416.

97 Teo, Y.C., Pan, Y., and Tan, C.H. (2013) *ChemCatChem*, **5**, 235–240.

98 Fan, W.,Yang, Q., Xu, F. and Li, P. (2014) *J. Org. Chem.*, **79**, 10588–10592.

99 Yadav, A.K., Srivastava, V.P., and Yadav, L.D.S. (2013) *New J. Chem.*, **37**, 4119–4124.

100 Srivastava, V.P., Yadav, A.K., and Yadav, L.D.S. (2013) *Synlett*, **24**, 465–470.

101 Xiao, T., Li, L., Lin, G. *et al.* (2014) *Green Chem.*, **16**, 2418–2421.

102 Yang, Z., Li, H., Zhang, L. *et al.* (2015) *Chem. Eur. J.*, **21**, 14723–14727.

103 Li, X., Gu, X., Li, Y. and Li, P. (2014) *ACS Catal.*, **4**, 1897–1900.

104 Zhao, Y., Zhang, C., Chin, K.F. *et al.* (2014) *RSC Adv.*, **4**, 30062–30067.

105 Liu, X., Ye, X., Bureš, F. *et al.* (2015) *Angew. Chem., Int. Ed.*, **54**, 11443–11447.

106 Samanta, S., Das, S. and Biswas, P. (2013) *J. Org. Chem.*, **78**, 11184–11193.

107 (a) Boens, N., Leen, V., and Dehaen, W. (2012) *Chem. Soc. Rev.*, **41**, 1130–1172; (b) Ulrich, G., Ziessel, R., and Harriman, A. (2008) *Angew. Chem., Int. Ed.*, **47**, 1184–1201.

108 Kamkaew, A., Lim, S.H., Lee, H.B. *et al.* (2013) *Chem. Soc. Rev.*, **42**, 77–88.

109 Kubo, Y., Eguchi, D., Matsumoto, A. *et al.* (2014) *J. Mater. Chem. A*, **2**, 5204–5211.

110 Huang, L. and Zhao, J. (2013) *RSC Adv.*, **3**, 23377–23388.

111 Huang, L. and Zhao, J. (2013) *Chem. Commun.*, **49**, 3751–3753.

112 Wang, X.-Z., Meng, Q.-Y., Zhong, J.-J., Gao, X.-W., Lei, T., Zhao, L.-M., Li, Z.-J., Chen, B., Tung, C.-H. and Wu, L.-Z. (2015) *Chem. Commun.*, **51**, 11256–11259.

113 For an ESR-spectroscopic proof of the $O_2^{\bullet-}$ radical anion in the presence of TBA eosin y, see: Liu, Q., Li, Y.-N., Zhang, H.-H., Chen, B., Tung, C.-H. and Wu, L.-Z. (2012) *Chem. Eur. J.*, **18**, 620–627.

114 Liu, Z., Zhang, Y., Cai, Z. *et al.* (2015) *Adv. Synth. Catal.*, **357**, 589–593.

115 Shen, Z.-C., Yang, P., and Tang, Y. (2016) *J. Org. Chem.*, **81**, 309–317.

116 Pitre, S. P., McTiernan, C. F., Ismaili, H. and Scaiano, J. C. (2013) *J. Am. Chem. Soc.*, **135**, 13286–13289.

117 Zou, Y.-Q., Chen, J.-R., Liu, X.-P. *et al.* (2012) *Angew. Chem., Int. Ed.*, **51**, 784–788.

118 Penders, I.G.T.M., Amara, Z., Horvath, R. *et al.* (2015) *RSC Adv.*, **5**, 6501–6504.

119 Hedstrand, D.M., Kruizinga, W.H., and Kellogg, R.M. (1978) *Tetrahedron Lett.*, **19**, 1255–1258.

120 (a) Neumann, M., Füldner, S., König, B., and Zeitler, K. (2011) *Angew. Chem., Int. Ed.*, **50**, 951–954; for the $Ru(bpy)_3^{2+}$-catalyzed dehalogenation process, see: (b) Narayanam, J., Tucker, J.W., and Stephenson, C.R.J. (2009) *J. Am. Chem. Soc.*, **131**, 8756–8757.

121 Pitre, S. P., McTiernan, C. D., Ismaili, H. and Scaiano, J. C. (2014) *ACS Catal.*, **4**, 2530–2535.

122 Jadhav, S. D., Bakshi, D. and Singh, A. (2015) *J. Org. Chem.*, **80**, 10187–10196.

123 Meyer, A. U., Slanina, T., Yao, C.-J. and König, B. (2016) *ACS Catal.*, **6**, 369–375.
124 Srivastava, V.P., Yadav, A.K., and Yadav, L.D. (2014) *Synlett*, **25**, 665–670.
125 Yadav, A.K., Srivastava, V.P., and Yadav, L.D. (2014) *RSC Adv.*, **4**, 4181–4186.
126 Yoshioka, E., Kohtani, S., Jichu, T. *et al.* (2015) *Synlett*, **26**, 265–270.
127 Lin, Q.-Y., Xu, X.H., and Qing, F.-L. (2015) *Org. Biomol. Chem.*, **13**, 8740–8749.
128 Yang, W., Yang, S., Li, P., and Wang, L. (2015) *Chem. Commun.*, **51**, 7520–7523.
129 Liu, H., Feng, W., Kee, C.W. *et al.* (2010) *Green Chem.*, **12**, 953–956.
130 Yang, D.-T., Meng, Q.-Y., Zhong, J.-J. *et al.* (2013) *Eur. J. Org. Chem.*, 7528–7532.
131 Gu, L., Jin, C., Liu, J. *et al.* (2014) *Chem. Commun.*, **50**, 4643–4645.
132 Cantillo, D., de Frutos, O., Rincón, J.A. *et al.* (2014) *Org. Lett.*, **16**, 896–899.
133 Hari, D.P., Schroll, P., and König, B. (2012) *J. Am. Chem. Soc.*, **134**, 2958–2961.
134 Maity, P., Kundu, D., and Ranu, B.C. (2015) *Eur. J. Org. Chem.*, 1727–1734.
135 Shaaban, S., Jolit, A., Petkova, D., and Maulide, N. (2015) *Chem. Commun.*, **51**, 13902–13905.
136 Xiao, T., Dong, X., Tang, Y., and Zhou, L. (2012) *Adv. Synth. Catal.*, **354**, 3195–3199.
137 Hari, D.P., Hering, T., and König, B. (2012) *Org. Lett.*, **14**, 5334–5337.
138 Majek, M. and Jacobi von Wangelin, A. (2015) *Angew. Chem., Int. Ed.*, **54**, 2270–2274.
139 Guo, W., Lu, L.-Q., Wan, Y. *et al.* (2015) *Angew. Chem., Int. Ed.*, **54**, 2265–2269.
140 Gu, L., Jin, C., and Liu, J. (2015) *Green Chem.*, **17**, 3733–3736.
141 Zhang, H.-T., Gu, L.-J., Huang, X.-Z. *et al.* (2016) *Chin. Chem. Lett.*, **27**, 256–260.
142 Hari, D.P., Hering, T., and König, B. (2014) *Angew. Chem., Int. Ed.*, **53**, 725–728.
143 Yao, C.-J., Sun, Q., Rastogi, N., and König, B. (2015) *ACS Catal.*, **5**, 2935–2938.
144 Yu, J., Zhang, L., and Yan, G. (2012) *Adv. Synth. Catal.*, **354**, 2625–2628.
145 Majek, M. and Jacobi von Wangelin, A. (2013) *Chem. Commun.*, **49**, 5507–5509.
146 Majek, M., Filace, F., and Jacobi von Wangelin, A. (2015) *Chem. Eur. J.*, **21**, 4518–4522.
147 (a) Zhou, J., Liu, Q., Feng, W. *et al.* (2015) *Chem. Rev.*, **115**, 395–465; (b) Ceroni, P. (2011) *Chem. Eur. J.*, **17**, 9560–9564.
148 (a) Majek, M., Faltermeier, W., Dick, B. *et al.* (2015) *Chem. Eur. J.*, **21**, 15496–15501; (b) Häring, M., Perez-Ruiz, R., Jacobi von Wangelin, A., and Díaz-Díaz, D. (2015) *Chem. Commun.*, **51**, 16848–16851.
149 Ghosh, I., Gosh, T., and König, B. (2014) *Science*, **346**, 725–728.
150 For reviews focussing on multicatalytic combinations with photocatalysis, see: (a) Neumann, M. and Zeitler, K. (2013) in *Chemical Photocatalysis* (ed.

B. König), de Gruyter, Berlin, pp. 151–168; (b) Skubi, K.L., Blum, T.R., and Yoon, T.P. (2016) *Chem. Rev.*, **116**. doi: 10.1021/acs.chemrev.6b00018
151 Pan, Y.; Kee, C. W.; Chen, L.; Tan, C.-H. (2011) *Green Chem.* 2011, **13**, 2682–2685
152 Neumann, M. and Zeitler, K. (2013) *Chem. Eur. J.*, **19**, 6950–6955.
153 (a) Neumann, M. and Zeitler, K. (2012) *Org. Lett.*, **14**, 2658–2661; for a similar approach also see: (b) Fidaly, K., Ceballos, C., Falguières, A. *et al.* (2012) *Green Chem.*, **14**, 1293–1297.
154 (a) Nicewicz, D. and MacMillan, D.W.C. (2008) *Science*, **322**, 77–80; (b) Nagib, D.A., Scott, M.E., and MacMillan, D.W.C. (2009) *J. Am. Chem. Soc.*, **131**, 10875–10877.
155 (a) Zhong, J.-J., Meng, Q.-Y., Liu, B. *et al.* (2014) *Org. Lett.*, **16**, 1988–1991; (b) Wu, C.-J., Zhong, J.-J., Meng, Q.-Y., Lei, T., Gao, X.-W., Tung, C.-H. and Wu, L.-Z. (2015) *Org. Lett.*, **17**, 884–887.
156 Zhong, J.-J., Wu, C.-J., Meng, Q.-Y. *et al.* (2014) *Adv. Synth. Catal.*, **356**, 2846–2852.
157 Xiang, M., Meng, Q.-Y., Li, J.-X. *et al.* (2015) *Chem. Eur. J.*, **21**, 18080–18084.
158 Zhou, C., Li, P., Zhu, X., and Wang, L. (2015) *Org. Lett.*, **17**, 6198–6201.
159 Hopkinson, M.N., Sahoo, B., and Glorius, F. (2014) *Adv. Synth. Catal.*, **356**, 2794–2800.
160 Luo, J. and Zhang, J. (2016) *ACS Catal.*, **6**, 873–877.
161 (a) Tellis, J. C., Primer, D. N. and Molander, G. A. (2014) *Science*, **345**, 433–436; (b) Zuo, Z., Ahneman, D. T., Chu, L., Terrett, J. A., Doyle, A. G. and MacMillan, D. W. C. (2014) *Science*, **345**, 437–440.
162 Mühldorf, B. and Wolf, R. (2016) *Angew. Chem., Int. Ed.*, **55**, 427–430.
163 Metternich, J.B. and Gilmour, R. (2016) *J. Am. Chem. Soc.*, **138**, 1040–1045.

7

Visible Light and Copper Complexes: A Promising Match in Photoredox Catalysis

Suva Paria and Oliver Reiser

[1] Institut für Organische Chemie, Universität Regensburg, Universitätsstraße 31, 93053 Regensburg, Germany

7.1 Introduction

Ever since in 1912 Ciamician [1] foresaw the potential of photochemistry by exploiting solar energy there have been considerable efforts for more than a century now to tap into this abundant resource as a mediator for chemical transformations [2, 3]. Since most organic compounds are not able to absorb visible light, catalysts are needed that can take this role and subsequently transfer energy or electrons to a substrate to start a chemical reaction. For electron transfer processes, at some point the catalyst needs to be regenerated either by back electron transfer from the substrate or from a sacrificial electron donor or acceptor. Consequently, a suitable photoredox catalyst should be able to efficiently absorb light in the visible light region, have a long life-time in the excited state, and have strong reduction and/or oxidation power to transfer or accept electrons from a substrate.

Most photoredox catalysts that meet these requirements are metal complexes based on ruthenium or iridium. Although advantageous in terms of stability and activity, these metals are scarce and have dramatically increased in price with the discovery of their versatile applicability as catalysts and also as components of light-emitting devices. For large-scale applications, more abundant photomaterials are desirable. As an alternative, inexpensive dyes as photocatalysts have been successfully applied [4], offering a metal-free approach for chemical transformations, but these suffer from a more rapid degradation when compared to metal complexes. The advantageous role of copper as a widely available source for light-induced processes has been also recognized, featuring strong reducing power, sufficient life-time and high luminescence in the excited state. Copper complexes have been applied in photocatalytic hydrogen production from water [5], as photosensitizers in photoelectrochemical cells [6], or as active components in organic light emitting diodes (OLED) [7]; however, their use as photoredox catalysts for organic synthesis has been rare until recently [8].

Visible Light Photocatalysis in Organic Chemistry, First Edition.
Edited by Corey R. J. Stephenson, Tehshik P. Yoon and David W. C. MacMillan.

7.2 Photophysical Properties of Copper Catalysts

Seminal studies by McMillin and coworkers [9] provided insightful information on the excited state properties of Cu^I-phenanthroline derivatives **5** (Table 7.2), suggesting them as viable alternatives to ruthenium- or iridium-based photocatalysts [13]. Such complexes with the general formula $[Cu(NN)_2]^+$ have a distorted tetrahedral geometry with a d^{10} configuration at the metal center. Upon irradiating with light, an electron from the metal-centered t_2 orbital (oxidation of Cu(I) to Cu(II)) is promoted to the lowest energy (LUMO) ligand-centered π^* orbital (reduction of the ligand within the metal complex), which is known as metal to ligand charge transfer (MLCT) (Eq. (7.1)) [14]. Intersystem crossing (ISC) from the singlet MLCT state occurs rapidly, leading to the stable triplet excited MLCT state (Scheme 7.1).

$$[Cu^I(NN)_2]^+ \xrightarrow{h\nu} [Cu^{II}(NN)(NN)^-]^{+*} \tag{7.1}$$

Though a strong reductant, this excited state of such homoleptic $Cu(NN)_2^+$ complexes is only a mild oxidant, and consequently, examples for reductive quenching of excited copper complexes are scarce. In contrast, in ruthenium or iridium complexes for example $[Ru(bpy)_3]^{2+}$ (bpy = 2,2′-bipyridine), the MLCT triplet excited state is a more potent reductant as well as oxidant than the ground state species. Nevertheless, the possibility to tune the electronic properties of Cu(I)phenanthroline complexes by substituting one phenanthroline ligand by a strong donor or acceptor ligand, thus moving from homo- to heteroleptic complexes, is increasingly being recognized as a means to broaden the scope of copper complexes both as photoreductants as well as oxidants.

In 1987, Sauvage *et al.* [11a] evaluated $[Cu(dap)_2]^+$ (**5e**, dap = 2,9-bis (4-anisyl)-1,10-phenanthroline) for the photochemical reductive coupling of 4-bromonitrobenzene that proceeded upon irradiation at 350 nm, capitalizing on the strong reducing power of excited $[Cu(dap)_2]^+$ **5e** (−1.43 V) compared to excited $[Ru(bpy)_3]^{2+}$ (−0.81 V) [15a] or $[Ir\{dF(CF_3)ppy\}_2(dtbbpy)]^+$ (−0.89 V; $dF(CF_3)ppy$ = 2-(2,4-difluorophenyl)-5-trifluoromethylpyridine, dtbbpy = 4,4-di-*tert*-butyl-2,2-dipyridyl) [15b], which is only surpassed by *fac*-$Ir(ppy)_3$ (−1.73 V, ppy = 2-phenylpyridine) [15d], which has a more negative potential

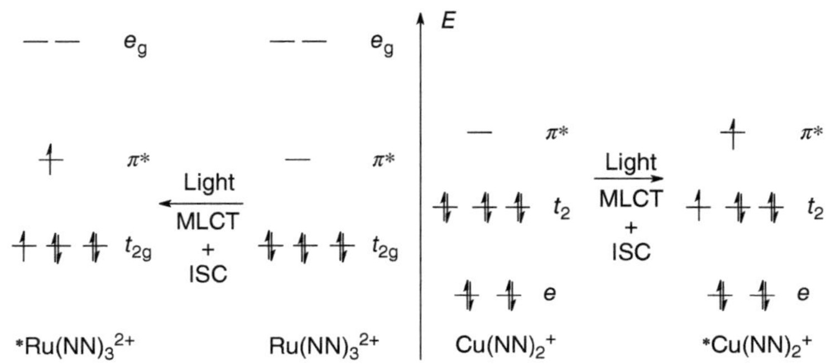

Scheme 7.1 Simplified molecular orbital depiction of low-spin d^6 Ru^{2+} and d^{10} Cu^+ complex.

Table 7.1 Comparison of redox potentials of photoexcited copper(I) catalyst with those of ruthenium(II) and iridium(III) catalysts.[a)]

Photocatalyst	$E_{1/2}$ (M^+/M^*) (V)	$E_{1/2}$ (M^+/M) (V)	$E_{1/2}$ (M^*/M^-) (V)	$E_{1/2}$ (M/M^-) (V)	Excitation λ_{max} (nm)	Excited state lifetime (ns)	References
$[Cu(dap)_2]^+$ (**5e**)	−1.43	+0.62			437	270 (130)	[11a,b]
$[Cu(dpp)(binc)]^+$ (**7**)	−1.88	+0.69			450[b)]	17 000[c)]	[11b]
$[Cu(dpdmp)((PPh_2)_2C_2B_9H_{10})]$ (**8**)			+1.01	−1.39, −1.60	449[b)]	1300	[11c]
$[Ru(bpy)_3]^{2+}$	−0.81	+1.29	+0.77	−1.33	452	1100	[15a]
$[Ir\{dF(CF_3)ppy\}_2(dtbbpy)]^+$	−0.89	+1.69	+1.21	−1.37	380	2300	[15b]
$[Ir(ppy)_2(dtbbpy)]^+$	−0.96	+1.21	+0.66			557	[15c]
fac-$Ir(ppy)_3$	−1.73	+0.77	+0.31	−2.19	375	1900	[15d]

a) Potentials are measured against saturated calomel electrode (SCE), all measurements were conducted in MeCN at room temperature unless otherwise noted.
b) In CH_2Cl_2.
c) In PMMA (poly(methyl methacrylate)).

than copper (Table 7.1). The excited state life-time, however, of homoleptic copper-phenanthrolines is shorter by a factor of 5–10 when compared to related ruthenium or iridium complexes (Figure 7.1 and Table 7.3), which is rationalized by the structural reorganization from the ground state tetrahedral to a square planner geometry in the excited state, resulting in exciplex quenching [19].

Introducing substituents at the 2,9-positions of phenanthrolines and moving to heteroleptic complexes, incorporating chelating phosphine or isonitrile ligands, increases life-time of the excited triplet state as well as photostability by preventing excited state structural relaxation and hence exciplex quenching. Following this rational, McMillin and coworkers introduced a new class of $[Cu(NN)(POP)]^+$ complexes **6** (POP = bis[2-(diphenylphosphino)phenyl]ether) [12], and more recently, Reiser *et al.* evaluated mixed copper–phenanthroline bisisonitrile complexes of type **7**, which both have excited life times in the microsecond region and thus even exceeding those of commonly used ruthenium and iridium photocatalysts (Tables 7.1 and 7.2) [11b]. Concurrent with a blue shifted absorption maximum, these complexes are also stronger reductants in the excited state compared to $[Cu(dap)_2]^+$ (**5e**). An exciting development was made by Chen *et al.* with the introduction of heteroleptic copper(I) complexes **8** containing phenanthroline and monoanionic nido-carborane-diphosphine ligands [11c]. In their long-lived photoexcited states, these complexes proved to be strong photo-oxidants, allowing cross-dehydrogenative coupling reactions via an oxidative quenching cycle (*vide infra*). In general, there is growing experimental evidence that photoexcited Cu(I) species can also be utilized as electron acceptors, and likewise, also photoexcited Cu(II) species can take this role.

Heteroleptic copper complexes **6–8** can be readily obtained by ligand exchange, for example from copper(I)bisphenanthroline complexes, contrasting octahedral

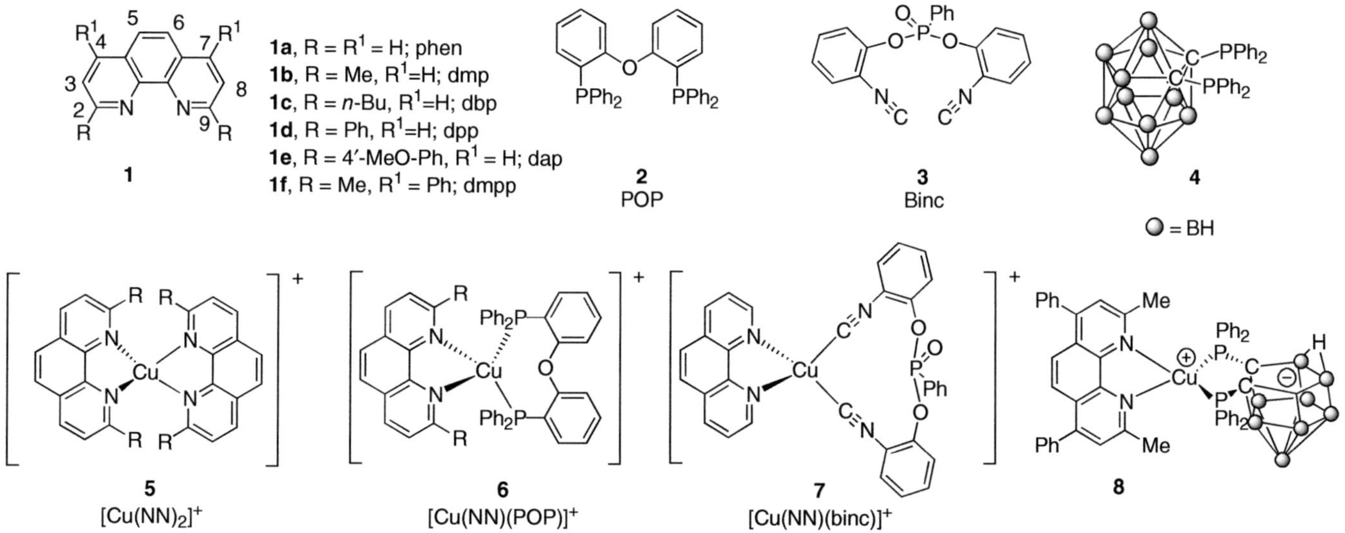

Figure 7.1 Ligands and copper(I) complexes derived thereof used as photoredox catalysts.

Table 7.2 Effect of ligand substitution for copper(I) complexes (*cf*. Figure 7.1).

Photocatalyst	Excitation λ_{max} (nm)	Excited state lifetime (ns)	References
$[Cu(phen)_2]^+$ (**5a**)	458	<20	[10]
$[Cu(dmp)_2]^+$ (**5b**)	454	85	[10]
$[Cu(dpp)_2]^+$ (**5d**)	448	250	[10]
$[Cu(dap)_2]^+$ (**5e**)	437	270	[11a]
$[Cu(phen)(POP)]^+$ (**6a**)	391	190	[12]
$[Cu(dmp)(POP)]^+$ (**6b**)	383	14 300	[12]
$[Cu(dbp)(POP)]^+$ (**6c**)	378	16 100	[12]
$[Cu(phen)(binc)]^+$ (7)	450	17 000	[11b]
$[Cu(dmpp)(C_2B_9H_{10})]$ (**8**)	449	1300	[11c]

ruthenium or iridium complexes of the type $[M(LL)_3]$, which are resistant to ligand exchange, especially in their ground state. Thus, photoactive copper complexes offer the advantage of tuning their properties by changing the nature of the chelating ligands to meet the requirements for a given photochemical process.

7.3 Application of Copper Based Photocatalysts in Organic Synthesis

The benefits of copper catalysis in combination with UV light have been recognized for some time in processes such as [2+2] cycloadditions [20] or atom transfer radical additions (ATRA) [21]. More recently, Fu *et al.* showed that the combination of Cu(I) and UV light is effective in the coupling of alkyl and aryl halides [22]. Application of copper catalysts in visible light photocatalysis however remained scarce, greatly contrasting with the use of $Ru(bpy)_3Cl_2$, whose potential for visible light photoredox processes was already recognized in the 1980s and early 1990s, owing to the pioneering work of Kellog, Pac, Deronzier, Tomioka, and Willner [3b], and was more recently brought again into the limelight by the groups of MacMillan, Yoon, and Stephenson, who demonstrated exciting new directions in organic synthesis by visible light photoredox catalysis [3].

The report of Sauvage *et al.* in 1987 disclosing the use of $[Cu(dap)_2]Cl$ (**3e**) for the C—Br bond activation of 4-nitrobenzyl bromide (**27**) (Scheme 7.2) [11a], marks the beginning of visible light copper photocatalysis. In its excited state, the copper(I) catalyst **3e** is a strong reductant, being able to directly transfer an electron to benzyl bromide **9**, which results in the generation of benzyl radical. This can undergo a formal dimerization to **10** or oxidation to **11**, the latter mediated by molecular oxygen. Triethylamine served as a stochiometric reductant to regenerate **3e** from the oxidized photocatalyst $[Cu(dap)_2]^{2+}$. Despite this promising utilization of **3e** for photocatalytic carbon–carbon bond forming reactions or oxidations, it took 25 years for the potential of this complex and related ones to be further explored.

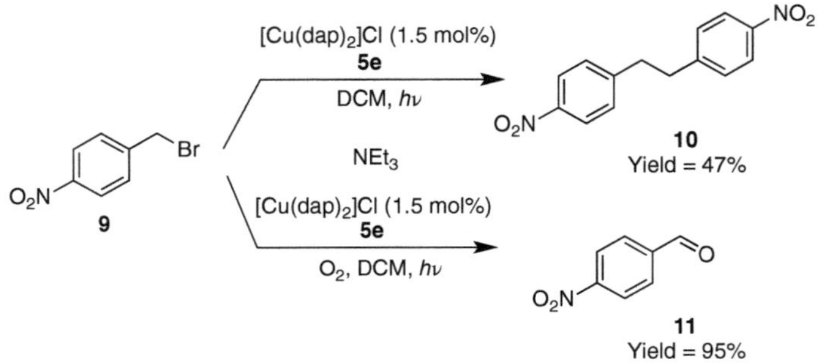

Scheme 7.2 Photochemical activation of 4-nitrobenzyl bromide (**9**) with [Cu(dap)$_2$]Cl (**5e**).

Complementing the elegant work of Stephenson *et al.* using ruthenium photocatalysts [23], Reiser *et al.* [24] demonstrated the potential of [Cu(dap)$_2$]Cl (**5e**) as a visible light photoredox catalyst for ATRA reactions of α-halocarbonyl compounds or perfluoralkyl iodides **13** to olefins **12** to give rise to **14** in good yields at low catalyst loading (Scheme 7.3). The reaction proceeded through the so-called oxidative quenching cycle, that is, no sacrificial electron donor such as a tertiary amine was needed in the process.

Notably, [Cu(dap)$_2$]Cl (**5e**) is powerful enough to allow ATRA reactions between electron-deficient benzyl halides **15** and electron-rich styrenes **16** or silyl enol ethers **18** (Scheme 7.4) [25].

If the resulting ATRA products **17** or **19** possess an *ortho*-nitro group they can be converted to biologically relevant tetrahydroquinolines **20** (Scheme 7.5).

Another class of suitable reagents for copper-catalyzed ATRA reactions with alkenes were found to be fluoroalkylsulfonylchlorides, revealing another notable feature of copper photoredox catalysts. In an insightful study, Dolbier *et al.* demonstrated net addition of fluoroalkyl and chlorine groups to *electron-deficient* alkenes catalyzed by [Cu(dap)$_2$]Cl (**5e**), where they were able to synthesize not only trifluoromethylated compounds but also mono- and difluoromethylated compounds (Scheme 7.6 top row, Table 7.3, entries 1–4, 15–17) [16a]. The latter could also be recently expanded to α,ϖ-aminoalkenes to result in an overall intramolecular aminodifluoromethylation [16b].

Reiser *et al.* observed that if the reaction is carried out between *electron neutral* alkenes with **21a** as substrate, trifluoromethylchloro*sufonylation* to alkenes

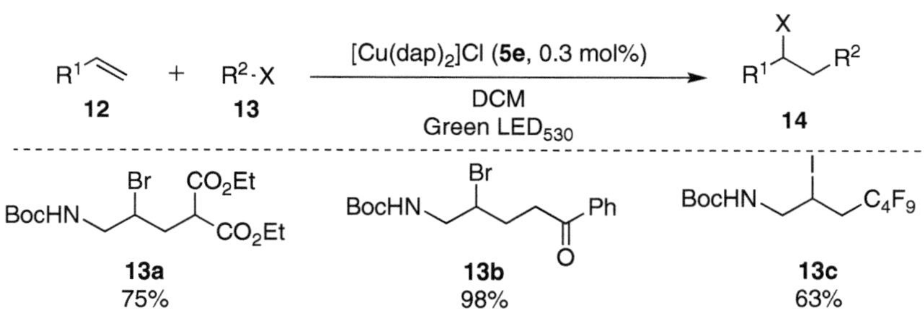

Scheme 7.3 [Cu(dap)$_2$]Cl (**5e**)-catalyzed ATRA reactions between alkenes **12** and α-halocarbonyl compounds or perfluoralkyl iodides **13**.

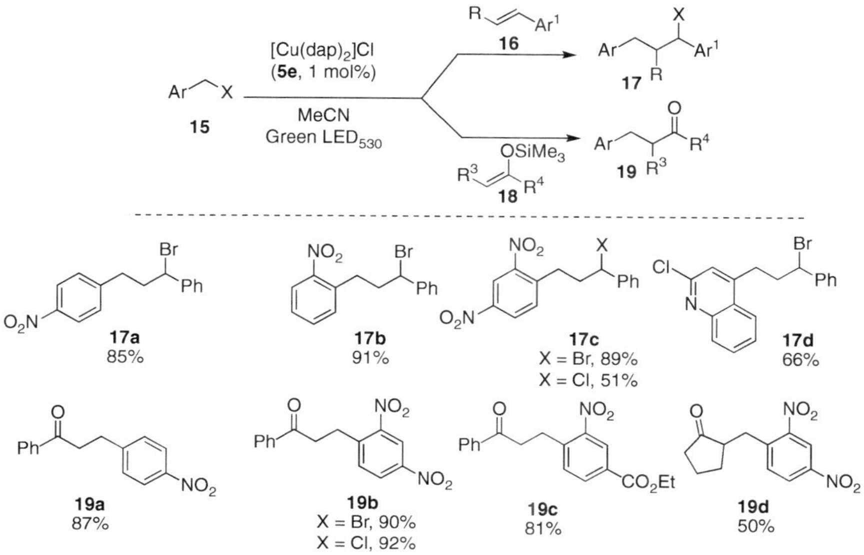

Scheme 7.4 [Cu(dap)$_2$]Cl (3e) catalyzed ATRA reactions between alkenes **16** or silylenolethers **18** and electron-deficient benzyl halides **15**.

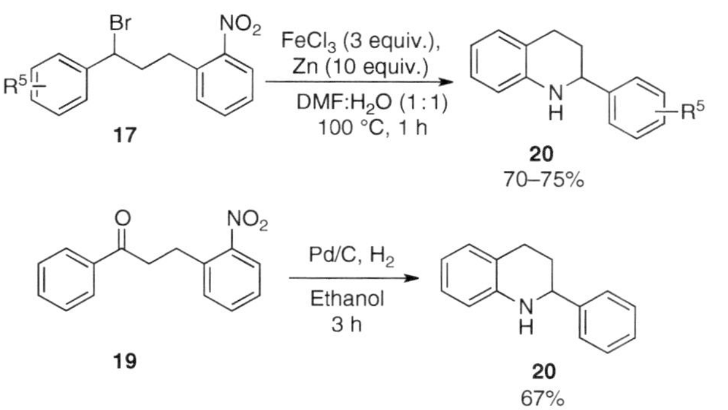

Scheme 7.5 Conversion of ATRA products to tetrahydroquinolines **20**.

is observed (Scheme 7.6, middle row; Table 7.3, entries 5, 8, 10–15), contrasting the course of the reaction these substrates take with [Ru(bpy)$_3$]Cl$_2$ under otherwise identical reaction conditions (Table 7.3, entries 6,9), for which *trifluoromethylchlorination* is observed (Scheme 7.6, bottom row) [17]. The latter was shown to be general by Han et al. with a wide range of alkenes when Ru(phen)$_3$Cl$_2$ (phen = 9,10-phenanthroline) is employed as catalyst [18].

To put forward a consistent mechanistic picture of these observations is not possible. The obtained products are consistent with an initial electron transfer from the copper(I) or Ru(II) catalyst to the fluoroalkylsulfonylchloride, upon which a fluoroalkyl radical is generated (Scheme 7.7). The latter adds to the alkene, generating radical **28**, but from this point on the reactions seem to take different pathways, depending on the electronic and steric environment

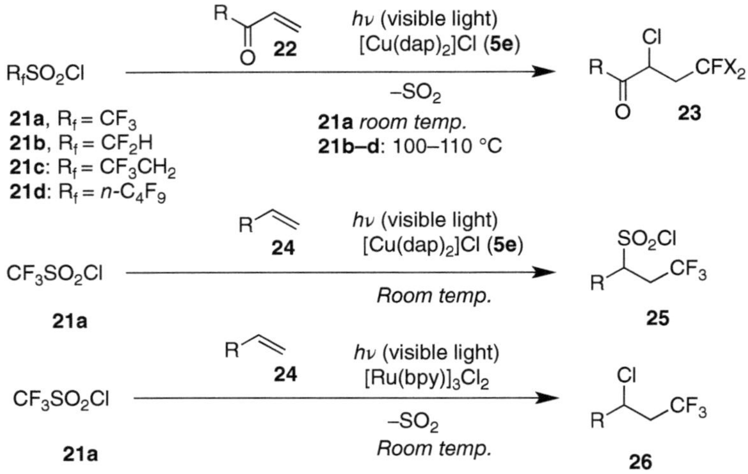

Scheme 7.6 ATRA reactions between alkenes and perfluorosulfonylchlorides.

of the alkene as well as on the type of catalyst. Between octahedral $Ru(NN)_3^{2+}$ and tetrahedral $Cu(NN)_2^+$ complexes the following differences can be noted: while $Ru(NN)_3^{2+}$ are more resistant to undergo ligand exchange especially in the ground state, in $Cu(NN)_2^+$ complexes substitution of a (NN)-ligand with nucleophiles can readily occur. Moreover, the oxidation potential of the copper(II)complex that is formed after electron transfer to the substrate is considerably lower than that of the corresponding Ru(III) complex. Studying the alkene- and catalyst-dependent courses of the reaction raises the question if the copper catalyst is implicated in the transfer of either the chlorosulfonyl or the chloro group, for example via an inner sphere mechanism calling for an intermediate like **27**.

Dolbier argued that the electrophilic radical **28** (R = electron withdrawing group) can neither be oxidized by Cu(II) nor can initiate a chain propagation with fluoroalkylsulfonylchloride [16a]. Thus, transfer of chloride from a species like **28** with concurrent back electron transfer to the copper catalyst is proposed. Following this line of argument, a more electron-rich radical **28** (R = alkyl) might still not be further oxidized by the Cu(II)-species, but might be sufficiently nucleophilic to initiate chain propagation with **21** by selective *attack on sulfur* with concurrent regeneration of perfluoroalkyl radicals. The stronger oxidation power of Ru(III) might lead to cation **30**, which will then recombine with chloride to the product. This mechanistic picture would rule out chain propagation of fluoroalkylsulfonylchlorides when ruthenium(II) photocatalysts are employed, while Cu(I) catalysts might differentiate between chain propagation and transfer of the nucleophile from a catalyst species like **27**, which obviously could also be a species like **27b** in which the SO_2Cl is still bound to copper [18]. Steric factors and remote strong donor atoms on the alkene also seem to play a role in the outcome of the reaction. Indeed, for CuCl/UV light-mediated ATRA reactions of prechlorinated alkanes to alkenes, Mitami *et al.* already speculated on the direct involvement of copper in the catalytic cycle as a chlorine delivering agent [21].

Table 7.3 Fluoroalkylsulfonylierungen und Fluoroalkylchlorierungen of alkenes.

Entry	R_fSO_2Cl	Alkene	Catalyst	ATRA product	Yield (%)	References
1	**21a**	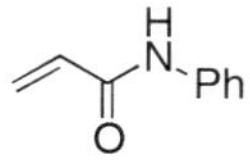	$[Cu(dap)_2]Cl$	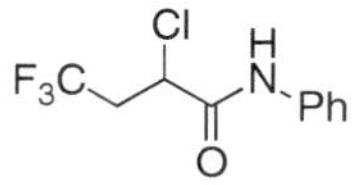	98	[16a]
2	**21a**	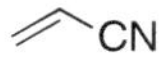	$[Cu(dap)_2]Cl$	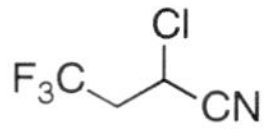	99	[16a]
3	**21a**	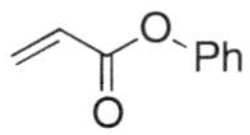	$[Cu(dap)_2]Cl$	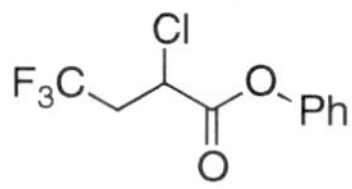	97	[16a]
4	**21a**	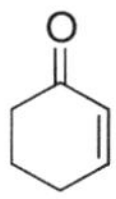	$[Cu(dap)_2]Cl$	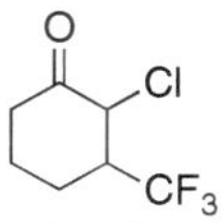	77	[16a]
5 6	**21a**	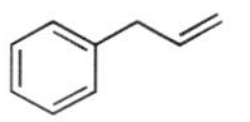	$[Cu(dap)_2]Cl$ $[Ru(bpy)_3]Cl_2$	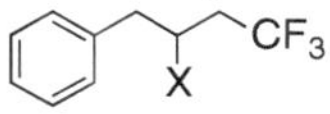	86 ($X = SO_2Cl$) 84 ($X = Cl$)	[17]
7	**21a**	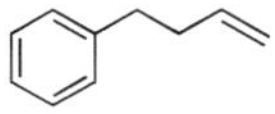	$[Ru(phen)_3]Cl_2$	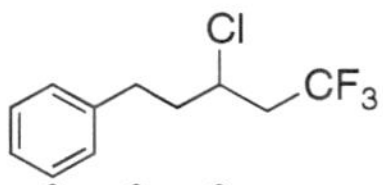	81	[18]
8 9	**21a**	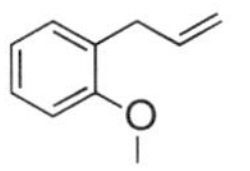	$[Cu(dap)_2]Cl$ $[Ru(bpy)_3]Cl_2$	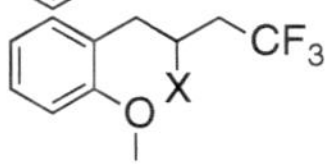	80 ($X = SO_2Cl$) 82 ($X = Cl$)	[17]
10	**21a**	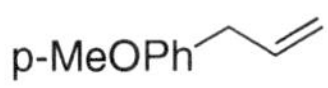	$[Cu(dap)_2]Cl$	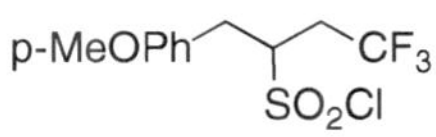	76	[17]
11	**21a**	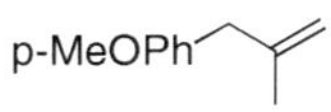	$[Cu(dap)_2]Cl$	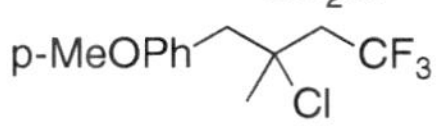	77	[17]
12	**21a**	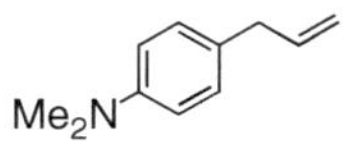	$[Cu(dap)_2]Cl$	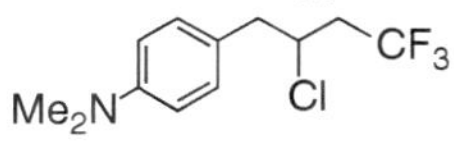	42	[17]
13	**21a**		$[Cu(dap)_2]Cl$	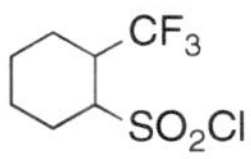	56	[17]
14	**21a**	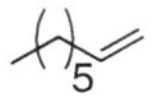	$[Cu(dap)_2]Cl$	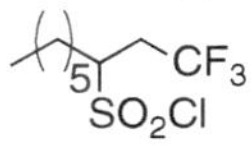	71	[17]
15	**21a**	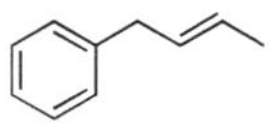	$[Cu(dap)_2]Cl$	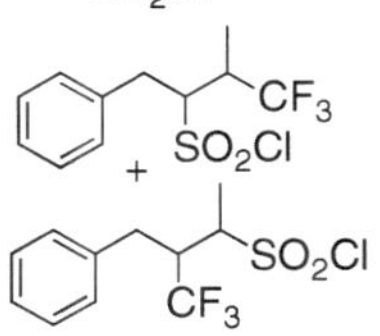 7 : 3 regioisomers	67	[17]
16	**21a**	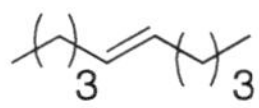	$[Ru(phen)_3]Cl_2$	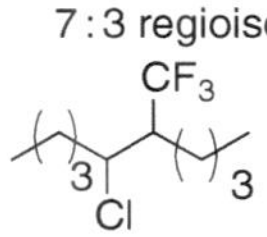	61	[18]
17 18 19	**21b** **21c** **21d**	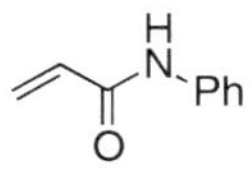	$[Cu(dap)_2]Cl$	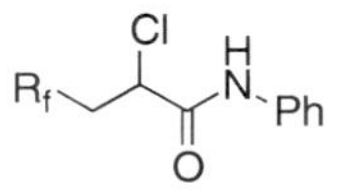	61 70 98	[16a]

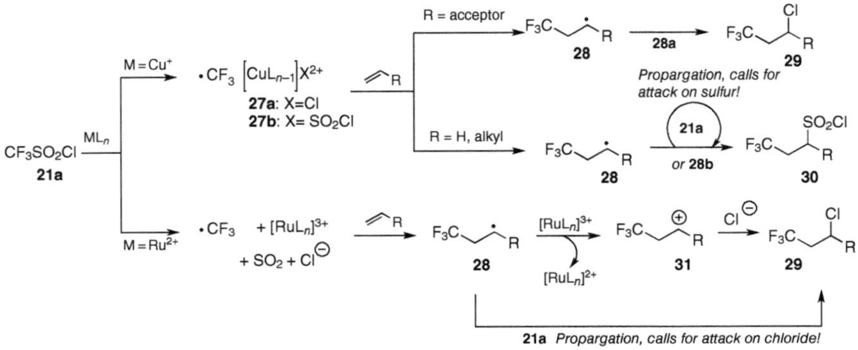

Scheme 7.7 Mechanistic proposal for the trifluoromethylchlorination or trifluoromethylchlorosulfonylation of alkenes.

A recent study on a light switchable, three-component also points to the importance of Cu(II)-species to be able to further oxidize a radical intermediate to a cation. Greaney *et al.* elegantly demonstrated that the Cu(dap)$_2$Cl-catalyzed azidination of styrenes-type double bonds takes a different course depending on the reaction being performed either in the dark or under visible light irradiation (Scheme 7.8) [26]. For example, upon treatment of styrenes **33** with the Zhdankin reagent, **31** can be converted to bisazidinated adducts **32** catalyzed by [Cu(dap)$_2$]Cl (**5e**) in the dark, while upon visible light irradiation adducts **34** that are derived from addition of azide and the solvent methanol are obtained. It is proposed that azide radicals are formed by single electron transfer (SET) from **31** mediated by electron transfer from the Cu(I) catalyst **5e** that add to styrenes forming radical **35**. While in a dark environment oxidation of **35** by Cu(II) does not take place and thus a radical chain reaction with **31** is not initiated to give bisazidination products **32**, under irradiation oxidation of **35**, presumably by Cu(II)*, it is possible to form cation **36**, which is then trapped by methanol that acts as a nucleophile.

The combination of Cu(I)phenantholine complexes and visible light was also found to be a successful liaison for various allylation reactions. Employing

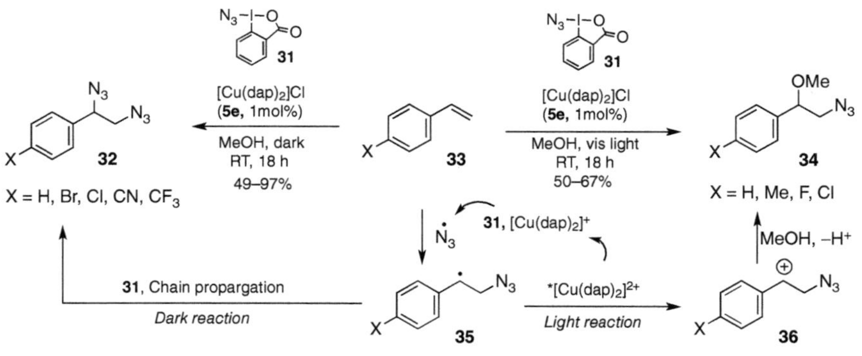

Scheme 7.8 Light switchable, [Cu(dap)$_2$Cl]-catalyzed azidination of styrenes.

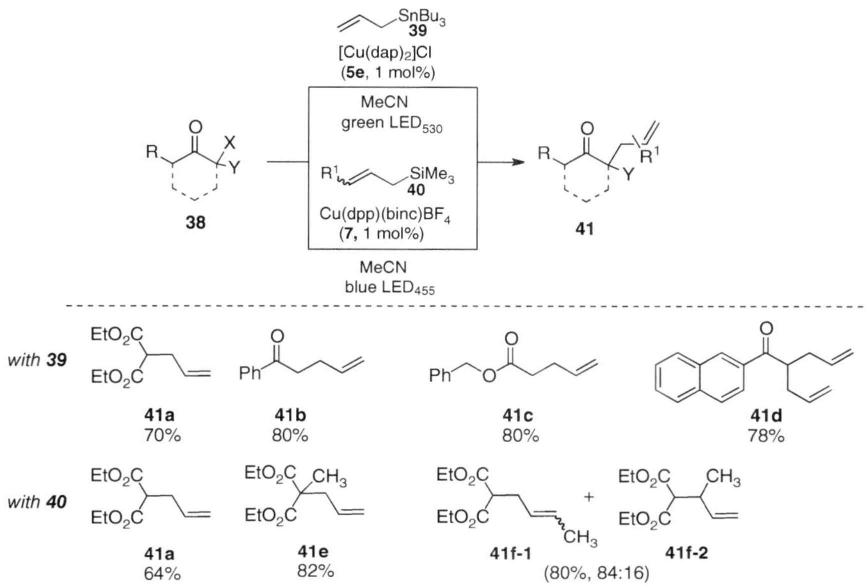

Scheme 7.9 Allylation between α-halo carbonyl compounds and allyltributyltin with $[Cu(dap)_2]Cl$ (**5e**) or allyltrimethylsilanes with $[Cu(dpp)(binc)]BF_4$ (**7**) as photoredox catalysts.

$[Cu(dap)_2]Cl$ (**5e**) the allylation of α-halo carbonyl compounds **38** with allyltributyltin **39** was demonstrated (Scheme 7.9) [24] as an alternative to previously reported radical conditions utilizing AIBN at 80°C or BEt_3 at ambient temperature or to photochemical conditions under UV irradiation in the absence of catalysts. With the more potent heteroleptic Cu(I) complex **7**, allylation becomes possible using allyltrimethylsilanes **40** rather than employing a tin-based reagent [11b]. Also, allyltosylates **43** are suitable coupling reagents, as demonstrated by Ollivier and coworkers with aryl radicals derived from diaryliodonium salts **42** using $[Cu(dpp)_2]PF_6$ (**5c**) as the photocatalyst (Scheme 7.10) [27].

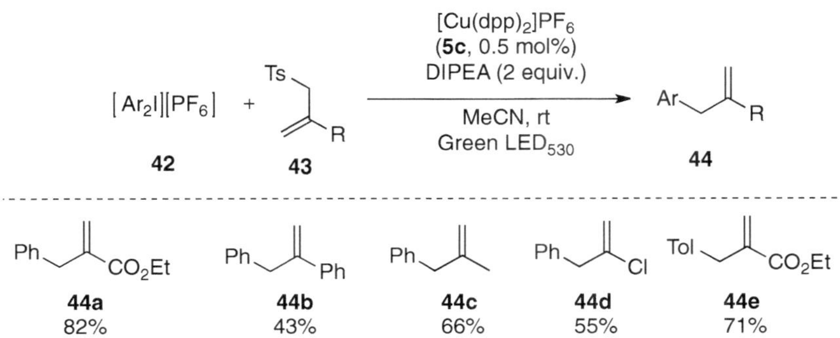

Scheme 7.10 Allylation between diaryliodonium salts **42** and allyl tosylates **43** with $[Cu(dpp)_2]PF_6$ (**5c**) as photoredox catalyst.

Similar to the ATRA process, it is assumed in all three processes that a SET between the excited Cu(I) catalyst and the organic halide is the key photochemical step of the reaction. After elimination of halide, the resulting radical then adds to the allyl acceptor, from which a tin, silyl, or tosyl radical is eliminated to form the allylation product. The couplings with allyltin and allylsilane were carried out in the absence of a sacrificial electron donor, assuming either propagation of the reaction by a chain mechanism or catalyst regeneration by the formation of trialkyltin or trialkylsilyl bromide with concurrent back electron transfer (Cu(II) to Cu(I)). In the reaction between diaryliodonium salts and allyl tosylates DIPEA was present as a stochiometric sacrificial electron donor, assuming the reduction of Cu(II) to Cu(I) in order to close the catalytic cycle. In all three cases, the mechanistic proposal put forward calls for a S_N2' pathway; however, all examples presented except one do not allow to rule out the possibility that products can also be formed by formal S_N2 substitutions. Quite strikingly, in the allylation between crotylsilane **40** ($R^1 = Me$) and diethyl α-bromomalonate a 5:1 mixture of **41f-1** (formal S_N2' product) and **41f-2** (formal S_N2 product) is obtained, which might suggest that under the photoredox conditions employed the allylsilane might undergo 1,3-migration, contrasting their stability for such processes under thermal conditions.

In all photoredox-catalyzed reactions with $Cu(dap)_2Cl$ and related complexes discussed so far, the strong reducing power of excited Cu(I)* allowed to start chemical transformations such as ATRA reactions or allylations by transferring an electron to an substrate. However, there is growing evidence that also Cu(I)-complexes as well as Cu(II)-complexes, possibly in their excited Cu(II)* state, might act as effective photo-oxidants to initiate photochemical processes, and the following examples show some recent developments that point to this reaction.

The advantage of a visible light mediated photoredox catalzyed process compared to the direct UV light-mediated pathway was demonstrated by Collins *et al.* with the cyclization of **45** for the synthesis of 5[helicene] **46** (Scheme 7.11) [28]. While the classical UV light-mediated 6π-electrocyclic ring

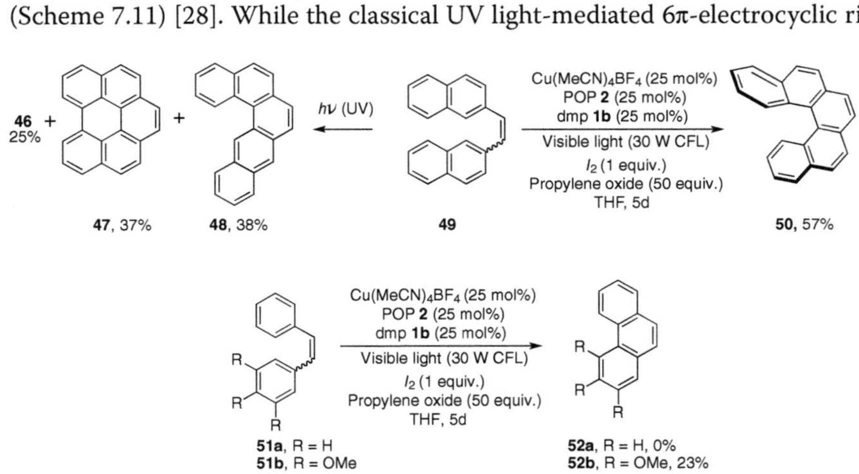

Scheme 7.11 Visible light mediated photoredox cyclization catalyzed by *in situ*-generated Cu-based complex **6b** versus the direct UV-mediated photocyclization process.

closure followed by rearomatization with iodine as a stochiometric oxidant shows a number of side reactions (formation of **47** and **48**), carrying out the reaction under visible light conditions in the presence of *in situ*-generated Cu(I)-photocatalyst **6b** (*cf.* Figure 7.1 and Table 7.2) proceeded cleanly and with improved yields to the desired **46**. Notably, the authors proposed an oxidation as the key photochemical step based on their studies with stilbenes **51** that need to be electron rich in order to undergo photocyclization mediated by **6b**.

By a related cyclization strategy utilizing a flow strategy that greatly reduced the reaction time, biologically relevant carbazoles **54** can be synthesized from di- or triarylamines **53** [29]. For the visible light mediated processes, *in situ*-generated, copper-based photocatalyst **6b** proved to be much superior with respect to $[Ru(bpy)_3](PF_6)_2$, but for this process, direct UV irradiation in the absence of a catalyst also proved to be effective. While, in general, identical products were obtained in the visible light- or UV light-mediated process, with triarylamines bearing *para*-substituents dramatical changes in the *endo/exo* ratio of the products were observed (Scheme 7.12) for reasons that were not understood at that time [30]. The iodine in the reaction mixture is thought to serve both as an oxidative quencher of the excited photocatalyst and as an oxidant for the final re-aromatization step.

In a conceptually related approach regarding the key photochemical step, the weak oxidation power of Cu(II) to Cu(I) was also recently utilized for the selective generation of aryl-stabilized α-aminoradicals that can be trapped by maleimides

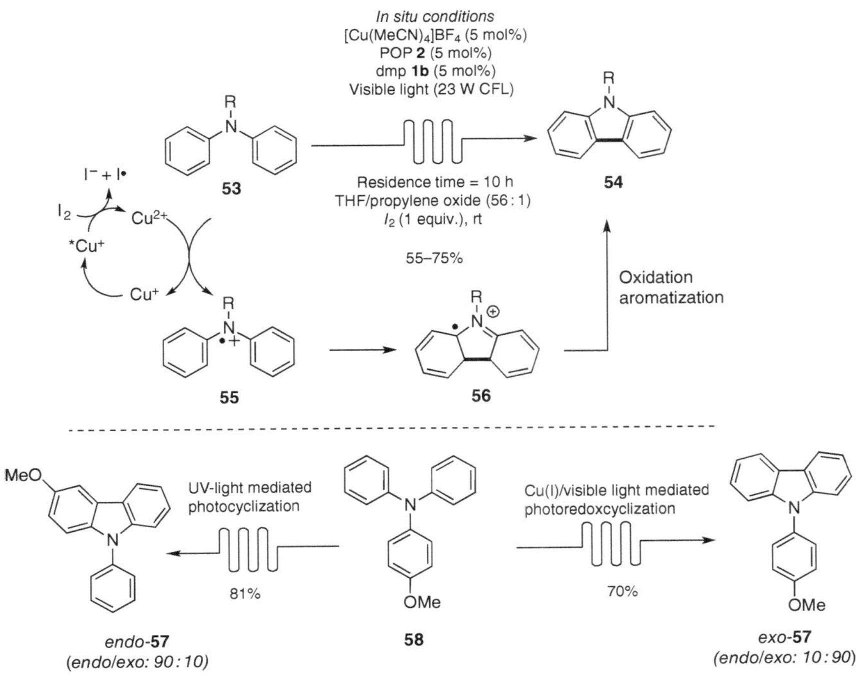

Scheme 7.12 Visible light mediated photoredox cyclization of di- and triarylamines catalyzed by *in situ*-generated Cu-based complex **6b** versus the direct UV-mediated photocyclization process.

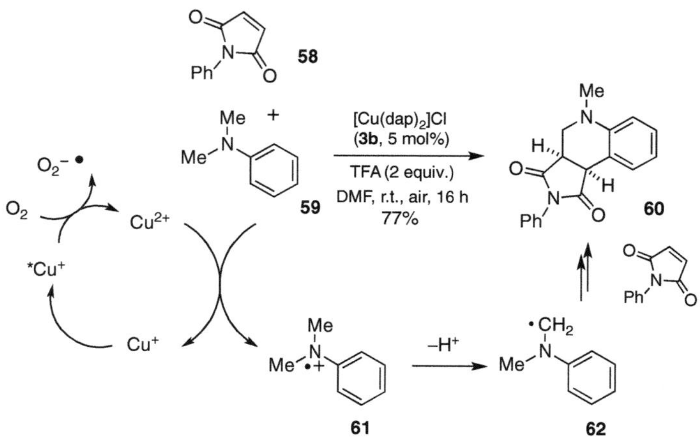

Scheme 7.13 Copper-catalyzed photo-oxidation of arylamines as key step in the anellation with maleimides.

to generate annulated heterocycles of type **60** [31]. The authors propose that $Cu(dap)_2Cl$ is oxidized by molecular oxygen upon visible light irradiation to generate Cu(II). Trifluoroacetic acid was found to be a crucial additive to promote this redox step (Scheme 7.13).

The photo-oxidation of *N*-aryltetraisoquinolines by the Cu(I)-complex **8** (*cf.* Table 7.2) to initiate a cross-dehydrogenative coupling with nitromethane shows another reaction mode in copper photoredox catalysis (Scheme 7.14) [11c].

Another remarkable feature of the combination of copper and visible light was discovered by Hwang *et al.* in the Sonogashira C–C cross coupling reaction (Scheme 7.15) [32], which proceeds at room temperature without requiring a palladium catalyst. As the decisive intermediate, a copper acetylide complexe **73** is

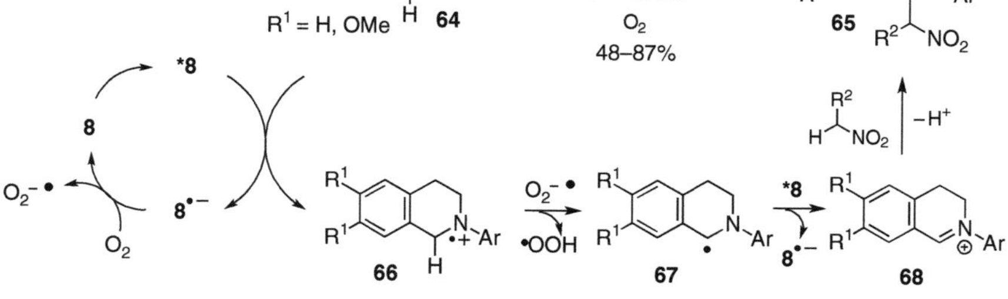

Scheme 7.14 Cross-dehydrogenative coupling between nitroalkanes and tetrahydroisoquinolines.

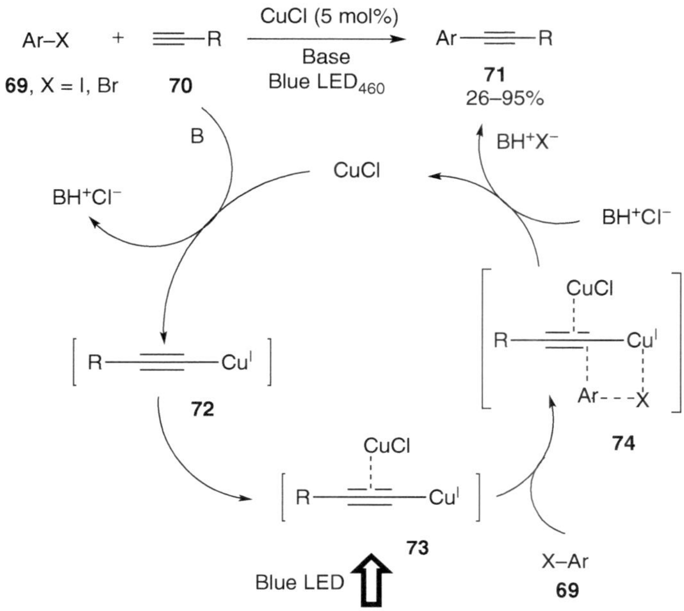

Scheme 7.15 Visible light mediated, copper(I)-catalyzed Sonogashira Coupling.

proposed, which is able to absorb light in the visible region (λ_{abs} = 425–485 nm). Photo excitation results in a ligand (acetylene) to metal (copper) charge transfer (LMCT), making the acetylene moiety electrophilic enough for nucleophilic attack of an aryl bromide or iodide **23**. Both electron-rich, but also electron-poor aryl halides were found to be suitable in this process.

Expanding on this concept, the oxidative C—N coupling of anilines **75** with terminal alkynes **70** was demonstrated to lead to α-ketoamides with a broad scope of substrates in very good yields [33]. Again, the excitation of a copper(I)–acetylide complex, that is, **72**, is proposed to be the decisive key photochemical step. Stepwise oxidation of this species via Cu(II)-acetylide **76** leads to the Cu(III)-species **77**, setting the stage for the reductive C–N-coupling step giving rise to **78**. Further oxidation of this species finally yields **76** with regeneration of the copper(I) catalyst (Scheme 7.16).

7.4 Outlook

Copper-based complexes have proven recently to be versatile catalysts in visible light mediated organic reactions [34]. This is not only due to the economical advantage copper offers compared to ruthenium or iridium complexes, but also to its role that might expand beyond acting as a simple electron transfer reagent. Copper(I) complexes are the photoexcitable species, being able to act as an electron donor, but recent developments suggest that such species can also accept an electron from a substrate. Likewise, photo-oxidations initiated by

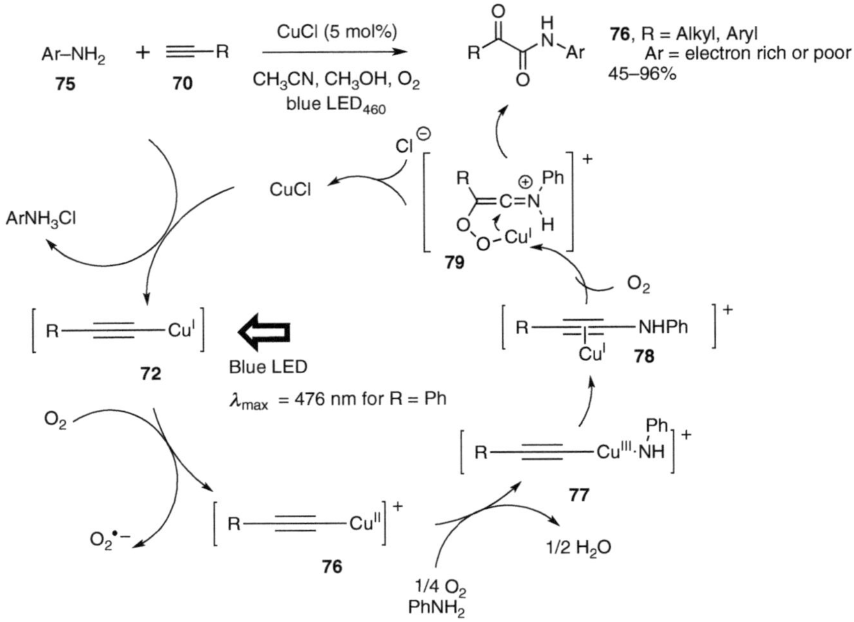

Scheme 7.16 Photoredox catalyzed, oxidative coupling or acetylenes and anilines.

Cu(II) complexes appear to be feasible. Moreover, the examples discussed for cross-coupling and ATRA reactions point to the possibility of coordination of the copper center with the substrate, leading to unique reactivities that have not been found with other metal-based photocatalysts. The much shorter life-time of excited Cu(I)*-phenanthroline complexes compared to analogous ruthenium or iridium complexes can be greatly extended in heteroleptic complexes using chelating bisphosphine, bisisonitrile, or borane ligands. Consequently, copper complexes might be able to establish themselves as viable alternatives to ruthenium or iridium complexes as visible light catalysts in many photoredox catalyzed reactions.

Acknowledgment

This work was supported by the DFG Graduiertenkolleg 1626 "Photocatalysis" and the DAAD.

References

1 Ciamician, G. (1912) *Science*, **36**, 385–394.
2 For selective reviews on important bond formations by photocatalysis see: (a) Fagnoni, M., Dondi, D., Ravelli, D., and Albini, A. (2007) *Chem. Rev.*, **107**, 2725–2756; (b) Hoffmann, N. (2008) *Chem. Rev.*, **108**, 1052–1103;

(c) Svoboda, J. and König, B. (2006) *Chem. Rev.*, **106**, 5413–5430; (d) T. Bach, *Synthesis* 1998, 683-703; For fine chemical synthesis with sunlight, see: (e) P. Esser, B. Pohlmann, H.-D. Scharf, *Angew. Chem.* 1994, **106**, 2093-2108; *Angew. Chem. Int. Ed.* **1994**, 33, 2009-2023.

3 For recent reviews on visible light photocatalysis see: (a) K. Zeitler, *Angew. Chem. Int. Ed. Engl.* 2009, **48**, 9785-9789; *Angew. Chem.* 2009, 121, 9969-9974; (b) F. Teply′, *Collect. Czech. Chem. Commun.* 2011, **76**, 859-917; (c) J. Xuan, W.-J. Xiao, *Angew. Chem.* 2012, **124**, 6934-6944; *Angew. Chem. Int. Ed.* **2012**, 51, 6828-6838; (d) Shi, L. and Xia, W. (2012) *Chem. Soc. Rev.*, **21**, 7687–7697; (e) Ischay, M.A. and Yoon, T.P. (2012) *Eur. J. Org. Chem.*, 3359–3372; (f) Tucker, J.W. and Stephenson, C.R.J. (2012) *J. Org. Chem.*, **77**, 1617–1622; (g) Xi, Y., Yi, H., and Lei, A. (2013) *Org. Biomol. Chem.*, **11**, 2387–2403; (h) Prier, C.K., Rankic, D.A., and MacMillan, D.W.C. (2013) *Chem. Rev.*, **113**, 5322–5363; (i) Xuan, J., Lu, L.-Q., Chen, J.-R., and Xiao, W.-J. (2013) *Eur. J. Org. Chem.*, 6755–6770; (j) Schultz, D.M. and Yoon, T.P. (2014) *Science*, **343**, 985.

4 (a) D. P. Hari, B. König, *Angew. Chem.* 2013, **125**, 4832-4842; *Angew. Chem. Int. Ed.* **2013**, 52, 4734-4743; (b) Ravelli, D. and Fagnoni, M. (2012) *ChemCatChem*, **4**, 169–171; (c) Nicewicz, D.A. and Nguyen, T.M. (2014) *ACS Catal.*, **4**, 355–360.

5 (a) Edel, A., Marnot, P.A., and Sauvage, J.P. (1984) *Nouv. J. Chim.*, **8**, 495–498; (b) Khnayzer, R.S., McCusker, C.E., Olaiya, B.S., and Castellano, F.N. (2013) *J. Am. Chem. Soc.*, **135**, 14068–14070; (c) S.-P. Luo, E. Mejía, A. Friedrich, A. Pazidis, H. Junge, A.-E. Surkus, R. Jackstell, S. Denurra, S. Gladiali, S. Lochbrunner, M. Beller, *Angew. Chem.* 2013, **125**, 437-441; *Angew. Chem. Int. Ed.* **2013**, 52, 419-423; (d) Mejía, E., Luo, S.-P., Karnahl, M. *et al.* (2013) *Chem. Eur. J.*, **19**, 15972–15978;For a highlight on noble metal free photocatalytic hydrogen productions see: (e) van den Bosch, B., Chen, H.-C., van der Vlugt, J.I. *et al.* (2013) *ChemSusChem*, **6**, 790–793.

6 (a) Breddels, P.A., Blasse, G., Casadonte, D.J., and McMillin, D.R. (1984) *Ber. Bunsen Ges. Phys. Chem.*, **88**, 572; (b) Alonso-Vante, N., Nierengarten, J.F., and Sauvage, J.P. (1994) *J. Chem. Soc. Dalton Trans.*, 1649–1654.

7 (a) Zhang, Q., Zhou, Q., Cheng, Y. *et al.* (2004) *Adv. Mater.*, **16**, 432–436; (b) Armaroli, N., Accorsi, G., Holler, M. *et al.* (2006) *Adv. Mater.*, **18**, 1313–1316.

8 *Leading reviews:* (a) M. Majek, A. Jacobi von Wangelin, *Angew. Chem.* 2013, **125**, 6033-6035; *Angew. Chem. Int. Ed.* **2013**, 52, 5919-5921; (b) Paria, S. and Reiser, O. (2014) *ChemCatChem*, **6**, 2477–2483.

9 (a) McMillin, D.R. and Ahn, B.T. (1978) *Inorg. Chem.*, **17**, 2253–2258; (b) McMillin, D.R., Buckner, M.T., and Ahn, B.T. (1977) *Inorg. Chem.*, **16**, 943–945;For review article see: (c) McMillin, D.R. and McNett, K.M. (1998) *Chem. Rev.*, **98**, 1201–1220.

10 Ruthkosky, M., Castellano, F.N., and Meyer, G.J. (1996) *Inorg. Chem.*, **35**, 6406–6412.

11 (a) Kern, J.-M. and Sauvage, J.-P. (1987) *J. Chem. Soc. Chem. Commun.*, 546–548; (b) Knorn, M., Rawner, T., Czerwieniec, R., and Reiser, O. (2015)

ACS Catal., **5**, 5186–5193; (c) Wang, B., Shelar, D.P., Han, X.Z. *et al.* (2015) *Chem Eur. J.*, **21**, 1184–1190.

12 Cuttell, D.G., Kuang, S.-M., Fanwick, P.E. *et al.* (2001) *J. Am. Chem. Soc.*, **124**, 6–7.

13 Armaroli, N. (2001) *Chem. Soc. Rev.*, **30**, 113–124.

14 Irving, H. and Williams, J.P. (1953) *J. Chem. Soc.*, 3192–3210.

15 (a) Kalyanasundaram, K. (1982) *Coord. Chem. Rev.*, **46**, 159–244; (b) Lowry, M.S., Goldsmith, J.I., Slinker, J.D. *et al.* (2005) *Chem. Mater.*, **17**, 5712–5719; (c) Slinker, J.D., Gorodetsky, A.A., Lowry, M.S. *et al.* (2004) *J. Am. Chem. Soc.*, **126**, 2763–2767; (d) Flamigni, L., Barbieri, A., Sabatini, C. *et al.* (2007) *Top. Curr. Chem.*, **281**, 143–203.

16 (a) Tang, X.J. and Dolbier, W.R. (2015) *Angew. Chem. Int. Ed.*, **54**, 4246–4249; (b) Zhang, Z., Tang, X., Thomoson, C.S., and Dolbier, W.R. (2015) *Org. Lett.*, **17**, 3528–3531.

17 Bagal, D.B., Kachkovskyi, G., Knorn, M. *et al.* (2015) *Angew. Chem. Int. Ed.*, **54**, 6999–7002.

18 Oh, S.H., Malpani, Y.R., Ha, N. *et al.* (2014) *Org. Lett.*, **16**, 1310–1313.

19 Lavie-Cambot, A., Cantuel, M., Leydet, Y. *et al.* (2008) *Coord. Chem. Rev.*, **252**, 2572–2584.

20 (a) Schwendiman, D.P. and Kutal, C. (1977) *J. Am. Chem. Soc.*, **99**, 5677–5682; (b) Grutsch, P.A. and Kutal, C. (1979) *J. Am. Chem. Soc.*, **101**, 4228–4233; (c) Hertel, R., Mattay, J., and Runsink, J. (1991) *J. Am. Chem. Soc.*, **113**, 657–665; (d) Langer, K. and Mattay, J. (1995) *J. Org. Chem.*, **60**, 7256–7266; (e) Banerjee, S. and Ghosh, S. (2003) *J. Org. Chem.*, **68**, 3981–3989; (f) Sarkar, N., Nayek, A., and Ghosh, S. (2004) *Org. Lett.*, **6**, 1903–1905.

21 (a) Mitani, M., Nakayama, M., and Koyama, K. (1980) *Tetrahedron Lett.*, **21**, 4457–4460; (b) Mitani, M., Kato, I., and Koyama, K. (1983) *J. Am. Chem. Soc.*, **105**, 6719–6721.

22 (a) Do, H.-Q., Bachman, S., Bissember, A.C. *et al.* (2014) *J. Am. Chem. Soc.*, **136**, 2162–2167; (b) Uyeda, C., Tan, Y., Fu, G.C., and Peters, J.C. (2013) *J. Am. Chem. Soc.*, **135**, 9548–9552; (c) Ziegler, D.T., Choi, J., Muñoz-Molina, J.M. *et al.* (2013) *J. Am. Chem. Soc.*, **135**, 13107–13112; (d) Ratani, T.S., Bachman, S., Fu, G.C., and Peters, J.C. (2015) *J. Am. Chem. Soc.*, **137**, 13902–13907.

23 Nguyen, J.D., Tucker, J.W., Konieczynska, M.D., and Stephenson, C.R.J. (2011) *J. Am. Chem. Soc.*, **133**, 4160–4163.

24 Pirtsch, M., Paria, S., Matsuno, T. *et al.* (2012) *Chem. Eur. J.*, **18**, 7336–7340.

25 Paria, S., Pirtsch, M., Kais, V., and Reiser, O. (2013) *Synthesis*, **45**, 2689–2698.

26 Fumagalli, G., Rabet, P.T.G., Boyd, S., and Greaney, M.F. (2015) *Angew. Chem. Int. Ed.*, **54**, 11481–11484.

27 Baralle, A., Fensterbank, L., Goddard, J.-P., and Ollivier, C. (2013) *Chem. Eur. J.*, **19**, 10809–10813.

28 Hernandez-Perez, A.C., Vlassova, A., and Collins, S.K. (2012) *Org. Lett.*, **14**, 2988–2991.

29 A. C. Hernandez-Perez, S. K. Collins, *Angew. Chem.* 2013, **125**, 12928-12932; *Angew. Chem. Int. Ed.* **2013**, 52, 12696-12700.

30 Hernandez-Perez, A.C., Caron, A., and Collins, S.K. (2015) *Chem. Eur. J.*, **21**, 16673–16678.
31 Nicholls, T.P., Constable, G.E., Robertson, J.C. *et al.* (2016) *ACS Catal.*, **6**, 451–457.
32 Sagadevan, A. and Hwang, K.C. (2012) *Adv. Synth. Catal.*, **354**, 3421–3427.
33 For additional reading of primary literature published after the transcription of this chapter(a) Kainz, Q.M., Matier, C.D., Bartoszewicz, A. *et al.* (2016) Asymmetric copper-catalyzed C-N cross-couplings induced by visible light. *Science*, **351** (6274), 681–684; (b) Rabet, P.T.G., Fumagalli, G., Boyd, S., and Greaney, M.F. (2016) Benzylic C—H azidation using the Zhdankin reagent and a copper photoredox catalyst. *Org. Lett.*, **18** (7), 1646–1649; (c) Nicholls, T.P., Constable, G.E., Robertson, J.C. *et al.* (2016) Brønsted acid cocatalysis in copper(i)-photocatalyzed α-amino C—H bond functionalization. *ACS Catal.*, **6** (1), 451–457; (d) Xiao, P., Dumur, F., Zhang, J. *et al.* (2014) Copper complexes in radical photoinitiating systems: applications to free radical and cationic polymerization upon visible LEDs. *Macromolecules*, **47** (12), 3837–3844; (e) Knorn, M., Rawner, T., Czerwieniec, R., and Reiser, O. (2015) Copper(phenanthroline)(bisisonitrile)] + -complexes for the visible-light-mediated atom transfer radical addition and allylation reactions. *ACS Catal.*, **5** (9), 5186–5193; (f) Hernandez-Perez, A.C. and Collins, S.K. (2016) Heteroleptic Cu-Based sensitizers in photoredox catalysis. *Acc. Chem. Res.*, **49** (8), 1557–1565; (g) Wang, B., Shelar, D.P., Han, X.-Z. *et al.* (2015) Long-lived excited states of zwitterionic copper(I) complexes for photoinduced cross-dehydrogenative coupling reactions. *Chem. Eur. J.*, **21** (3), 1184–1190; (h) Yang, F., Koeller, J., and Ackermann, L. (2016) Photoinduced copper-catalyzed C–H arylation at room temperature. *Angew. Chem. Int. Ed.*, **55** (15), 4759–4762; (i) Reiser, O. (2016) Shining light on copper: unique opportunities for visible-light-catalyzed atom transfer radical addition reactions and related processes. *Acc. Chem. Res.*, **49** (9), 1990–1996; (j) Rawner, T., Knorn, M., Lutsker, E. *et al.* (2016) Synthesis of trifluoromethylated sultones from alkenols using a copper photoredox catalyst. *J. Org. Chem.*, **81** (16), 7139–7147; (k) Pagire, S.K., Paria, S., and Reiser, O. (2016) Synthesis of β-hydroxysulfones from sulfonyl chlorides and alkenes utilizing visible light photocatalytic sequences. *Org. Lett.*, **18** (9), 2106–2109.
34 Sagadevan, A., Ragupathi, A., Lin, C.C. *et al.* (2015) *Green Chem.*, **17**, 1113–1119.

8

Arene Functionalization by Visible Light Photoredox Catalysis

Durga Prasad Hari, Thea Hering, and Burkhard König

University of Regensburg, Institute of Organic Chemistry, Department of Chemistry and Pharmacy, D-93040 Regensburg, Germany

8.1 Introduction

Photocatalyzed functionalization of arenes and heteroarenes is becoming an increasingly viable alternative to the traditional transition metal-catalyzed methods regularly adopted for C—H activation [1, 2]. In contrast to metal-catalyzed reactions, photocatalytic reactions mostly proceed via a radical pathway giving access to a different class of reactivity [3]. Thus photocatalysis enables transformations that would not proceed under thermal conditions. Since substituted arenes are such a ubiquitous motif in organic molecules, especially in pharmaceuticals and dyes, and serve as attractive building blocks for the elaboration of complex molecules, new and efficient methods for arene functionalization are of great interest [4–7,47k]. This chapter provides an overview over the recent developments and the early examples of functionalizing arenes by visible light induced processes.

Photocatalytic arene functionalizations typically involve the formation of a reactive aryl radical that is highly reactive toward a variety of compounds and functional groups [8–11]. The chemistry of aryl radicals is well established and forms the basis for many named reactions including the Sandmeyer reaction, the Pschorr cyclization, the Gomberg–Bachmann reaction, and the Meerwein arylation [12, 13]. In the past decade many elegant methodologies to functionalize aromatic compounds by the use of visible light and a suitable photocatalyst have been developed [1].

8.1.1 Aryl Diazonium Salts

Although this area flourished recently, the first examples of photocatalytic generation of aryl radicals have been reported as early as 1984 [14], when Deronzier and Cano–Yelo achieved the synthesis of phenanthrene derivatives by a photocatalytic Pschorr reaction using $[Ru(bpy)_3]^{2+}$ as a photoredox catalyst and aryl diazonium salt **1** as an oxidative quencher (Scheme 8.1). The classical copper-catalyzed Pschorr reaction [15, 16] starts with reduction of the diazonium salt **1** followed by intramolecular cyclization. A similar mechanism

Visible Light Photocatalysis in Organic Chemistry, First Edition.
Edited by Corey R. J. Stephenson, Tehshik P. Yoon and David W. C. MacMillan.

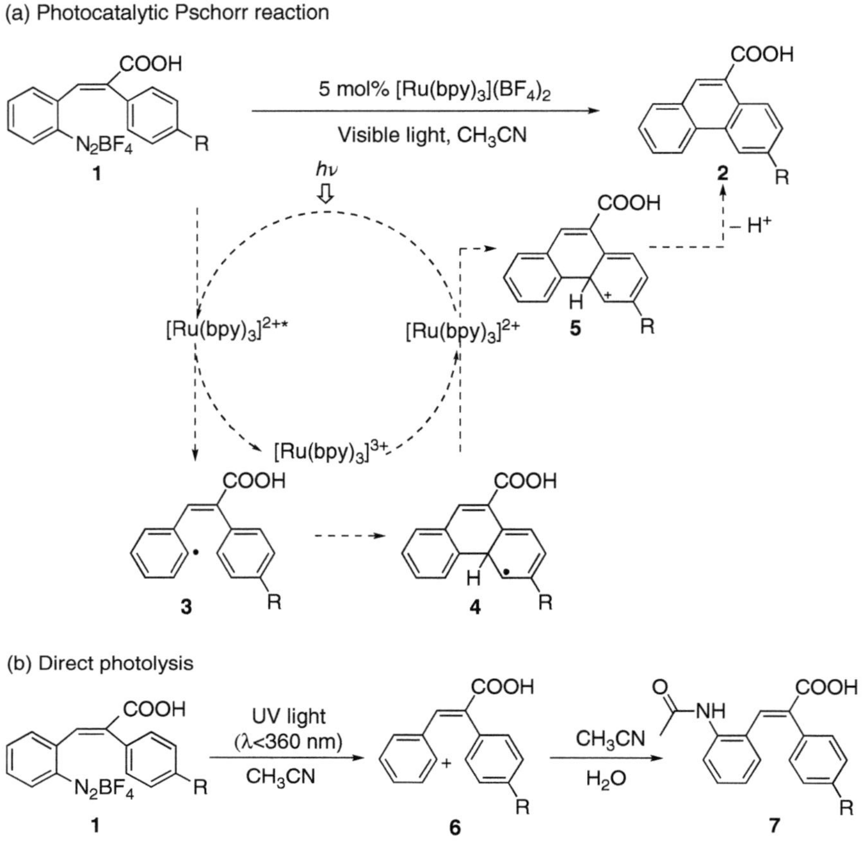

Scheme 8.1 Mechanism of the photocatalyzed Pschorr reaction as reported by Deronzier (a) compared to the direct photolysis under UV irradiation (b).

was proposed for the photocatalytic version: upon excitation by visible light ($\lambda > 410$ nm) $[Ru(bpy)_3](BF_4)_2$ transfers an electron to the diazonium salt **1**, which is cleaved to the aryl radical species **3** and nitrogen. The aryl radical **3** cyclizes to give the radical intermediate **4** that is subsequently oxidized to the cationic intermediate **5** by the strong oxidant $[Ru(bpy)_3]^{3+}$. Deprotonation restores the aromaticity and phenanthrene **2** is obtained in quantitative yields. It is important to note that this product cannot be prepared by direct photolysis of diazonium salt **1** since it mainly leads to acetanilide **7** with only small quantities of the desired phenanthrene **2** (Scheme 8.1). Direct photolysis of **1** by UV light does not yield the aryl radical as obtained by photocatalytic reduction: An aryl cation **6** is formed instead, which is directly attacked by the solvent acetonitrile to give **7**.

Based on the successful Photo-Pschorr reaction Cano–Yelo and Deronzier intended to extend their methodology to synthesize fluorenone, fluorene, and dibenzofuran from the corresponding diazonium salts (Scheme 8.2) [17]. However, these reactions turned out to be less efficient only yielding the cyclization

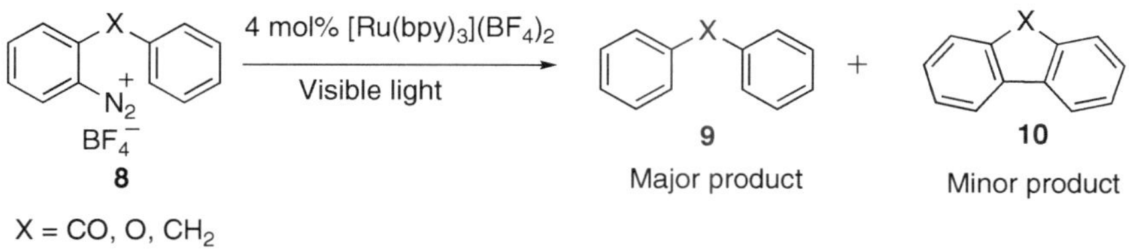

Scheme 8.2 Extension of the photocatalytic Pschorr reaction for the synthesis of fluorenone, fluorene and dibenzofuran.

product **10** in poor yields. A finding which can be explained by the more rigid structure of the diazonium salt **8** and the smaller gain in aromatic stabilization energy compared to phenanthrene.

Transition metal-catalyzed reactions for C—H arylation have been studied and have many advantages including broad substrate scope and functional group tolerance. However, most of the arylation methods still require high temperatures. Recently, Sanford and coworkers reported ligand-directed C—H arylation by merging photoredox catalysis with palladium catalysis using aryl diazonium salts as the aryl radical source at room temperature [18]. This method has several attractive features in comparison to other C—H arylation methods including a non-acidic solvent, low temperature, and the generation of easily removable HBF_4. Irradiation of a mixture of diazonium salt, substrate, $Pd(OAc)_2$, and $Ru(bpy)_3Cl_2 \cdot 6H_2O$ using a 26 W light bulb in methanol afforded the corresponding products in good yields (Scheme 8.3). This dual catalytic method is compatible with a variety of functional groups, directing groups including amides, pyrazoles, pyrimidines, and oxime ethers.

The mechanism of the dual catalytic reaction is depicted in Scheme 8.4. After photoexcitation of $[Ru(bpy)_3]^{2+}$ its excited state is oxidatively quenched by aryl diazonium salt **12**, giving an aryl radical **14** and the strong oxidant $[Ru(bpy)_3]^{3+}$. Addition of the very reactive aryl radical to palladacycle **15**, which is generated by C—H activation of the substrate, affords the intermediate **16.** Oxidation of the Pd^{III} intermediate **16** by $[Ru(bpy)_3]^{3+}$ regenerates the photocatalytic cycle while producing the intermediate **17**. Finally, the Pd^{IV} intermediate **17** undergoes reductive elimination to release the arylated product **13** while regenerating the Pd^{II} catalyst.

König and his coworkers utilized the aryl radical for direct C—H arylation of heteroarenes with aryl diazonium salts in green light using the organic dye eosin Y as a photoredox catalyst (Scheme 8.5) [19]. In comparison to other reported heteroarene C—H arylation methods, this transformation avoids the use of ligands, bases, metals, and elevated temperatures. Diazonium salts with electron-neutral or -withdrawing groups are smoothly converted into the corresponding products in good to excellent yields. Furthermore, this photoreaction was applied to prepare dithiophenes, which have found applications in materials science for example, preparing field effect transistors. No significant yields were observed when the reaction was conducted without a photocatalyst or without light, confirming the photocatalytic nature of the reaction.

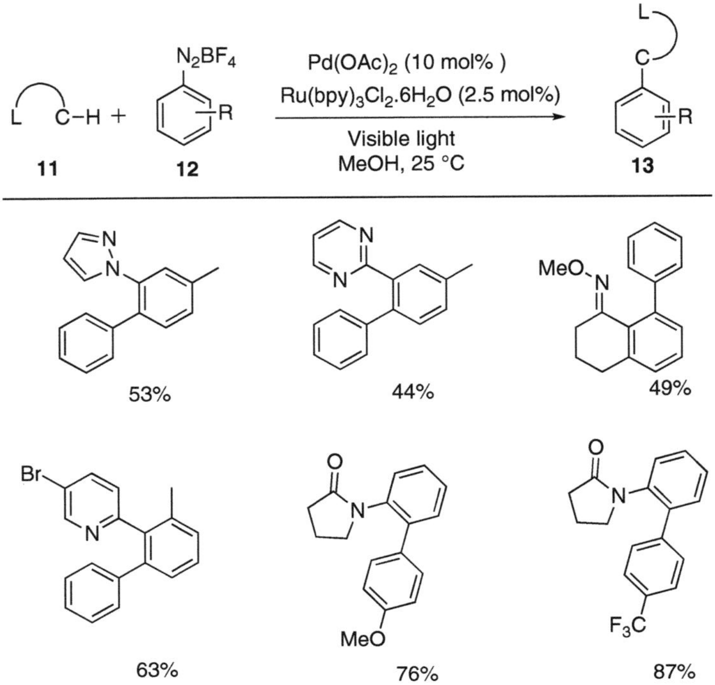

Scheme 8.3 Room temperature Pd/Ru catalyzed C—H arylation.

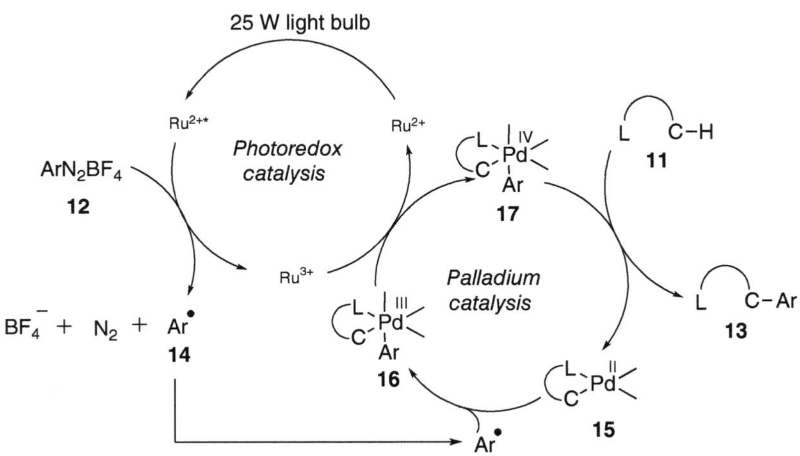

Scheme 8.4 Suggested mechanism for the Pd/Ru catalyzed C—H arylation.

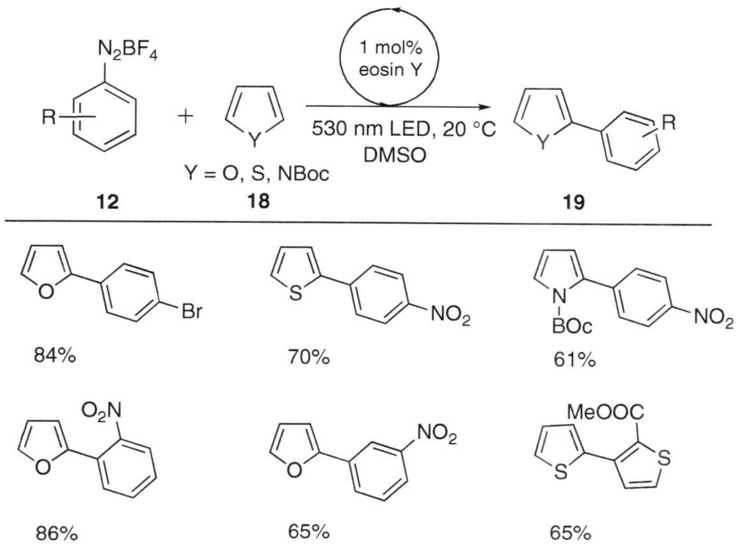

Scheme 8.5 Photocatalytic direct C—H arylation of heteroarenes.

A plausible mechanism of this metal-free photocatalytic reaction starts with a single electron transfer (SET) from the excited state of eosin Y to the aryl diazonium salt **12**, thus forming the cation radical of eosin Y and an aryl radical, which then subsequently adds to the heteroarene **18** to give another radical intermediate **20**. Further oxidation of the aryl radical intermediate by the eosin Y cation radical affords carbocation **21** and closes the catalytic cycle. Alternatively, **20** is oxidized by aryl diazonium salt **12** in a chain transfer mechanism. Deprotonation of carbocation **21** gives the desired product **19**. Radical intermediates were trapped with TEMPO (**22**, **23**) confirming the radical nature of this photoreaction (Scheme 8.6).

In an effort extending photocatalytic direct C—H arylation methods, König and coworkers applied the method to the synthesis of benzothiophenes, but mixtures of regioisomers were formed in very poor yields. However, a radical annulation visible light photoredox catalysis solved the problem (Scheme 8.7) [20]. Terminal and internal alkynes smoothly reacted in this transformation. Furthermore, the synthetic application of this photocatalytic annulation reaction was demonstrated by preparing the key intermediate **29** in the synthesis of the Raloxifene **30** (Scheme 8.8).

Based on the observation made in this study as well as on earlier reports, the authors proposed a tentative mechanism (Scheme 8.9). One electron transfer from excited state eosin Y to diazonium salt **24** provides an aryl radical **31** and the radical cation of eosin Y. Addition of the aryl radical to alkynes affords the vinyl radical intermediate **32**, which further undergoes homolytic substitution at the sulfur atom to give sulfuranyl radical **33**. Oxidation of the sulfuranyl radical **33** by the eosin Y radical cation or the aryl diazonium salt **24** in a chain transfer fashion furnishes the cation **34**. Finally, the cation intermediate **34** transfers

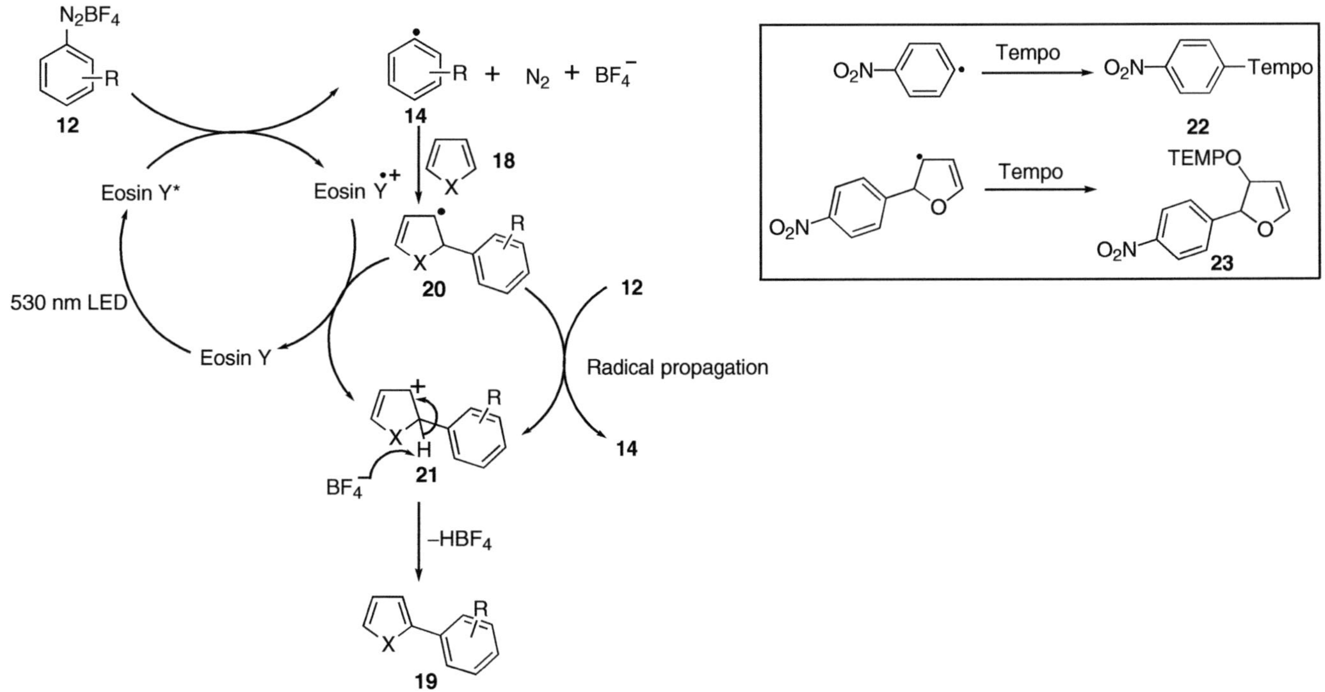

Scheme 8.6 Suggested mechanism for the direct C—H arylation of heteroarenes in visible light.

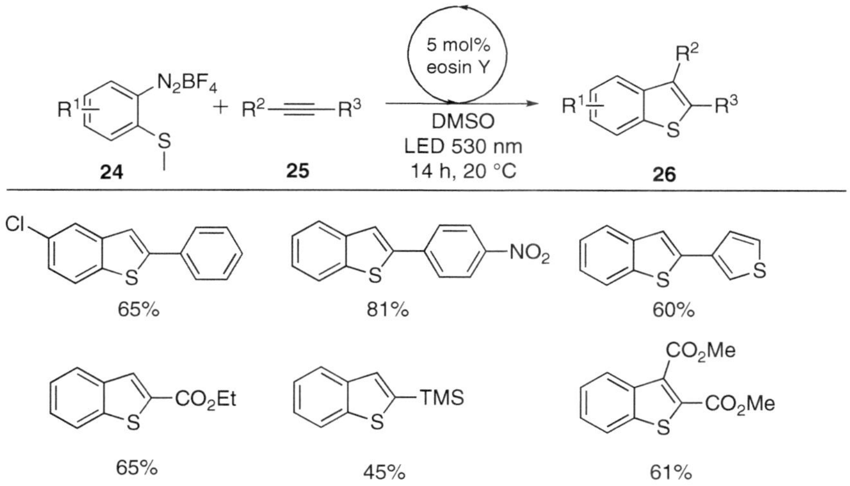

Scheme 8.7 Photocatalyzed annulations method for synthesis of benzothiophenes.

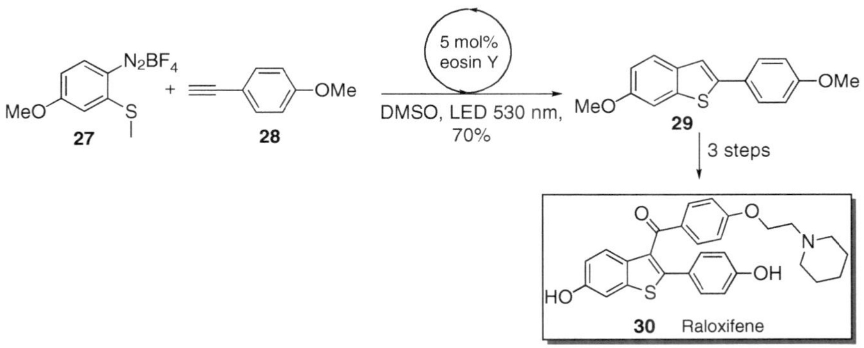

Scheme 8.8 Application of the photoreaction in the preparation of the key intermediate in the synthesis of the drug Raloxifene.

a methyl group to the solvent molecule DMSO giving the reaction product. TEMPO trapped products support the existence of radical intermediates during the photoreaction.

Meerwein developed an arylation method for unsaturated compounds with aryl diazonium salts in 1939, but the broader synthetic application of this reaction in organic synthesis was prevented by low yields, high catalyst loading, aqueous reaction media, and undesired side products [21]. Recently, a photocatalytic variant of the Meerwein reaction was reported, which further improves the process (Scheme 8.10) [22]. Control experiments confirmed that light and catalyst are essential for significant conversion of the arylated products. Halogen bearing diazonium salts have been used in the photoreaction leaving the carbon halogen bond intact and allowing for further synthetic elaboration. However, the reaction is mainly applicable to unsaturated substrates including styrene, coumarins, quinones, and phenyl acetylenes. Later, the Meerwein arylation

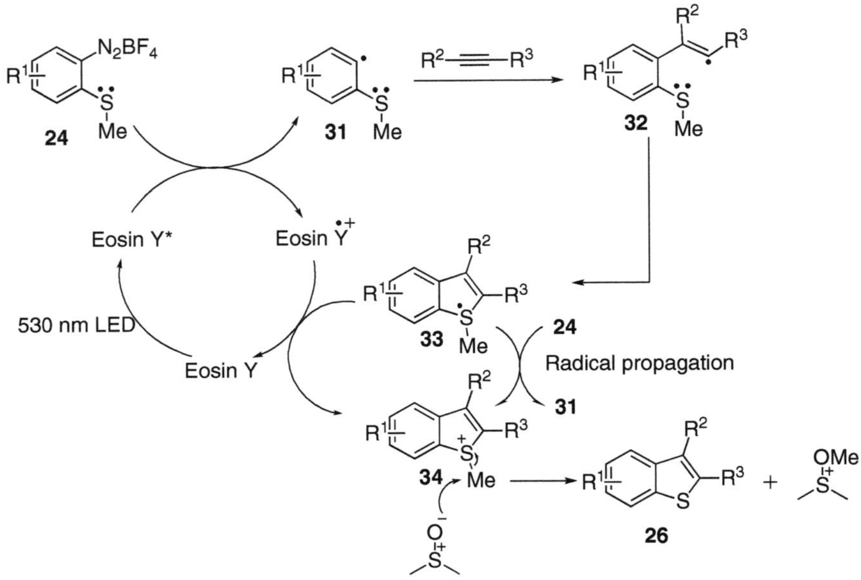

Scheme 8.9 Mechanism of the photocatalyzed annulation reaction for the synthesis of the benzothiophenes.

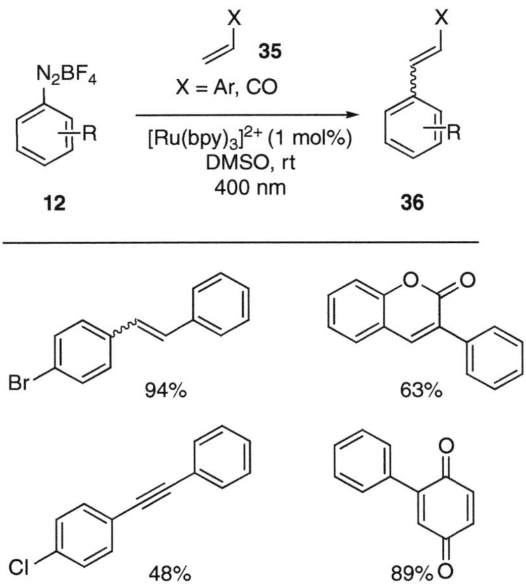

Scheme 8.10 The photocatalyzed version of intermolecular Meerwein arylation.

reaction was extended to enamides and enecarbamtes using aryl diazonium salts and blue light by the Yu group [23]. The mechanism of the photocatalytic Meerwein reaction starts with quenching of the excited catalysts by the diazonium salt to form the aryl radical. The aryl radical adds to the unsaturated compound to give a radical intermediate, which is further undergoes oxidation followed by deprotonation to yield the product.

Recently a photocatalytic method for the α-arylation of enol acetates using aryl diazonium salts in blue light was reported [24] (Scheme 8.11). Various photocatalysts and solvents were screened for this reaction; optimal yields were observed in DMF and DMSO when $Ru(bpy)_3Cl_2$ was employed as photocatalyst. The transformation proceeds at mild reaction conditions, has a broad substrate scope regarding diazonium salts and enol acetates, and avoids strong bases, expensive ligands, and toxic metal enolates. Furthermore, the authors employed this methodology for the synthesis of functionalized indoles in an efficient two-step process without isolation of intermediates (Scheme 8.12).

As expected for the significant conversion of the products both light and photocatalyst were necessary.

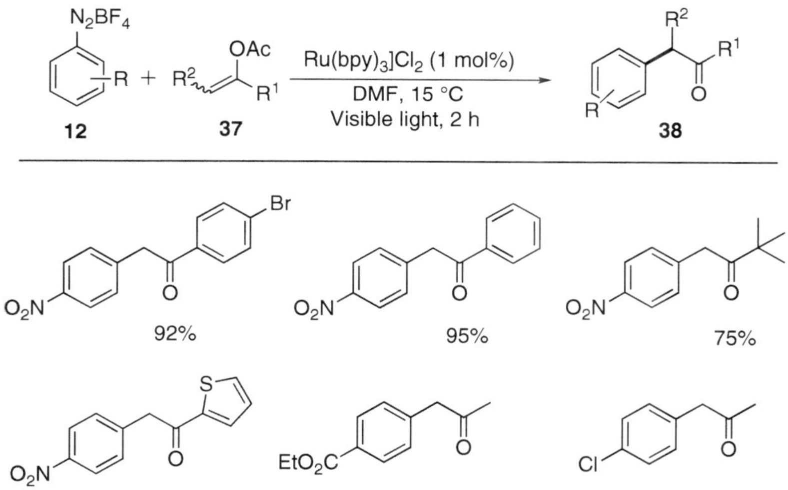

Scheme 8.11 α-Arylation of enol acetates using aryl diazonium salts in visible light.

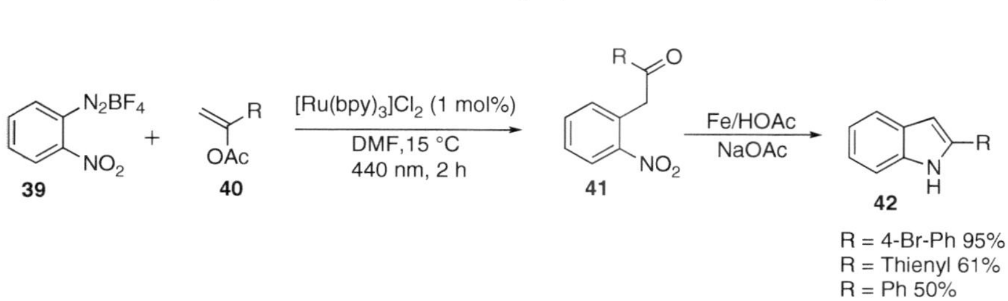

Scheme 8.12 Preparation of precursors of functionalized indoles by photoredox catalysis.

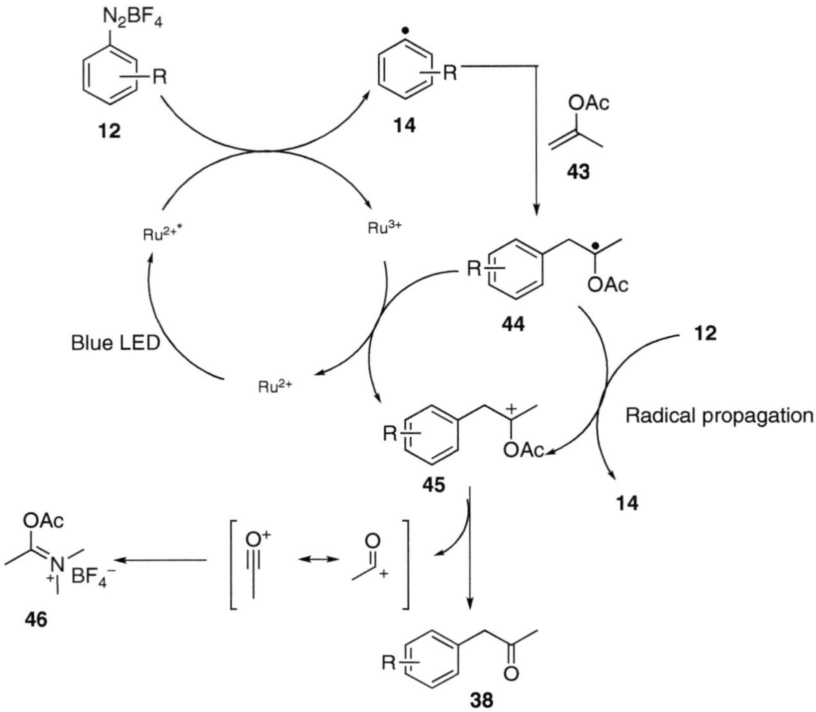

Scheme 8.13 A plausible mechanism for the α-arylation of enol acetates by photoredox catalysis.

The mechanistic pathway (Scheme 8.13) for the α-arylation of enol acetates proposed by the authors involves initial blue light excitation of $Ru(bpy)_3^{2+}$ to generate the excited state $Ru(bpy)_3^{2+*}$, which transfers a single electron to diazonium salt **12** to give aryl radical **13** and $Ru(bpy)_3^{3+}$. Addition of the aryl radical to the enol double bond **43** furnishes another radical intermediate **44**, which is further oxidized either by $Ru(bpy)_3^{3+}$ or by diazonium salt **12** to provide cation **45**. Reduction of $Ru(bpy)_3^{3+}$ closes the catalytic cycle. Finally, the desired product is obtained by transfer of an acyl cation to DMF giving the stable salt **46**, which is the likely intermediate in the first step of the Vilsmeier–Haack reaction.

Phenanthrenes are important structural motifs of organic compounds due to their wide presence in natural products, and have evinced considerable interest with respect to pharmaceutical and material chemistry. However, a common synthetic approach for the synthesis of phenanthrenes is photocyclization of stilbenes under UV-light irradiation. The UV light-mediated method may induce the formation of side products and requires very expensive quartz glassware, UV lamps, and protective eyewear. Recently, Zhou and his coworkers reported a visible light photocatalytic synthesis of phenanthrenes from diazonium salts and alkynes using eosin Y [25]. Noteworthy, addition of base to the reaction decreases the yield, due to a direct reaction with the diazonium salt.

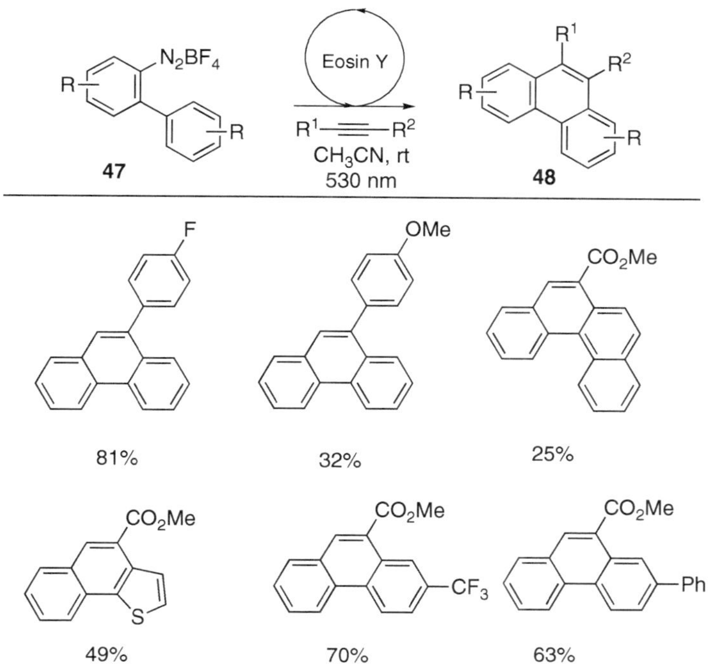

Scheme 8.14 Synthesis of phenanthrenes by visible light photocatalysis.

A variety of functional groups including nitro, keto, methoxy, halogen, and ester groups are tolerated well in this reaction (Scheme 8.14). Reduction of diazonium salts by eosin Y* produces a biaryl radical, which further adds to the alkyne to form a vinyl radical intermediate and the intramolecular addition to the aromatic ring affords the cyclic radical intermediate. Finally, the cyclic radical intermediate undergoes oxidation followed by deprotonation to give the desired product **48**.

The photoredox-catalyzed Meerwein arylation addition reaction was reported by König and his coworkers for amino arylation of alkenes in visible light (Scheme 8.15). The influence of water on this reaction was significant: a larger amount of water yields 1,2-diphenylethanol as a by-product. This photocatalytic method utilizes mild and efficient conditions to access a variety of amides and it was compatible with a wide range of functional groups including ester, nitro, halide, ether, alkyl groups. The substrate scope with respect to diazonium salts, nitriles, and alkenes is large and many photoreaction products have the potential for further synthetic elaboration.

The proposed mechanism of the reaction is depicted in Scheme 8.16. Initial SET from the excited state of the photocatalyst $Ru(bpy)_3^{2+*}$ to diazonium salt **12** gives an aryl radical **14**, which adds to the alkene **49** to give the corresponding radical intermediate **51**, which is further oxidized to the carbenium intermediate **52**. Finally, the intermediate **52** is trapped by a nitrile and followed by hydrolysis to give the required product **50**. Radical intermediate **51** is oxidized either by the

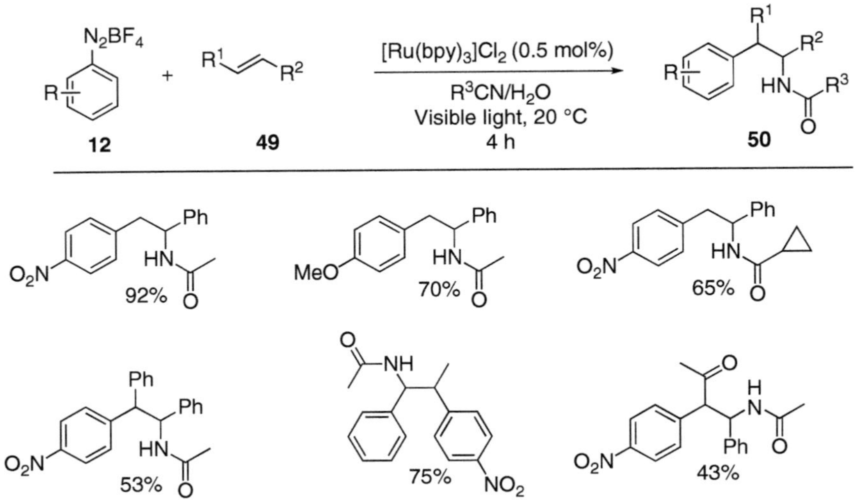

Scheme 8.15 The photoredox-catalyzed Meerwein arylation addition reaction.

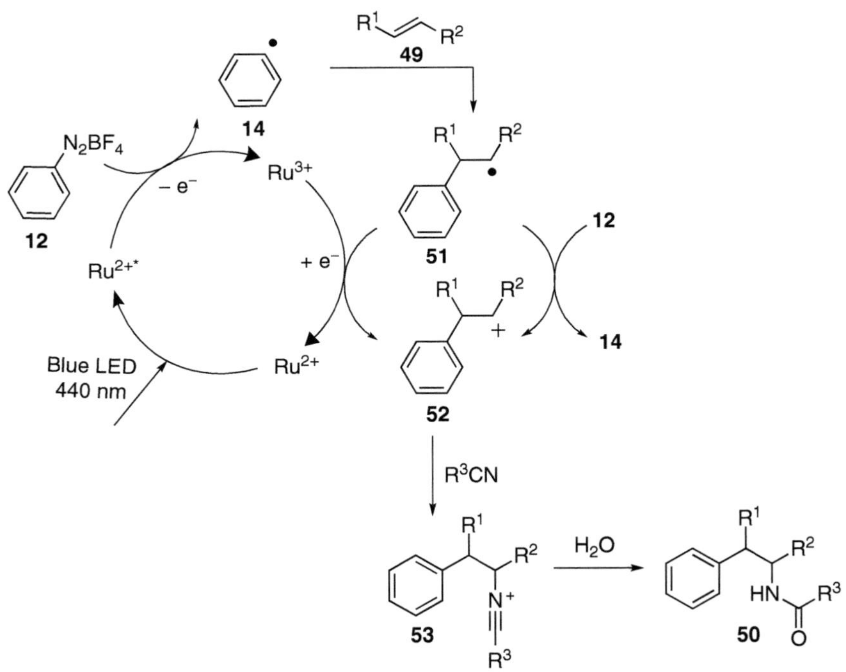

Scheme 8.16 Suggested mechanism for the photoredox-catalyzed Meerwein arylation addition reaction.

$Ru(bpy)_3{}^{3+}$ to regenerate the catalytic cycle or by the diazonium salt **12** in a chain transfer mechanism.

Oxindoles are found as important structural components in natural products and pharmaceutical chemistry. Palladium-catalyzed oxidative difunctionalization of alkenes in *N*-arylacrylamides is an efficient method for their synthesis and an alternative method is the radical-mediated cyclization of *N*-arylacrylamides. Zou and coworkers reported an efficient method for the preparation of oxindoles using photoredox-catalyzed diarylation of *N*-arylacrylamide **54** with aryl diazonium salt **12** (Scheme 8.17) [26]. *N*-protected substrates including methyl, butyl, benzyl, and isopropyl were effective in this transformation.

Recently, Glorius and coworkers reported a dual catalytic method for intramolecular oxy and aminoarylation of alkenes with aryl diazonium salts by combining gold and photoredox catalysis in visible light (Scheme 8.18) [27]. This dual catalytic process involves the formation of C—C and C—Nu bonds across the alkene at room temperature and does not require any external strong oxidants. The dual catalytic nature of the reaction was confirmed by control experiments. The absence of the photocatalyst or light led to very poor yields, while no product was identified in the absence of the gold catalyst. The scope of the reaction was broad with respect to diazonium salts as well as alkenes.

A tentative mechanism of the reaction is shown in Scheme 8.19. Initially, the alkene substrate **58** reacts with the cationic Au(I) catalyst **59** to afford the intermediate **60** resulting from *anti*-selective cyclization. A SET from the excited photocatalyst to the diazonium salt **12** generates an aryl radical, which adds to the intermediate **60** to afford the Au(II) intermediate **61** bearing both coupling partners. Oxidation of the unstable intermediate **61** by $[Ru(bpy)_3]^{3+}$ regenerates the photocatalyst while producing the Au(III) intermediate **62**, which then undergoes reductive elimination to yield the cyclized product **63** and the Au(I) catalyst.

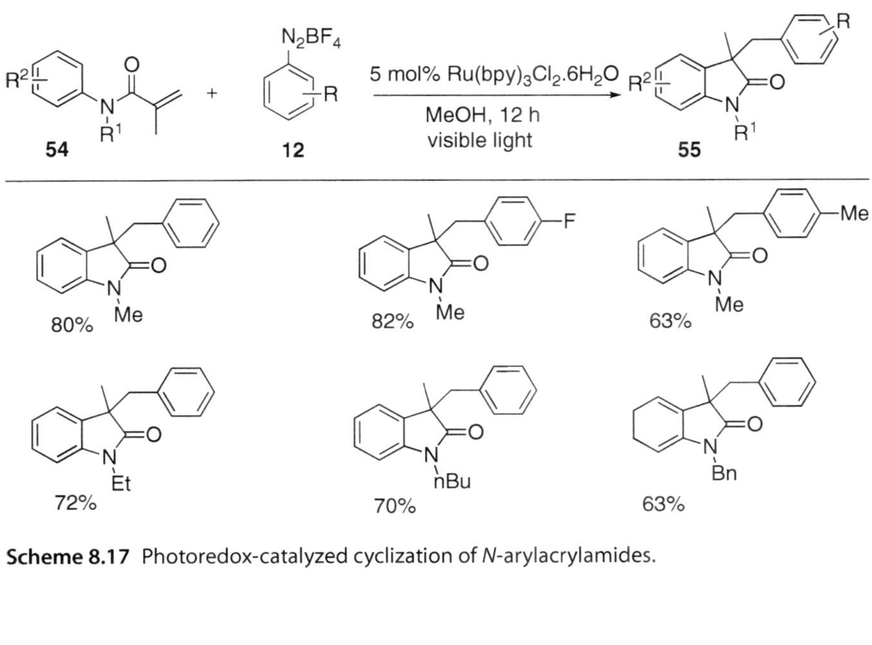

Scheme 8.17 Photoredox-catalyzed cyclization of *N*-arylacrylamides.

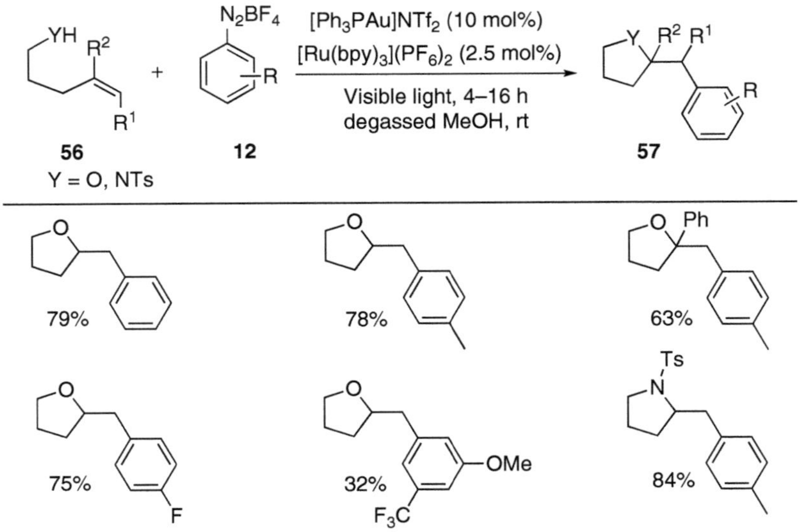

Scheme 8.18 Dual catalytic system for intramolecular oxy and amino arylation.

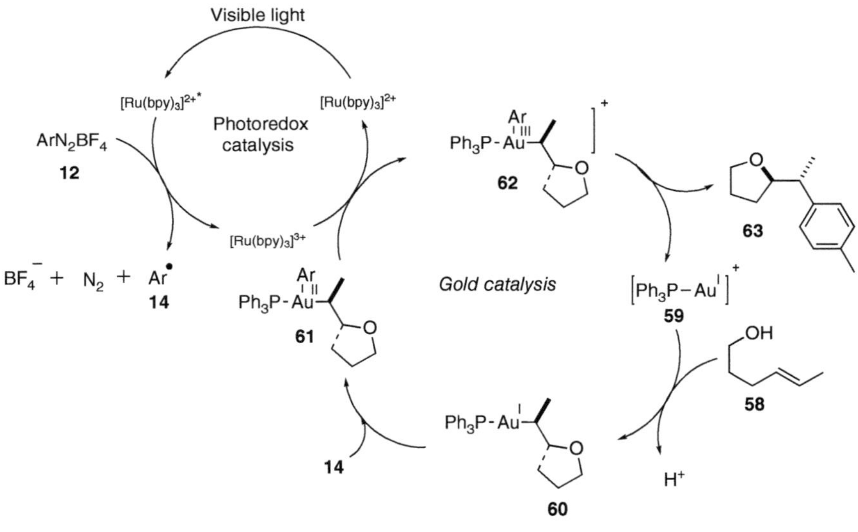

Scheme 8.19 Proposed mechanism of the dual catalytic system.

Aryl sulfides are important structural motifs in nature and they are synthesized from readily available diazonium salts with thiols under basic or neutral reaction conditions, but the intermediate that is formed during the reaction is a potent explosive. To overcome this hazard, Jacobi and coworkers recently developed a synthesis of aryl sulfides from aryl diazonium salts in the presence of eosin Y [28]. The photoreaction is highly tolerable to different functional groups and has the broad substrate scope. The proposed mechanism of the reaction starts with a SET from eosin Y to the diazonium salt forming the aryl radical, which then reacts with the disulfide to give a trivalent sulfur radical. A SET oxidation of the sulfur radical by the eosin Y radical cation produces an electrophilic species, while regenerating the photocatalyst. Finally, the electrophilic species undergoes substitution in the presence of DMSO to afford the target aryl sulfide.

The Stadler–Zielgler reaction allows the formation of C—S bonds. Aryl amines are converted into the corresponding diazonium salts followed by reaction with thiolates to yield aryl sulfides. To avoid the isolation of the diazonium salts and to increase the efficiency of the reaction, Noël and coworkers reported a mild, one pot Stadler–Zielgler reaction by photoredox catalysis for the synthesis of aryl sulfides with minimum formation of diazosulfides (Scheme 8.20) [29]. In addition, the authors have developed a continuous-flow protocol addressing safety concerns associated with the explosive intermediates during the reaction. Notably, alkylthiols show similar reactivities compared to arylthiols and *N*-heterocyclic substrates afforded lower yields in this photoreaction.

All of the reactions described so far dealt with C—C and C—S bond formation utilizing the oxidative quenching of the photocatalyst using aryl diazonium salts. Very recently, the Yan group reported a carbon–borylation reaction in visible light using eosin Y (Scheme 8.21) [30]. Aryl diazonium salts bearing both electron- donating and electron-withdrawing groups reacted smoothly to give the corresponding products in moderate to good yields. Heteroaromatic diazonium salts do not react well in this photoreaction. The proposed mechanism of the reaction starts with a SET from the excited state of eosin Y to the aryl diazonium salt **12** to give an aryl radical, which adds to the complex **69**, which is formed *in situ* by the interaction between **67** and the BF_4^-, to afford the borylated product **68** and radical anion **70**. One electron oxidation of the reaction intermediate **70** by the eosin Y radical cation closes the catalytic cycle (Scheme 8.22).

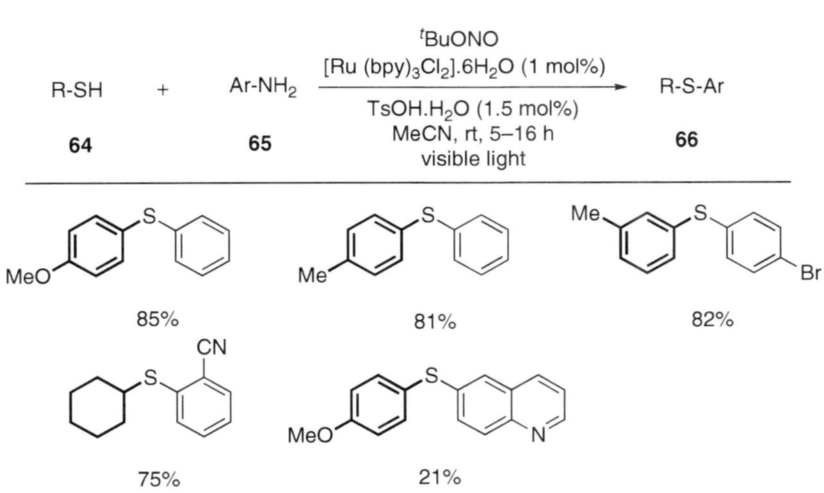

Scheme 8.20 Photocatalytic Stadler–Zielgler reaction for synthesis of aryl sulfide.

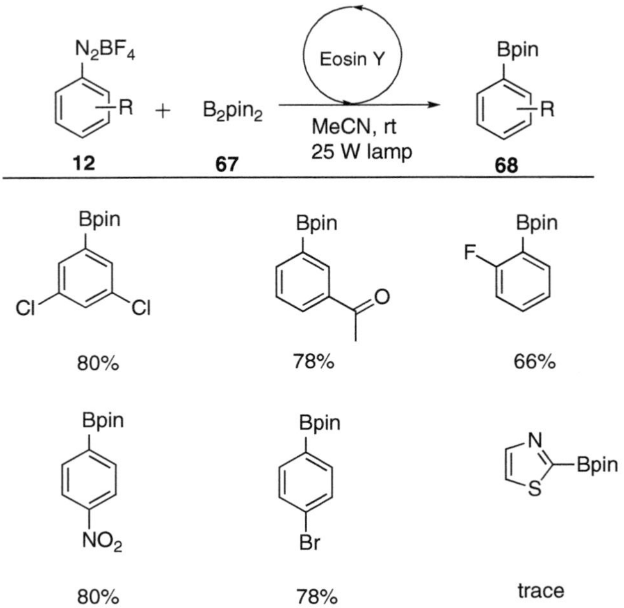

Scheme 8.21 Borylation of aryl diazonium salts in visible light.

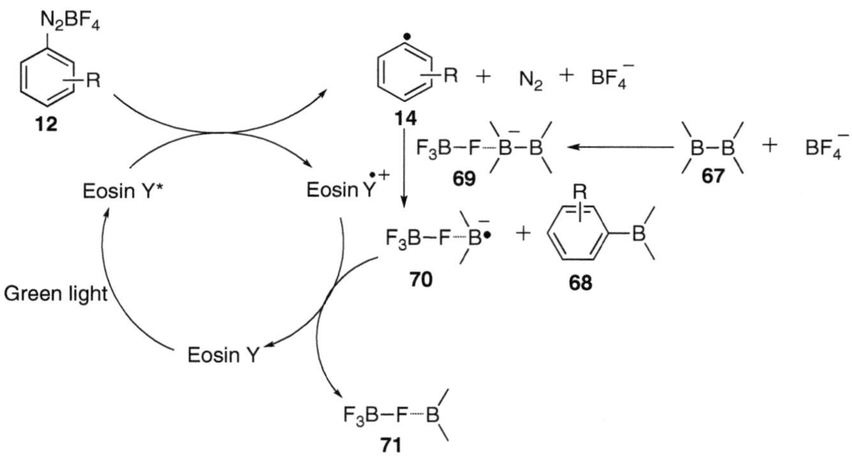

Scheme 8.22 Proposed mechanism for borylation reaction in visible light.

8.1.2 Diaryl Iodonium Salts

Diaryl iodonium salts are an alternative source of aryl radicals and they have been used as photoinitiators for cationic polymerization, surface functionalization, and also as arylating reagents, but only a few applications in visible light mediated organic synthesis have been reported. Recently, Sanford described a dual catalytic method for the C—H arylation with iodonium salts at room temperature [31]. The necessity of a photocatalyst and visible light, the inhibition by radical scavengers, and the obtained chemioselectivity support a radical mechanism for this transformation.

Later, Xiao and his coworkers reported the direct C—H arylation of heteroarenes and arenes using diaryl iodonium salts as radical precursors in visible light with $Ru(bpy)_3Cl_2$ as a photoredox catalyst (Scheme 8.23) [32]. The counter anion of the diphenyl iodonium salts has an impact on the reaction yield and the authors found that trifluoromethanesulfonate gave the best yields. The substrate scope toward diaryl iodonium salts is large and the desired products were isolated in good yields with electron-withdrawing groups. In addition, unsymmetrical iodonium salts were tested in this photoreaction, and it was found that the sterically more hindered aromatic group is preferentially transferred to the heteroarene. The proposed mechanism of the reaction is similar to the previously reported mechanism. Initially, a SET from the excited $[Ru(bpy)_3]^{2+*}$ to the diaryl iodonium salt takes place, giving an aryl radical and the strong oxidant $[Ru(bpy)_3]^{3+}$. Addition of the aryl radical to the arene gives another radical intermediate, which is oxidized by $[Ru(bpy)_3]^{3+}$ to a carbocation while regenerating the photocatalyst. Finally, proton elimination of the carbocation intermediate affords the product.

Another photocatalytic reduction of diaryl iodonium salts using $[Cu(dpp)_2]$ $[PF_6]$ for allylation reaction with allylsulfones has been reported by Ollivier and his coworkers (Scheme 8.24) [33]. Compared to photocatalysts such as Ru– or Ir– polypyridine complexes, no significant improvement was observed with $[Cu(dpp)_2][PF_6]$. However, it demonstrates that the copper complex is an inexpensive alternative to known photocatalysts. The control experiments showed that the copper catalyst, visible light, and an electron donor were necessary for this reaction. Advantages of this photo reaction are mild reaction conditions and the broad scope of diaryl iodonium salts and allylsulfones.

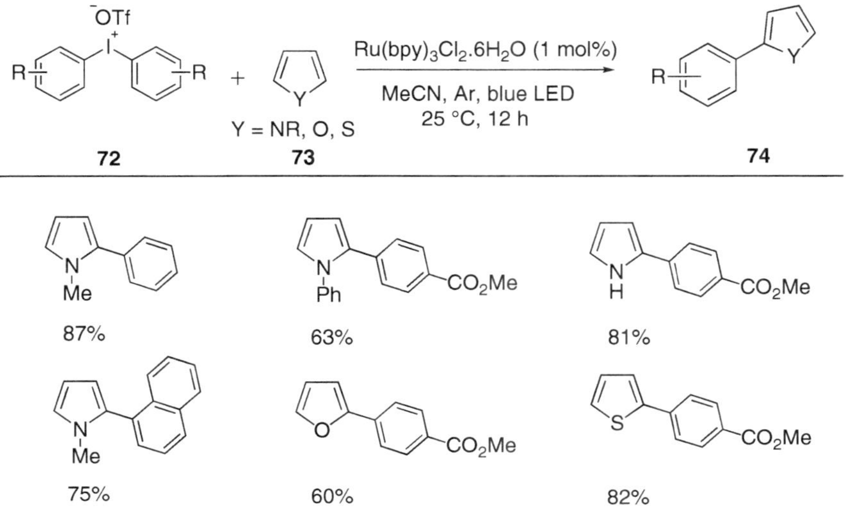

Scheme 8.23 Direct arylation of heteroarenes with diaryl iodonium salts.

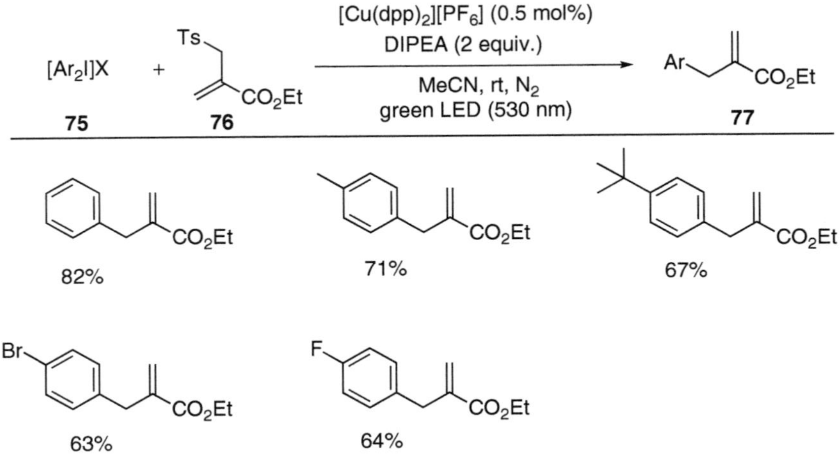

Scheme 8.24 Allylation of aryl radicals by photoredox catalysis.

To gain further details of the reaction mechanism, the authors monitored the reaction by ^{1}H NMR and found the Cu (II) species during the reaction. The signals of the $[Cu(dpp)_2]^+$ were recovered without any significant modification by adding two equivalents of DIPEA. Based on these results and literature reports, the authors proposed a tentative mechanism (Scheme 8.25). Upon green light irradiation, the excited state of the photocatalyst is oxidatively quenched by diaryl iodonium salt **75** to produce $[Cu(dpp)_2]^{2+}$ and the iodanyl radical intermediate **78**, which further undergoes fragmentation to furnish the aryl radical. Further on, the aryl radical adds to the allylsulfone **76** to afford the allylation product **77**. Finally, the catalyst $[Cu(dpp)_2]^+$ is regenerated by DIPEA.

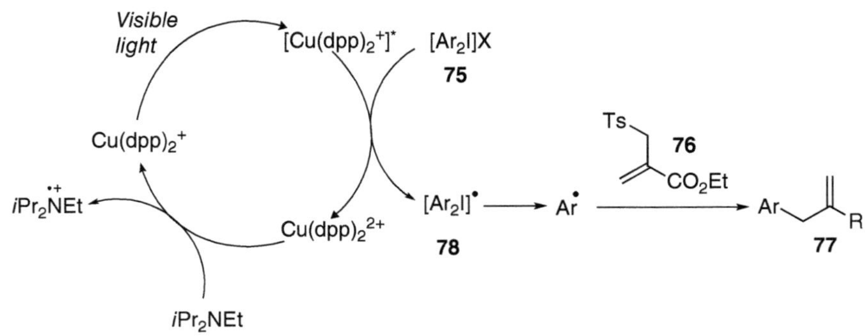

Greaney and coworkers further utilized the diaryl iodonium salts for oxy and aminoarylation of styrenes using photoredox catalysis (Scheme 8.26) [34]. Visible light irradiation of 1 mol% of $Ir(ppy)_3$, styrene (1 equiv.), Ph_2IBF_4 (2 equiv.), and 20 mol% of $Zn(OAC)_2$ in methanol provided the corresponding oxyarylated product in 70% yield. The authors investigated the scope of the reaction by employing a variety of iodonium salts. It was found that alkyl, ester, halogen, trifluoromethyl groups containing iodonium salts reacted smoothly

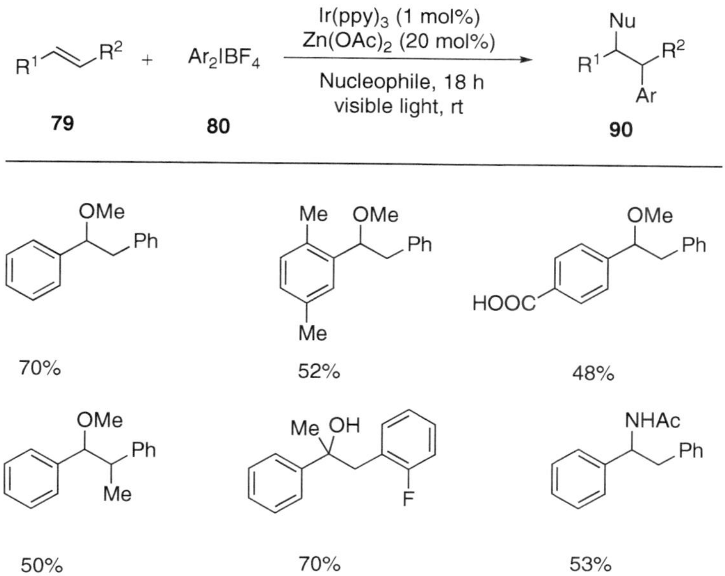

Scheme 8.25 Photoreaction for oxy and aminoarylation.

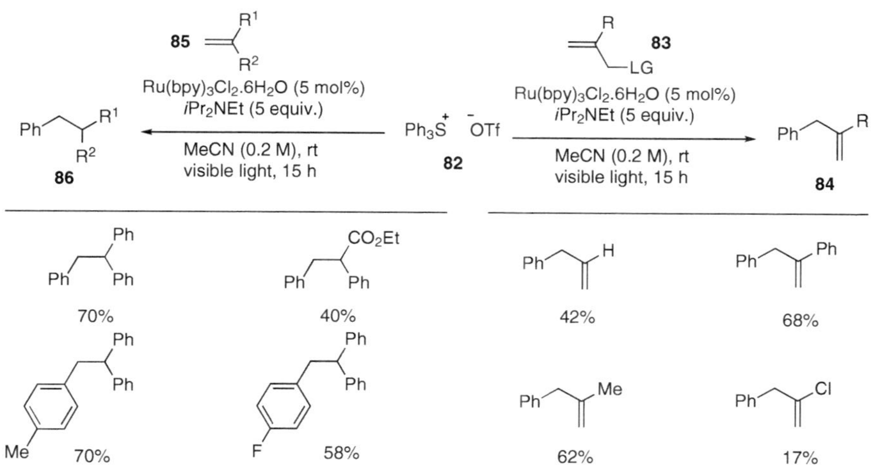

Scheme 8.26 Photocatalytic reduction of triaryl sulfonium salts.

to afford the corresponding products in good yields. Diaryl iodonium salts having tetrafluoroborate as counter anion were found to be more efficient than the analogous triflate salts. In addition, the authors also investigated the other nucleophiles such as water and nitriles in this mutli-component reaction. The suggested mechanism of the reaction starts with oxidative quenching of the excited state photocatalyst by the diaryliodonium salt furnishing an aryl radical and Ir(IV). The aryl radical adds to the styrene forming another benzylic radical intermediate, which then undergoes further oxidation by Ir(IV) to produce the cation while regenerating the photocatalyst. Finally, the cation was trapped by various nucleophiles to give the corresponding products.

8.1.3 Triaryl Sulfonium Salts

Sulfonium salts have found many applications in organic synthesis. They can be reduced electrochemically, photochemically, or by pulse radiolysis. In nature, the biochemical reduction of the sulfonium salt of S-adenosyl-L-methionine is possible by Fe-S cluster containing enzymes [35, 36]. However, the application of sulfonium salts as aryl radicals has received only little attention so far. Recently, Ollivier and his coworkers reported the reduction of triaryl sulfonium salts using photoredox catalysis and the subsequently formed aryl radical was utilized for carbon–carbon bond formation (Scheme 8.27) [37]. The photoreaction of 1 equiv. of the triphenyl sulfonium salts with phenylsulfone (10 equiv.), Hünig's base (5 equiv.) using 1 mol% $Ru(bpy)_3Cl_2$ catalyst under blue light irradiation provided the corresponding product in 68% yield after 15 h. In addition, the authors also investigated other acceptors such as styrenes, acrylate, methyl acrylate, acrylonitrile and screened sulfonium salts with different counter anions. No significant effect of the counter anion on the reaction yields was noted.

In contrast to other onium salt involving mechanisms, this reaction (Scheme 8.28) starts presumably with the reductive quenching of the photocatalyst $Ru(bpy)_3{}^{2+}$ by DIPEA to give a strong reductant Ru(I) that transfers an electron to the triphenyl sulfonium triflate **75**. The generated sulfuranyl radical decomposes into a phenyl radical and diphenyl sulfide. Finally, the phenyl radical adds either to **76** or **78**. The radical intermediate undergoes a ß-fragmentation to produce the allylated product **77** or hydrogen abstraction from the amine

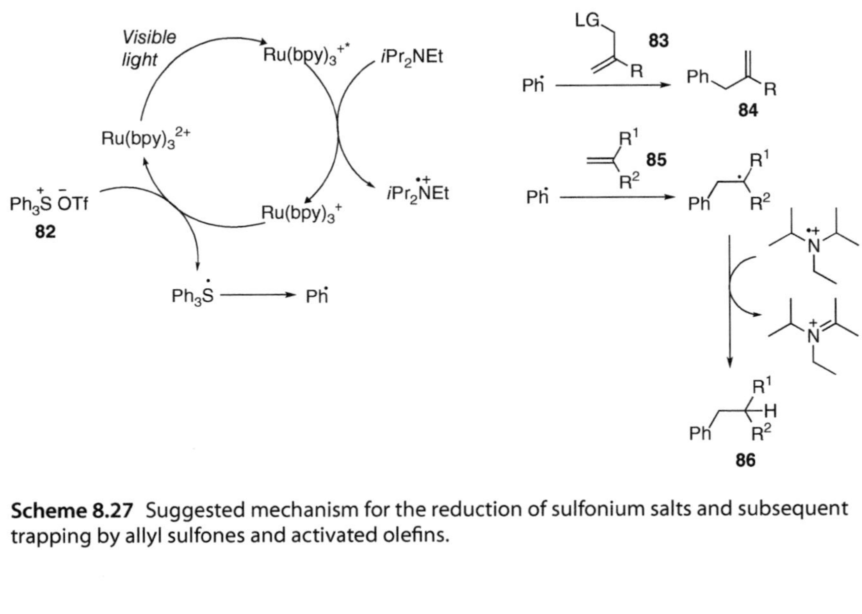

Scheme 8.27 Suggested mechanism for the reduction of sulfonium salts and subsequent trapping by allyl sulfones and activated olefins.

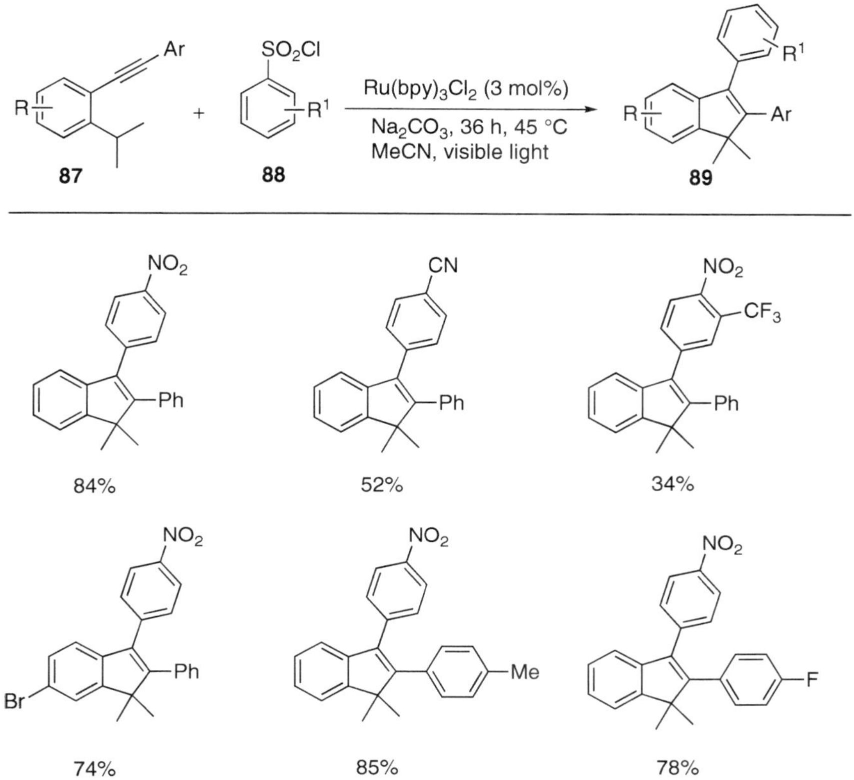

Scheme 8.28 Photocatalyzed reaction for the synthesis of 1*H*-indenes.

radical cation to afford the reduced compound **79**. Notably, the diphenyl sulfide side product does not react under the photoreductive conditions.

8.1.4 Aryl Sulfonyl Chlorides

Aryl radicals were also generated from the corresponding aryl sulfonyl chlorides. Recently, Li and coworkers utilized aryl sulfonyl chloride for constructing functionalized 1*H*-indenes through radical arylative cyclization of *ortho*-alkyl arylalkynes using visible light photoredox catalysis (Scheme 8.29) [38]. Notably, in this transformation the photocatalyst can be recovered and reused without any loss of activity.

The proposed mechanism of the reaction starts with a SET from $Ru(bpy)_3{}^{2+*}$ to aryl sulfonyl chloride generating an aryl radical and a strong oxidant. An addition of the aryl radical to the alkyne gives another radical intermediate, which is oxidized by $Ru(bpy)_3{}^{3+}$ to afford the cation while closing the catalytic cycle. Finally, deprotonation followed by carbocyclization furnishes the desired 1*H*-indene. The presence of a base accelerates the reaction by removing HCl, which is formed during the reaction.

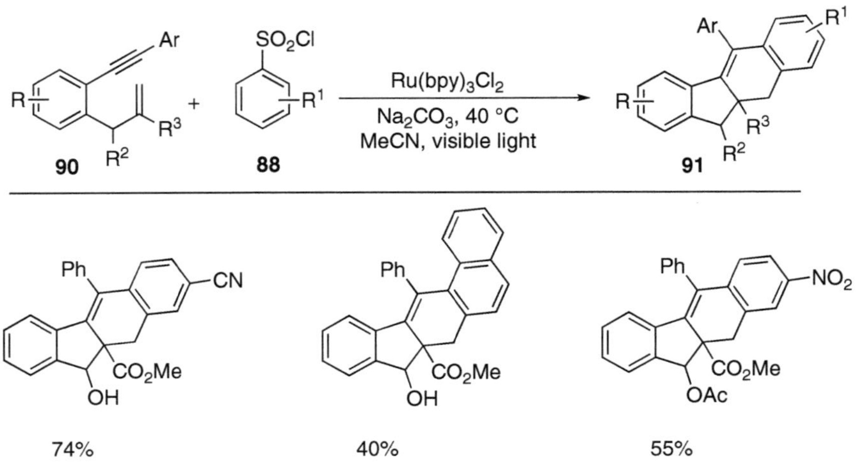

Scheme 8.29 Tandem cyclization of 1,6-enynes with arylsulfonyl chlorides using visible light photoredox catalysis.

Tandem cyclization is an important strategy in organic chemistry to construct substituted polycyclic compounds. Li and coworkers reported an efficient tandem cyclization of 1,6-enynes with arylsulfonyl chlorides using visible light photoredox catalysis (Scheme 8.29) [39]. The reaction of **90** with **88**, $Ru(bpy)_3Cl_2 \cdot 6\ H_2O$, Na_2CO_3, and 36 W compact fluorescent light irradiation in MeCN at 40 °C gave the desired product **91** in 12%. Interestingly, after replacing the light source by a 5 W blue LED, the yield of **91** was dramatically increased to 79%. In addition, the nature of the base has an influence on this reaction, and better yields were observed with inorganic bases as compared to that with organic bases.

8.2 Applications of Aryl Diazonium Salts

As described in the preceding sections photocatalytic arylation reactions in visible light are a viable and versatile tool to functionalize arene moieties in homogeneous solution. Recently, photocatalytically generated aryl radicals from aryl diazonium salts have also been employed for the covalent functionalization of solid surfaces like gold, polynitrophenylene, and cellulose. The use of aryl radicals in surface modification is well established for electrochemical or metal-catalyzed methods and light-induced processes may provide additional advantages. Pinson and coworkers described the modification of a gold surface with aryl diazonium salts and visible light absorbing photocatalysts like eosin Y or $[Ru(bpy)_3]^{2+}$ (Scheme 8.30) [40].

An even more elaborate example was reported by the Berthelot group using a combination of inkjet printing and photocatalysis for creating covalently bound polymer structures on polynitrophenylene (Scheme 8.31) [41]. A mixture of the diazonium salt, acrylic acid, and the photocatalyst $[Ru(bpy)_3]^{2+}$ is placed on the

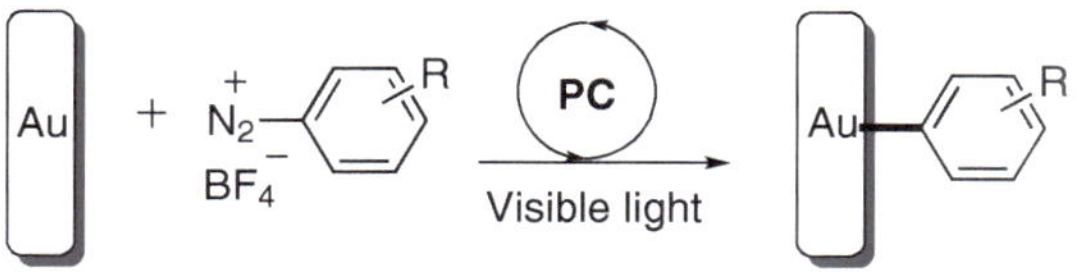

Scheme 8.30 Photocatalytic arylation of gold surfaces by visible light. PC = photocatalyst.

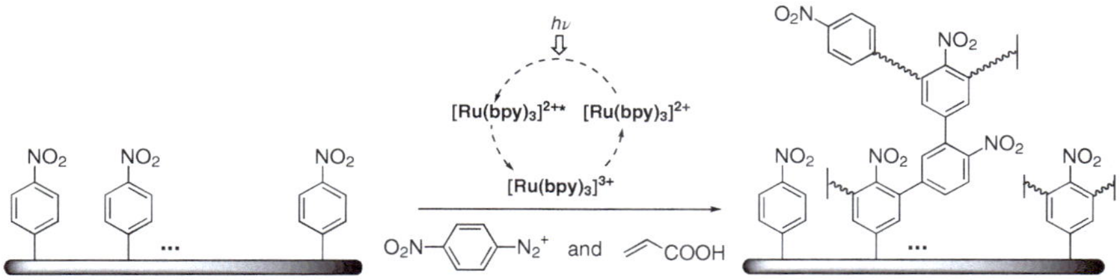

Scheme 8.31 Schematic picture of the surface modification by a combination of aryl radical coupling to a polymeric surface (surface functionalization not shown; mechanism as in scheme 8.30) and reaction in homogeneous solution with acrylic acid.

surface and the diazonium salt is reduced to the aryl radical, which in part reacts directly with the surface nitrophenyl moieties while also forming small oligomers cross-linked by polyacrylic acid strings, which eventually react with the surface. This methodology was applied to pattern different materials and is a promising technique for applications in the synthesis of flexible electronic devices.

The most recent example demonstrates how the photocatalytic arene functionalization on surfaces can be used to achieve spatial resolution [42]. A cellulose sheet covalently linked with a coumarin derivative is soaked in a solution of the diazonium salt and the photocatalyst in DMSO and covered with a photomask before irradiation. As can be seen in Scheme 8.32 the reaction does only take place in the light exposed areas.

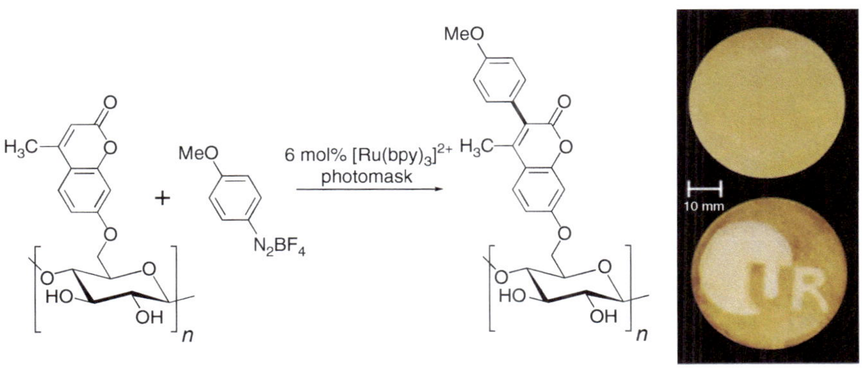

Scheme 8.32 Patterning a coumarin-functionalized cellulose sheet by spatially controlled photoreaction. The picture on the right shows the cellulose sheet before (above) and after irradiation (bottom).

8.3 Photoinduced Ullmann C—N Coupling

The Ullmann reaction for the construction of C—N bonds was first reported in 1903 for the synthesis of arylamines from readily available aryl halides and amines using stoichiometric amounts of copper at elevated temperatures. During the past years, several elegant methods have been developed for C—N coupling reactions. However, understanding the mechanism for these reactions has evolved only slowly. Recently, Peters and coworkers reported experimental data for the reaction of aryl halides with a Cu(I) amido complex in order to prove the radical mechanism for the Ullmann coupling reaction [43, 44]. Noteworthy, this study introduces a photoinduced variant of the Ullmann coupling for C—N bond formation.

Irradiation of a copper carbazolide complex and iodobenzene in CH_3CN using a 13 W compact fluorescent light bulb at room temperature for 10 h afforded the corresponding product in 77% yield (Scheme 8.33). Bromobenzene and fluorobenzene also undergo Ullmann coupling reaction to produce the corresponding products in 40%, and 5% yields, respectively. The control experiments have proven that both light and copper catalysts were necessary for a significant conversion. A higher yield was obtained with CD_3CN compared to CH_3CN which may be explained by a kinetic isotopic effect for the abstraction of hydrogen/deuterium from the solvent either by a phenyl radical or by the radical cation of the copper complex. The authors also developed a catalytic procedure for C—N bond formation. Irradiation of lithium carbazolide and iodobenzene in the presence of 10 mol% copper carbazolide complex provides the corresponding product in 64% yield.

The mechanism of the photoinduced Ullmann coupling reaction is depicted in Scheme 8.34. Irradiation of a copper carbazolide complex initiates an electron transfer to the aryl halide to produce a radical anion, which decomposes to afford the aryl radical. The aryl radical reacts with the copper complex to furnish the C—N coupling product. To distinguish between a concerted oxidative addition and a pathway involving SET, the authors carried out a

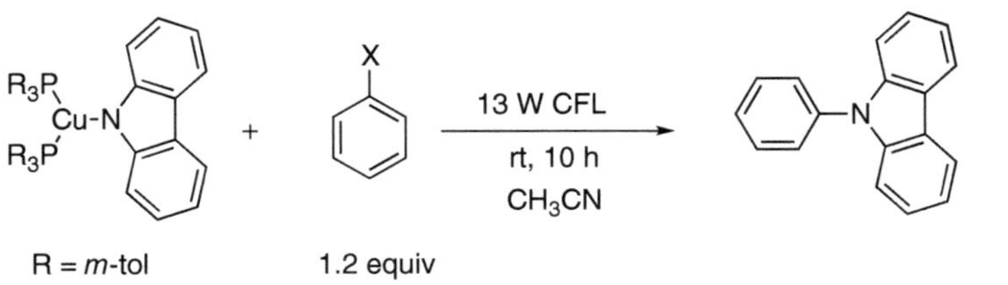

92 **93** **94**

X	Yield (%)
I	77
Br	40
F	5

Scheme 8.33 Photoinduced Ullmann coupling for C—N bond formation.

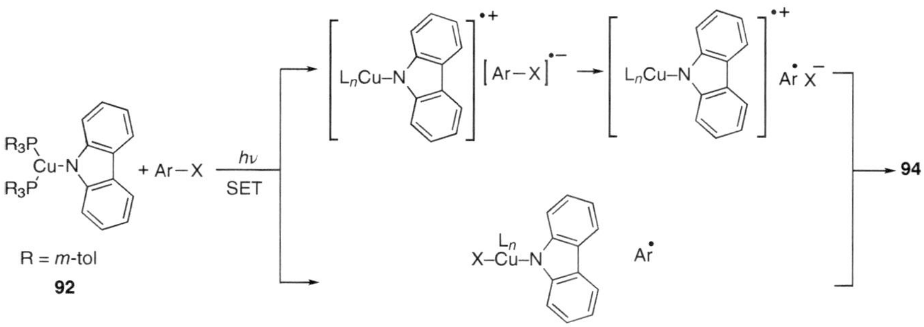

Scheme 8.34 Mechanism for photoinduced Ullmann coupling for C—N bond formation.

series of experiments comparing relative reactivities. If the C—X bond cleavage proceeds via a concerted oxidative addition the main product is the coupling of 1-bromonapthalene; if the reaction proceeds via a SET then coupling of 4-chlorobenzene is favorable. When the copper complex was irradiated with a 1 : 1 mixtures of 1-bromonapthalene and 4-chlorobenzene, the Ullmann coupling product derived from 4-chlorobenzene was predominant. This observation supports the SET pathway for C—N bond formation.

Later, Fu and coworkers expanded the scope of the photoinduced Ullmann C—N coupling reaction for *N*-arylation to a variety of nitrogen nucleophiles such as indoles, benzimidazoles, and imidazoles at room temperature using CuI as catalyst without a ligand co-additive (Scheme 8.35) [45]. In addition,

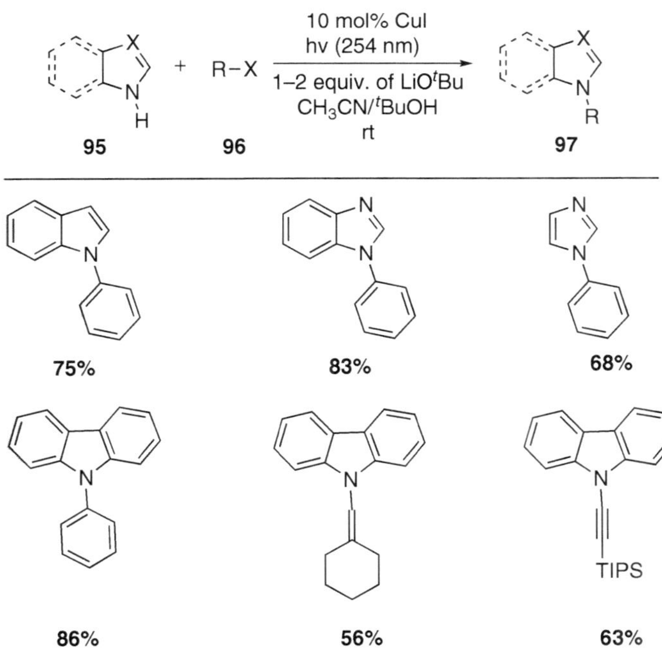

Scheme 8.35 Photo induced copper-catalyzed *N*-arylation of aromatic nitrogen nucleophiles.

this method is applicable for a variety of electrophilic partners including aryl halides, alkenyl halides, and alkenyl bromides. The authors replaced the CuI catalyst by $Ru(bpy)_3Cl_2$ as photocatalyst for this transformation, but no product was observed. Control experiments confirmed that both light and CuI were necessary for this photoreaction. Recently, Kuo Chu Hwang and coworkers reported a visible light mediated Sonogashira cross-coupling reaction for C—C bond formation using CuCl as a catalyst. This method exhibits a wide substrate scope and high functional group tolerance [46].

8.4 Conclusion

The examples on arene functionalization by visible light photoredox catalysis described in this chapter demonstrate the great potential of the transformation in organic synthesis and for applications in material chemistry [47]. The field has developed rapidly in the last years, but many opportunities and challenges remain. A variety of different aryl radical precursor, such as diaryliodonium salts, aryl diazonium salts, triarylsulfonium salts, aryl sulfonyl chlorides or simple aryl halides, have been utilized. The scope of the accessible products and reactions ranges from arene-(heteroarene) coupling conventionally achieved by Suzuki reaction and arene-olefin couplings giving Heck-type products, over Ullman-type couplings to the synthesis of heterocycles, α-arylated ketones and phenanthrenes. Moreover, new and highly interesting reactions became possible by merging photoredox catalysis with palladium, gold, or copper catalysis. These dual catalytic approaches will open up avenues to even more advanced transformations in the future. Arene functionalization using visible light photoredox catalysis has developed into a viable tool for organic synthesis.

References

1 Prier, C.K., Rankic, D.A., and Macmillan, D.W. (2013) Visible light photoredox catalysis with transition metal complexes: applications in organic synthesis. *Chem. Rev.*, **113**, 5322–5363.

2 Narayanam, J.M.R. and Stephenson, C.R.J. (2011) Visible light photoredox catalysis: applications in organic synthesis. *Chem. Soc. Rev.*, **40**, 102–113.

3 Reckenthäler, M. and Griesbeck, A.G. (2013) Photoredox catalysis for organic syntheses. *Adv. Synth. Catal.*, **355**, 2727–2744.

4 Nakamura, I. and Yamamoto, Y. (2004) Transition-metal-catalyzed reactions in heterocyclic synthesis. *Chem. Rev.*, **104**, 2127–2198.

5 Alberico, D., Scott, M.E., and Lautens, M. (2007) Aryl – aryl bond formation by transition-metal-catalyzed direct arylation. *Chem. Rev.*, **107**, 174–238.

6 Lyons, T.W. and Sanford, M.S. (2010) Palladium-catalyzed ligand-directed C—H functionalization reactions. *Chem. Rev.*, **110**, 1147–1169.

7 Beletskaya, I.P. and Cheprakov, A.V. (2000) The heck reaction as a sharpening stone of palladium catalysis. *Chem. Rev.*, **100**, 3009–3066.

8 Hari, D.P. and König, B. (2013) The photocatalyzed Meerwein arylation: classic reaction of aryl diazonium salts in a new light. *Angew. Chem. Int. Ed.*, **52**, 4734–4743.
9 Hari, D.P., Hering, T., and König, B. (2013) Synthetic applications of aryl diazonium salts enabled by visible light photoredox catalysis. *Chem. Today*, **31**, 59–62.
10 Mo, F., Dong, G., Zhang, Y., and Wang, J. (2013) Recent applications of arene diazonium salts in organic synthesis. *Org. Biomol. Chem.*, **11**, 1582–1593.
11 Hari, D.P., Hering, T., and Konig, B. (2013) Synthetic applications of aryl diazonium salts enabled by visible light photoredox catalysis. *Chim. Oggi-Chem. Today*, **31**, 59–62.
12 Pratsch, G. and Heinrich, M. (2012) in *Radicals in Synthesis III*, vol. **320** (eds M. Heinrich and A. Gansäuer), Springer, Berlin Heidelberg, pp. 33–59.
13 Galli, C. (1988) Radical reactions of arenediazonium ions: an easy entry into the chemistry of the aryl radical. *Chem. Rev.*, **88**, 765–792.
14 Cano-Yelo, H. and Deronzier, A. (1984) Photocatalysis of the Pschorr reaction by tris-(2,2[prime or minute]-bipyridyl)ruthenium(II) in the phenanthrene series. *J. Chem. Soc. Perkin Trans.*, **2**, 1093–1098.
15 Pschorr, R. (1896) Neue synthese des Phenanthrens und seiner Derivate. *Ber. Deutsch. Chemischen Ges.*, **29**, 496–501.
16 Leake, P.H. (1956) The Pschorr synthesis. *Chem. Rev.*, **56**, 27–48.
17 Cano-Yelo, H. and Deronzier, A. (1987) Photocatalysis of the Pschorr reaction by Ru(bpy)32+. *J. Photochem.*, **37**, 315–321.
18 Kalyani, D., McMurtrey, K.B., Neufeldt, S.R., and Sanford, M.S. (2011) Room-temperature C–H arylation: merger of Pd-catalyzed C–H functionalization and visible-light photocatalysis. *J. Am. Chem. Soc.*, **133**, 18566–18569.
19 Hari, D.P., Schroll, P., and König, B. (2012) Metal-free, visible-light-mediated direct C–H arylation of heteroarenes with aryl diazonium salts. *J. Am. Chem. Soc.*, **134**, 2958–2961.
20 Hari, D.P., Hering, T., and König, B. (2012) Visible light photocatalytic synthesis of benzothiophenes. *Org. Lett.*, **14**, 5334–5337.
21 Meerwein, H., Büchner, E., and van Emster, K. (1939) Über die Einwirkung aromatischer Diazoverbindungen auf α,β-ungesättigte Carbonylverbindungen. *J. Prakt. Chem.*, **152**, 237–266.
22 Schroll, P., Hari, D.P., and König, B. (2012) Photocatalytic arylation of alkenes, alkynes and enones with diazonium salts. *Chem. Open*, **1**, 130–133.
23 Jiang, H., Huang, C., Guo, J. *et al.* (2012) Direct C–H functionalization of enamides and enecarbamates by using visible-light photoredox catalysis. *Chem. Eur. J.*, **18**, 15158–15166.
24 Hering, T., Hari, D.P., and König, B. (2012) Visible-light-mediated α-arylation of enol acetates using aryl diazonium salts. *J. Org. Chem.*, **77**, 10347–10352.
25 Xiao, T., Dong, X., Tang, Y., and Zhou, L. (2012) Phenanthrene synthesis by eosin Y-catalyzed, visible light-induced [4+2] benzannulation of biaryldiazonium salts with alkynes. *Adv. Synth. Catal.*, **354**, 3195–3199.
26 Fu, W., Xu, F., Fu, Y. *et al.* (2013) Synthesis of 3,3-disubstituted oxindoles by visible-light-mediated radical reactions of aryl diazonium salts with N-arylacrylamides. *J. Org. Chem.*, **78** (23), 12202–12206.

27 Sahoo, B., Hopkinson, M.N., and Glorius, F. (2013) Combining gold and photoredox catalysis: visible light-mediated oxy- and aminoarylation of alkenes. *J. Am. Chem. Soc.*, **135**, 5505–5508.

28 Majek, M. and von Wangelin, A.J. (2013) Organocatalytic visible light mediated synthesis of aryl sulfides. *Chem. Commun. (Camb. UK)*, **49**, 5507–5509.

29 Wang, X., Cuny, G.D., and Noël, T. (2013) A mild, one-pot Stadler–Ziegler synthesis of arylsulfides facilitated by photoredox catalysis in batch and continuous-flow. *Angew. Chem. Int. Ed.*, **52**, 7860–7864.

30 Yu, J., Zhang, L., and Yan, G. (2012) Metal-free, visible light-induced borylation of aryldiazonium salts: a simple and green synthetic route to arylboronates. *Adv. Synth. Catal.*, **354**, 2625–2628.

31 Neufeldt, S.R. and Sanford, M.S. (2012) Combining transition metal catalysis with radical chemistry: dramatic acceleration of palladium-catalyzed C–H arylation with diaryliodonium salts. *Adv. Synth. Catal.*, **354**, 3517–3522.

32 Liu, Y.-X., Xue, D., Wang, J.-D. *et al.* (2013) Room-temperature arylation of arenes and heteroarenes with diaryl-iodonium salts by photoredox catalysis. *Synlett*, **24**, 507–513.

33 Baralle, A., Fensterbank, L., Goddard, J.-P., and Ollivier, C. (2013) Aryl radical formation by copper(I) photocatalyzed reduction of diaryliodonium salts: NMR evidence for a CuII/CuI mechanism. *Chem. A Eur. J.*, **19**, 10809–10813.

34 Fumagalli, G., Boyd, S., and Greaney, M.F. (2013) Oxyarylation and aminoarylation of styrenes using photoredox catalysis. *Org. Lett.*, **15**, 4398–4401.

35 Farrar, C.E., Siu, K.K.W., Howell, P.L., and Jarrett, J.T. (2010) Biotin synthase exhibits burst kinetics and multiple turnovers in the absence of inhibition by products and product-related biomolecules. *Biochemistry*, **49**, 9985–9996.

36 Wang, S.C. and Frey, P.A. (2007) S-adenosylmethionine as an oxidant: the radical SAM superfamily. *Trends Biochem. Sci.*, **32**, 101–110.

37 Donck, S., Baroudi, A., Fensterbank, L. *et al.* (2013) Visible-light photocatalytic reduction of sulfonium salts as a source of aryl radicals. *Adv. Synth. Catal.*, **355**, 1477–1482.

38 Xia, J.-D., Deng, G.-B., Zhou, M.-B. *et al.* (2012) Reusable visible light photoredox catalysts; catalyzed benzylic C(sp3)–H functionalization/carbocyclization reactions. *Synlett*, **23**, 2707–2713.

39 Deng, G.-B., Wang, Z.-Q., Xia, J.-D. *et al.* (2013) Tandem cyclizations of 1,6-enynes with arylsulfonyl chlorides by using visible-light photoredox catalysis. *Angew. Chem. Int. Ed.*, **52**, 1535–1538.

40 Bouriga, M., Chehimi, M.M., Combellas, C. *et al.* (2013) Sensitized photografting of diazonium salts by visible light. *Chem. Mater.*, **25**, 90–97.

41 Garcia, A., Hanifi, N., Jousselme, B. *et al.* (2013) Polymer grafting by inkjet printing: a direct chemical writing toolset. *Adv. Funct. Mater.*, **23**, 3668–3674.

42 Schroll, P., Fehl, C., Dankesreiter, S., and Konig, B. (2013) Photocatalytic surface patterning of cellulose using diazonium salts and visible light. *Org. Biomol. Chem.*, **11**, 6510–6514.

43 Creutz, S.E., Lotito, K.J., Fu, G.C., and Peters, J.C. (2012) Photoinduced Ullmann C–N coupling: demonstrating the viability of a radical pathway. *Science*, **338**, 647–651.

44 Majek, M. and Jacobi von Wangelin, A. (2013) Ambient-light-mediated copper-catalyzed C–C and C–N Bond formation. *Angew. Chem. Int. Ed.*, **52**, 5919–5921.

45 Ziegler, D.T., Choi, J., Muñoz-Molina, J.M. *et al.* (2013) A versatile approach to Ullmann C–N couplings at room temperature: new families of nucleophiles and electrophiles for photoinduced, copper-catalyzed processes. *J. Am. Chem. Soc.*, **135**, 13107–13112.

46 Sagadevan, A. and Hwang, K.C. (2012) Photo-induced sonogashira C–C coupling reaction catalyzed by simple copper(I) chloride salt at room temperature. *Adv. Synth. Catal.*, **354**, 3421–3427.

47 For additional reading of primary literature that has emerged since transcription of this chapter see:(a) Senaweera, S. and Weaver, J.D. (2016) Dual C–F, C–H functionalization via photocatalysis: access to multifluorinated biaryls. *J. Am. Chem. Soc.*, **138** (8), 2520–2523; (b) Tang, X.-J. and Dolbier, W.R. (2015) Efficient Cu-catalyzed atom transfer radical addition reactions of fluoroalkylsulfonyl chlorides with electron-deficient alkenes induced by visible light. *Angew. Chem. Int. Ed.*, **54** (14), 4246–4249; (c) Zhang, Y.-P., Feng, X.-L., Yang, Y.-S., and Cao, B.-X. (2016) Metal-free, C–H arylation of indole and its derivatives with aryl diazonium salts by visible-light photoredox catalysis. *Tetrahedron Lett.*, **57** (21), 2298–2302; (d) Meyer, A.U., Berger, A.L., and König, B. (2016) Metal-free C–H sulfonamidation of pyrroles by visible light photoredox catalysis. *Chem. Commun.*, **52** (72), 10918–10921; (e) Hering, T. and König, B. (2016) Photocatalytic activation of N-chloro compounds for the chlorination of arenes. *Tetrahedron*, **72** (48), 7821–7825; (f) Singh, A., Kubik, J.J., and Weaver, J.D. (2015) Photocatalytic C–F alkylation; facile access to multifluorinated arenes. *Chem. Sci.*, **6** (12), 7206–7212; (g) Zheng, Y.-W., Chen, B., Ye, P. *et al.* (2016) Photocatalytic hydrogen-evolution cross-couplings: benzene C–H amination and hydroxylation. *J. Am. Chem. Soc.*, **138** (32), 10080–10083; (h) Jiang, H., Cheng, Y., Wang, R. *et al.* (2014) Synthesis of isoquinolines via visible light-promoted insertion of vinyl isocyanides with diaryliodonium salts. *Chem. Commun.*, **50** (46), 6164–6167; (i) Prasad Hari, D., Hering, T., and König, B. (2014) The photoredox-catalyzed Meerwein addition reaction: intermolecular amino-arylation of alkenes. *Angew. Chem. Int. Ed.*, **53** (3), 725–728; (j) Xia, X.-D., Lu, L.-Q., Liu, W.-Q. *et al.* (2016) Visible-light-driven photocatalytic activation of inert sulfur ylides for 3-acyl oxindole synthesis. *Chem. Eur. J.*, **22** (25), 8432–8437; (k) Ghosh, I., Marzo, L., Das, A. *et al.* (2016) Visible light mediated photoredox catalytic arylation reactions. *Acc. Chem. Res.*, **49** (8), 1566–1577; (l) Arora, A. and Weaver, J.D. (2016) Visible light photocatalysis for the generation and use of reactive azolyl and polyfluoroaryl intermediates. *Acc. Chem. Res.*, **49** (10), 2273–2283.

9

Visible-Light Photocatalysis in the Synthesis of Natural Products

Gregory L. Lackner, Kyle W. Quasdorf, and Larry E. Overman

University of California, Irving, 4042A Frederick Reines Hall, Irvine, CA 92697, USA

Visible-light photocatalysis is expanding the applications of photochemistry in organic synthesis. Although the foundations of this chemistry were laid nearly 40 years ago [1], many important advances in this area of catalysis were recorded only within the past 6 years [2]. Nonetheless, the ability of visible-light photocatalysis to solve substantial problems in organic synthesis is already signaled by its use as a central step in recent total syntheses of several structurally diverse natural products.

To date, the widest application of visible-light photocatalysis in natural product total synthesis has been to generate carbon radicals. In 2011, Stephenson and coworkers exploited this tactic in their total synthesis of (+)-gliocladin C (**1**) [3]. Taking a lead from reports in the 1970s and 1980s that carbon radicals could be generated using visible-light photocatalysis, the Stephenson group first developed synthetically attractive conditions to form carbon radicals from halide precursors and employed these intermediates in reductive dehalogenation and C—C bond-forming processes [4].

The central challenge in the synthesis of (+)-gliocladin C (**1**) is formation of the C3 quaternary-carbon stereocenter that links the tetracyclic cyclotryptophan fragment to the β-carbon (C3′) of the indole substituent (Scheme 9.1). The Stephenson group envisaged accomplishing this objective by the addition of the tertiary benzylic radical generated from cyclotryptophan bromide **3** to the β-carbon of an indole derivative **2**. Stereoselection was expected to be high in this step, as earlier studies had verified that tertiary radicals of this type react with carbon electrophiles from the *Re* face to form new quaternary carbon stereocenters [5].

In order to evaluate this strategy for assembling the C3–C3′ linkage of (+)-gliocladin C, the coupling of readily available cyclotryptamine bromide **9** was studied (Scheme 9.2). Upon visible-light irradiation (blue light-emitting diodes (LEDs), $\lambda_{max} = 454$ nm), the coupling of bromide **9** with *N*-methylindole (**10**) in the presence of 1 mol% of [Ru(bpy)$_3$Cl$_2$] took place exclusively at the C2′ position of the indole to form the C3–C2′-coupled product **11** in 52% yield. By blocking C2′ with a carbomethoxy group, selectivity was reversed, allowing the desired C3–C3′-linked product **13** to be formed cleanly in 58% yield. Ultimately, it was

Visible Light Photocatalysis in Organic Chemistry, First Edition.
Edited by Corey R. J. Stephenson, Tehshik P. Yoon and David W. C. MacMillan.

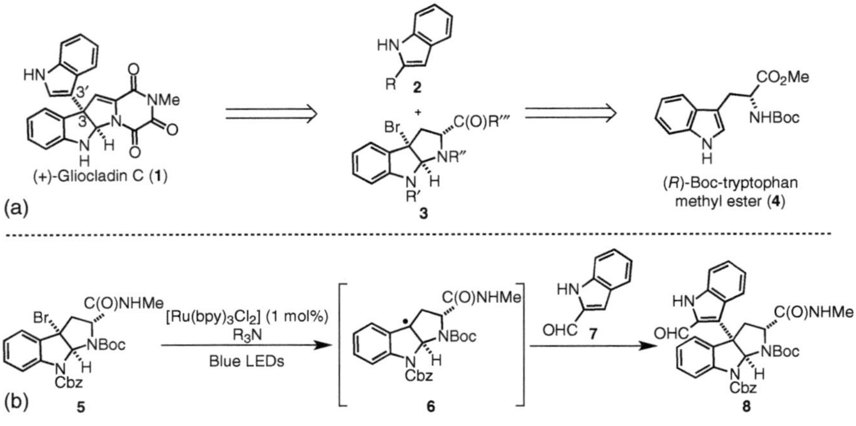

Scheme 9.1 Retrosynthetic analysis of (+)-gliocladin C (a). Visible-light photoredox coupling step utilized in the total synthesis (b).

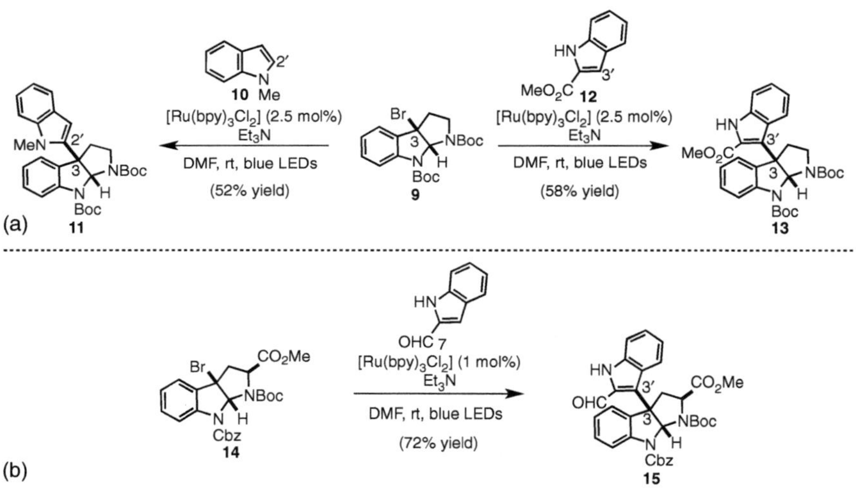

Scheme 9.2 Conditions for selective C2′ and C3′ coupling (a). Optimized coupling conditions (b).

found that introduction of a formyl substituent at the C2′ position of the indole fragment increased the yield of the photocatalytic-coupling step to 72% yield.

With the reaction conditions for the coupling step established, the total synthesis of (+)-gliocladin C commenced from (*R*)-Boc-tryptophan methyl ester (**4**) (Scheme 9.3). Protection of the indole nitrogen of **4**, bromocyclization, and conversion of the ester to an amide delivered the bromide substrate **5** for the key visible-light photoredox-coupling step in 79% yield over the three steps. Under optimized conditions, this critical step proceeded smoothly to give product **8** in 82% yield. To compensate for competitive reductive debromination of **5**, 5 equiv. of 2-formylindole (**7**) were employed. In addition, the use of Bu_3N instead of Et_3N as a reductive quencher proved to be advantageous in gram-scale reactions.

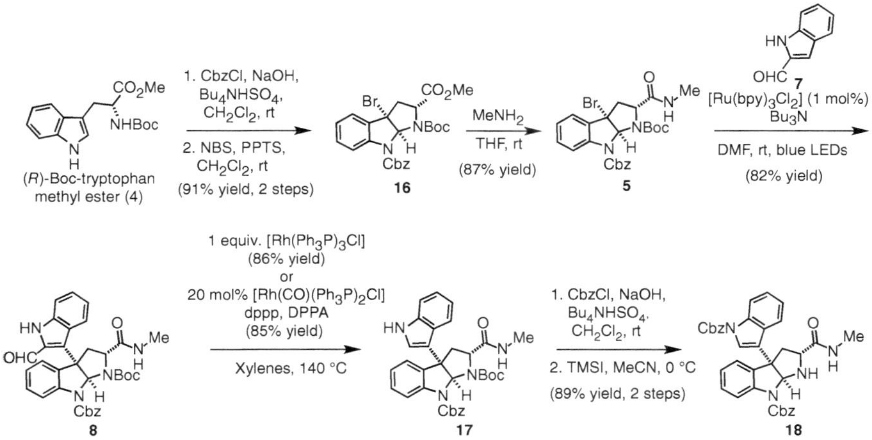

Scheme 9.3 Synthesis of **18** and the photoredox coupling step.

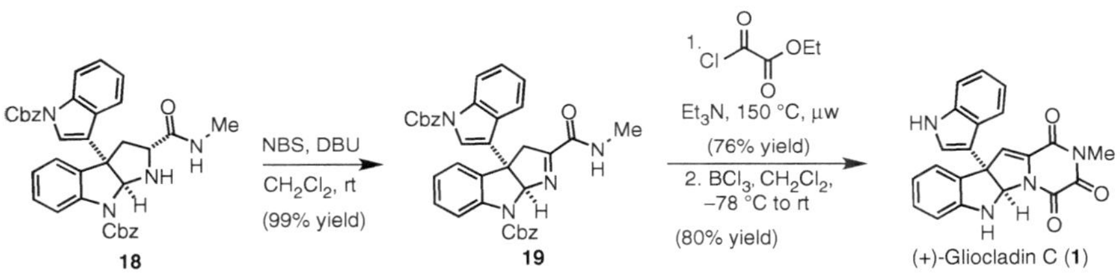

Scheme 9.4 Completion of Stephenson's total synthesis of (+)-gliocladin C.

Removal of the superfluous formyl group was achieved most efficiently by reaction of **8** with a stoichiometric amount of Wilkinson's catalyst, $[Rh(PPh_3)_3Cl]$. Protection of the indoline nitrogen of product **17**, followed by removal of the Boc group, delivered intermediate **18** in 89% yield over the two steps.

In three additional steps, intermediate **18** was elaborated to (+)-gliocladin C (Scheme 9.4). Overall, this synthesis proceeded in 10 steps from (*R*)-Boc-tryptophan methyl ester (**4**) and constituted the first use of visible-light photocatalysis to form carbon–carbon bonds in the total synthesis of a complex natural product.

As five total syntheses of (+)-gliocladin C have now been recorded, it is instructive to compare and contrast the strategy of Stephenson's synthesis with that of the other syntheses. In the inaugural total synthesis of (+)-gliocladin C by Overman and Shin in 2007 [6], and an improved second-generation approach disclosed by this group in 2011 [7], the critical C3–C3′ linkage was established early in the synthesis (Scheme 9.5). This general strategy was employed also in Gong's recent total synthesis [8] and a formal synthesis by Hao [9]. The use of a free-radical coupling to unite the cyclotryptophan and indole fragments late in the synthesis is the signature feature of Stephenson's approach to (+)-gliocladin C, and nicely illustrates the utility of visible-light photocatalysis to accomplish selective chemistry with reactants containing significant functionality [10].

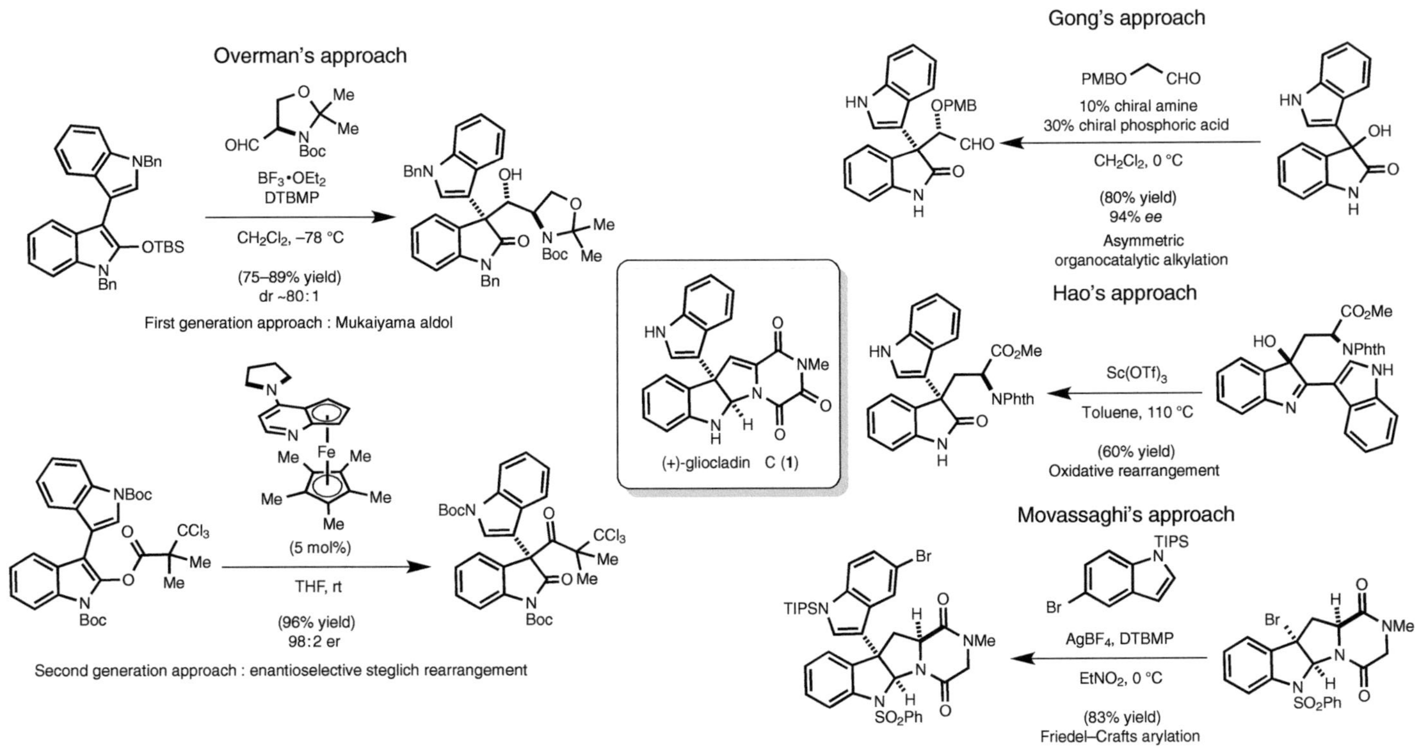

Scheme 9.5 Key steps in other total syntheses of (+)-gliocladin C.

Taking inspiration from Stephenson's total synthesis of (+)-gliocladin C and Overman's total synthesis of (−)-aplyviolene (*vide infra*) [11], Li and coworkers reported in 2013 total syntheses of drimentines A (**20**), F (**21**), and G (**22**) using the coupling of a cyclotryptophan tertiary radical generated by visible-light photocatalysis as the pivotal step [12]. These authors envisaged addition of the nucleophilic tertiary benzylic radical generated from bromide precursor **24** to the exomethylene double bond of enone **23** (Scheme 9.6). Enone **23** was seen arising from the inexpensive sesquiterpene lactone (+)-sclareolide (**25**).

The two reactants needed to evaluate the pivotal visible-light photoredox-coupling step were assembled from simple starting materials (Scheme 9.7). Cyclotryptamine bromide **24** was available in high yield by bromocyclization of (*S*)-tryptophan derivative **26** [13]. Beginning from (+)-sclareolide (**25**), enone fragment **23** was prepared in five steps in 55% overall yield.

With bromide **24** and enone **23** in hand, various conditions were examined to accomplish the fragment coupling (Table 9.1). Classical methods for radical chain addition reactions of alkyl halides, such as exposure of bromide **24** and an excess of enone **23** to azobisisobutyronitrile (AIBN), Bu_3SnH, or $(TMS)_3SiH$ at elevated temperatures, furnished none of the desired coupled products. The use of Et_3B, O_2, and Bu_3SnH also proved ineffective. In all of these cases, debromination of **24** was observed, with enone **23** remaining intact. Movassaghi's conditions for homocoupling of similar tertiary benzylic bromides upon reaction with

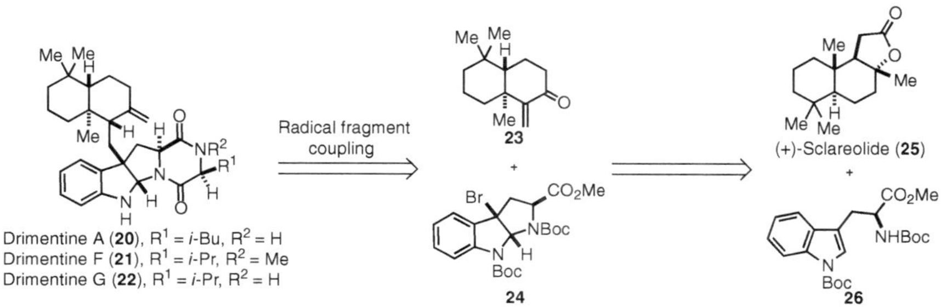

Scheme 9.6 Retrosynthetic analysis of drimentines A, F, and G.

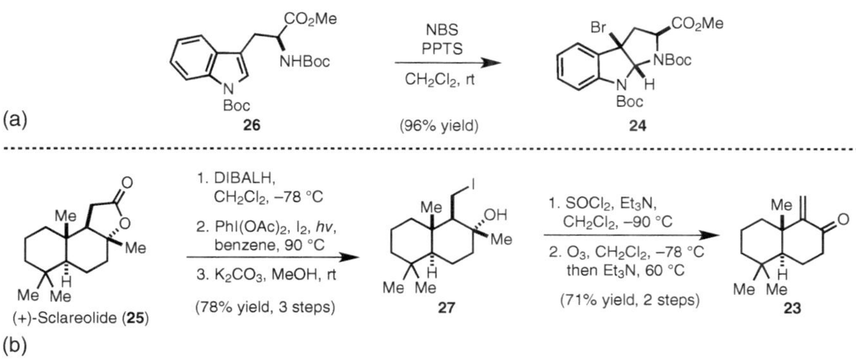

Scheme 9.7 Preparation of precursors for photoredox fragment coupling.

Table 9.1 Optimization of the photoredox fragment coupling.

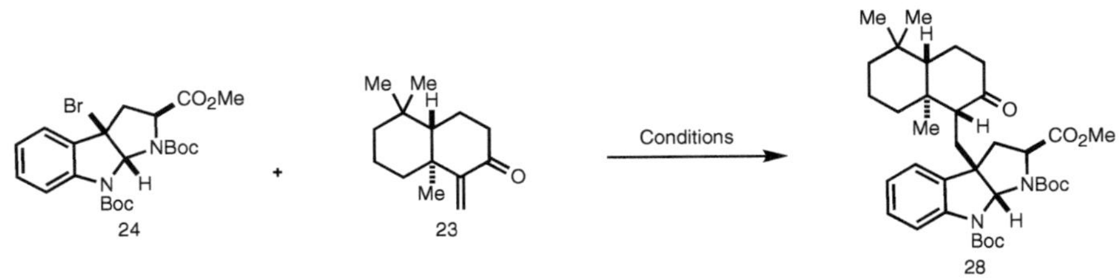

Entry	Conditions	Temperature	Ratio 23 : 24	Yield (%)
1	AIBN, Bu_3SnH, or $(TMS)_3SiH$, toluene	80 °C	4 : 1	0
2	Et_3B, O_2, Bu_3SnH, THF	rt	4 : 1	0
3	[Co(PPh_3)Cl], acetone	rt	4 : 1	0
4	Bu_3SnH (slow addition), benzene	80 °C	10 : 1	58
5	[Ru(bpy)$_3$Cl$_2$] • 6H$_2$O (2.5%), blue LEDs, Et_3N, DMF	rt	4 : 1	51
6	[Ir(ppy)$_2$(dtbbpy)PF$_6$] (2.5%), blue LEDs, Et_3N, DMF	rt	4 : 1	89
7	[Ir(ppy)$_2$(dtbbpy)PF$_6$] (2.5%), blue LEDs, Et_3N, DMF	rt	1 : 1.5	91
8	Blue LEDs, Et_3N, DMF	rt	4 : 1	12

[Co(PPh_3)Cl] were also explored [14]; however, only the homodimerization product of **24** was observed, even in the presence of 4 equiv. of the enone. Eventually, the authors found that slow addition of Bu_3SnH and conducting the reaction at high dilution (0.005 M in benzene) proved beneficial, allowing the desired coupled product to be formed in 58% yield. With the importance of generating low concentrations of the radical species in a controlled manner suggested by this result, Li and coworkers explored visible-light photocatalysis to facilitate the transformation. Utilizing [Ru(bpy)$_3$Cl$_2$]-catalyzed photoredox conditions employed by Stephenson for reductive dehalogenation [4a], the desired alkene coupling was achieved in 51% yield. Upon further reaction optimization, it was found that the use of [Ir(ppy)$_2$(dtbbpy)PF$_6$] as the photocatalyst was superior to [Ru(bpy)$_3$Cl$_2$]·6H$_2$O. Of particular note, yields approaching 90% (based on enone **23**) for the catalytic photoredox-coupling step were obtained using only a 50% excess of radical precursor **24**. Control experiments established that the photocatalyst was crucial to the success of the reaction, as its omission led to the formation of **28** in much diminished yield.

Having defined an efficient sequence for preparing pentacyclic intermediate **28**, the total syntheses of alkaloids **20–22** were accomplished in five additional steps (Scheme 9.8). After removing the Boc group of **28** with TFA, standard peptide coupling conditions were used to append one of three Boc-protected α-amino acids. These products were then converted to diketopiperazines **31a–c** in good overall yield over three steps. Reaction of these intermediates with MeMgBr/$CeCl_3$ followed by dehydration of the resulting tertiary alcohols with $SOCl_2$ in pyridine completed the total syntheses of (−)-drimentine A (**20**), F (**21**), and G (**22**) in 26%, 18%, and 31% yields, respectively, for the final steps.

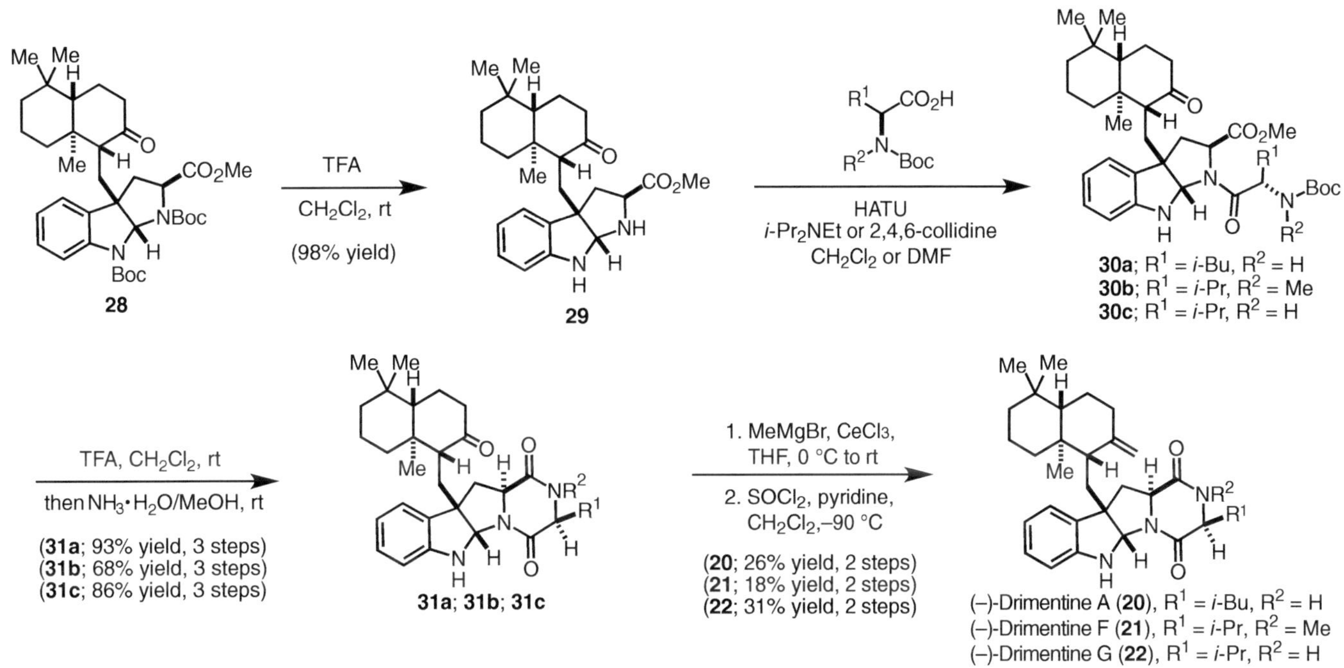

Scheme 9.8 Completion of total syntheses of (−)-drimentines A, F, and G.

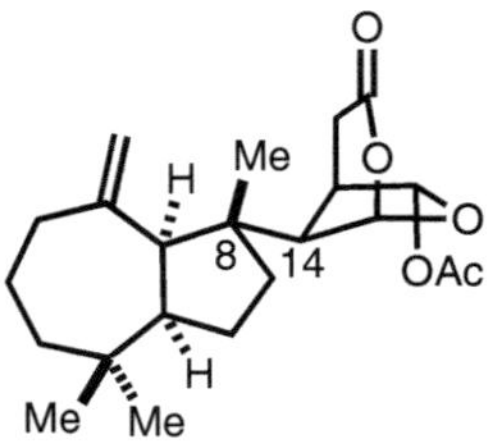

(−)-Aplyviolene (**32**)

Figure 9.1 (−)-Aplyviolene (**32**).

In 2012, Overman and Schnermann reported the use of visible-light photocatalysis to accomplish the coupling of a trialkyl tertiary radical with a cyclopentenone acceptor as the pivotal step of their total synthesis of (−)-aplyviolene (**32**) [11]. The central challenge in the synthesis of aplyviolene and structurally related rearranged spongian diterpenes [15] is forming the C8–C14 σ-bond that connects the quaternary and tertiary stereocenters C8 and C14 and unites the *cis*-perhydroazulene and bicyclic lactone fragments (Figure 9.1).

The synthetic strategy that was followed evolved from this group's earlier total synthesis of (−)-aplyviolene (**32**) [16] in which the two cyclic fragments of this diterpene were joined by conjugate addition of the thermodynamic enolate generated from *cis*-perhydroazulenone **33** with bromocyclopentenone **34** (Scheme 9.9) [17]. This union took place from the convex face of the *cis*-perhydroazulene nucleophile and from the face opposite the four-carbon side chain of the Michael acceptor to correctly set the relative configuration of the contiguous C8 and C14 stereocenters. The coupled product **35** was subsequently advanced to (−)-aplyviolene (**32**) by reductive silylation of the α-bromoketone and removal of the extraneous hydroazulene carbonyl group to form **36**. In five additional steps, the cyclopentene fragment of this intermediate was elaborated to 2-hydroxytetrahydrofuran **37**, setting the stage for cyclization to form the distinctive dioxobicyclo[3.2.1]octanone fragment of (−)-aplyviolene. Lactonization to form the bridged-bicyclic fragment was accomplished by the method of Mukaiyama to give intermediate **38** in high yield [18]. Baeyer–Villiger oxidation of **38** then delivered (−)-aplyviolene (**32**), thereby completing the first total synthesis of a member of this family of structurally elaborate rearranged spongian diterpenes.

While this enolate/Michael addition strategy united the two chiral fragments efficiently with high stereocontrol, the authors reasoned that direct incorporation of the *cis*-perhydroazulene moiety in the correct oxidation state would streamline the synthesis considerably. The coupling of a tertiary organocuprate

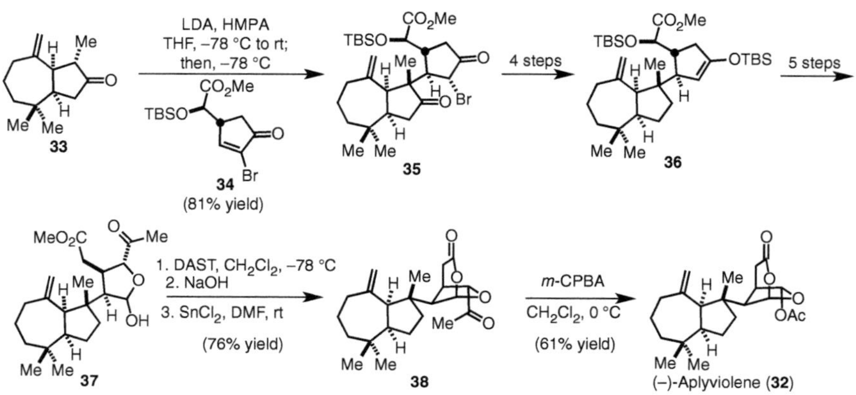

Scheme 9.9 First-generation enolate-coupling approach to (−)-aplyviolene.

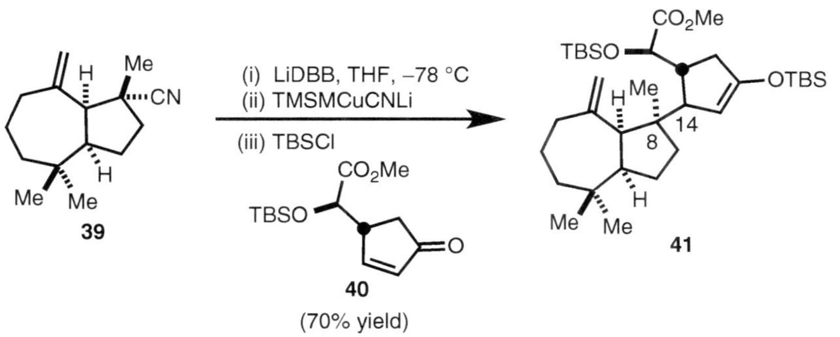

Scheme 9.10 Tertiary organometallic conjugate addition to form the C8 – C14 bond.

intermediate, formed by reductive lithiation of nitrile **39** and transmetalation, with enone **40** was first investigated to accomplish this aim (Scheme 9.10) [19]. To the authors' surprise, this union yielded product **41** having the incorrect relative configuration of the newly formed quaternary carbon stereocenter, indicating that bond formation had occurred from the more-hindered, concave face of the *cis*-perhydroazulene nucleophile.

As the unexpected stereochemical outcome of the cuprate coupling depicted in Scheme 9.10 was attributed to the carbon–metal bond of the *cis*-perhydroazulene nucleophile [19], the authors turned to examine the coupling of the corresponding tertiary *cis*-perhydroazulene radical. To generate this nucleophilic radical, the authors chose the method reported by Okada, in which visible light, a dihydropyridine, and catalytic $Ru(bpy)_3^{2+}$ are employed to generate carbon radicals by reductive fragmentation of (*N*-acyloxy)phthalimide intermediates (Scheme 9.11) [20]. One attraction of this method is the ready formation of the (*N*-acyloxy)phthalimide intermediate from the corresponding carboxylic acid. In addition, the stability and crystallinity of these radical precursors made them ideal for generating carbon radicals from structurally complex synthetic intermediates.

After some experimentation, it was found that the coupling of enone **43** and the tertiary radical generated from (*N*-acyloxy)phthalimide **42** (prepared from (+)-fenchone in nine steps) was best accomplished in dichloromethane using $[Ru(bpy)_3(BF_4)_2]$, *i*-Pr_2NEt and Hantzsch dihydropyridine **44** upon irradiation with blue LEDs (Scheme 9.12). As anticipated, the carbon radical conjugate addition occurred exclusively from the less-hindered convex face of the *cis*-perhydroazulene unit to provide addition product **45** in 61% yield. The yield

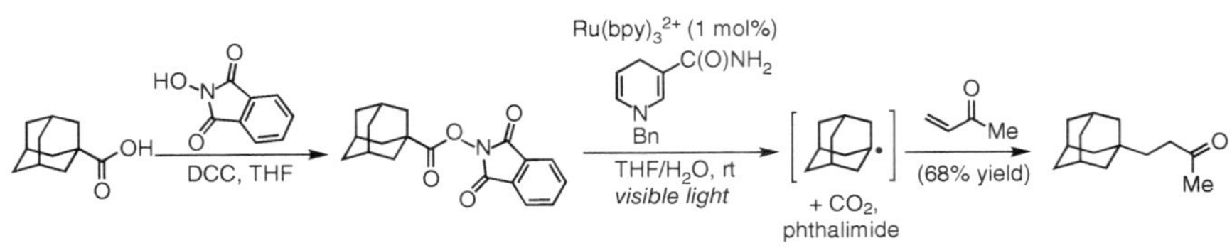

Scheme 9.11 Okada's report of the coupling of the tertiary adamantal radical with methyl vinyl ketone.

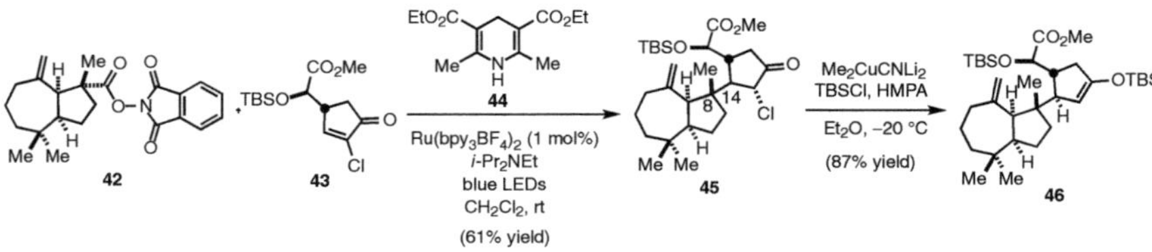

Scheme 9.12 Photoredox-catalyzed radical coupling to form the C8 – C14 bond of (−)-aplyviolene.

of the fragment-coupling step was even higher, as the major by-product formed resulted from reductive dechlorination of product **45**, a known conversion under related visible-light photocatalysis conditions [4a,21]. Reductive enolsilylation of coupled product **45** secured **46**, thereby completing a formal total synthesis of (−)-aplyviolene.

This improved second-generation total synthesis of (−)-aplyviolene via a critical radical coupling step showcases the utility of visible-light photocatalysis in generating trialkyl tertiary radical intermediates that are inaccessible by most traditional methods. For example, the potential bromide precursor of the *cis*-perhydroazulene radical could not be prepared because of facile elimination of this tertiary homoallylic bromide to form a 1,3-diene. This synthesis also illustrates the capability of visible-light photocatalysis to construct demanding σ-bonds that connect two contiguous stereocenters.

A second area where visible-light photocatalysis has already impacted natural product total synthesis is in the area of cycloaddition reactions. The ability of visible-light photocatalysis to promote cycloadditions by single-electron reduction, single-electron oxidation, or triplet-energy transfer provides new opportunities for using cycloaddition reactions to construct structurally complex cyclic and polycyclic molecules. Visible-light photocatalytic cycloadditions are particularly attractive in this arena, as they typically avoid elevated reaction temperatures and circumvent the use of high-energy ultraviolet light [2, 22]. In addition, the regiochemical and stereochemical outcomes of cycloadditions carried out by visible-light photocatalysis can be complementary to those obtained by established methods.

Photochemical [2 + 2]-cycloadditions brought about by photolytic sensitization, typically with ultraviolet light, play an indispensible role in the construction of cyclobutane-containing natural products [23]. Reports of the use of visible-light photocatalysis to form cyclobutanes, specifically intramolecular and intermolecular [2 + 2]-cycloadditions of aromatic enones promoted by one-electron reduction, first appeared from the Yoon group in 2008 and 2009 [24]. The complementary use of oxidative visible-light photocatalysis to realize intramolecular [2 + 2]-cycloadditions of electron-rich styrenes, and crossed intermolecular [2 + 2]-cycloadditions of styrenes and alkenes, were also described by this group [25]. Success in these latter transformations requires choosing a catalytic photooxidant with an excited-state potential sufficient to oxidize the alkene substrate, but not the cycloadduct. Another approach for achieving this selectivity in visible-light photocatalytic reactions was introduced by Nicewicz and coworker [26]. In this method, a tunable relay oxidant is oxidized by the photocatalyst and it subsequently selectively oxidizes the styrene to promote cyclodimerization.

Nicewicz illustrated the utility of this method in the syntheses of two symmetrical lignan natural products, magnosalin (**49**) and pellucidin A (**51**) (Scheme 9.13). In the former case, (*E*)-asarone (**47**) was dimerized in a head-to-head manner to give the C_2-symmetric product, (±)-magnosalin (**49**), in 50% yield by irradiation with visible light in the presence of the pyrylium photocatalyst **48** and the relay oxidant anthracene. In stark contrast, an analogous cycloaddition mediated by ultraviolet light furnished only the *meso* isomer in low yield [27]. In a similar manner, stereoselective head-to-head dimerization of 2,4,5-trimethoxystyrene (**50**) delivered (±)-pellucidin A (**51**) in 37% yield. The use of *N*,*N*-diethylaniline as the relay oxidant and low temperature were necessary to prevent cycloreversion of cycloadduct **51**.

An intramolecular [2 + 2]-cycloaddition is the salient feature of the concise total synthesis of the cannabinoid cannabiorcicyclolic acid (**56**) reported by Lu and Yoon (Scheme 9.14) [28]. The distinctive feature of this visible-light photocatalytic cycloaddition is its promotion by triplet-energy transfer from the photoexcited catalyst to the diene substrate, rather than electron transfer. This synthesis began by condensation of resorcinol **52** and citral (**53**) to form 2*H*-chromene (**54**). This diene intermediate was subsequently photocyclized in the presence of visible light and the photocatalyst

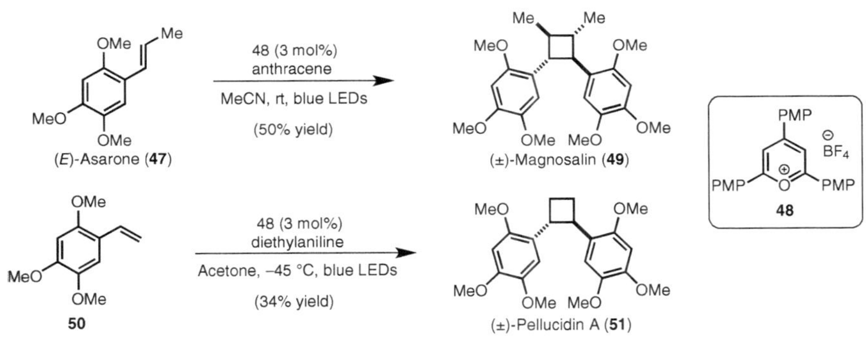

Scheme 9.13 Photoredox-catalyzed [2 + 2]-cycloadditions in total syntheses of (±)-magnosalin and (±)-pellucidin A.

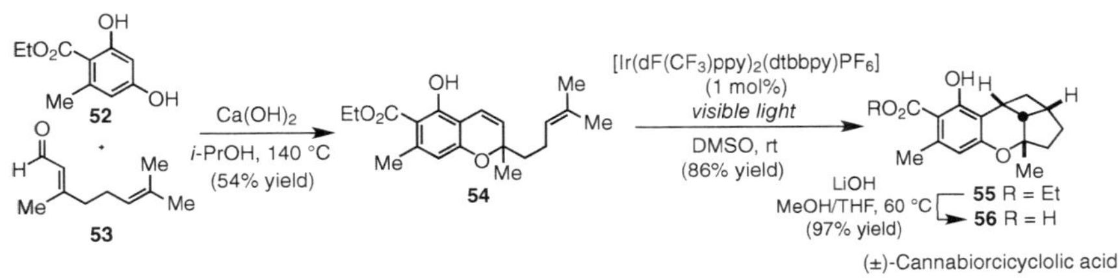

Scheme 9.14 Photoredox-catalyzed [2 + 2]-cycloaddition in the total synthesis of (±)-cannabiorcicyclolic acid.

$[Ir(dF(CF_3)ppy)_2(dtbbpy)PF_6]$ at room temperature to provide tetracyclic product **55** in 86% yield. Saponification of the ester substituent of **55** completed a short synthesis of (±)-cannabiorcicyclolic acid (**56**).

Visible-light photocatalysis also holds considerable promise for facilitating cycloadditions that form rings other than cyclobutanes. In their synthesis of (±)-heitziamide A (**59**), Yoon and coworkers illustrate unique opportunities provided by carrying out [4 + 2]-cycloadditions in this way [29]. Heitziamide A is suggested to be biosynthesized by [4 + 2]-cycloaddition (Diels–Alder reaction) of the co-isolated natural products fagaramide (**52**) and myrcene (**53**) [30]. Although this disconnection is synthetically attractive, it was anticipated—and verified by the authors—that thermal Diels–Alder reaction of **52** and **53** did not provide heitziamide A (**59**), but rather regioisomer **60** (Scheme 9.15).

In examining the scope of related photoredox-catalyzed [4 + 2]-cycloadditions brought about by visible light, the Yoon group discovered that the regioselectivity of the cycloaddition of styrenes and 1,3-dienes is complementary to that of the corresponding thermal reactions. Thus, exposure of styrene **61** and myrcene (**66**) to 2 mol% $[Ru(bpz)_3(PF_6)_2]$, visible light, and air produced exclusively cycloadduct **62**. This reaction is believed to be promoted by single-electron oxidation of styrene **61** by photogenerated $Ru(bpz)_3^{2+*}$. Subsequent removal of the silyl ether protecting group from cycloadduct **62**, and oxidation of the resulting primary alcohol to the carboxylic acid gave intermediate **63** in high yield. Formation of the isobutyl amide proceeded in 81% yield to complete the synthesis of (±)-heitziamide A (**67**) (Scheme 9.16).

This expedient total synthesis of (±)-heitziamide A (**59**) demonstrates the capability of visible-light photoredox catalysis to couple electronically mismatched dienes and dienophiles and showcases the complementary regioselectivity obtained by conducting cycloadditions in this way.

Several themes emerge from the total syntheses discussed in this chapter. For generating reactive free-radical intermediates or promoting cycloaddition reactions by one-electron reduction or oxidation, visible-light photocatalysis offers an attractive green method to accomplish these transformations without the need for stoichiometric reductants or oxidants. Moreover, in several of the total syntheses discussed, visible-light photocatalysis was successful where more

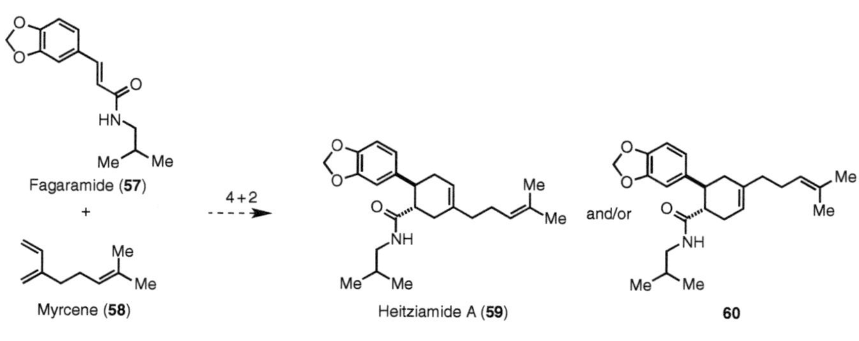

Scheme 9.15 Proposed biosynthesis of heltziamide and the expected cycloadduct 60 of thermal cycloaddition.

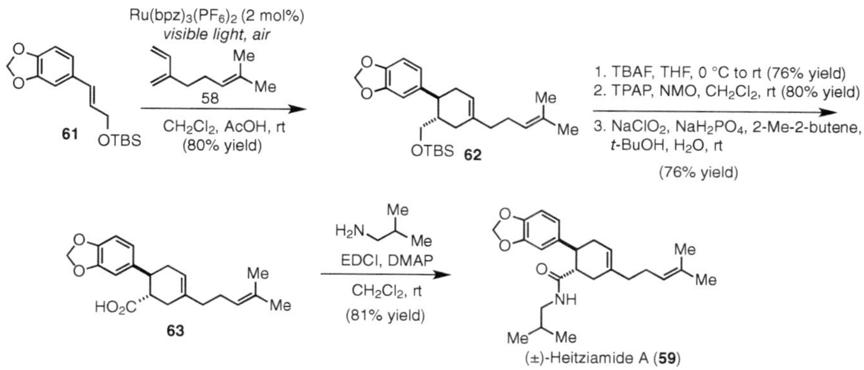

Scheme 9.16 Photoredox-catalyzed [4 + 2]-cycloaddition for the total synthesis of (±)-heitziamide.

conventional methods failed. To date, the applications of visible-light photocatalysis to total syntheses have been limited to the promotion of free-radical coupling reactions and [2 + 2]- and [4 + 2]-cycloadditions. We are certain to see this list expand to include forming a wider set of reactive intermediates, promoting additional pericyclic processes, and other important transformations.

One time-tested approach for determining the scope and limitations of new chemistry is to evaluate its utility for the synthesis of architecturally complex, polyfunctional natural products [31]. Many advances in the use of visible-light photocatalysis in organic synthesis were registered only within the past few years [32]. Nonetheless, the natural product total syntheses discussed in this chapter illustrate the key role these transformations can play in constructing structurally complex molecules. As progress in this field continues to advance, it is certain that visible-light photocatalysis will play an increasing role in future syntheses of complex natural products.

References

1 Teplý, F. (2011) *Collect. Czech. Chem. Commun.*, **76**, 859–917.
2 Prier, C.K., Rankic, D.A., and MacMillan, D.W.C. (2013) *Chem. Rev.*, **113**, 5322–5363.
3 Furst, L., Narayanam, J.M.R., and Stephenson, C.R.J. (2011) *Angew. Chem. Int. Ed.*, **50**, 9655–9659.
4 (a) Narayanam, J.M.R., Tucker, J.W., and Stephenson, C.R.J. (2009) *J. Am. Chem. Soc.*, **131**, 8756–8757; (b) Tucker, J.W., Narayanam, J.M.R., Krabbe, S.W., and Stephenson, C.R.J. (2010) *Org. Lett.*, **12**, 368–371; (c) Tucker, J.W., Nguyen, J.D., Narayanam, J.M.R. *et al.* (2010) *Chem. Commun.*, **46**, 4985–4987; (d) Furst, L., Matsuura, B.S., Narayanam, J.M.R. *et al.* (2010) *Org. Lett.*, **12**, 3104–3107.
5 Bruncko, M., Crich, D., and Samy, R. (1994) *J. Org. Chem.*, **59**, 5543–5549.
6 Overman, L.E. and Shin, Y. (2007) *Org. Lett.*, **9**, 339–341.

7 DeLorbe, J.E., Jabri, S.Y., Mennen, S.M. *et al.* (2011) *J. Am. Chem. Soc.*, **133**, 6549–6552.
8 Song, J., Guo, C., Adele, A. *et al.* (2013) *Chem. Eur. J.*, **19**, 3319–3323.
9 Sun, M., Hao, X.-Y., Liu, S., and Hao, X.-J. (2013) *Tetrahedron Lett.*, **54**, 692–694.
10 The total synthesis of (+)-gliocladin C reported by Boyer and Movassaghi in 2012 also constructed the C3–C3′ bond late in the synthesis, see:Boyer, N. and Movassaghi, M. (2012) *Chem. Sci.*, **3**, 1798–1803.
11 Schnermann, M.J. and Overman, L.E. (2012) *Angew. Chem. Int. Ed.*, **51**, 9576–9580.
12 Sun, Y., Li, R., Zhang, W., and Li, A. (2013) *Angew. Chem. Int. Ed.*, **52**, 9201–9204.
13 López, C.S., Pérez–Balado, C., Rodriguez-Graña, P., and de Lera, A.R. (2008) *Org. Lett.*, **10**, 77–80.
14 Movassaghi, M. and Schmidt, M.A. (2007) *Angew. Chem. Int. Ed.*, **46**, 3725–3728.
15 Keyzers, R.A., Northcote, P.T., and Davies-Coleman, M.T. (2006) *Nat. Prod. Rep.*, **23**, 321–331.
16 Schnermann, M.J. and Overman, L.E. (2011) *J. Am. Chem. Soc.*, **133**, 16425–16427.
17 Lebsack, A.D., Overman, L.E., and Valentekovich, R.J. (2001) *J. Am. Chem. Soc.*, **123**, 4851–4852.
18 Mukaiyama, T., Murai, Y., and Shoda, S. (1981) *Chem. Lett.*, **10**, 431–432.
19 Schnermann, M.J., Untiedt, N.L., Jiménez-Osés, G. *et al.* (2012) *Angew. Chem. Int. Ed.*, **51**, 9581–9586.
20 Okada, K., Okamoto, K., Morita, N. *et al.* (1991) *J. Am. Chem. Soc.*, **113**, 9401–9402.
21 (a) van Bergen, T.J., Hedstrand, D.M., Kruizinga, W.H., and Kellogg, R.M. (1979) *J. Org. Chem.*, **44**, 4953–4962; (b) Mashraqui, S.J. and Kellogg, R.M. (1985) *Tetrahedron Lett.*, **26**, 1453–1456.
22 Fundamental aspects of these recent developments are summarized in other chapters in this volume, see:Tucker, J.W. and Stephenson, R.J. (2012) *J. Org. Chem.*, **77**, 1617–1622.
23 (a) Hoffman, N. (2008) *Chem. Rev.*, **108**, 1052–1103; (b) Bach, T. and Hehn, J.P. (2011) *Angew. Chem. Int. Ed.*, **50**, 1000–1045.
24 (a) Ischay, M.A., Anzovino, M.E., Du, J., and Yoon, T.P. (2008) *J. Am. Chem. Soc.*, **130**, 12886–12887; (b) Du, J. and Yoon, T.P. (2009) *J. Am. Chem. Soc.*, **131**, 14604–14605.
25 (a) Ischay, M.A., Lu, Z., and Yoon, T.P. (2010) *J. Am. Chem. Soc.*, **132**, 8572–8574; (b) Ischay, M.A., Ament, M.S., and Yoon, T.P. (2012) *Chem. Sci.*, **3**, 2807–2811.
26 Riener, M. and Nicewicz, D.A. (2013) *Chem. Sci.*, **4**, 2625–2629.
27 Yamamura, S., Niwa, M., Yukimasa, T., and Nonoyama, M. (1982) *Bull. Chem. Soc. Jpn.*, **55**, 3573–3579.
28 Lu, Z. and Yoon, T.P. (2012) *Angew. Chem. Int. Ed.*, **51**, 10329–10332.
29 Lin, S., Ischay, M.A., Fry, C.G., and Yoon, T.P. (2011) *J. Am. Chem. Soc.*, **133**, 19350–19353.

30 Mbaze, L.M., Lado, J.A., Wansi, J.D. *et al.* (2009) *Phytochemistry*, **70**, 1442–1447.

31 Cannon, J.S. and Overman, L.E. (2012) *Angew. Chem. Int. Ed.*, **51**, 4288–4311.

32 For additional reading of primary literature published after the transcription of this manuscript see: (a) Nicholls, T.P., Leonori, D., and Bissember, A.C. (2016) Applications of visible light photoredox catalysis to the synthesis of natural products and related compounds. *Nat. Prod. Rep.*, **33** (11), 1248–1254; (b) Ma, Z., Wang, X., Wang, X. *et al.* (2014) Asymmetric syntheses of sceptrin and massadine and evidence for biosynthetic enantiodivergence. *Science*, **346** (6206), 219–224; (c) Müller, D.S., Untiedt, N.L., Dieskau, A.P. *et al.* (2015) Constructing quaternary stereogenic centers using tertiary organocuprates and tertiary radicals. Total synthesis of trans-clerodane natural products. *J. Am. Chem. Soc.*, **137** (2), 660–663; (d) Chen, J.-Q., Wei, Y.-L., Xu, G.-Q. *et al.* (2016) Intramolecular 1,5-H transfer reaction of aryl iodides through visible-light photoredox catalysis: a concise method for the synthesis of natural product scaffolds. *Chem. Commun.*, **52** (38), 6455–6458; (e) Blum, T.R., Zhu, Y., Nordeen, S.A., and Yoon, T.P. (2014) Photocatalytic synthesis of dihydrobenzofurans by oxidative [3 + 2] cycloaddition of phenols. *Angew. Chem. Int. Ed.*, **53** (41), 11056–11059; (f) Zhang, H., Liu, P.-F., Chen, Q. *et al.* (2016) Synthesis and absolute configuration assignment of albucidin: a late-stage reductive deiodination by visible light photocatalysis. *Org. Biomol. Chem.*, **14** (13), 3482–3485; (g) Orgren, L.R., Maverick, E.E., and Marvin, C.C. (2015) Synthesis of (±)-tetrabenazine by visible light photoredox catalysis. *J. Org. Chem.*, **80** (24), 12635–12640; (h) Chatterjee, T., Cho, J.Y., and Cho, E.J. (2016) Synthesis of substituted oxazoles by visible-light photocatalysis. *J. Org. Chem.*, **81** (16), 6995–7000; (i) Tao, D.J., Slutskyy, Y., and Overman, L.E. (2016) Total synthesis of (−)-chromodorolide B. *J. Am. Chem. Soc.*, **138** (7), 2186–2189. (j) Slutskyy, Y., Jamison, C.R., Zhao, P., Lee, J., Rhee, Y.H. and Overman, L.E., (2017) Versatile construction of 6-substituted *cis*-2,8-dioxabicyclo[3.3.0]octan-3-ones. Short enantioselective total syntheses of cheloviolenes A and B and dendrillolide C. *J. Am. Chem. Soc.*, **139** (21), 7192–7195.

10

Dual Photoredox Catalysis: The Merger of Photoredox Catalysis with Other Catalytic Activation Modes

Christopher K. Prier and David W. C. MacMillan

Princeton University, Merck Center for Catalysis at Princeton, NJ 08544, USA

10.1 Introduction

Recently, much research in the field of photoredox catalysis has focused on the merger of photoredox catalyst systems with other modes of small-molecule activation. While photoredox catalysis is highly effective at promoting many reaction types in a monocatalytic sense, its combination with other activation modes has enabled remarkable chemical transformations that are not possible using either catalyst system on its own. Often, these transformations proceed via synergistic catalysis, in which the photocatalyst and the co-catalyst simultaneously activate separate reaction partners toward a coupling event [1]. Alternatively, the two catalysts may act together to activate a single substrate, or act at disparate points in a particular mechanistic pathway. While the two catalysts perform distinct roles, their catalytic cycles often remain intricately interwoven, with each of the catalysts interacting with the substrates and with each other at multiple points in a given reaction mechanism.

Dual catalysis strategies in photoredox catalysis offer numerous advantages over a monocatalytic approach. A functional group that is typically inert under photoredox conditions, for instance, may be activated by a co-catalyst to engage in single-electron transfer. Furthermore, while photoredox catalysis allows for the generation of unique (typically open-shell) reactive intermediates, it does not provide a means of controlling their subsequent reactivity. Intercepting these intermediates with a second catalyst allows for their reactivity to be modulated or the selectivity of subsequent reactions to be controlled. In certain cases, the co-catalyst can dramatically alter the inherent reactivity of an intermediate generated via photoredox; the addition of a radical to a transition metal, for instance, generates an organometallic intermediate that has distinctly different properties and reactivity compared to the radical itself.

The ability to perform photoredox reactions with enantioselectivity is a particularly important feature of these synergistic strategies. As the photocatalyst is not covalently or otherwise associated with the reactive intermediates it generates, it lacks the capacity to impart selectivity in any bond-forming steps; in fact, with limited exceptions, enantioselective photoredox transformations primarily

Visible Light Photocatalysis in Organic Chemistry, First Edition.
Edited by Corey R. J. Stephenson, Tehshik P. Yoon and David W. C. MacMillan.

rely on the action of a complementary chiral catalyst [2]. Dual catalytic systems are also highly modular, allowing changes to be made to one catalyst without affecting the efficiency of the other, thus facilitating reaction optimization. In an asymmetric synergistic photoredox reaction, for instance, the catalyst controlling enantioselectivity in the bond-forming step can often be extensively modified without affecting the efficiency of electron-transfer events performed by the photocatalyst.

Perhaps more so than other fields of catalysis, photoredox catalysis has proved particularly amenable to merger with complementary activation modes. This feature may arise in part from the nature of typical photoredox conditions (room temperature, visible light irradiation), which do not preclude the use of most other catalyst systems. Photoredox catalysts are also tolerant of a wide range of reaction conditions, solvents, and additives, and the substrates for these reactions are typically relatively inert, and thus unlikely to engage in deleterious reactivity with the second catalyst. Furthermore, in the case of the typical transition metal-based photocatalysts ($Ru(bpy)_3^{2+}$, *fac*-$Ir(ppy)_3$, etc.), the metal coordination sphere is fully saturated, preventing the catalyst from engaging in any type of inner-sphere interaction with reaction substrates or other catalysts; that is, the photocatalyst acts *exclusively* through outer-sphere mechanisms (*i.e.* electron or energy-transfer). The predictability of redox events based on known reduction potentials also aids in the rational selection of catalysts and substrates that will be either inert or susceptible to single-electron transfer with the photocatalyst, depending on the needs of the particular application.

This chapter first discusses reactions that have been enabled by the merger of photoredox catalysis with organocatalytic activation modes. Next, examples of synergistic photoredox catalysis with Brønsted and Lewis acids are reviewed. The chapter concludes with chemistry enabled by the combination of photoredox and transition metal catalysts.

10.2 Merger of Photoredox Catalysis with Organocatalysis

Given the broad compatibility of photoredox catalysis with diverse organic functionality, it is not surprising that photoredox has been extensively employed in concert with organocatalytic activation modes. Early work was particularly focused on the merger of photoredox catalysis with enamine catalysis. In this general activation mode, condensation of a primary or secondary amine with a carbonyl compound catalytically generates an enamine intermediate, a species activated at the α-position of the carbonyl toward a range of electrophilic functionalization reactions. Our laboratory and others have extensively investigated the use of chiral secondary amine catalysts to achieve enantioselective α-functionalization reactions of aldehydes and ketones [3].

In 2008, the author's laboratory reported the merger of photoredox catalysis with enamine organocatalysis to perform the enantioselective α-alkylation of aldehydes, a particularly challenging transformation that was, to a large degree,

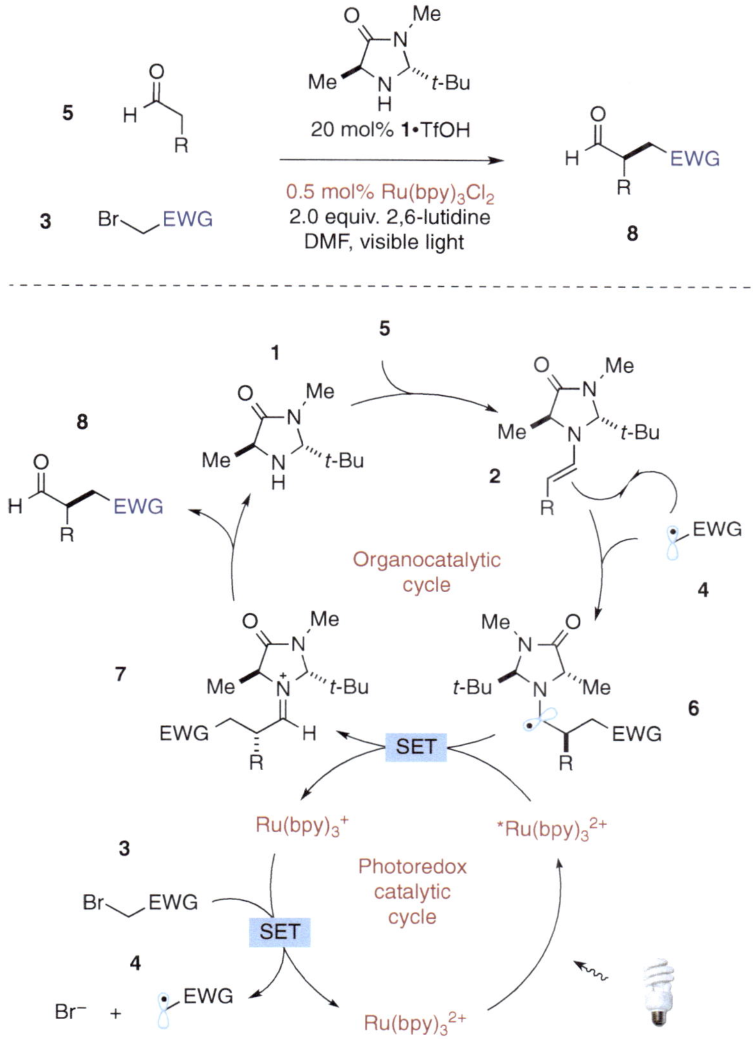

Scheme 10.1 Merger of photoredox and enamine catalysis: the asymmetric α-alkylation of aldehydes.

elusive using amine catalysis alone (Scheme 10.1) [4]. Given the electron-rich nature of the enamine π-system, it was anticipated that these intermediates would couple with electrophilic carbon-centered radicals generated via the action of a photoredox catalyst. Furthermore, the use of a chiral imidazolidinone catalyst **1** was expected to control the facial approach of the radical and provide a means of achieving an enantioselective coupling. The proposed mechanism, involving distinct photocatalytic and organocatalytic cycles, is shown in Scheme 10.1. Initiation of the reaction requires quenching of the photocatalyst $^{*}Ru(bpy)_3^{2+}$ by a sacrificial amount of enamine **2** to provide $Ru(bpy)_3^{+}$ (not shown). The $Ru(bpy)_3^{+}$ intermediate is sufficiently reducing ($E_{1/2}^{II/I} = -1.33$ V vs

SCE (saturated calomel electrode)) [5] to transfer an electron to the alkyl bromide radical precursor **3** (e.g., $E_{1/2}^{\text{red}} = -0.49$ V vs SCE for phenacyl bromide) [6]. This reduction of the alkyl halide induces mesolytic cleavage to afford bromide and the electron-deficient radical **4**. Meanwhile, condensation of the simple aldehyde **5** with the imidazolidinone organocatalyst **1** gives chiral enamine **2**. Addition of the electrophilic radical to the accessible *Si* face of the enamine then generates α-amino radical **6**, which may undergo single-electron oxidation by $^*\text{Ru(bpy)}_3^{2+}$ ($E_{1/2}^{*\text{II/I}} = +0.77$ V vs SCE, compared to −0.92 to −1.12 V vs SCE for α-amino radicals) [5, 7]. In this electron-transfer step, the two catalytic cycles intersect to provide Ru(bpy)_3^+ as well as iminium ion **7**. Hydrolysis of the iminium then releases the enantioenriched α-alkylated aldehyde **8** and regenerates the organocatalyst. Photoredox catalysis thus not only generates the reactive radical species but also performs a key oxidation in the organocatalytic cycle.

This strategy was successfully applied to the α-alkylation of aldehydes using a range of α-bromocarbonyls **9** as radical precursors (Scheme 10.2a) [4]. Using

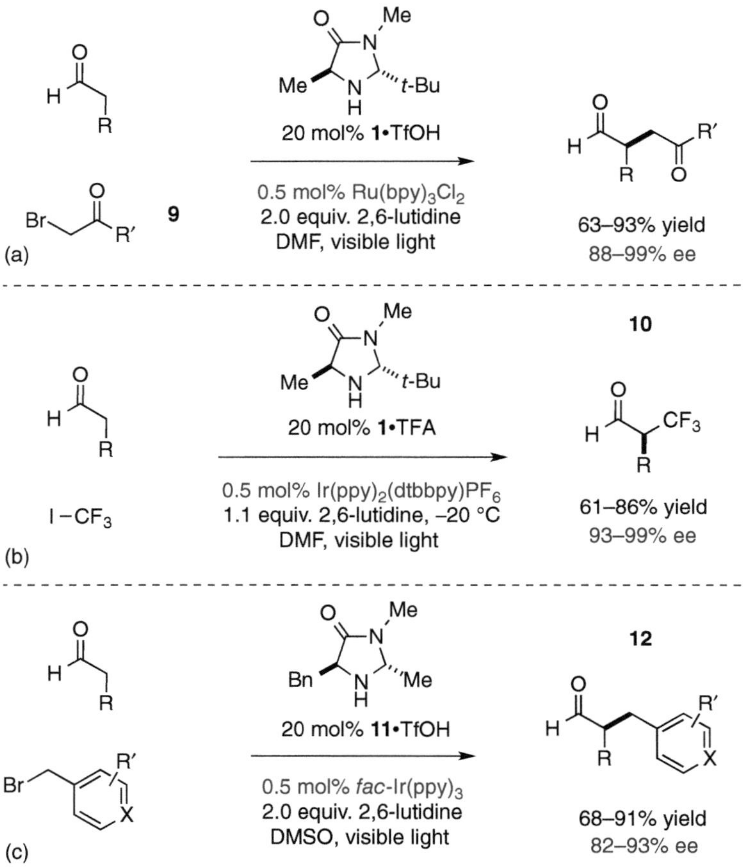

Scheme 10.2 Enantioselective α-functionalization reactions achieved via photoredox and enamine catalysis. (a) α-Alkylation of aldehydes. (b) α-Trifluoromethylation of aldehydes. (c) α-Benzylation of aldehydes.

$Ru(bpy)_3^{2+}$ as the photocatalyst and imidazolidinone **1** as the amine catalyst, a range of aldehydes couple with radicals derived from α-bromomalonates, ketones, and esters in high yields and enantioselectivities. The merger of photoredox and amine catalysis has subsequently been employed to achieve α-functionalization reactions using a diverse set of electrophilic radicals. For instance, generation of trifluoromethyl radicals under this reaction manifold has enabled the enantioselective α-trifluoromethylation of aldehydes (Scheme 10.2b) [8]. These reactions employ the iridium photocatalyst $Ir(ppy)_2(dtbbpy)PF_6$, which in its Ir(II) oxidation state is sufficiently reducing ($E_{1/2}^{III/II} = -1.51$ V vs SCE) [9] to transfer an electron to the radical precursor trifluoromethyl iodide (CF_3I, $E_{1/2}^{red} = -1.22$ V vs SCE) [10]. Mesolytic cleavage provides the trifluoromethyl radical, which adds with high facial selectivity to the chiral enamine derived from imidazolidinone **1**, providing access to a range of α-trifluoromethyl aldehydes **10**. Perfluoroalkyl iodides may also be employed as substrates to provide access to enantioenriched α-perfluoroalkyl aldehydes.

Another transformation enabled by photoredox organocatalysis is the asymmetric α-benzylation of aldehydes (Scheme 10.2c) [11]. It was found that electron-deficient benzyl halides could be induced to undergo coupling (via generation of the corresponding benzylic radicals) using the iridium photocatalyst *fac*-$Ir(ppy)_3$ and imidazolidinone organocatalyst **11** to give α-benzyl aldehyde products **12**. In contrast to the photocatalytic cycle displayed in Scheme 10.1, in which a reductive quenching cycle is operative, the α-benzylation reaction is proposed to proceed via an oxidative quenching pathway in which the photoexcited species *fac*-*$Ir(ppy)_3$ ($E_{1/2}^{IV/*III} = -1.73$ V vs SCE) performs single-electron reduction of the benzyl halide. The reaction was found to require highly electron-deficient benzyl halides; the presence of electron-withdrawing groups on the arene presumably both facilitates single-electron reduction and renders the resulting benzylic radical more electrophilic. The reaction is also amenable to the coupling of heterobenzyl moieties incorporating a range of electron-deficient heteroaromatics such as pyridines, pyrazines, pyrimidines, and benzimidazoles.

While these examples all employ transition metal-based photoredox catalysts, this general approach to aldehyde α-functionalization has also been performed using the organic photocatalyst eosin Y by Zeitler and coworkers [12, 13]. The α-alkylation of aldehydes has additionally been performed via semiconductor catalysis; König and coworkers have identified surface-modified titanium dioxide and the heterogeneous material $PbBiO_2Br$ as efficient catalysts for the reaction under visible-light irradiation [14]. The coupling of chiral enamines with electrophilic radicals has also been applied to the enantioselective α-amination of aldehydes via the generation of electrophilic nitrogen-centered radicals; in this case, however, the use of a photoactive aminating reagent obviates the requirement for a photoredox catalyst [15].

Koike and Akita have merged amine and photoredox catalysis to perform an α-oxyamination of aldehydes [16]. In this transformation, the amine catalyst morpholine condenses with the aldehyde **13** to give the enamine intermediate **14** (Scheme 10.3). Instead of undergoing coupling at this stage, however, a single-electron transfer event between the enamine and *$Ru(bpy)_3^{2+}$ generates the

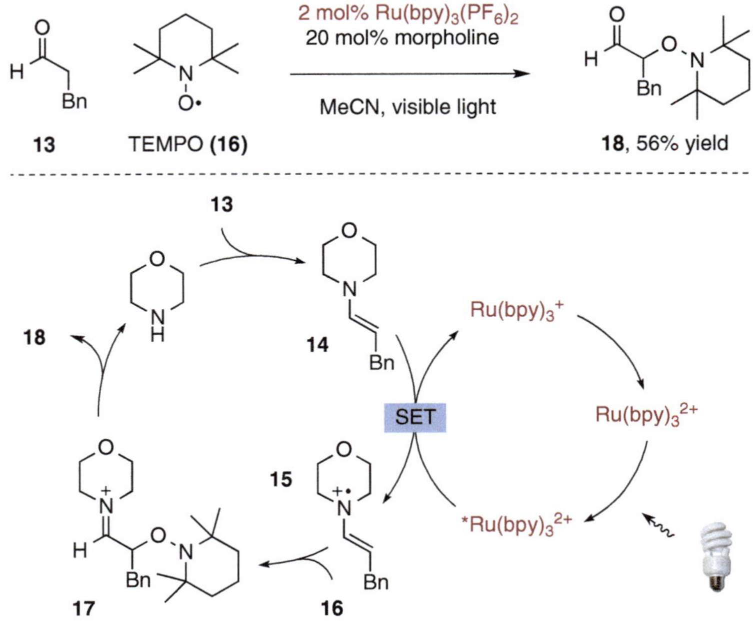

Scheme 10.3 Photoredox and amine catalysis: α-oxyamination of aldehydes with TEMPO.

enamine radical cation **15**; this species is an electrophilic radical possessing significant spin density at the carbon atom β to nitrogen [17, 18]. A proposed radical–radical coupling between radical cation **15** and 2,2,6,6-tetramethylpiperidine-1-oxyl (TEMPO, **16**) provides iminium ion **17**, hydrolysis of which generates α-oxyaminated aldehyde **18**. The stoichiometric oxidant operative in this process is potentially adventitious oxygen [19, 20].

In the reactions discussed thus far, amine and photoredox catalysis have enabled the α-functionalization of carbonyl compounds through the coupling of either enamine or enamine radical cation intermediates. Recently, a dual catalytic system of amine and photoredox catalysts has been shown to activate the more remote β-position of saturated carbonyl compounds. In the first example of this chemistry, *fac*-$Ir(ppy)_3$ and amine catalyst **19** were employed to achieve the β-arylation of aldehydes and ketones with cyanoarenes (Scheme 10.4) [21]. The key to this transformation is the transient generation of the β-enaminyl radical **20**, a 5π-electron intermediate that possesses radical character at the carbonyl β-position. In the mechanism of this reaction, the photoexcited catalyst $^*Ir(ppy)_3$ ($E_{1/2}^{IV/*III} = -1.73$ V vs SCE) [22] first transfers an electron to a benzonitrile such as 1,4-dicyanobenzene (**21**, $E_{1/2}^{red} = -1.6$ V vs SCE) [23] to give the arene radical anion **22** and the oxidized photocatalyst $Ir^{IV}(ppy)_3{}^+$. Concurrent condensation of a saturated aldehyde **23** with secondary amine **19** provides enamine **24**. This species is sufficiently electron-rich to undergo single-electron oxidation by $Ir^{IV}(ppy)_3{}^+$ ($E_{1/2}^{IV/III} = +0.77$ V vs SCE) [22], generating the enamine radical cation **25** and turning over the photocatalytic cycle. Enamine oxidation significantly lowers the

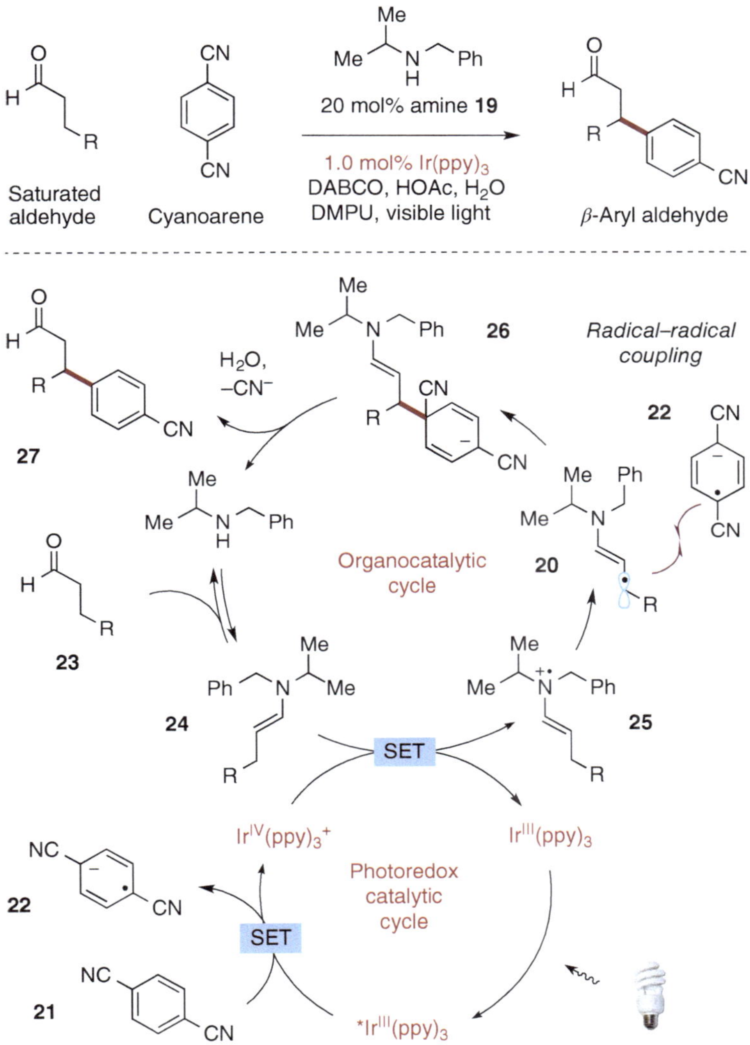

Scheme 10.4 Photoredox and amine catalysis enable the direct β-arylation of saturated aldehydes.

pK_a of the allylic C—H bonds (originally at the carbonyl β-position), allowing deprotonation at this position by weak base to deliver the key β-enaminyl radical intermediate **20**. Radical–radical coupling between this species and the arene radical anion **22** is proposed to forge the new carbon–carbon bond and provide dienyl anion **26**. Aromatization via loss of cyanide and hydrolysis of the enamine then provides the β-aryl aldehyde product **27** while regenerating the amine catalyst **19**. In scope studies, this synergistic catalytic system was found to enable the coupling of diverse electron-deficient cyanoarenes, including cyano-substituted heteroarenes, with a broad range of saturated aldehydes.

Cyclic ketones were also found to be amenable to β-arylation, and azepane was identified as a superior amine catalyst for these substrates.

A remarkable feature of the β-arylation reaction is the manner in which two radicals, both generated in catalytic quantities, selectively undergo radical–radical coupling. The success of this transformation has been attributed to the long-lived and persistent nature of the arene radical anion coupling partner [24]. In seeking other potential partners for radical–radical coupling reactions, ketyl radicals were next identified as stabilized radicals with the potential to engage in coupling with the β-enaminyl radical. Indeed, a synergistic combination of $Ir(ppy)_3$ and azepane (**28**) as the amine catalyst was found to promote the β-coupling of saturated ketones with aryl ketones to provide γ-hydroxyketone products (Scheme 10.5a) [25]. As in the β-arylation reaction, the merged photo- and organocatalyst system generates the activated 5π-electron intermediate **29**. Simultaneous generation of the ketyl radical **30** proceeds via single-electron reduction of a diaryl ketone ($E_{1/2}^{red} = -1.83$ V vs SCE for benzophenone) [26] by the excited state of the photoredox catalyst. Radical–radical coupling followed by hydrolysis of the enamine then delivers the γ-hydroxyketone adduct **31**. Remarkably, this reaction achieves the selective coupling of two different ketones: a cyclohexanone capable of forming an enamine and an aryl ketone capable of undergoing single-electron reduction. This protocol may also be applied to the coupling of ketyl radicals derived from acetophenones and other aryl–alkyl ketones; in these cases, the photocatalyst $Ir(p\text{-MeO-ppy})_3$ was found to be effective.

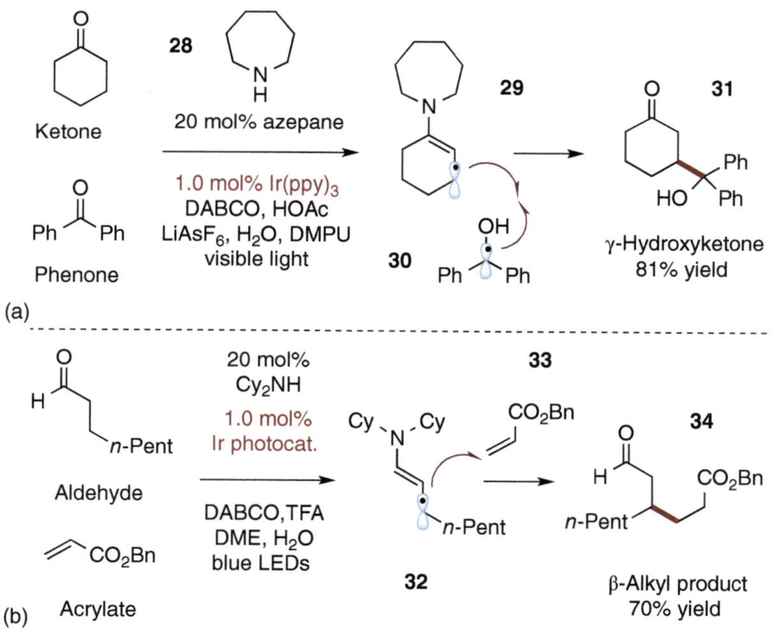

Scheme 10.5 β-Aldol and β-alkylation reactions proceeding via β-enaminyl radicals. (a) β-Aldol coupling of saturated ketones. (b) β-Alkylation of saturated aldehydes.

Highly electron-rich radicals such as the β-enaminyl radical are strongly nucleophilic; thus, in addition to engaging in radical–radical coupling with other open-shell species, the 5π-electron intermediate may react with closed-shell electrophiles such as electron-deficient π-systems. This reactivity of the β-enaminyl radical has been exploited to achieve the β-alkylation of saturated aldehydes with Michael acceptors (Scheme 10.5b) [27]. In this transformation, the β-enaminyl radical **32** undergoes addition to radical acceptors such as benzyl acrylate (**33**); the sterically encumbered organocatalyst dicyclohexylamine (Cy_2NH) is employed to prevent 1,4-conjugate addition of the organocatalyst to the Michael acceptor. The C—C bond-forming step then generates an intermediate α-carbonyl radical (not shown), which may undergo single-electron reduction followed by protonation and enamine hydrolysis to deliver the β-alkylated aldehyde adduct **34**. This reaction is amenable to the construction of quaternary carbon centers at the aldehyde β-carbon, and may also be carried out in an intramolecular manner.

Photoredox catalysis has been extensively applied to the development of amine α-functionalization reactions, often via the conversion of these compounds to their corresponding iminium ions. As in the chemistry discussed earlier, amine organocatalysts have been employed alongside photoredox catalysts to expand the utility of these transformations. For instance, Rueping and coworkers have shown that a combination of L-proline and $Ru(bpy)_3^{2+}$ efficiently catalyzes an oxidative Mannich coupling of *N*-phenyltetrahydroisoquinolines (Scheme 10.6a) [28]. The reaction is proposed to proceed via addition of an enamine (generated from the condensation of L-proline with acetone) to the iminium ion **35** (generated by the photoredox catalyst via two single-electron oxidations), providing the α-alkylated amine **36**. Despite the use of a chiral amine catalyst in this reaction, however, only low enantioselectivities were observed. In a separate

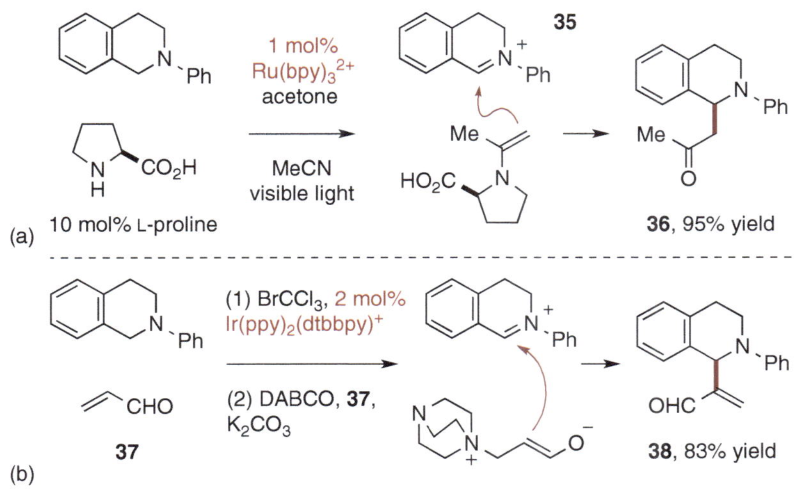

Scheme 10.6 Uses of amine organocatalysis to perform coupling reactions with photoredox-generated iminium ions. (a) Oxidative Mannich coupling of amines. (b) α-Acroleination of amines with DABCO organocatalyst.

reaction, Lu and Xiao have employed 1,4-diazabicyclo[2.2.2]octane (DABCO) as a nucleophilic catalyst to perform the coupling of iminium ions with acrolein (**37**, Scheme 10.6b) [29]. In their two-step protocol, the amine is first converted to the iminium ion **35** via photoredox catalysis under visible light irradiation. Next, in the absence of light, acrolein and DABCO are added and react with the iminium to provide the coupled product **38** via the Morita–Baylis–Hillman pathway. The chiral amine β-isocupreidine was found to catalyze the formation of adduct **38** with a moderate level of enantioinduction (83 : 17 e.r.). In another example of nucleophilic catalysis merged with photoredox, Jiao and coworkers have used 4-methoxypyridine as a nucleophilic catalyst in the oxidation of α-haloesters to α-ketoesters [30].

Despite the broad range of amine α-functionalization reactions enabled by photoredox catalysis, achieving highly enantioselective variants of these reactions has proved challenging. In one approach to this problem, DiRocco and Rovis merged photoredox catalysis with *N*-heterocyclic carbene (NHC) catalysis [31] to perform an asymmetric α-acylation of amines (Scheme 10.7) [32]. In this reaction, a chiral NHC catalyst functions to generate a chiral acyl anion equivalent for addition into iminium ions generated via photoredox catalysis. In the proposed photoredox cycle, oxidative quenching of $^*Ru(bpy)_3{}^{2+}$ by *meta*-dinitrobenzene (*m*-DNB, **39**, $E_{1/2}^{red} = -0.90$ V vs SCE) [33] generates $Ru(bpy)_3{}^{3+}$ and the arene radical anion **40**. $Ru(bpy)_3{}^{3+}$ may then oxidize the tetrahydroisoquinoline **41** to its radical cation, which after loss of a proton and another electron yields the iminium ion **35**. Simultaneously, the aminoindanol-derived NHC catalyst **42** reacts with the simple aldehyde *n*-butanal (**43**) to form the nucleophilic Breslow intermediate **44**. Attack of this activated intermediate on the iminium **35** forges the new C—C bond, with the chiral information on the NHC backbone effectively controlling the configuration at the newly formed stereocenter. The adduct **45** may then suffer elimination of the product α-acylamine **46**, regenerating the carbene catalyst. A range of aliphatic aldehydes may be reacted under these conditions to give α-acylamine products with high enantioselectivity. As in the asymmetric α-alkylation of aldehydes, the merger of photoredox catalysis with an organocatalytic method (either amine or NHC catalysis) is key to the development of an enantioselective transformation.

In the NHC-catalyzed α-acylation of amines, enantioinduction is achieved via the formation of a chiral nucleophile, which adds to the achiral iminium ion electrophile. In contrast, Stephenson and Jacobsen have developed an asymmetric α-alkylation of amines using a chiral catalyst that binds the iminium ion and controls the approach of an achiral nucleophile [34]. A chiral thiourea is employed as an anion-binding catalyst [35]; the thiourea binds the chloride anion of an iminium chloride salt (generated via photoredox catalysis), providing the stereo-defined ion pair **47** (Scheme 10.8). Studies on this system identified the chiral thiourea **48** as capable of promoting the highly enantioselective reaction of a tetrahydroisoquinoline-derived iminium ion with a silyl keteneacetal nucleophile to provide α-alkylated amine **49**. As in the α-acroleination reaction, a two-step reaction protocol is employed: in the first step, the amine is stoichiometrically converted to the iminium ion **35** in acetonitrile; the solvent

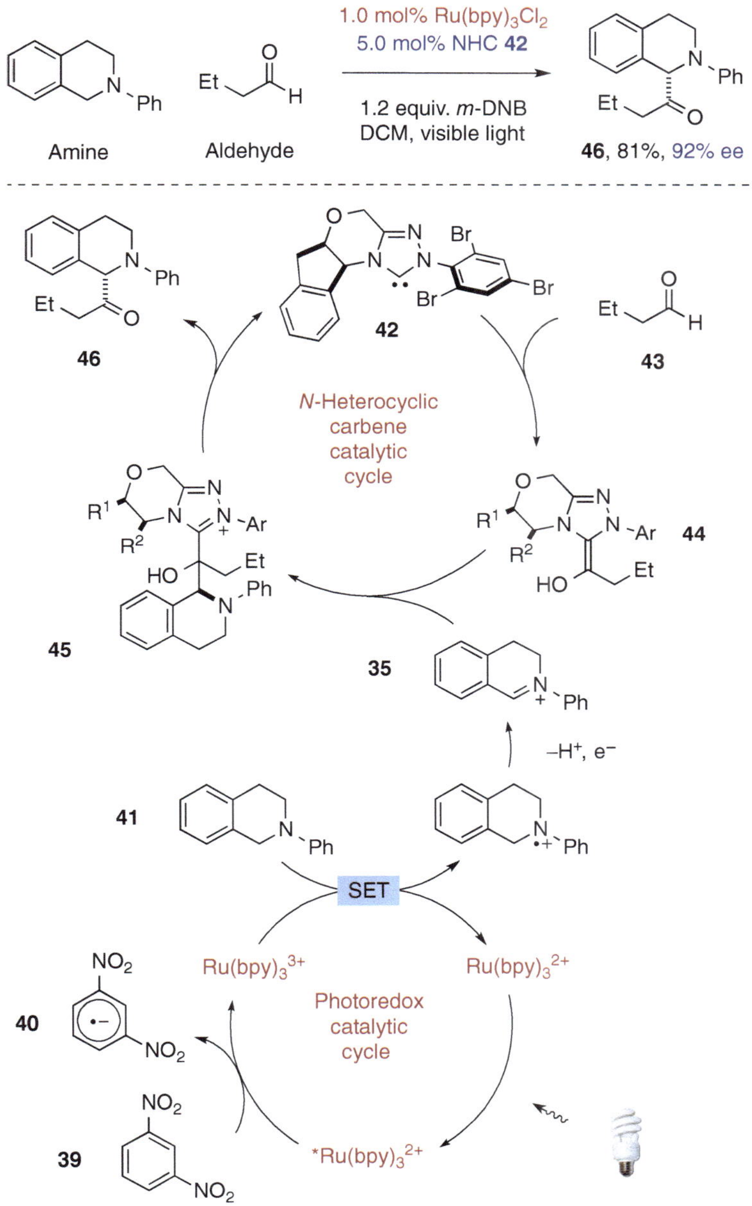

Scheme 10.7 Merger of photoredox catalysis with *N*-heterocyclic carbene catalysis: the enantioselective α-acylation of amines.

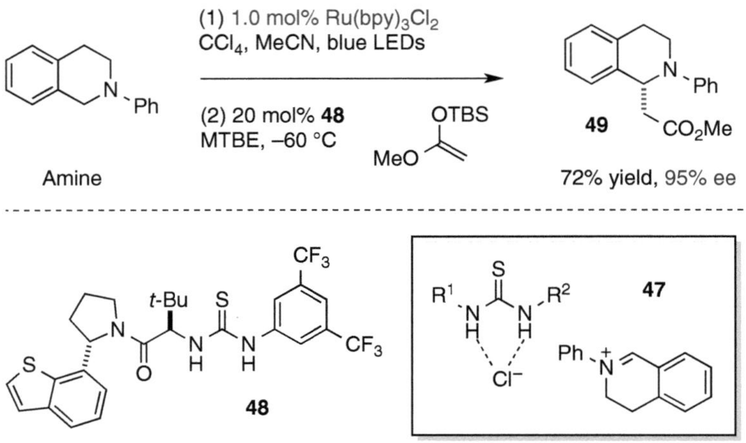

Scheme 10.8 Photoredox and anion-binding catalysis enable an enantioselective α-alkylation of amines.

is subsequently switched to methyl *tert*-butyl ether (MTBE) for the second step, the addition of the silyl ketene acetal nucleophile. While polar solvents are required for conversion in the photoredox step, nonpolar solvents are required for high levels of enantioselectivity to be obtained in the anion-binding step.

An inherent limitation of photoredox catalysis is the fact that these catalysts perform only the transfer of single electrons. In certain scenarios where single-electron transfer is not facile, however, the transfer of an electron and a proton together—as a hydrogen atom (H•)—has been identified as a powerful strategy for catalysis. For instance, Nicewicz has employed hydrogen atom transfer (HAT) catalysts in combination with photoredox catalysts to enable the anti-Markovnikov hydroetherification of alkenes (Scheme 10.9) [36]. In this reaction, alkenols such as **50** are converted to cyclic ethers such as tetrahydrofuran **51** via a mechanism involving single-electron oxidation of the olefin to its radical cation. As unactivated alkenes possess high redox potentials (typically above +1.7 V vs SCE), Nicewicz made use of the highly oxidizing organic photocatalyst 9-mesityl-10-methylacridinium perchlorate pioneered by Fukuzumi [37]. Upon photoexcitation, the 9-mesityl-10-methylacridinium ion (Acr^+–Mes) is converted to the strongly oxidizing electron-transfer state $Acr^•$–$Mes^{•+}$ ($E_{1/2}^{red}$ = +2.06 V vs SCE) [38], which removes an electron from alkenol **50** ($E_{1/2}^{red}$ = +1.95 V vs SCE) to give the alkene radical cation **52**. This intermediate is proposed to form the three-membered adduct **53** between the radical cation and the oxygen nucleophile, followed by cleavage of one of the C—O bonds to give the cyclized radical **54** [39]. The observed anti-Markovnikov selectivity is predicated on formation of the more stable radical intermediate. Subsequently, the tertiary radical **54** must accept an electron and a proton to be converted to the product tetrahydrofuran **51**. The electron could potentially be delivered by the reduced photocatalyst intermediate $Acr^•$–Mes ($E_{1/2}^{red}$ = −0.57 V vs SCE) [38].

In reactions employing the acridinium Acr^+–Mes as the sole catalyst, however, low yields of product were obtained. Speculating that the reduction

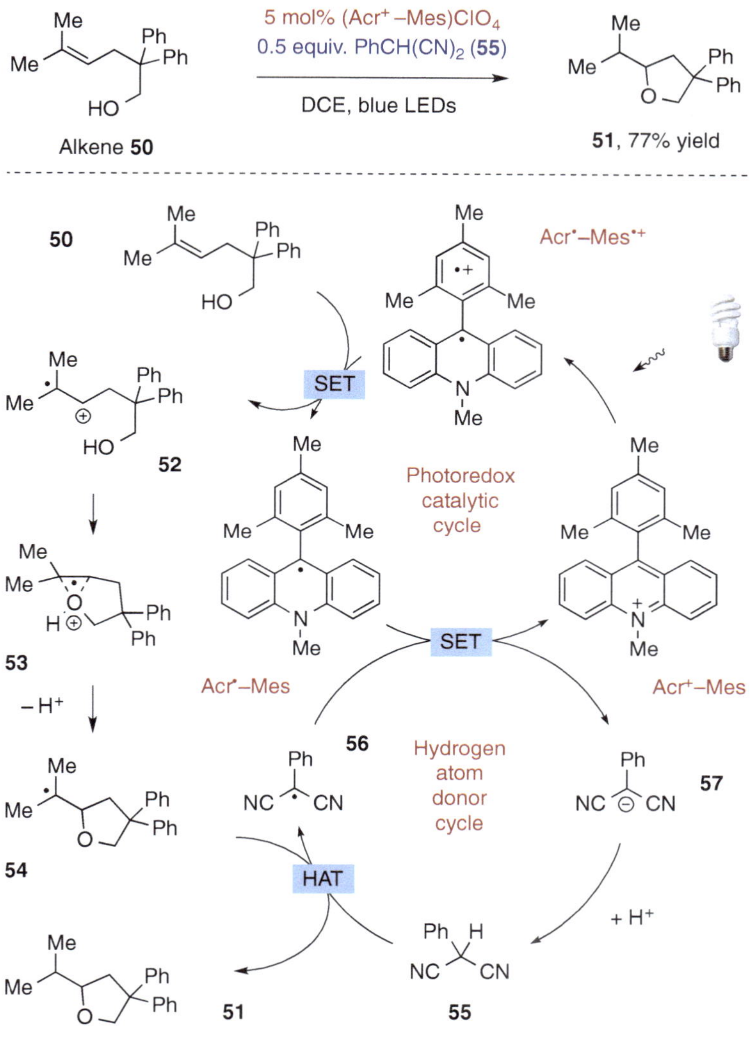

Scheme 10.9 Intramolecular hydroetherification of alkenes enabled by the merger of photoredox and hydrogen atom donor catalysts.

step was limiting product formation, Nicewicz explored the use of hydrogen atom donors to convert radical **54** to the product **51**. It was found that the addition of 2-phenylmalononitrile (**55**) significantly improved yields of the hydroetherification product. This species features a relatively weak C—H bond (bond dissociation energies, BDE = 77 kcal mol^{-1}) [40] and is proposed to undergo HAT with **54** to give the tertiary radical **56**. This species may undergo single-electron reduction by Acr•–Mes to complete the photocatalytic cycle. Upon protonation, the anion **57** is converted back to **55**, completing the hydrogen atom donor catalytic cycle. In donating a hydrogen atom to the tertiary radical **54**, this process converts a radical that is difficult to reduce (**54**) to one that is more readily reduced (**56**) by the photocatalyst (**56**). Using this dual catalytic system, the photoredox hydroetherification reaction was shown to act on a range of alkenols to provide the products of 5-*exo*, 6-*exo*, or 7-*exo* cyclization; in all cases, these reactions give exclusively the anti-Markovnikov regioisomer.

The dual photoredox and H-atom transfer catalyst system has subsequently been employed to develop a range of alkene hydrofunctionalization reactions. Intramolecular anti-Markovnikov hydroamination may be accomplished when the alkene substrate is tethered to a protected amine; this reaction provides access to a range of pyrrolidine and piperidine ring systems (Scheme 10.10a) [41]. Thiols such as thiophenol were found to act as highly efficient HAT catalysts for this transformation; abstraction of a hydrogen atom provides a thiyl radical that may be readily reduced ($E_{1/2}^{red}$ = +0.16 V vs SCE for PhS•) [42] and protonated to turn over the thiol catalyst. Intermolecular hydroamination of alkenes may also be achieved using trifluoromethanesulfonamide ($TfNH_2$) and various nitrogen heterocycles as the nitrogen reactant [43]. Other classes of suitable nucleophiles include carboxylic acids, which again undergo coupling with alkenes with complete anti-Markovnikov selectivity (Scheme 10.10b) [44]. Benzene sulfinic acid was identified as the preferred H-atom donor for this reaction. A similar catalyst system may be employed to achieve the hydrotrifluoromethylation of alkenes using the Langlois reagent (**58**) (Scheme 10.10c) [45]. In this case, however, the mechanism is postulated to proceed not via oxidation of the alkene, but rather via single-electron oxidation of the Langlois reagent. The resulting electrophilic trifluoromethyl radical may then add directly to the alkene substrate; the radical

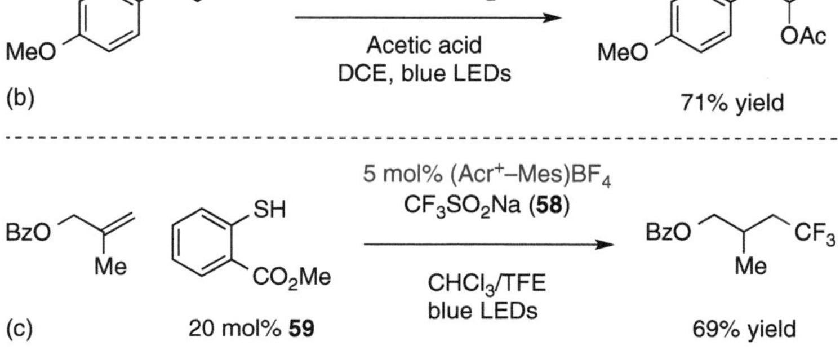

Scheme 10.10 Anti-Markovnikov alkene functionalization reactions promoted by photoredox and hydrogen atom donor catalysts. (a) Anti-Markovnikov hydroamination of alkenes. (b) Anti-Markovnikov hydroacetoxylation of alkenes. (c) Anti-Markovnikov hydrotrifluoromethylation of alkenes.

product of this addition may then accept a hydrogen atom from either methyl thiosalicylate (**59**) or 2,2,2-trifluoroethanol (TFE).

Further studies have been directed at exploiting this dual catalyst system to perform complexity-building cycloaddition reactions. Upon exposure to the mesityl-acridinium photocatalyst, HAT catalyst **55**, and visible light, allylic alcohols and alkenes were found to engage in [3+2] cycloadditions to provide substituted tetrahydrofurans (Scheme 10.11) [46]. This reaction proceeds via a polar-radical-crossover mechanism: in the polar step, anti-Markovnikov addition of the alcohol nucleophile to the radical cation **60** (generated via single-electron oxidation) forges a C—O bond and gives adduct **61**. In the subsequent one-electron step, radical addition to the pendent alkene forms a C—C bond and generates primary radical **62**. Hydrogen atom abstraction from the catalyst **55** furnishes the tetrahydrofuran product. While the C2 and C3 stereocenters are typically formed with high *trans* selectivity, lower diastereoselectivity is observed with respect to the relationship between the C3 and C4 substituents.

In the alkene hydrofunctionalization chemistry, thiols are particularly effective hydrogen atom donors, undergoing loss of H• to give readily reduced thiyl radicals. In the reverse sense, however, these thiyl radicals are also capable of performing the *abstraction* of hydrogen atoms from weak C—H bonds. This reactivity of thiyl radicals has been exploited in the α-arylation of benzylic ethers and alcohols using a combination of the thiol organocatalyst methyl 2-mercaptoacetate (**63**) and the iridium photocatalyst $Ir(ppy)_3$ (Scheme 10.12) [47]. As discussed earlier, electron-deficient benzonitriles (such as 1,4-dicyanobenzene **21**) may accept an electron from $^*Ir(ppy)_3$ to give arene radical anions capable of engaging in radical–radical coupling with other open-shell intermediates. Subsequently, single-electron oxidation and deprotonation of thiol catalyst **63**, potentially via proton-coupled electron transfer (PCET, see subsequent text), then generates the key thiyl radical **64**. This radical may perform the abstraction of a hydrogen atom from the benzylic C—H bond of the ether substrate **65**, a step expected to be favorable on the basis of the S—H

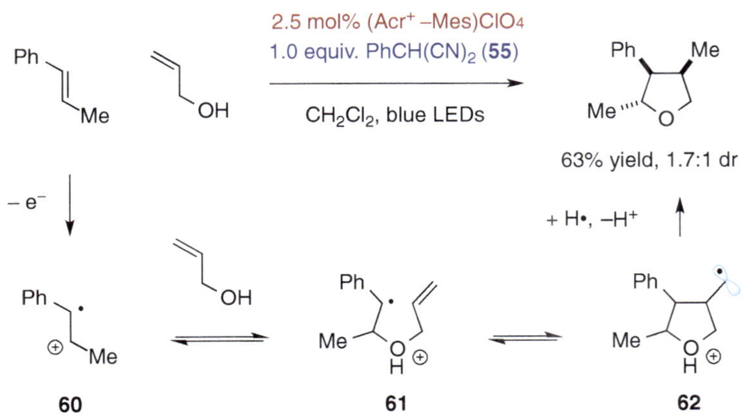

Scheme 10.11 Polar-radical-crossover mechanism of a [3+2] cycloaddition between alkenes and allylic alcohols.

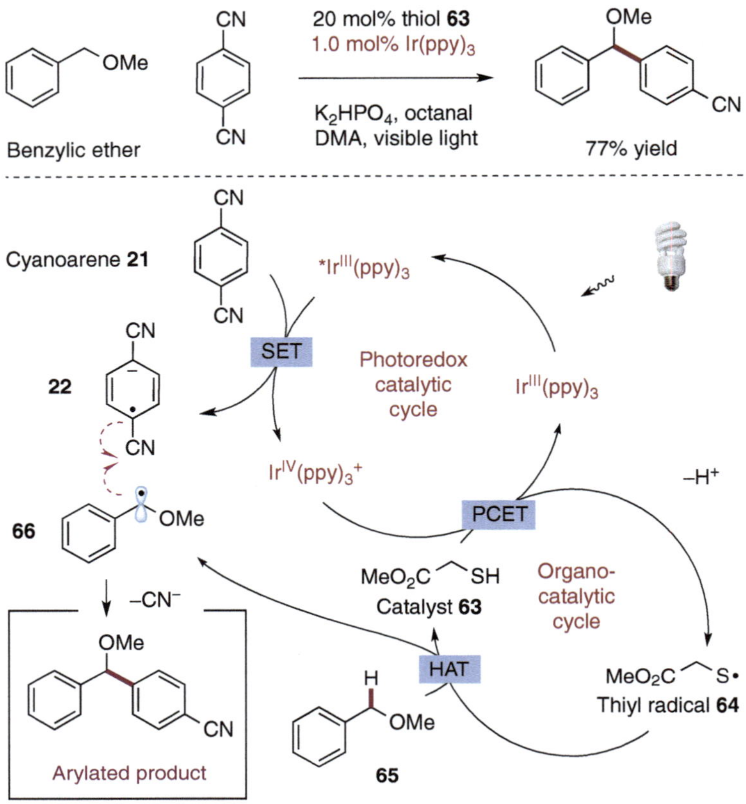

Scheme 10.12 Photoredox and thiol catalysts mediate the α-arylation of benzylic ethers and alcohols.

bond strength (BDE = 86.8–87.2 kcal mol^{-1} for MeO_2CCH_2S—H) [48] compared to that of the benzylic C—H bond (BDE = 85.8 kcal mol^{-1} for PhCH(—H)OMe) [49]. The hydridic nature of the α-alkoxy C—H bond also favors its abstraction by the electrophilic thiyl radical [50]. The HAT step regenerates the thiol catalyst **63** and provides the benzylic radical **66**; radical–radical coupling between the benzylic radical and the radical anion **22**, followed by loss of cyanide, delivers the α-arylated ether product. A particularly attractive feature of this system is the predictable selectivity of the reaction for arylation at the weakest substrate C—H bond. Both benzylic ethers and benzylic alcohols undergo coupling with a variety of cyanoarenes under these conditions [51].

10.3 Merger of Photoredox Catalysis with Acid Catalysis

10.3.1 Photoredox Catalysis and Brønsted Acid Catalysis

Protonation may be one of the simplest methods of activating organic molecules, and a number of reports have appeared documenting the advantages of employing Brønsted acid co-catalysts in photoredox processes. In one example of this type of dual activation, Yoon has shown that formic acid and $Ru(bpy)_3^{2+}$ promote a reductive cyclization of bis(enones) to give cyclopentane products [52]. As shown in Scheme 10.13, the reaction is initiated by reductive quenching of the photocatalyst by Hünig's base to generate a $Ru(bpy)_3^{+}$ intermediate. Protonation of the bis(enone) **67** by formic acid, to give protonated species **68**, is then postulated to activate the substrate toward single-electron reduction by $Ru(bpy)_3^{+}$. The resulting neutral radical **69** undergoes 5-*exo*-trig radical cyclization to form α-keto radical **70** predominantly as the *trans* diastereomer; abstraction of a hydrogen atom from the radical cation of Hünig's base then furnishes the cyclopentane diketone product **71**. A range of aryl as well as aliphatic bis(enones) undergo cyclization to *trans*-substituted cyclopentanes under these conditions. No product is obtained in the absence of the Brønsted acid, highlighting the importance of catalytic acid in activating the organic substrate toward single-electron transfer. Interestingly, activation of the same bis(enone) substrates with a Lewis acid rather than a proton promotes divergent reactivity and leads to the products of [2+2] cycloaddition (see subsequent text). In addition, Zeitler has shown that the reductive cyclization of bis(enones) may also be accomplished using a hydrogen-bonding thiourea catalyst to promote the single-electron reduction of enone **67** [53].

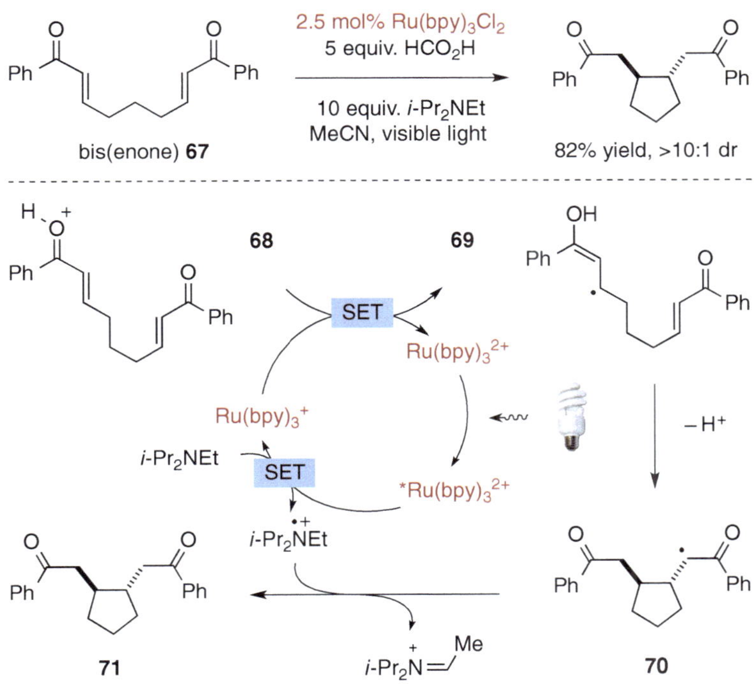

Scheme 10.13 Reductive cyclization of bis(enones) promoted by photoredox and Brønsted acid catalysts.

Knowles has employed Brønsted acids to promote challenging single-electron transfers in the context of an intramolecular ketyl-olefin coupling reaction, in which the acrylate-tethered aryl ketone **72** is converted to cyclopentane product **73** (Scheme 10.14) [54]. This reaction requires the single-electron reduction of the ketone to its corresponding ketyl radical, a step that is significantly disfavored on the basis of the redox potentials of the substrate($E_{1/2}^{red} = -2.10$ V vs SCE for acetophenone) [55] and the photoredox reductant ($Ru(bpy)_3^+$, $E_{1/2}^{II/I} = -1.33$ V vs SCE). To overcome this barrier, a system that enables the required reduction to proceed via PCET was designed. PCET describes the independent transfer of a proton and an electron in a single elementary step; the concerted nature of this process can lower the kinetic barrier relative to sequential pathways involving discrete proton-transfer and electron-transfer steps. In the ketyl-olefin cyclization reaction , $Ru(bpy)_3^+$ and diphenyl phosphoric acid were found to constitute an effective electron donor/proton donor pair, enabling the formation of ketyl radical **74**. This process is driven by the increased basicity of the ketyl **74** relative to the parent ketone; PCET is potentially also operative in Yoon's

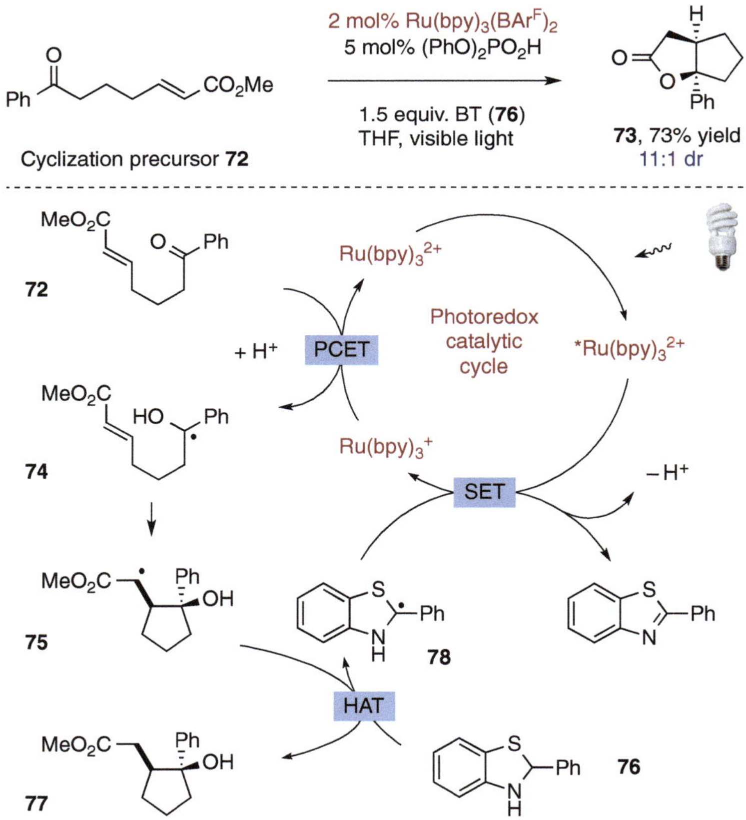

Scheme 10.14 Ketyl-olefin coupling enabled by proton-coupled electron transfer (PCET) using Brønsted acid and photoredox catalysts.

reductive bis(enone) cyclization. Following generation of ketyl radical **74**, intramolecular radical cyclization yields α-carbonyl radical **75**. Abstraction of a hydrogen atom from the reductant 2-phenyl-dihydrobenzothiazole (BT, **76**) then generates the cyclopentane **77**, which may lactonize to give the observed product **73**. Hydrogen atom abstraction from BT generates the stabilized radical **78**, which may reduce $^{*}Ru(bpy)_3^{2+}$ to turn over the photocatalytic cycle. Oxidation of radical **78** also releases a proton, regenerating the catalytic acid that was first consumed in the PCET step. A range of alkyl–aryl ketones undergo intramolecular cyclization using this co-catalytic system; in addition to acrylates, the acceptor moiety may also be a styrene or vinyl nitrile.

Knowles and coworkers then questioned whether a chiral Brønsted acid could induce asymmetry in reactions involving ketone-derived ketyl radicals. In such a reaction, the chiral acid co-catalyst would be required to facilitate single-electron reduction of the ketone and then remain hydrogen bonded to the resulting ketyl during a subsequent C—C bond-forming step. Ultimately, it was found that the chiral phosphoric acid **79** can indeed promote an aza-pinacol cyclization with high levels of yield and enantioselectivity (Scheme 10.15) [56]. Concerted electron and proton transfer to the ketohydrazone **80** delivers catalyst-bound ketyl radical **81**, which may undergo radical cyclization and subsequent reduction by Hantzsch dihydropyridine (HEH) to deliver the 1,2-amino alcohol derivative **82**. The reaction was found to be tolerant of various modifications to the aryl ring as well as the carbon tether.

In another example of merging photoredox with acid catalysis, Yoon has shown that trifluoroacetic acid (TFA) significantly improves the efficiency of the photoredox-catalyzed α-alkylation of amines with Michael acceptors, a reaction that proceeds via the addition of α-amino radicals such as **83** to enones, enals, or acrylates (Scheme 10.16) [57]. Evidence was obtained suggesting that the acid co-catalyst accelerates the C—C bond-forming step, presumably via protonation

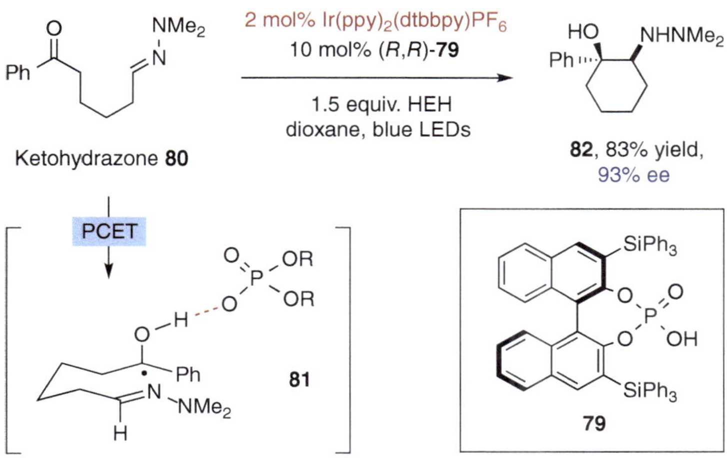

Scheme 10.15 Enantioselective aza-pinacol cyclization promoted by a chiral Brønsted acid.

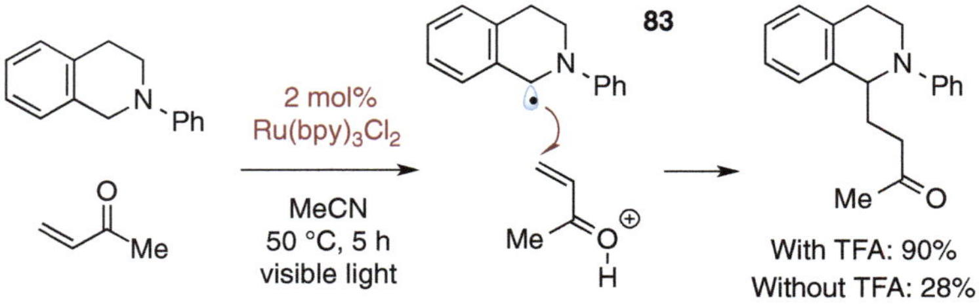

Scheme 10.16 Trifluoroacetic acid accelerates the α-alkylation of amines with Michael acceptors.

of the electrophile, an event that renders this component more reactive toward coupling with the nucleophilic α-amino radical. Thus, in contrast to the above-mentioned examples, the Bronsted acid does not facilitate an electron-transfer step but rather promotes the subsequent radical reactivity.

10.3.2 Photoredox Catalysis and Lewis Acid Catalysis

In addition to Brønsted acids, Lewis acids have also found significant utility as co-catalysts in photoredox reactions. Yoon has shown that lithium salts may be employed to catalyze photoredox transformations of bis(enones); in contrast to the Brønsted acid-promoted reductive cyclization of these compounds (see Scheme 10.13), Lewis acid co-catalysis leads to the generation of cyclobutane products (Scheme 10.17a) [58]. In the reaction, coordination of the lithium cation to the aryl enone **67** raises the reduction potential of the substrate, enabling single-electron reduction by $Ru(bpy)_3{}^+$. The resulting lithium-bound radical anion **84**, displaying markedly different reactivity compared to its neutral radical analog **69**, then undergoes [2+2] cycloaddition to yield ketyl radical **85**. Loss of an electron furnishes the cyclobutane product **86**. While these intramolecular cyclizations afford *cis*-dione products, intermolecular [2+2] cycloadditions were shown to give *trans*-diones. This use of a lithium salt as a Lewis acid to promote enone reduction has been used to achieve a range of other [2+2] and [4+2] cycloaddition reactions [59, 60]. The lithium salt is typically employed in (super)stoichiometric quantities but in some cases may be used at catalytic loadings.

Other Lewis acids have been employed to facilitate similar single-electron reduction events. Samarium(III) triflate has been used to promote the reductive dimerization of chalcones (Scheme 10.17b) [61], and lanthanum(III) triflate has been employed in the [3+2] cycloaddition of aryl cyclopropyl ketones (Scheme 10.17c) [62]. Lewis acids have additionally been employed in photoredox transformations of *N*-aryl glycine esters, with Rueping using zinc acetate to promote α-arylation with indole nucleophiles [63] and Wu using copper(II) triflate to promote oxidative Mannich reactions [64]. Magnesium perchlorate has also been employed to promote the photoredox reduction of nitro compounds to oximes [65].

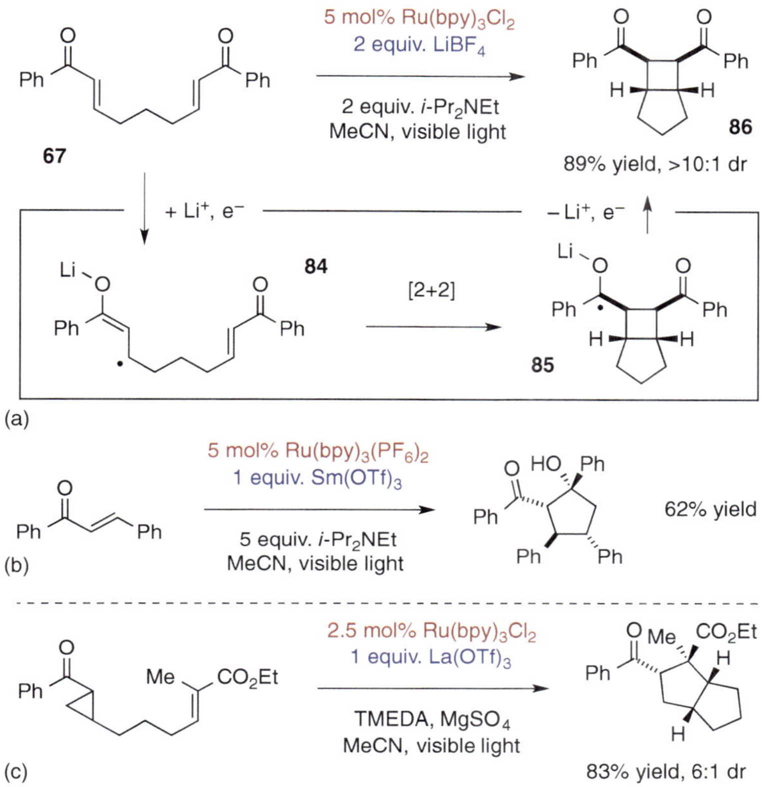

Scheme 10.17 Cyclization reactions enabled by the combination of photoredox and Lewis acid catalysis. (a) Lithium-promoted bis(enone) [2+2] cycloaddition. (b) Samarium(III)-promoted chacone dimerization. (c) Lanthanum(III)-promoted [3+2] cycloaddition.

A critical feature of the bis(enone) [2+2] cycloaddition is that single-electron reduction of the enone moiety occurs exclusively when it is bound to the lithium Lewis acid. Yoon recognized that this gating mechanism presents an opportunity for the development of an asymmetric reaction employing chiral Lewis acids. Efforts toward this goal led to the identification of chiral europium complexes, generated from the combination of $Eu(OTf)_3$ with chiral dipeptide ligands, that catalyze the highly enantioselective [2+2] intermolecular cycloaddition of aryl enones with aliphatic enones (Scheme 10.18) [66]. Furthermore, the Schiff base ligand **87** was found to preferentially provide the *trans*-dione diastereomer of the cyclobutane product, while its reduced analog **88** predominantly delivers the *cis*-dione product. Electronically diverse aryl enones may be employed. This photoredox approach to asymmetric cycloadditions—in which catalytic activation is achieved via a difference in reduction potentials between the catalyst-bound and catalyst-unbound substrate—contrasts strongly with traditional photochemical approaches to cycloadditions, in which the substrate itself must undergo photoexcitation and few methods for catalytic activation exist.

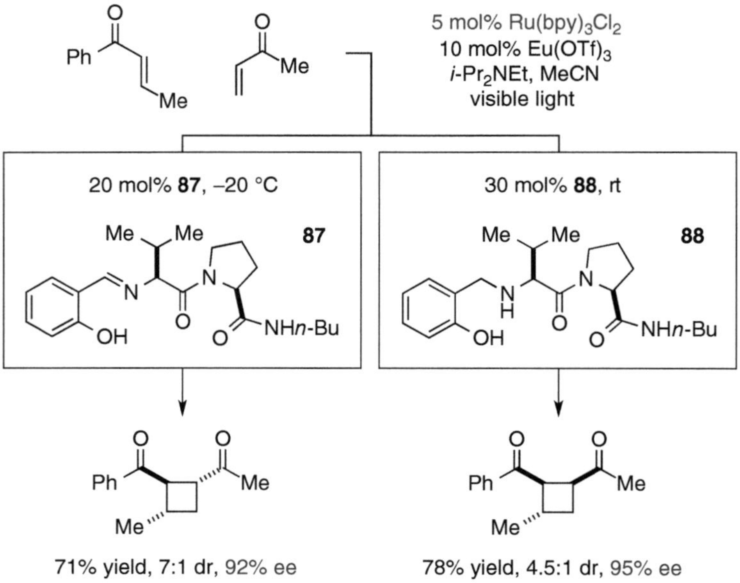

Scheme 10.18 Enantioselective [2+2] cycloadditions of enones promoted by chiral Lewis acid complexes.

10.4 Merger of Photoredox Catalysis with Transition Metal Catalysis

Many examples of synergistic photoredox catalysis employ transition metal catalysts. Transition metals may interface with photoredox catalysis in a variety of ways, but one important reaction manifold is their combination with organic radicals to give organometallic intermediates. Photoredox catalysis can thus allow commonplace functional groups to be transformed *in situ* into useful organometallic intermediates. At the same time, transition metals are typically redox active and will engage in single-electron transfer; the ability of the photoredox catalyst to alter the oxidation state of metals is also a powerful means of modulating the metal's reactivity.

An early and seminal report in this area is Sanford's merger of photoredox catalysis with palladium catalysis to perform C—H arylation reactions with aryldiazonium salts [67]. Extensive work from the Sanford group in the area of high-valent palladium catalysis has demonstrated that a variety of Lewis basic directing groups may ligate palladium(II) and facilitate the activation of proximal C—H bonds; this reaction manifold has been employed to develop a wide range of C—O, C—S, C—X, C—N, and C—C bond-forming reactions [68]. Most of these C—H functionalization reactions require the use of strong two-electron oxidants to convert Pd(II) to the Pd(IV) oxidation state, and often harsh reaction conditions [69]. To overcome these limitations, Sanford targeted one-electron chemistry as a means of accessing high oxidation state metal species under relatively mild conditions.

A strategy involving the generation of aryl radicals via photoredox chemistry was found to be effective; upon visible light irradiation of a mixture of 2-arylpyridine **89**, phenyldiazonium salt **90**, palladium(II) acetate, and $Ru(bpy)_3^{2+}$ at room temperature, the C—H arylation product **91** is obtained in good yield (Scheme 10.19) [67]. According to the proposed mechanism for this reaction, the photoexcited species $^*Ru(bpy)_3^{2+}$ donates an electron to phenyldiazonium salt **90**, generating the phenyl radical **92** with loss of dinitrogen. Simultaneously, C—H activation of 2-arylpyridine **89** generates Pd(II) intermediate **93**. The aryl radical is proposed to add to this species, oxidizing the metal by one electron to generate Pd(III) intermediate **94**. At this point, the two catalytic cycles merge, with the highly oxidizing intermediate $Ru(bpy)_3^{3+}$ removing an electron from **94**, giving the Pd(IV) species **95**. The required Pd(IV) intermediate is thus accessed via two separate one-electron oxidation events,

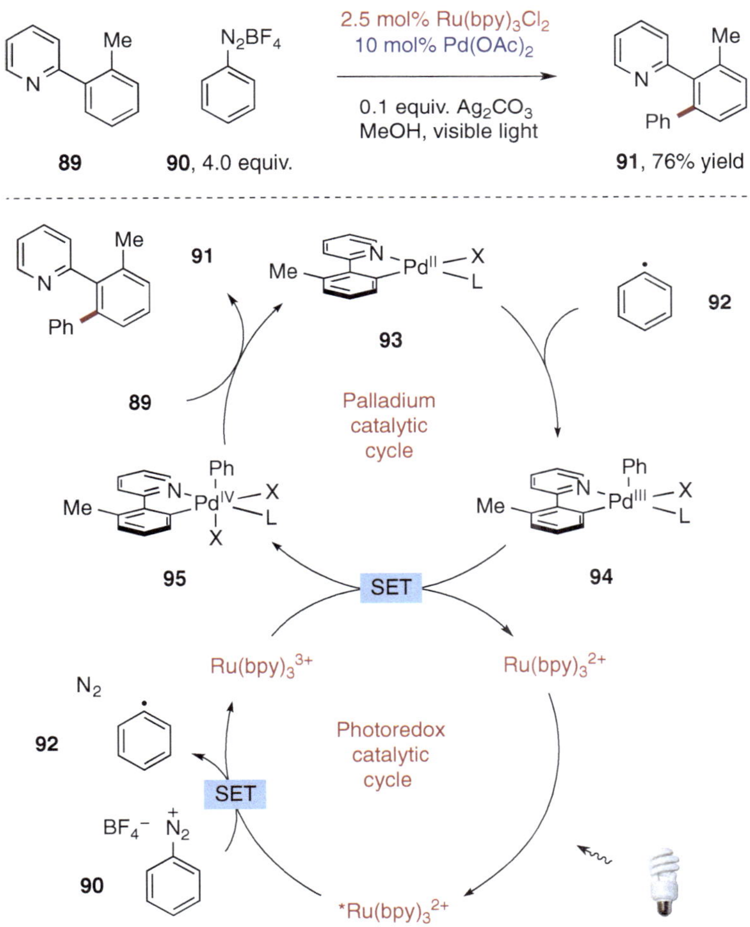

Scheme 10.19 Merger of palladium and photoredox catalysis enables arene C—H arylation with aryldiazonium salts.

rather than a single two-electron oxidation as occurs in most other examples of Pd(IV) catalysis. Reductive elimination from this high-valent palladium complex generates the biphenyl product **91** and, following C—H activation of another arylpyridine **89**, regenerates Pd(II) complex **93**. In addition to pyridines, other Lewis basic functionality including amides, pyrazoles, pyrimidines, and oximes may be employed as directing groups, and a range of electronically diverse aryl-diazonium salts are competent reaction partners. Sanford has also accomplished photoredox- and palladium-mediated C—H arylation using aryliodonium salts as precursors to aryl radicals [70].

Ye and Sanford have leveraged a similar dual photoredox/transition metal catalyst system to achieve the trifluoromethylation of arylboronic acids [71]. Recognizing that copper catalysts had previously been exploited to perform this transformation using either nucleophilic [72] or electrophilic [73] sources of the trifluoromethyl group, they hypothesized that an alternative radical approach employing copper and photoredox catalysts could be viable. Indeed, a combination of $Ru(bpy)_3{}^{2+}$ and copper(I) acetate was found to successfully promote the conversion of phenylboronic acid (**96**) to trifluoromethylbenzene (**97**) using trifluoromethyl iodide (**98**) as the $\bullet CF_3$ source (Scheme 10.20). The reaction gives only trace product in the absence of the copper catalyst, demonstrating that the reaction does not proceed via the direct interaction of the trifluoromethyl radical with the boronic acid. The reaction is proposed to proceed first via single-electron oxidation of the copper(I) species **99** ($E^{red}_{1/2} = -0.08$ V vs SCE for Cu(I)) by $^*Ru(bpy)_3{}^{2+}$ to give a copper(II) species **100** and $Ru(bpy)_3{}^{+}$. Single-electron reduction of trifluoromethyl iodide by $Ru(bpy)_3{}^{+}$ then generates the trifluoromethyl radical ($\bullet CF_3$, **101**) and regenerates the ground-state $Ru(bpy)_3{}^{2+}$. Addition of $\bullet CF_3$ to the copper(II) species **100** oxidizes the metal by a single electron to give the high-valent trifluoromethyl–copper(III) species **102**. Subsequent transmetallation with phenylboronic acid provides the aryl–copper(III) intermediate **103**, which may undergo reductive elimination to release the trifluoromethylated arene **97** and regenerate the copper(I) species. The authors alternatively suggest that the order of trifluoromethyl radical addition and transmetallation may be reversed, with transmetallation between **100** and the boronic acid giving rise to a copper(II)–aryl species which could be converted to copper(III) species **103** upon addition of the CF_3 radical. In this reaction as well as the directed C—H arylation, the photocatalyst thus plays a dual role, both generating the key radical species and also, by performing a single-electron oxidation of the metal, playing an integral part in the transition metal catalytic cycle. Both reactions also highlight the ability of synergistic photoredox/metal catalyst systems to access unusual oxidation states—such as Pd(IV) and Cu(III)—of broadly useful transition metals and thereby enable unique reactivity. In the case of the boronic acid trifluoromethylation, the dual catalytic system was found to successfully trifluoromethylate both electron-rich and electron-deficient arylboronic acids as well as boronic acids derived from heteroarenes such as pyridine and quinoline. Perfluoroalkylation may be achieved using perfluoroalkyl iodides as radical precursors.

The arylation and trifluoromethylation reactions developed by Sanford furthermore highlight the ability of dual photoredox/metal catalyst systems to

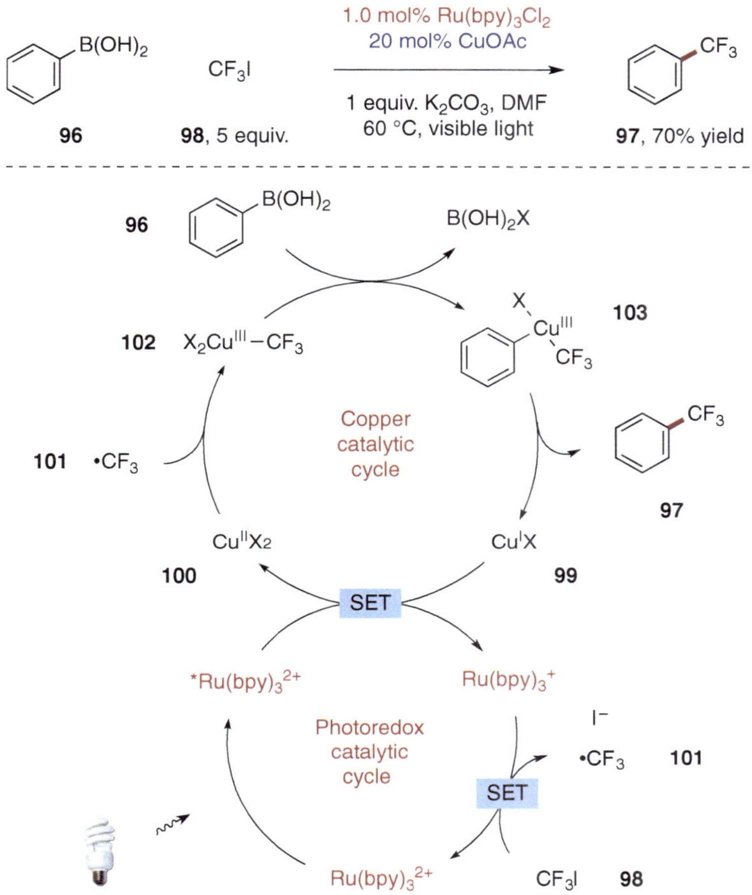

Scheme 10.20 Merger of photoredox and high-valent copper catalysis: trifluoromethylation of arylboronic acids.

perform cross-coupling reactions between an organic radical on the one hand and, on the other hand, a more established coupling partner for transition metal catalysis (boronic acid, C—H bond proximal to a directing group). In a similar manner, MacMillan and Doyle [74] as well as Molander [75] have recently employed photoredox catalysis to engage organic radicals in nickel-catalyzed cross-coupling with aryl halides. In the report from MacMillan and Doyle, photoredox and nickel catalysts were used to achieve the sp^3–sp^2 cross-coupling of α-amino acids with aryl halides (Scheme 10.21). In a representative reaction, the iridium photocatalyst $Ir[dF(CF_3)ppy]_2(dtbbpy)PF_6$ and $NiCl_2$• glyme promote the decarboxylative arylation of *N*-Boc-proline (Boc-Pro-OH) with *p*-iodotoluene (**104**) to provide benzylic amine **105**. The first step in the putative reaction pathway is single-electron oxidation of the carboxylic acid ($E_p^{red} = +0.95$ V vs SCE for Boc-Pro-OCs) [76] by the photoexcited iridium catalyst ($E_{1/2}^{*III/II} = +1.21$ V vs SCE) [77]; the resulting carboxyl radical then loses

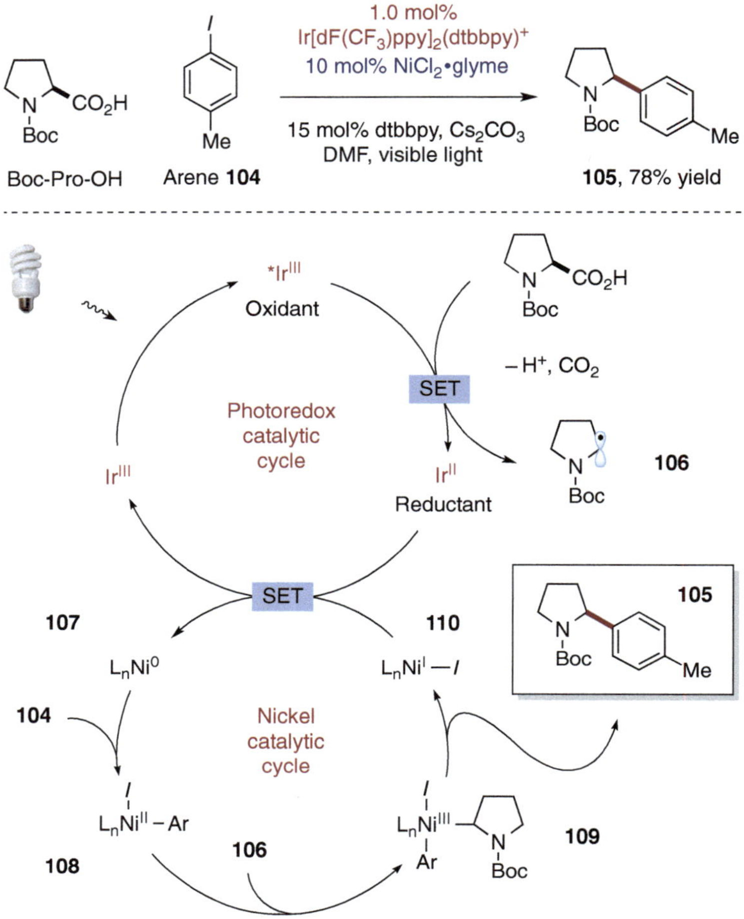

Scheme 10.21 Merger of photoredox and nickel catalysis in the decarboxylative arylation of α-amino acids.

carbon dioxide to give α-amino radical **106**. Concurrently, oxidative addition of the aryl iodide **104** to the Ni(0) complex **107** (generated via reduction of the Ni(II) catalyst precursor) furnishes Ni(II)–aryl complex **108**. Combination of α-amino radical **106** with this complex then generates the Ni(III)–aryl–alkyl species **109**, which is primed for reductive elimination of the product benzylic amine **105**. This step also releases Ni(I) species **110**, which may be reduced to Ni(0) by the Ir(II) state of the photocatalyst to turn over both the photoredox and nickel catalytic cycles. A range of aryl iodides and bromides were found to be amenable to coupling under these conditions, as were certain heteroaryl chlorides. In addition to α-amino acids, α-oxy carboxylic acids may also be employed, providing access to benzylic ether products. Furthermore, oxidizable tertiary amines may undergo arylation at an α-C—H bond; these species are converted to α-amino radicals via an oxidation/α-deprotonation pathway.

Molander's nickel/photoredox system was developed for the cross-coupling of benzylic trifluoroborates with aryl bromides, another sp^2–sp^3 cross-coupling reaction (Scheme 10.22) [75]. Conventional cross-coupling reactions of these sp^3-hybridized organometallic nucleophiles typically suffer from a slow and rate-limiting transmetallation step; a radical approach was therefore pursued that would offer an alternative one-electron mechanism of transmetallation. Organotrifluoroborates in particular were employed given their known potential to undergo single-electron oxidation to yield organic radicals [78]; thus, single-electron transfer from potassium benzyltrifluoroborate (**111**, $E_p^{red} = +1.1\,V$ vs SCE) [78] to the excited state of Ir[dF(CF_3)ppy]$_2$(bpy)PF_6 ($E_{1/2}^{*III/II} = +1.32\,V$ vs SCE) [79] generates the benzylic radical **112**. Concurrent oxidative addition of the aryl bromide **113** to the Ni(0) catalyst provides Ni(II) complex **114**, which may react with the benzylic radical to furnish diorgano–Ni(III) complex **115**. Reductive elimination provides the diaryl

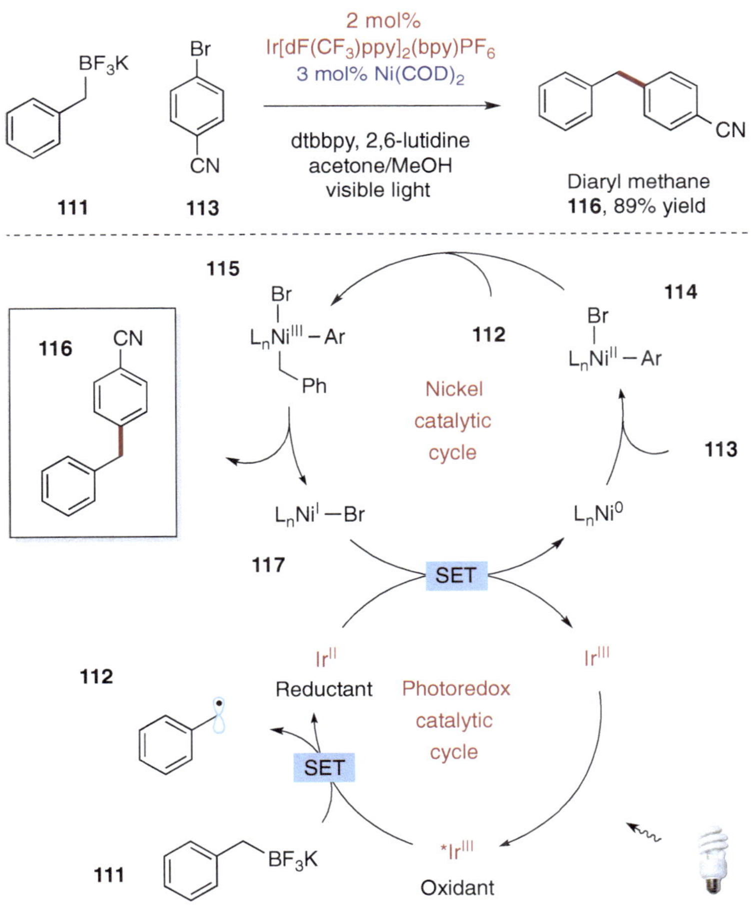

Scheme 10.22 Cross-coupling of benzylic trifluoroborates with aryl bromides promoted by nickel and photoredox catalysts.

methane product **116** and Ni(I) species **117**, which is reduced back to Ni(0) by the Ir(II) photocatalyst intermediate. Broad electronic variation is tolerated on both the trifluoroborate and aryl bromide coupling partners, and various nitrogen-containing heteroaryl bromides also couple with good efficiency. Furthermore, an attractive feature of these transition metal-catalyzed cross-coupling reactions is the potential for asymmetric induction to be achieved through the use of chiral ligands on the metal. Indeed, Molander demonstrated that a 2,2′-bis(2-oxazoline) ligand provides a moderate level of enantioselectivity (75 : 25 e.r.) in the coupling of a racemic secondary benzylic trifluoroborate salt.

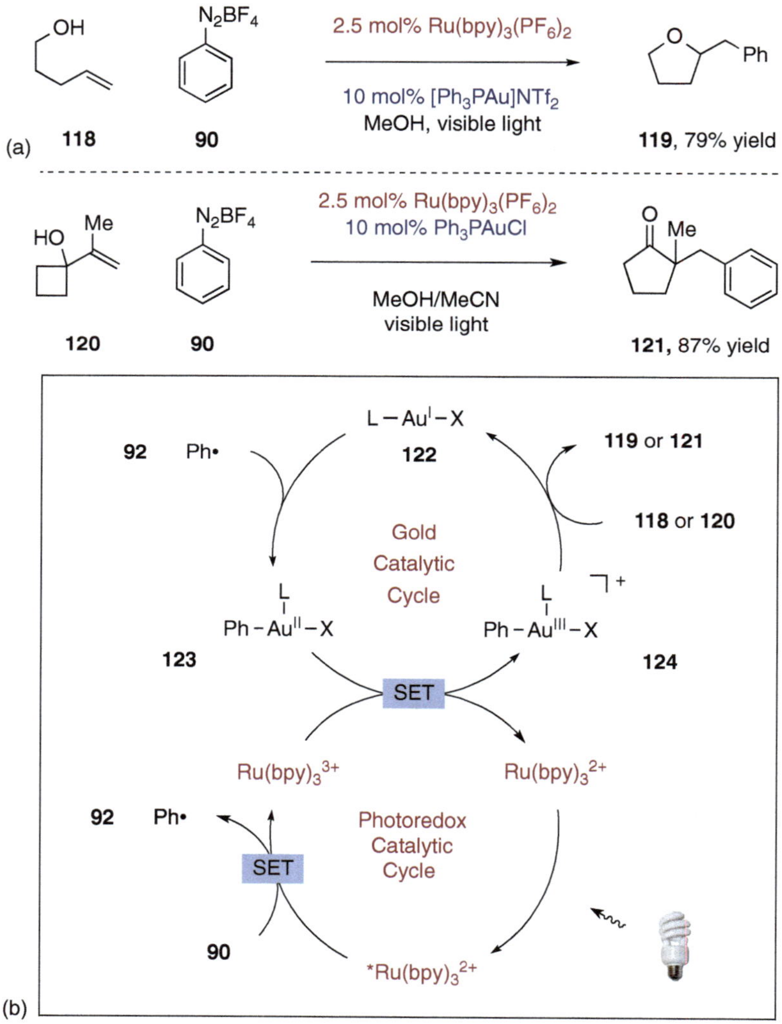

Scheme 10.23 Arylative reactions of alkenes achieved through gold and photoredox co-catalysis. (a) Oxyarylation of alkenes with aryldiazonium salts. (b) Arylative ring expansion of vinylcyclobutanols.

The stereoconvergent nature of this coupling provides further evidence for the intermediacy of a benzylic radical.

Glorius and Toste have merged the unique chemistry of gold catalysis with photoredox to achieve two different arylative transformations of alkenes with aryldiazonium salts. First, Glorius reported the oxyarylation of alkenes using $Ru(bpy)_3^{2+}$ and $[Ph_{3P}Au]NTf_2$; in a representative reaction, the alkenol **118** reacts with phenyldiazonium salt **90** to give tetrahydrofuran **119**, the product of intramolecular cyclization and intermolecular arylation [80] (Scheme 10.23a). Amine-tethered alkenes also undergo aminoarylation under these conditions to give pyrrolidine products. In Toste's reaction, under similar conditions, the vinylcyclobutanol **120** undergoes an arylative ring expansion to provide cyclopentanone **121** (Scheme 10.23b) [81]. Both of these reactions potentially proceed via the generic dual catalytic mechanism shown in Scheme 10.23. As in Sanford's C—H arylation reaction, the aryldiazonium salt **90** is reduced by $^*Ru(bpy)_3^{2+}$ to generate the aryl radical **92**. Addition of this radical to the Au(I) catalyst **122** then gives gold(II) species **123**, which may be further oxidized to cationic aryl–gold(III) complex **124** by $Ru(bpy)_3^{3+}$. This electrophilic and π-Lewis acidic gold complex may then coordinate the alkene substrate (**118** or **120**), promoting either intramolecular attack of a nucleophile (in the oxyarylation reaction) or cyclobutanol ring expansion. The resulting alkyl–aryl–gold(III) species (not shown) may then suffer reductive elimination of the organic substituents to give the products (**119** or **121**) as well as regenerate the Au(I) catalyst **122**.

Lastly, Stephenson and Rueping have both reported the α-alkynylation of tertiary amines using photoredox and copper catalysts (Scheme 10.24a) [82]. These reactions are proposed to proceed via addition of an *in situ*-generated copper acetylide to an iminium ion (generated via oxidative photocatalysis). While Stephenson employed a two-step procedure, in which the copper catalyst

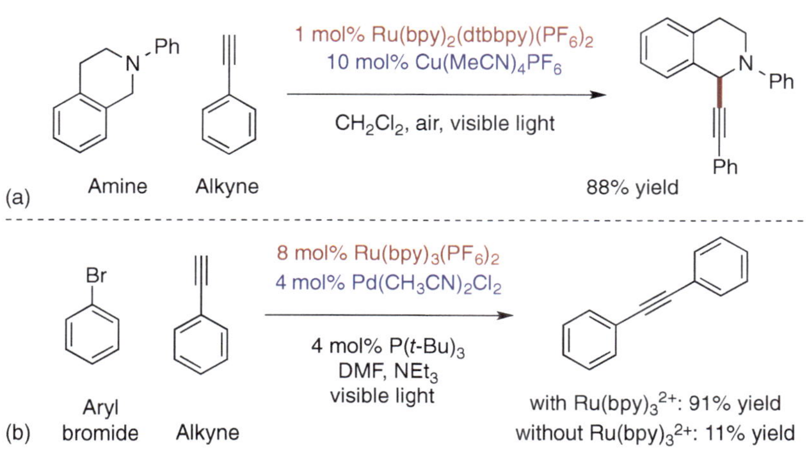

Scheme 10.24 Merger of photoredox and transition metal catalysis in alkynylation reactions. (a) Photoredox and Cu-catalyzed α-alkynylation of amines. (b) $Ru(bpy)_3^{2+}$ and Pd-catalyzed Sonogashira coupling.

and alkyne are added only after the iminium ion is formed, Rueping achieved the reaction in a single step using oxygen as the stoichiometric oxidant. This α-alkynylation has additionally been achieved using rose bengal as the photocatalyst [83]. A combination of $Ru(bpy)_3{}^{2+}$ and a palladium catalyst has also been shown to promote a copper-free Sonogashira coupling of aryl bromides with alkynes (Scheme 10.24b) [84]. The rate and efficiency of the reaction was shown to be improve upon the inclusion of the photocatalyst and light irradiation, although the mechanism of photoredox activation in this reaction has yet to be clearly determined.

10.5 Conclusions

The combination of photoredox catalysis with complementary activation modes has achieved an astonishing range of notable chemical transformations. [85–89] These reactions have demonstrated that photoredox catalysis can significantly benefit from the merger with organocatalysis, acid catalysis, and transition metal catalysis. Photoredox catalysts have even been used in concert with enzymes; in one example, alcohol dehydrogenase was employed to recycle the NADH reductant used in a photoredox reductive dehalogenation reaction [90, 91]. The examples of dual photoredox catalysis discussed in this chapter have illustrated the ability of these co-catalysts to impart enantioselectivity, activate typically inert functionality, fundamentally alter the reactivity of radical intermediates, and enable previously elusive chemical reactions. Given the wide complementarity between photoredox catalysis and other activation modes, it is likely that this topic will be a major focus of research in photoredox catalysis for years to come.

References

1 Allen, A.E. and MacMillan, D.W.C. (2012) *Chem. Sci.*, **3**, 633–658.
2 For select examples of enantioselective photoredox transformations that use 'chiral-at-metal' catalysts as both the photocatalyst and Lewis acid, see: (a) Huo, H., Shen, X., Wang, C. *et al.* (2014) *Nature*, **515**, 100–103; (b) Huo, H., Wang, C., Harms, K., and Meggers, E. (2015) *J. Am. Chem. Soc.*, **137**, 9551–9554; (c) Tan, Y., Yuan, W., Gong, L., and Meggers, E. (2015) *Angew. Chem., Int. Ed.*, **54**, 13045–13048; (d) Wang, C., Qin, J., Shen, X. *et al.* (2016) *Angew. Chem., Int. Ed.*, **55**, 685–688.
3 (a) MacMillan, D.W.C. (2008) *Nature*, **455**, 304–308; (b) Mukherjee, S., Yang, J.W., Hoffmann, S., and List, B. (2007) *Chem. Rev.*, **107**, 5471–5569.
4 Nicewicz, D.A. and MacMillan, D.W.C. (2008) *Science*, **322**, 77–80.
5 Kalyanasundaram, K. (1982) *Coord. Chem. Rev.*, **46**, 159–244.
6 Tanner, D.D. and Singh, H.K. (1986) *J. Org. Chem.*, **51**, 5182–5186.
7 Wayner, D.D.M., Dannenberg, J.J., and Griller, D. (1986) *Chem. Phys. Lett.*, **131**, 189–191.

8 Nagib, D.A., Scott, M.E., and MacMillan, D.W.C. (2009) *J. Am. Chem. Soc.*, **131**, 10875–10877.
9 Slinker, J.D., Gorodetsky, A.A., Lowry, M.S. *et al.* (2004) *J. Am. Chem. Soc.*, **126**, 2763–2767.
10 Bonesi, S.M. and Erra-Balsells, R. (2000) *J. Chem. Soc. Perkin Trans. 2*, 1583–1595.
11 Shih, H.-W., Vander Wal, M.N., Grange, R.L., and MacMillan, D.W.C. (2010) *J. Am. Chem. Soc.*, **132**, 13600–13603.
12 Neumann, M., Füldner, S., König, B., and Zeitler, K. (2011) *Angew. Chem. Int. Ed.*, **50**, 951–954.
13 For an example of this transformation promoted by rose bengal, see: Fidaly, K., Ceballos, C., Falguières, A. *et al.* (2012) *Green Chem.*, **14**, 1293–1297.
14 Cherevatskaya, M., Neumann, M., Füldner, S. *et al.* (2012) *Angew. Chem. Int. Ed.*, **51**, 4062–4066.
15 Cecere, G., König, C.M., Alleva, J.L., and MacMillan, D.W.C. (2013) *J. Am. Chem. Soc.*, **135**, 11521–11524.
16 Koike, T. and Akita, M. (2009) *Chem. Lett.*, **38**, 166–167.
17 Um, J.M., Gutierrez, O., Schoenebeck, F. *et al.* (2010) *J. Am. Chem. Soc.*, **132**, 6001–6005.
18 For examples of asymmetric catalysis via enamine radical cation intermediates, see: Beeson, T.D., Mastracchio, A., Hong, J.-B. *et al.* (2007) *Science*, **316**, 582–585.
19 For a related reaction employing photoredox catalysis with stoichiometric enamines, see: Yasu, Y., Koike, T., and Akita, M. (2012) *Chem. Commun.*, **48**, 5355–5357.
20 For a similar α-oxyamination reaction performed in tandem with a Michael addition, see: Yoon, H.-S., Ho, X.-H., Jang, J. *et al.* (2012) *Org. Lett.*, **14**, 3272–3275.
21 Pirnot, M.T., Rankic, D.A., Martin, D.B.C., and MacMillan, D.W.C. (2013) *Science*, **339**, 1593–1596.
22 Flamigni, L., Barbieri, A., Sabatini, C. *et al.* (2007) *Top. Curr. Chem.*, **281**, 143–203.
23 Wakasa, M., Sakaguchi, Y., Nakamura, J., and Hayashi, H. (1992) *J. Phys. Chem.*, **96**, 9651–9656.
24 For reviews on the persistent radical effect: (a) Fischer, H. (2001) *Chem. Rev.*, **101**, 3581–3610; (b) Studer, A. (2001) *Chem. Eur. J.*, **7**, 1159–1164; (c) Focsaneanu, K.-S. and Scaiano, J.C. (2006) *Helv. Chim. Acta*, **89**, 2473–2482.
25 Petronijević, F.R., Nappi, M., and MacMillan, D.W.C. (2013) *J. Am. Chem. Soc.*, **135**, 18323–18326.
26 Wagner, P.J., Truman, R.J., Puchalski, A.E., and Wake, R. (1986) *J. Am. Chem. Soc.*, **108**, 7727–7738.
27 Terrett, J.A., Clift, M.D., and MacMillan, D.W.C. (2014) *J. Am. Chem. Soc.*, **136**, 6858–6861.
28 Rueping, M., Vila, C., Koenigs, R.M. *et al.* (2011) *Chem. Commun.*, **47**, 2360–2362.

29 Feng, Z.-J., Xuan, J., Xia, X.-D. *et al.* (2014) *Org. Biomol. Chem.*, **12**, 2037–2040.
30 Su, Y., Zhang, L., and Jiao, N. (2011) *Org. Lett.*, **13**, 2168–2171.
31 (a) Moore, J.L. and Rovis, T. (2010) *Top. Curr. Chem.*, **291**, 118–144; (b) Enders, D., Niemeier, O., and Henseler, A. (2007) *Chem. Rev.*, **107**, 5606–5655.
32 DiRocco, D.A. and Rovis, T. (2012) *J. Am. Chem. Soc.*, **134**, 8094–8097.
33 Bock, C.R., Connor, J.A., Gutierrez, A.R. *et al.* (1979) *J. Am. Chem. Soc.*, **101**, 4815–4824.
34 Bergonzini, G., Schindler, C.S., Wallentin, C.-J. *et al.* (2014) *Chem. Sci.*, **5**, 112–116.
35 For reviews on anion-binding catalysis, see: (a) Brak, K. and Jacobsen, E.N. (2013) *Angew. Chem. Int. Ed.*, **52**, 534–561; (b) Phipps, R.J., Hamilton, G.L., and Toste, F.D. (2012) *Nature Chem.*, **4**, 603–614.
36 Hamilton, D.S. and Nicewicz, D.A. (2012) *J. Am. Chem. Soc.*, **134**, 18577–18580.
37 Fukuzumi, S. and Ohkubo, K. (2013) *Chem. Sci.*, **4**, 561–574.
38 Ohkubo, K., Mizushima, K., Iwata, R. *et al.* (2010) *Chem. Commun.*, **46**, 601–603.
39 Arnold, D.R., Chan, M.S.W., and McManus, K.A. (1996) *Can. J. Chem.*, **74**, 2143–2166.
40 Bordwell, F.G., Cheng, J.-P., Ji, G.-Z. *et al.* (1991) *J. Am. Chem. Soc.*, **113**, 9790–9795.
41 Nguyen, T.M. and Nicewicz, D.A. (2013) *J. Am. Chem. Soc.*, **135**, 9588–9591.
42 Larsen, A.G., Holm, A.H., Roberson, M., and Daasbjerg, K. (2001) *J. Am. Chem. Soc.*, **123**, 1723–1729.
43 Nguyen, T.M., Manohar, N., and Nicewicz, D.A. (2014) *Angew. Chem. Int. Ed.*, **53**, 6198–6201.
44 Perkowski, A.J. and Nicewicz, D.A. (2013) *J. Am. Chem. Soc.*, **135**, 10334–10337.
45 Wilger, D.J., Gesmundo, N.J., and Nicewicz, D.A. (2013) *Chem. Sci.*, **4**, 3160–3165.
46 Grandjean, J.-M.M. and Nicewicz, D.A. (2013) *Angew. Chem. Int. Ed.*, **52**, 3967–3971.
47 Qvortrup, K., Rankic, D.A., and MacMillan, D.W.C. (2014) *J. Am. Chem. Soc.*, **136**, 626–629.
48 Escoubet, S., Gastaldi, S., Vanthuyne, N. *et al.* (2006) *J. Org. Chem.*, **71**, 7288–7292.
49 Tumanov, V.E., Kromkin, E.A., and Denisov, E.T. (2002) *Russ. Chem. Bull. Int. Ed.*, **51**, 1641–1650.
50 Roberts, B.P. (1999) *Chem. Soc. Rev.*, **28**, 25–35.
51 For a related reaction in which a stoichiometric reagent performs hydrogen atom abstraction, see: Dai, C., Meschini, F., Narayanam, J.M.R., and Stephenson, C.R.J. (2012) *J. Org. Chem.*, **77**, 4425–4431.
52 Du, J., Espelt, L.R., Guzei, I.A., and Yoon, T.P. (2011) *Chem. Sci.*, **2**, 2115–2119.

53 Neumann, M. and Zeitler, K. (2013) *Chem. Eur. J.*, **19**, 6950–6955.
54 Tarantino, K.T., Liu, P., and Knowles, R.R. (2013) *J. Am. Chem. Soc.*, **135**, 10022–10025.
55 Martens, F.M., Verhoeven, J.W., Gase, R.A. *et al.* (1978) *Tetrahedron*, **34**, 443–446.
56 Rono, L.J., Yayla, H.G., Wang, D.Y. *et al.* (2013) *J. Am. Chem. Soc.*, **135**, 17735–17738.
57 Espelt, L.R., Wiensch, E.M., and Yoon, T.P. (2013) *J. Org. Chem.*, **78**, 4107–4114.
58 Ischay, M.A., Anzovino, M.E., Du, J., and Yoon, T.P. (2008) *J. Am. Chem. Soc.*, **130**, 12886–12887.
59 (a) Du, J. and Yoon, T.P. (2009) *J. Am. Chem. Soc.*, **131**, 14604–14605; (b) Hurtley, A.E., Cismesia, M.A., Ischay, M.A., and Yoon, T.P. (2011) *Tetrahedron*, **67**, 4442–4448; (c) Tyson, E.L., Farney, E.P., and Yoon, T.P. (2012) *Org. Lett.*, **14**, 1110–1113.
60 Yasu *et al.* have similarly employed lithium tetrafluoroborate to promote single-electron reduction of duroquinone by $^{*}Ru(bpy)_3{}^{2+}$; see reference 18.
61 Zhao, G., Yang, C., Guo, L. *et al.* (2012) *J. Org. Chem.*, **77**, 6302–6306.
62 Lu, Z., Shen, M., and Yoon, T.P. (2011) *J. Am. Chem. Soc.*, **133**, 1162–1164.
63 Zhu, S. and Rueping, M. (2012) *Chem. Commun.*, **48**, 11960–11962.
64 Gao, X.-W., Meng, Q.-Y., Xiang, M. *et al.* (2013) *Adv. Synth. Catal.*, **355**, 2158–2164.
65 Cai, S., Zhang, S., Zhao, Y., and Wang, D.Z. (2013) *Org. Lett.*, **15**, 2660–2663.
66 Du, J., Skubi, K.L., Schultz, D.M., and Yoon, T.P. (2014) *Science*, **344**, 392–396.
67 Kalyani, D., McMurtrey, K.B., Neufeldt, S.R., and Sanford, M.S. (2011) *J. Am. Chem. Soc.*, **133**, 18566–18569.
68 (a) Lyons, T.W. and Sanford, M.S. (2010) *Chem. Rev.*, **110**, 1147–1169; (b) Hickman, A.J. and Sanford, M.S. (2012) *Nature*, **484**, 177–185.
69 For instance, arene C–H arylation with aryldiazonium salts proceeds at 100 °C: (a) Kalyani, D., Deprez, N.R., Desai, L.V., and Sanford, M.S. (2005) *J. Am. Chem. Soc.*, **127**, 7330–7331; (b) Deprez, N.R. and Sanford, M.S. (2009) *J. Am. Chem. Soc.*, **131**, 11234–11241.
70 Neufeldt, S.R. and Sanford, M.S. (2012) *Adv. Synth. Catal.*, **354**, 3517–3522.
71 Ye, Y. and Sanford, M.S. (2012) *J. Am. Chem. Soc.*, **134**, 9034–9037.
72 (a) Chu, L. and Qing, F.-L. (2010) *Org. Lett.*, **12**, 5060–5063; (b) Senecal, T.D., Parsons, A.T., and Buchwald, S.L. (2011) *J. Org. Chem.*, **76**, 1174–1176; (c) Jiang, X., Chu, L., and Qing, F.-L. (2012) *J. Org. Chem.*, **77**, 1251–1257.
73 (a) Xu, J., Luo, D.-F., Xiao, B. *et al.* (2011) *Chem. Commun.*, **47**, 4300–4302; (b) Liu, T. and Shen, Q. (2011) *Org. Lett.*, **13**, 2342–2345.
74 Zuo, Z., Ahneman, D., Chu, L. *et al.* (2014) *Science*. doi: 10.1126/science.1255525
75 Tellis, J.C., Primer, D.N., and Molander, G.A. (2014) *Science*, **345**, 433–436.
76 Zuo, Z. and MacMillan, D.W.C. (2014) *J. Am. Chem. Soc.*, **136**, 5257–5260.
77 Lowry, M.S., Goldsmith, J.I., Slinker, J.D. *et al.* (2005) *Chem. Mater.*, **17**, 5712–5719.
78 Yasu, Y., Koike, T., and Akita, M. (2012) *Adv. Synth. Catal.*, **354**, 3414–3420.

79 Hanss, D., Freys, J.C., Bernardinelli, G., and Wenger, O.S. (2009) *Eur. J. Inorg. Chem.*, 4850–4859.

80 Sahoo, B., Hopkinson, M.N., and Glorius, F. (2013) *J. Am. Chem. Soc.*, **135**, 5505–5508.

81 Shu, X.-z., Zhang, M., He, Y. *et al.* (2014) *J. Am. Chem. Soc.*, **136**, 5844–5847.

82 (a) Freeman, D.B., Furst, L., Condie, A.G., and Stephenson, C.R.J. (2012) *Org. Lett.*, **14**, 94–97; (b) Rueping, M., Koenigs, R.M., Poscharny, K. *et al.* (2012) *Chem. Eur. J.*, **18**, 5170–5174.

83 Fu, W., Guo, W., Zou, G., and Xu, C. (2012) *J. Fluorine Chem.*, **140**, 88–94.

84 Osawa, M., Nagai, H., and Akita, M. (2007) *Dalton Trans.*, 827–829.

85 For additional reading of primary literature on about the mechanism of Ni-Ir metallophotoredox catalysis see: (a) Skubi, K.L., Blum, T.R., and Yoon, T.P. (2016) *Chem. Rev.*, **116**, 10035–10074; (b) Levin, M.D., Kim, S., and Toste, F.D. (2016) *ACS Cent. Sci.*, **2**, 293–301; (c) Hopkinson, M.N., Sahoo, B., Li, J.-L., and Glorius, F. (2014) *Chem. Eur. J.*, **20**, 3874–3886; (d) Lang, X., Zhao, J., and Chen, X. (2016) *Chem. Soc. Rev.*, **45**, 3026–3038; (e) Tóth, B.L., Tischler, O., and Novák, Z. (2016) *Tetrahedron Lett.*, **57**, 4505–4513.

86 For additional reading regarding the mechanism of Ni-Ir metallaphotoredox catalysis, see: (a) Gutierrez, O., Tellis, J.C., Primer, D.N. *et al.* (2015) *J. Am. Chem. Soc.*, **137**, 4896–4899; (b) Oderinde, M.S., Varela-Alvarez, A., Aquila, B. *et al.* (2015) *J. Org. Chem.*, **80**, 7642–7651; (c) Terrett, J.A., Cuthbertson, J.D., Shurtleff, V.W., and MacMillan, D.W.C. (2015) *Nature*, **524**, 330–334; (d) Heitz, D.R., Tellis, J.C., and Molander, G.A. (2016) *J. Am. Chem. Soc.*, **138**, 12715–12718; (e) Shields, B.J. and Doyle, A.G. (2016) *J. Am. Chem. Soc.*, **138**, 12719–12722; (f) Welin, E.R., Le, C., Arias-Rotondo, D.M. *et al.* (2017) *Science*, **355**, 380–385.

87 For additional reading on new methodologies employing Ni-Ir metallaphotoredox catalysis see: (a) Noble, A., McCarver, S.J., and MacMillan, D.W.C. (2015) *J. Am. Chem. Soc.*, **137**, 624–627; (b) Primer, D.N., Karakaya, I., Tellis, J.C., and Molander, G.A. (2015) *J. Am. Chem. Soc.*, **137**, 2195–2198; (c) Chu, L., Lipshultz, J.M., and MacMillan, D.W.C. (2015) *Angew. Chem. Int. Ed.*, **54**, 7929–7933; (d) Karakaya, I., Primer, D.N., and Molander, G.A. (2015) *Org. Lett.*, **17**, 3294–3297; (e) Tasker, S.Z. and Jamison, T.F. (2015) *J. Am. Chem. Soc.*, **137**, 9531–9534; (f) Arendt, K.M. and Doyle, A.G. (2015) *Angew. Chem. Int. Ed.*, **54**, 9876–9880; (g) "Chip" Le, C.C. and MacMillan, D.W.C. (2015) *J. Am. Chem. Soc.*, **137**, 11938–11941; (h) Yamashita, Y., Tellis, J.C., and Molander, G.A. (2015) *Proc. Natl. Acad. Sci.*, **112**, 12026–12029; (i) El Khatib, M., Serafim, R.A.M., and Molander, G.A. (2016) *Angew. Chem. Int. Ed.*, **55**, 254–258; (j) Ryu, D., Primer, D.N., Tellis, J.C., and Molander, G.A. (2016) *Chem. Eur. J.*, **22**, 120–123; (k) Jouffroy, M., Primer, D.N., and Molander, G.A. (2016) *J. Am. Chem. Soc.*, **138**, 475–478; (l) Luo, J. and Zhang, J. (2016) *ACS Catal.*, **6**, 873–877; (m) Zuo, Z., Cong, H., Li, W. *et al.* (2016) *J. Am. Chem. Soc.*, **138**, 1832–1835; (n) Amani, J., Sodagar, E., and Molander, G.A. (2016) *Org. Lett.*, **18**, 732–735; (o) Jouffroy, M., Kelly, C.B., and Molander, G.A. (2016) *Org. Lett.*, **18**, 876–879; (p) Patel, N.R., Kelly, C.B., Jouffroy, M., and Molander, G.A. (2016) *Org. Lett.*, **18**, 764–767; (q) Joe, C.L. and Doyle, A.G. (2016) *Angew. Chem. Int. Ed.*, **55**, 4040–4043;

(r) Jouffroy, M., Davies, G.H.M., and Molander, G.A. (2016) *Org. Lett.*, **18**, 1606–1609; (s) Shaw, M.H., Shurtleff, V.W., Terrett, J.A. *et al.* (2016) *Science*, **352**, 1304–1308; (t) Karimi-Nami, R., Tellis, J.C., and Molander, G.A. (2016) *Org. Lett.*, **18**, 2572–2575; (u) Tellis, J.C., Amani, J., and Molander, G.A. (2016) *Org. Lett.*, **18**, 2994–2997; (v) Corcoran, E.B., Pirnot, M.T., Lin, S. *et al.* (2016) *Science*, **353**, 279–283; (w) Johnston, C.P., Smith, R.T., Allmendinger, S., and MacMillan, D.W.C. (2016) *Nature*, **536**, 322–325; (x) Patel, N.R. and Molander, G.A. (2016) *J. Org. Chem.*, **81**, 7271–7275; (y) Zhang, X. and MacMillan, D.W.C. (2016) *J. Am. Chem. Soc.*, **138**, 13862–13865; (z) Ahneman, D.T. and Doyle, A.G. (2016) *Chem. Sci.*, **7**, 7002–7006; (aa) Gutiérrez-Bonet, Á., Tellis, J.C., Matsui, J.K. *et al.* (2016) *ACS Catal.*, **6**, 8004–8008; (ab) Vara, B.A., Jouffroy, M., and Molander, G.A. (2017) *Chem. Sci.*, **8**, 530–535; (ac) Amani, J. and Molander, G.A. (2017) *J. Org. Chem.*, **82**, 1856–1863; (ad) Matsui, J.K. and Molander, G.A. (2017) *Org. Lett.*, **19**, 436–439.

88 For additional reading on dual gold-photoredox catalysis see: (a) He, Y., Wu, H., and Toste, F.D. (2015) *Chem. Sci.*, **6**, 1194–1198; (b) Patil, D.V., Yun, H., and Shin, S. (2015) *Adv. Synth. Catal.*, **357**, 2622–2628; (c) Kim, S., Rojas-Martin, J., and Toste, F.D. (2016) *Chem. Sci.*, **7**, 85–88; (d) Tlahuext-Aca, A., Hopkinson, M.N., Sahoo, B., and Glorius, F. (2016) *Chem. Sci.*, **7**, 89–93; (e) Um, J., Yun, H., and Shin, S. (2016) *Org. Lett.*, **18**, 484–487; (f) Zhang, Q., Zhang, Z.-Q., Fu, Y., and Yu, H.-Z. (2016) *ACS Catal.*, **6**, 798–808; (g) Tlahuext-Aca, A., Hopkinson, M.N., Garza-Sanchez, R.A., and Glorius, F. (2016) *Chem. Eur. J.*, **22**, 5909–5913.

89 For additional reading on dual photoredox/transition metal catalysis involving palladium, rhodium or ruthenium see: (a) Lang, S.B., O'Nele, K.M., and Tunge, J.A. (2014) *J. Am. Chem. Soc.*, **136**, 13606–13609; (b) Xuan, J., Zeng, T.-T., Feng, Z.-J. *et al.* (2015) *Angew. Chem. Int. Ed.*, **54**, 1625–1628; (c) Ghosh, T., Slanina, T., and König, B. (2015) *Chem. Sci.*, **6**, 2027–2034; (d) Fabry, D.C., Zoller, J., Raja, S., and Rueping, M. (2014) *Angew. Chem. Int. Ed.*, **53**, 10228–10231; (e) Fabry, D.C., Ronge, M.A., Zoller, J., and Rueping, M. (2015) *Angew. Chem. Int. Ed.*, **54**, 2801–2805.

90 Maidan, R. and Willner, I. (1986) *J. Am. Chem. Soc.*, **108**, 1080–1082.

91 For related uses of photoredox in enzymatic catalysis, see: (a) Mandler, D. and Willner, I. (1984) *J. Am. Chem. Soc.*, **106**, 5352–5353; (b) Mandler, D. and Willner, I. (1986) *J. Chem. Soc. Perkin Trans. 2*, 805–811; (c) Mandler, D. and Willner, I. (1986) *J. Chem. Soc. Chem. Commun.*, 851–853.

11

Enantioselective Photocatalysis

Susannah C. Coote and Thorsten Bach

Technische Universität München, Lehrstuhl für Organische Chemie I, Lichtenbergstraße 4, 85747 Garching bei München, Germany

11.1 Introduction

The ability to carry out chemical reactions using only a light source offers numerous advantages to the synthetic chemist, especially from the point of view of sustainable synthesis: waste is minimized, and light is readily available. However, progress in the field of enantioselective photochemistry [1, 2] has lagged far behind the great advances made in ground-state asymmetric synthesis. In particular, the subdiscipline of enantioselective photocatalysis has, until recently, received very little attention [3]. The dearth of catalytic asymmetric photoreactions is especially unfortunate, since photochemical reactions often offer complementary reactivity to that of ground-state processes, allowing access to complex targets that are not accessible using "classical" methods [4, 5]. Although chiral auxiliaries have been widely used in photochemical reactions for the control of absolute configuration, this method is somewhat inefficient in that extra synthetic steps are required to introduce and remove the auxiliary. Equally, the most efficient asymmetric syntheses would ideally employ only substoichiometric amounts of a chiral source, as part of a catalytic process in which substantial chiral amplification can be expected.

Catalytic enantioselective photoreactions are especially challenging, for a number of reasons. In general, the inherent difficulties in controlling the conformations of excited-state intermediates have not yet been adequately addressed. In order to develop a successful enantioselective photocatalytic reaction, one must ensure that light is selectively absorbed by the desired substrate/intermediate, and not by other species that may be present (e.g., uncomplexed substrate, or reaction products). Equally, excited-state intermediates are characterized by high reactivity, short lifetimes, and weak inter- and intramolecular interactions, which means that effective chirality transfer is often less efficient in photochemical reactions than in the corresponding ground-state methods. In the high-energy excited states involved in photochemical reactions, nonadiabatic processes can run through surface crossings unaffected by very small differences in energy. Thus, the selective production of energetically differentiated

Visible Light Photocatalysis in Organic Chemistry, First Edition.
Edited by Corey R. J. Stephenson, Tehshik P. Yoon and David W. C. MacMillan.

diastereoisomeric transition states is a considerable challenge; and for a long time, it seemed that *catalytic* photoreactions that proceed with high levels of enantioselection were simply not possible. Added to the intrinsic difficulties mentioned, the majority of practicing organic chemists lack appropriate education and/or experience in photochemical methods. As a result, photochemistry has remained somewhat overlooked as a discipline, such that relatively few researchers have been inspired to move into the area of asymmetric photocatalysis. However, a number of recent groundbreaking advances in enantioselective photocatalysis mean that these methods are now becoming competitive with the corresponding ground-state processes, such that more widespread adoption of enantioselective photochemical reactions can now be envisaged.

Herein, we provide an overview of the state of the art in enantioselective photocatalysis from the perspective of the synthetic organic chemist. In the first part, we provide a summary of the most important pioneering contributions to enantioselective photocatalysis, followed by a discussion of more recent developments in the field, in which we have placed particular emphasis on processes that show potential for wider use in asymmetric synthesis. We have restricted our account to enantioselective photocatalysis in solution, focusing on catalysts that combine chirality with photocatalytic activity. Other associated processes, in which the photocatalytic and enantiodifferentiating steps are decoupled, are described elsewhere in this book. For discussions of advances in other related areas (e.g., the use of supramolecular assemblies or biological catalysts, or solid-state photocatalysis), the interested reader is directed to the wide range of reviews and monographs dedicated to the various areas of enantioselective photochemistry [1, 2].

11.2 The Twentieth Century: Pioneering Work

One of the most attractive approaches to asymmetric photocatalysis involves the application of circularly polarized light (CPL), since, in theory, no extra chiral information would be required to induce asymmetric induction. Such *absolute asymmetric synthesis* [6] enjoyed a great deal of attention during the 1970s [7] amid widespread interest in the proposed role of CPL in the origin of optically active biomolecules in Nature. However, attempts to employ CPL to induce asymmetric induction in synthetically useful reactions led invariably to extremely low enantioselectivity, as a consequence of the generally very small differences in absorption of (+)-CPL and (−)-CPL by organic molecules. This method is therefore no longer considered to hold potential in the context of synthetic enantioselective photochemistry, and is not discussed in further detail.

In 1965, Hammond and Cole studied the isomerization of *trans*-diphenylcyclopropane **1**, sensitized by an optically active naphthalene derivative **2** [8]. This study represented the first catalytic enantioselective photoreaction, and paved the way for a number of further studies of this system. Starting with racemic *trans*-diphenylcyclopropane **1**, irradiation in benzene in the presence of amide **2** established the photostationary state, at which a *cis*:*trans* ratio of 1 : 1.03 was observed, with the chiral *trans* isomer displaying an enantiomeric excess (ee) of 7% (Scheme 11.1). The isomerization was thought to proceed through

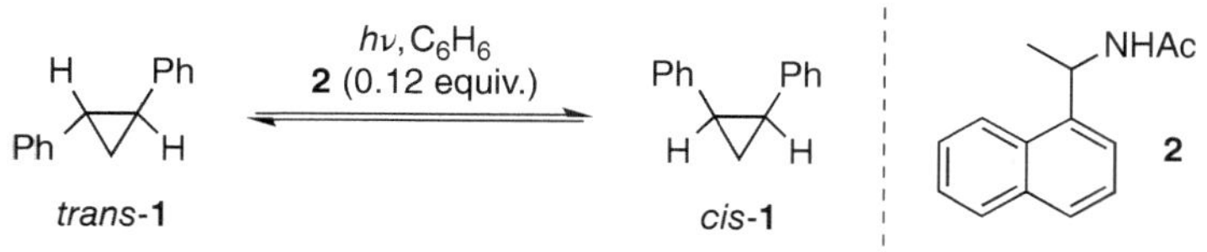

Scheme 11.1 Isomerization of *trans*-diphenylcyclopropane (**1**).

diastereoisomeric exciplexes, formed from the excited state of amide **2** (which was shown to act as a singlet sensitizer) and the diphenylcyclopropane, with the enantioselectivity arising as a result of the diastereoisomeric exciplexes having different lifetimes.

Since this seminal report, a number of other groups studied the reaction in more depth, culminating in a clearer understanding of the processes involved. Ouannès *et al.* employed chiral ketone **3** (Figure 11.1) as a triplet sensitizer [9], noting that *trans*-diphenylcyclopropane has an unusually low triplet energy (ca. 222 kJ mol^{-1}) [10]. Employing 33 mol% **3**, the authors observed results broadly similar to those obtained by Hammond and Cole, with an enantioselectivity of 3% ee observed under the optimal conditions. In a similar manner, Vondenhof and Mattay reported the use of 1,1′-binaphthalene-2,2′-dicarbonitrile (**4**) (Figure 11.1) as a chiral photoinduced electron transfer (PET) sensitizer for the same reaction [11]. Starting with *cis*-**1**, the photostationary state was reached after 19 h of irradiation in the presence of 10 mol% **4**, generating a 67 : 33 mixture of achiral *cis*- and chiral *trans*-diphenylcyclopropanes. In this case, the isomerization was proposed to proceed through a radical-ion pair produced via PET. In common with the related attempts described, only a low enantioselectivity of 4% ee was observed. More recently, Inoue and coworkers discovered that employing menthyl pyromellitate **5** as a photosensitizer, at room temperature in pentane, a slightly higher enantioselectivity (10.5% ee) can be induced in this isomerization reaction [12].

Along similar lines, Kagan and coworkers studied the optical photoactivation of racemic sulfoxide **6** in the presence of naphthalene derivative **2** [13]. It had been previously demonstrated that sulfoxides may be photoracemized in the presence of naphthalene [14]; thus, it was hoped that irradiating racemic **6** in the presence of a chiral sensitizer might lead to the enrichment of one of the sulfoxide enantiomers. In the event, irradiation of a solution of **6** in diethyl ether containing 20 mol% **2** furnished **6** in 4% ee (Scheme 11.2). The deracemization was thought

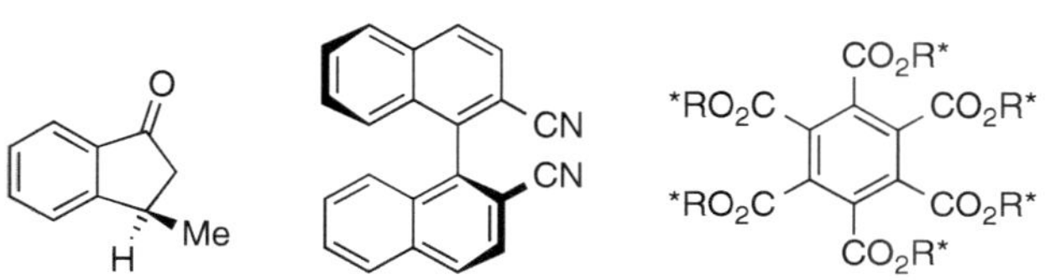

3 **4** **5** (R* = (−)-menthyl)

Figure 11.1 Asymmetric photosensitizers used for the isomerization of *trans*-diphenylcyclopropane.

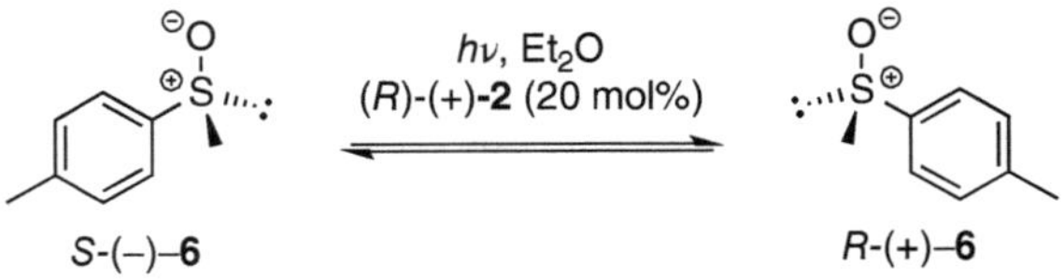

Scheme 11.2 Photoderacemization of 1-methyl-4-(methylsulfinyl)benzene (**6**).

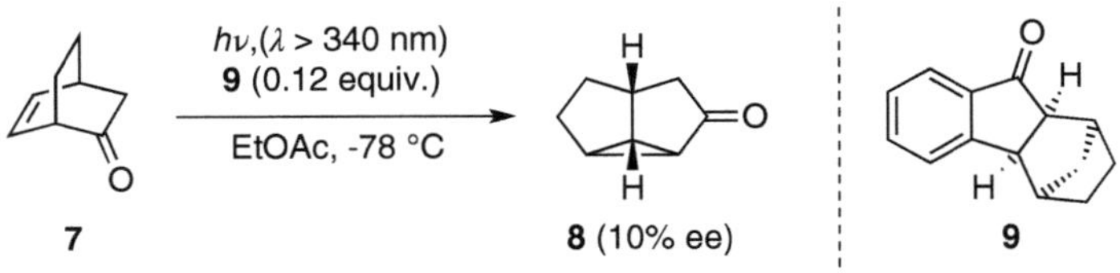

Scheme 11.3 Enantioselective oxadi-π-methane utilizing chiral ketone (**9**) as the catalyst.

to proceed via diastereoisomeric exciplexes formed from the excited singlet state of the chiral sensitizer **2** and the ground-state sulfoxide substrate.

Demuth *et al* described the first enantioselective oxadi-π-methane rearrangement, using a chiral ketone **9** as a triplet sensitizer [15]. Employing a catalytic amount of **9** (0.12 equiv.) at low temperature, the expected tricyclic photoproduct **8** was formed from the precursor **7** with 10% ee (±3%; Scheme 11.3). At room temperature, a lower enantioselectivity was observed.

The photoisomerization of a chiral *Z*-cyclooctene into chiral *E*-cyclooctene has enjoyed a great deal of attention, particularly in the Inoue research group. In 1978, Inoue *et al.* published the first example of the use of aromatic ester singlet sensitizers for the isomerization of a simple cyclic alkene [16]. Starting with *Z*-cyclooctene **10**, direct irradiation at 185 nm generated an approximately 1 : 1 mixture of the achiral *Z*-**10** and the chiral *E*-**10**, whereas in the presence of aromatic esters, a sensitized reaction occurred upon irradiation at 254 nm, yielding approximately 4 : 1 *Z*-**10**:*E*-**10** at the photostationary state. In the initial reports [16], the optimal results were achieved using di-(−)-menthylisophthalate as sensitizer (at best, an asymmetric induction of 4% ee was observed), although subsequent investigations improved on these results significantly (*vide infra*). The authors proposed a singlet exciplex mechanism involving energy transfer from the excited aromatic ester to the cyclooctene, which is thought to produce a twisted, excited singlet cyclooctene. Thus, upon formation of the exciplex, the cyclooctene moiety can rotate around the double bond, generating two discrete diastereoisomeric exciplexes, which then decay to the achiral *Z*-**10** or the chiral *E*-**10**. Subsequent reports focused on the investigation of a wider range of chiral aromatic esters as sensitizers for the same process [17]. After considerable experimentation, the optimal conditions involved sensitization by tetra-(−)-bornyl pyromellitate **11** at −88 °C, after which an ee value of 41% for (*R*)-*E*-**10** was recorded (Scheme 11.4). Subsequently, the optimum enantioselectivity was increased to 53% ee through further screening of aromatic ester sensitizers [18].

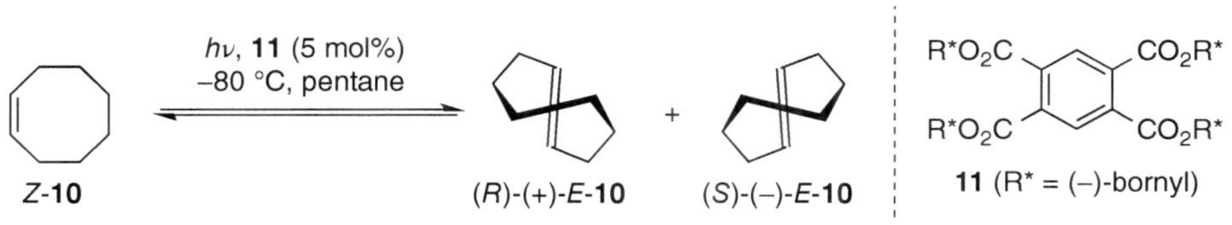

Scheme 11.4 Enantioselective photoisomerization of (Z)-cyclooctene (**10**).

These studies [17] also uncovered an intriguing temperature-switching effect in the product chirality: at a critical temperature (dependent on the sensitizer employed and the reaction conditions) the direction of the product's optical rotation inverted. Thus, above this critical temperature, the enantioselectivity of the product *E*-**10** increased with temperature, while below this temperature, the other enantiomer of *E*-**10** was formed preferentially, with the enantioselectivity increasing with decreasing temperature. For example, using (−)-menthyl pyromellitate **5** as sensitizer, this critical temperature was −19 °C, thus occurring within the experimental temperature range. The effect was attributed to temperature-dependent populations of the two diastereoisomeric sensitizer-cyclooctene exciplexes, with lower temperatures favoring the exciplex leading to one enantiomer, and higher temperatures favoring the other exciplex, thus the other enantiomer of the product. These unexpected findings were observed only when using orthodicarboxylates as sensitizers. Other aromatic esters did not display a similar temperature-dependent chirality inversion, and it was suggested that steric interactions between the two chiral alcohol units play an important role in the curious behavior observed.

Hoffmann and Inoue subsequently reported a similar study focusing on the enantiodifferentiating *Z-E* photoisomerization of cycloheptene under similar conditions [19]. In this case, the highly strained *E*-cycloheptene could not be isolated due to thermal instability, but could be trapped at low temperature by reaction with osmium(VIII) oxide or a reactive diene. Photolysis of *Z*-**12** led to lower conversion to *E*-**12** at the photostationary state than for *Z*-cyclooctene, presumably due to the particularly strained nature of the smaller cycle. The optimum conditions involved photolysis (40 min) in hexane at −80 °C in the presence of 20 mol% sensitizer **11** followed by trapping with diphenyl isobenzofuran, which generated the corresponding cycloadduct **13** in 77% ee (Scheme 11.5). At that time, this result represented the highest asymmetric induction ever obtained for a photosensitization.

In 1990, Kim and Schuster reported an enantioselective version [20] of their previously developed "triplex Diels Alder" reaction [21]; the photochemical [4+2] cycloaddition of electron-rich dienes with electron-rich dienophiles catalyzed by electron-deficient arene sensitizers. Two different chiral sensitizers were investigated, with the best results obtained using (−)-1,1′-bis(2,4-dicyanonaphthalene) (**15**) (Scheme 11.6). At −65 °C in toluene, employing only 1 mol% of sensitizer, the triplex Diels–Alder reaction of trans-β-methylstyrene with cyclohexadiene generated the desired product **14** with 19% ee (±2%). Slight increases in enantioselectivity could be achieved at lower concentrations of cyclohexadiene, and the degree of enantioselectivity was found to be temperature dependent: conducted

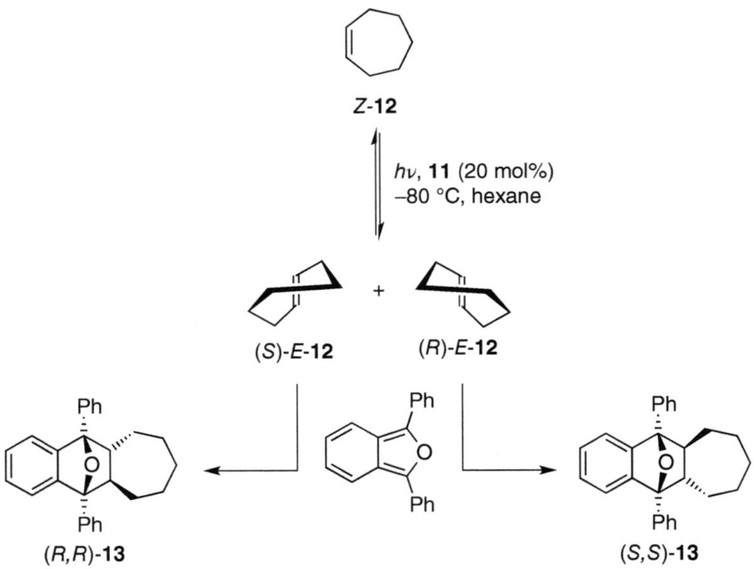

Scheme 11.5 Asymmetric photoisomerization of cycloheptene and subsequent trapping with diphenylisobenzofuran.

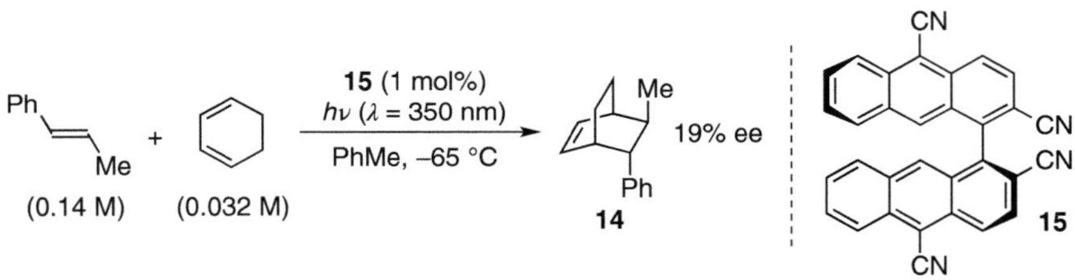

Scheme 11.6 "Triplex Diels–Alder" reaction.

at −15 °C, the same reaction yielded product **14** with 14% ee (±2%), while no enantioselectivity was generally observed at room temperature.

Based on extensive mechanistic studies, the reaction was proposed to proceed through a two-step sequence. Upon irradiation, the sensitizer is excited to its singlet excited state and forms an exciplex with the electron-rich dienophile. The exciplex subsequently reacts with cyclohexadiene (through a termolecular complex) to generate the [4+2] cycloadduct. Since the stabilization and lifetime of the exciplex are expected to be strongly affected by steric and electronic factors, it follows that the diastereoisomeric complexes that are generated from the chiral sensitizer and the prochiral dienophile might display different properties. It is thought that the diastereoisomeric complexes undergo rapid interconversion at room temperature but more slowly at low temperatures, and that selective trapping of one of the isomeric exciplexes by cyclohexadiene is responsible for the observed enantioselectivity. Experiments conducted in acetonitrile led only to racemic products, presumably due to the propensity of exciplexes to dissociate rapidly into ionic intermediates in polar media.

These early results presented focused on a variety of different photochemical reactions, including photoisomerizations (cyclopropanes and cyclooctene), oxadi-π-methane rearrangements, and deracemization of sulfoxides and photocycloadditions. Although the enantioselectivites obtained using the methods described generally remained low, these important contributions proved that catalytic enantioselective photoreactions were indeed possible, and inspired hope that continued effort would yield significant improvements in the levels of enantioselectivity achievable in such processes. These innovative reports provided a fundamental understanding of enantioselective photochemical processes and paved the way for further developments in this field, the most important of which are discussed in the next sections.

11.3 The Twenty-First Century: Contemporary Developments

Building on the pioneering research described in the previous section, the field of enantioselective photocatalysis has continued to inspire researchers, with vast improvements in the levels of enantioselectivity now attainable. In this section, more recent work toward catalytic enantioselective photoreactions is discussed, with particular emphasis on processes that are of potential use in synthetic applications. The most successful approaches can be conveniently divided into three distinct groups, namely, large-molecule chiral hosts, small-molecule chiral photosensitizers, and activation by chiral Lewis acids. These three approaches are described in detail in the following sections.

11.3.1 Large-Molecule Chiral Hosts

Inoue and coworkers have intensively investigated the photodimerization of 2-anthracenecarboxylic acid **16** in a variety of different media. Four products **17–20** are formed in the dimerization (**18** and **19** are chiral, while **17** and **20** are achiral; Scheme 11.7) and the dimerization is thought to proceed via singlet excited states. Previously, it was shown that in the presence of 25 mol%

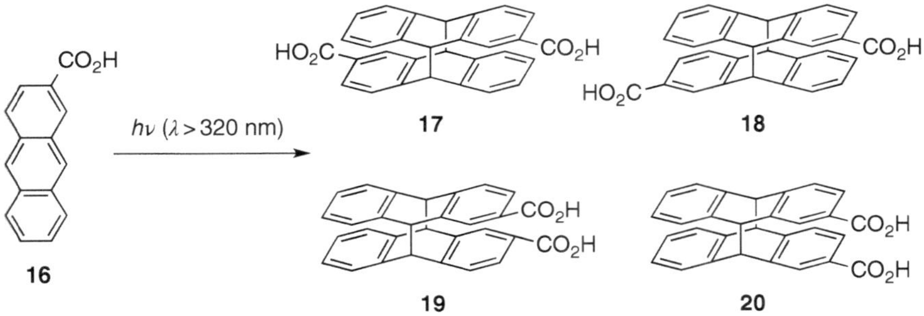

Scheme 11.7 Photodimerization of 2-antracenecarboxylic acid (**16**).

γ-cyclodextrin, **18** was produced in 43% yield and 28% ee, while **19** was produced in 8% yield and 1% ee [22].

It was found that the presence of γ-cyclodextrin allowed an acceleration of the photodimerization, which was attributed to the γ-cyclodextrin cavity size being ideal for the inclusion of two molecules of **16.** In the presence of native γ-cyclodextrin, only low asymmetric induction could be achieved, with regioselectivity and enantioselectivity being dependent on the amount of cyclodextrin employed. In an attempt to improve the selectivities observed, Inoue and coworkers designed new, chemically modified γ-cyclodextrin derivatives bearing diamino side chains [23]. Although the new derivatives themselves did not drastically improve the results, the addition of copper(II) salts was found to be very beneficial in that both the chemical yield and the enantioselectivity of head-to-head product **19** were improved. Under the optimal conditions, which employed only 10 mol% of modified γ-cyclodextrin **21** along with 50 mol% copper(II) salt, **19** was obtained with 64% ee at 83% conversion (Scheme 11.8). It is thought that Cu(II) coordinates with the diamino side chain of the modified γ-cyclodextrin, and with the carboxyl groups of two molecules of **16**, thus improving the regioselectivity of the reaction.

Since it was also shown that in the bulk solution (in the absence of a γ-cyclodextrin host) **16** tends to form aggregates that are very unreactive to photodimerization, it was possible to maintain the enantioselectivity while employing only a substoichiometric amount of chiral host, even though the host does not act as a sensitizer for the photoreaction.

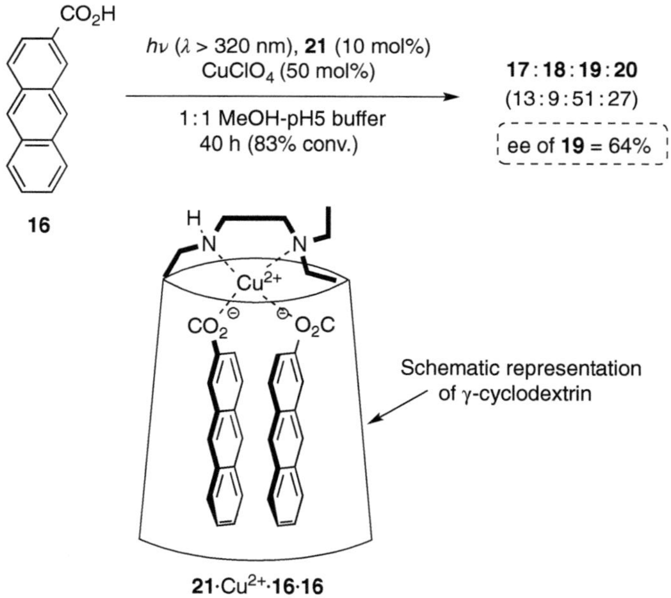

Scheme 11.8 Enantioselectivity induced by the modified copper-γ-cyclodextrin in the photodimerization of **16**.

The method is certainly an attractive entry into enantioselective photocatalysis, since γ-cyclodextrin is inexpensive, readily available, and can be derivatized easily to produce a large range of modified γ-cyclodextrins. However, to date, the method has only been applied to the specific case of photodimerization of 2-anthracenecarboxylic acid (**16**), for which the cavity of γ-cyclodextrin is particularly well suited. It remains to be seen whether this method can be adapted for use in other systems as a more general approach to photocatalysis.

11.3.2 Small-Molecule Chiral Photosensitizers

In a recent extension to a long-standing interest in photoisomerization (Section 1.2), Inoue and coworkers investigated the use of planar-chiral paracyclophane sensitizers [24] in the enantiodifferentiating *Z-E* isomerization of cyclooctene [25]. Using paracyclophane sensitizer **22**, a moderate enantioselectivity of 42% ee was achieved upon irradiation at −110 °C, a result similar to that obtained previously using point-chiral benzene(poly)carboxylate sensitizers (Section 1.2). However, applying conditions similar to the enantiodifferentiating photoisomerization of *Z,Z*-1,5-cyclooctadiene **23** led to rather more impressive levels of asymmetric induction. The best results were obtained at low temperature in isopentane/methylcyclohexane and employing 20 mol% of chiral sensitizer **22**, which generated *R*-(−)-*E,Z*-**23** in 87% ee (Scheme 11.9). The superior results obtained using *Z,Z*-**23** instead of *Z*-cyclooctene (*Z*-**10**) were attributed to the likelihood of *Z,Z*-**23** forming a better stabilized exciplex benefiting from closer interactions, due to its higher electron-donating ability.

It should be noted that photoisomerization from *Z*-**10** or *Z,Z*-**23** into the corresponding *E*-isomers sensitized by **22** is much less efficient than when sensitized by benzene(poly)carboxylates (at the photostationary state, *E*:*Z* ratios reach approximately 0.01 when employing **22**). However, the authors stress that upon circulating the photolysate continuously through a column loaded with silver(I) nitrate and silica, the *E*-isomer can be selectively removed from the mixture; thus, useful amounts of *E*-isomer for synthetic purposes could potentially be synthesized using this modification.

Bach and coworkers have developed a series of small-molecule photocatalysts capable of inducing excellent enantioselectivity in a variety of photochemical reactions. The catalysts have been developed on the basis of chiral template **24** [26], which was introduced in 1999 as a stoichiometric complexing agent to enable enantioselective photoreactions of prochiral lactam-containing substrates. Compound **24** comprises an oxazolotetrahydronaphthalene steric shield, as well as a rigid scaffold derived from Kemp's acid, which is capable of forming two hydrogen bonds with lactam-containing substrates. Complexation of such

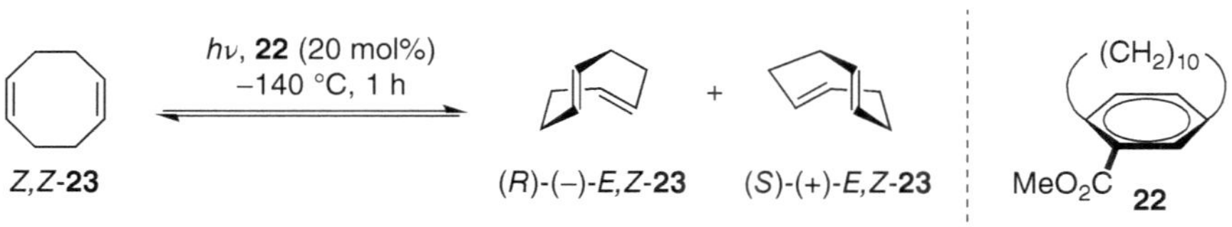

Scheme 11.9 Photoisomerization of Z,Z-cyclooctadiene with paracyclophane sensitizer (**22**).

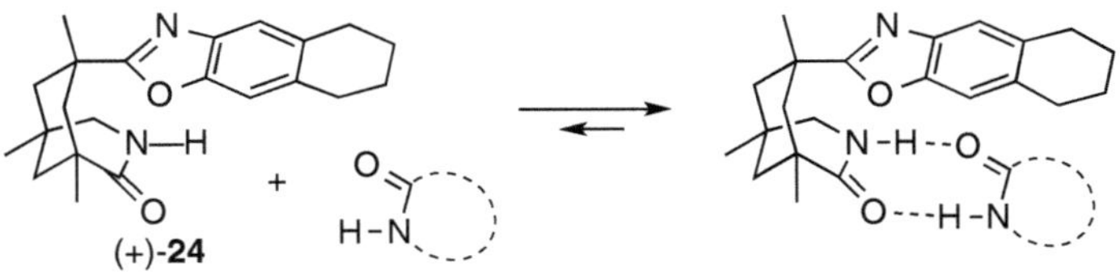

Scheme 11.10 Bach's chiral molecular template for stereocontrol in photochemical reactions.

substrates to chiral template **24** results in a stable, specific supramolecular complex in which one face of the substrate is shielded, thus allowing highly selective reaction on the open face of the substrate (Scheme 11.10).

Template **24** has been successfully employed in a number of different enantioselective photochemical reactions, including [2+2] photocycloadditions (both intermolecular [27] and intramolecular [28]), [4+4] photocycloadditions [29], electrocyclic [4π] [29] or [6π] [30] photocyclizations, Norrish–Yang reactions [31], Diels–Alder reactions of photochemically generated *ortho*-quinodimethane [32], and intermolecular [4+2] photocycloadditions involving singlet oxygen promoted by visible light [33]. As prochiral lactam-containing substrates, quinolones, isoquinolones, pyridones, and imidazolidinones have been employed. Reactions are generally conducted in nonpolar solvents such as toluene at low temperatures, since complexation between the substrate and the template is maximized under these conditions. For example, the enantioselective [2+2] photocycloaddition of substituted quinolone **25** with alkene **26** was employed as a key step in the total synthesis of the *Melodinus* alkaloid (+)-meloscine [27b,c]. Thus, in the presence of 2.5 equiv. of the chiral template **24**, the complex photoadduct **27** was obtained as a single isomer in excellent yield and good enantioselectivity (79% ee; Scheme 11.11). Upon subsequent HPLC separation, enantiopure material was obtained in 76% yield of isolated product.

Similarly, isoquinolone **28** undergoes efficient [2+2] photocycloaddition with a variety of electron-deficient alkenes, generating the desired functionalized cyclobutanes **29** in excellent yields and enantioselectivities (Scheme 11.12) [27d].

Despite the excellent results obtained using chiral template **24**, the need to employ stoichiometric or superstoichiometric amounts of a chiral source is a distinct disadvantage, even if, as in these cases, the template can be recovered after

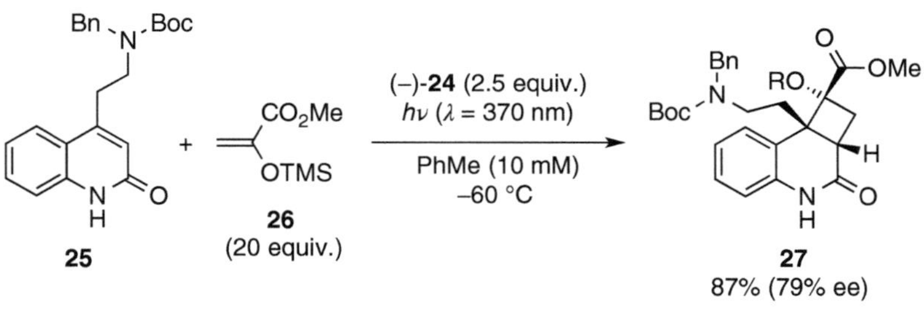

Scheme 11.11 Enantioselective [2+2] photocycloaddition using Bach's molecular template ((–)-**24**).

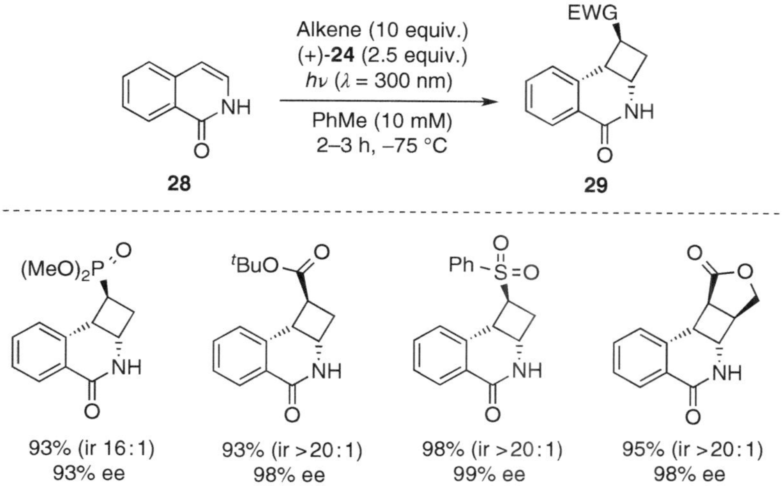

Scheme 11.12 [2+2]-Photocycloaddition of isoquinoline **28**, using Bach's molecular template ((+)-**24**).

the reaction. If the bulky oxazolotetrahydronaphthalene unit in template **24** could be modified such that it could act not only as a steric shield but also as a suitable photosensitizer, the photoreaction would be expected to occur only when the substrate is bound to the chiral host. Thus, assuming that the photoproduct dissociates from the chiral host after the photoreaction is complete, chiral multiplication could be expected, leading to photocatalysts that could be employed in substoichiometric amounts. Working from this premise, Bach and coworkers have developed a series of next-generation versions of **24** that can be employed in catalytic amounts. To date, three catalysts bearing different sensitizing units (benzophenone, xanthone, and thioxanthone) have been developed.

In 2005, Bach and coworkers introduced benzophenone-containing photocatalyst **32** (Scheme 11.14) and demonstrated its use in enantioselective photoreactions driven by PET [34, 35]. Thus, upon irradiation in the presence of **32**, quinolone **30**, bearing a pendant pyrrolidine moiety, is transformed into chiral spirocyclic pyrrolizidine **31**. The best results were obtained using 30 mol% of catalyst **32** in toluene at −60 °C, producing the desired product in 64% yield and 70% ee (Scheme 11.13). Lower catalyst loadings resulted in decreased

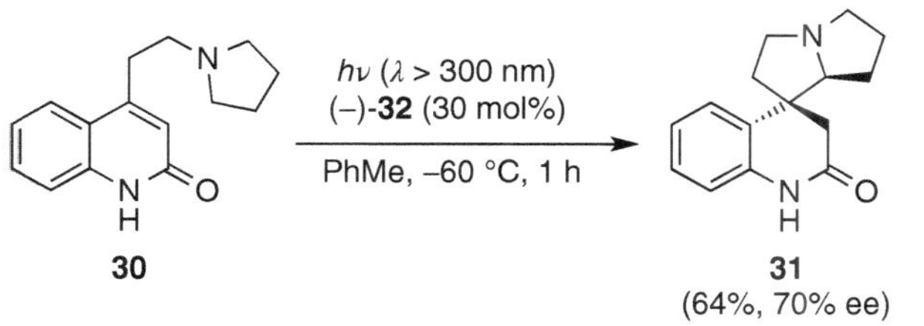

Scheme 11.13 Enantioselective benzophenone photocatalyst for the asymmetric radical cyclization of **30**.

enantioselectivity (e.g., using 20 mol% **32**, the desired product was obtained with 57% ee; using 10 mol% **32**, a 38% ee value was observed). These results represented a breakthrough in enantioselective photochemistry, and showed that photochemical routes have real potential in general asymmetric synthesis.

Upon UV-B irradiation, the catalyst is thought to accept an electron from the amine moiety of the substrate, resulting in the production of a radical-ion pair (Scheme 11.14). Next, a proton is transferred from the carbon atom adjacent to the electron-deficient nitrogen atom to the benzophenone radical anion, generating a radical pair. The nucleophilic α-amino radical undergoes a 1,4-addition to the quinolone. The catalytic cycle is thought to be completed by a back electron transfer with subsequent protonation of the resulting enolate. Throughout the process, the chiral catalyst enforces a chiral environment on the substrate, meaning that the pendant pyrrolidine ring remains on the nonblocked face of the substrate. Hence, efficient enantiodifferentiation is ensured.

In a similar manner, Krische and coworkers designed a new "sensitizing receptor" **33** and applied it to the enantioselective catalysis of the intramolecular [2+2] photocycloaddition of quinolone **34** [36]. The catalyst comprises a binding motif to coordinate **34** as well as a sensitizing benzophenone unit. In this case, sensitization occurs through triplet energy transfer instead of the PET described in the previous example. Intramolecular photocycloaddition of **34** was expected

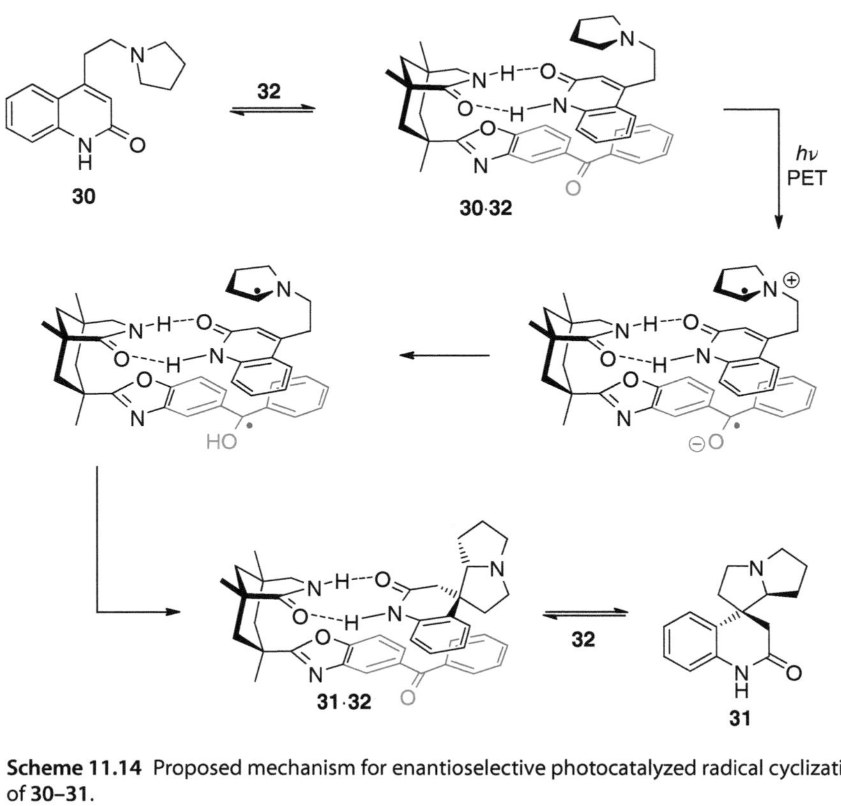

Scheme 11.14 Proposed mechanism for enantioselective photocatalyzed radical cyclization of **30–31**.

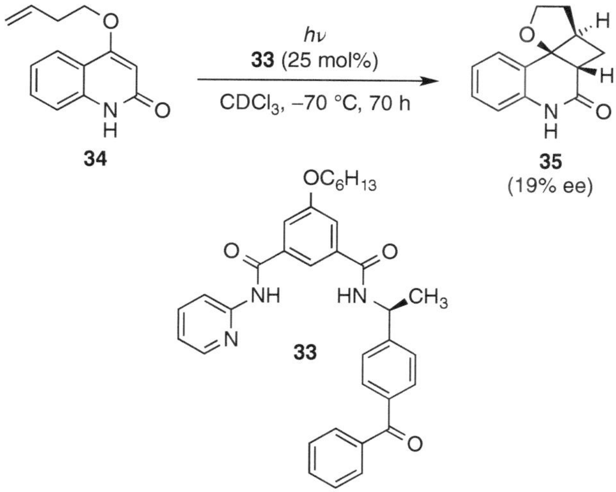

Scheme 11.15 Enantioselective [2+2] photocycloaddition of **34** using photocatalyst **33**, developed by Krische.

to take place in the chiral microenvironment of the host–guest complex, with the expectation that the sensitized reaction would take place more rapidly than the nonsensitized process. A mixture of quinolone **34** and chiral sensitizer **33** (25 mol%) was irradiated at −70 °C, resulting in quantitative conversion to the desired product **35** (Scheme 11.15). Low enantioselectivity was observed (19% ee), which did not improve upon increasing the loading of **33** (using 200 mol% **33**, **35** was obtained with 21% ee). The low enantioselectivity observed suggests that the catalyst confers only low facial differentiation; a significant background reaction of uncomplexed substrate seems less likely, since the catalytic effect of receptor **33** on the photocycloaddition was proved in separate experiments.

The benzophenone photocatalyst **32** developed by the Bach group was also investigated in the context of the same [2+2] photocycloaddition reaction [37]. The intramolecular [2+2] photocycloaddition of quinolone **34** in the presence of **32** generated two product regioisomers: the straight isomer *ent*-**35** and the crossed isomer **36** (Scheme 11.16). Although a rate acceleration was observed in the presence of only 10 mol% **32**, the enantioselectivities observed were disappointing (39% ee for *ent*-**35** and 17% ee for **36**). The low ee values were attributed to the comparatively low triplet energy of photocatalyst **32** as well as its relatively short-wavelength absorption. In addition, since the benzophenone moiety is not rigid but rather flexible, it is likely that it is less effective as a steric shield as the oxazolotetrahydronaphthalene unit in chiral template **24**.

The disappointing results employing photocatalyst **30** prompted the design of a new photocatalyst **37**, in which the sensitizing benzophenone moiety was replaced by a xanthone unit (Figure 11.2) [37, 38]. The triplet energy of xanthone (≈310 kJ mol^{-1}) [39] is significantly higher than that of benzophenone (≈287 kJ mol^{-1}); [39] it was anticipated that photocatalyst **37** would act as a more effective triplet sensitizer than its benzophenone analog **30**, since effective

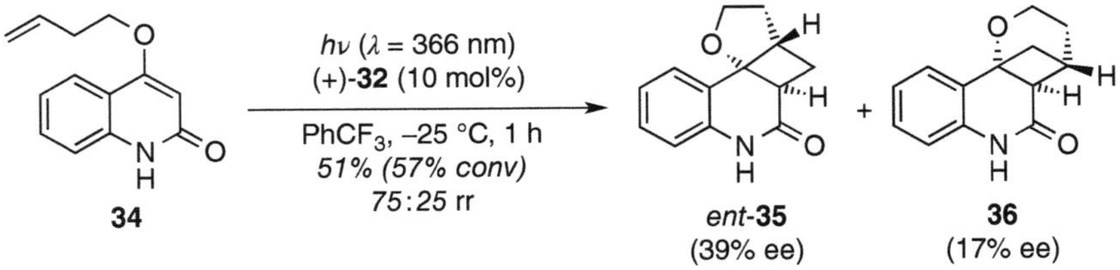

Scheme 11.16 Enantioselective [2+2] photocycloaddition of **34** using photocatalyst **32**, developed by Bach.

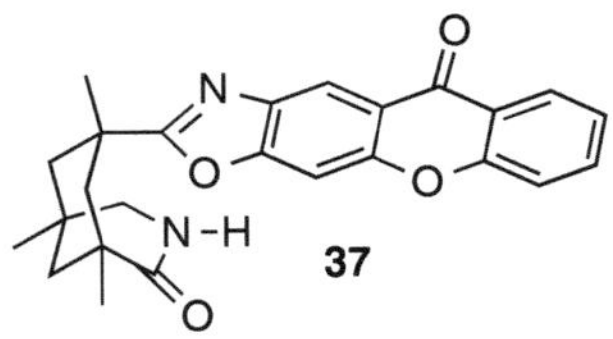

Figure 11.2 Next-generation enantioselective photocatalyst with xanthone photosensitizing motif.

sensitization (involving rapid energy transfer) requires that the triplet energy of the sensitizer is higher than that of the substrate. In addition, the xanthone moiety was expected to display superior abilities as a steric shield, since it is an inflexible, planar unit. The benzophenone moiety is much more flexible and thus less likely to be able to confer a rigid chiral environment on the substrate.

Indeed, employing xanthone photocatalyst **37** resulted in a dramatic increase in enantioselectivity, and both product isomers were obtained with excellent enantioselectivity (94% ee; Scheme 11.17). The optimal results were obtained using 20 mol% photocatalyst **37** in trifluorotoluene, irradiating at 366 nm at −25 °C. Trifluorotoluene was preferred over toluene as the reaction solvent, since photocatalyst decomposition is much faster in toluene. Catalyst decomposition is thought to occur through hydrogen-atom abstraction by the triplet excited state of the photocatalyst; restricting the availability of labile C—H bonds would therefore be expected to limit such degradation.

Mechanistically, it is thought that selective excitation of the xanthone chromophore in **37** is followed by rapid energy transfer to the quinolone substrate, exciting it to its triplet excited state. Subsequent intramolecular [2+2] photocycloaddition occurs, in which the xanthone moiety also functions as a steric shield, ensuring that photocycloaddition occurs only on the open face of the quinolone substrate (Scheme 11.18).

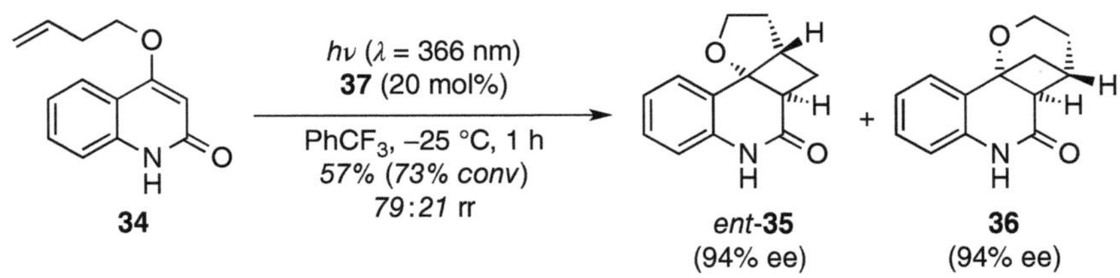

Scheme 11.17 Photochemical [2+2] cycloaddition of **34** using xanthone photocatalyst **37**.

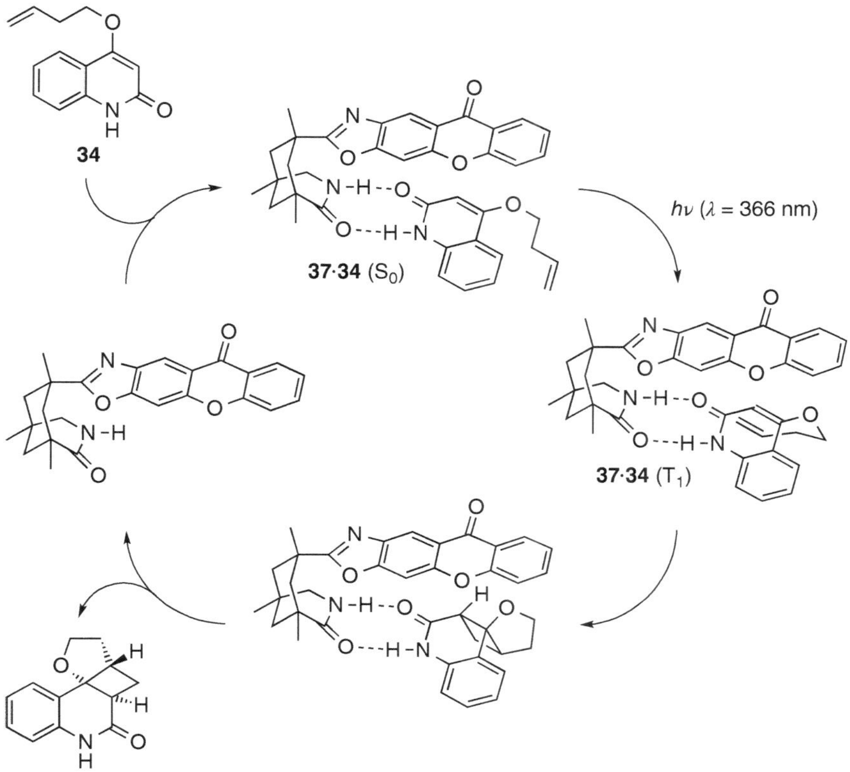

Scheme 11.18 Proposed mechanism for xanthone (**37**) catalyzed photochemical 2+2 cycloaddition.

In this case, lowering the catalyst loading leads to similar enantioselectivities but longer reaction times. Since the photocatalyst **37** undergoes significant decomposition upon prolonged irradiation, it was judged preferable to stop the reaction before complete conversion, in order to limit catalyst degradation. Interestingly, catalyst degradation was invariably accompanied by destruction of the minor regioisomer of the product. This phenomenon was attributed to the catalyst abstracting a hydrogen atom from the product, a process that appears to be more favorable for the minor product regioisomers than the major regioisomers.

Bach and coworkers subsequently studied more extensively the use of photocatalyst **37** for intramolecular enantioselective [2+2] photocycloadditions of quinolones, including the investigation of several new quinolone substrates [38]. First, the intramolecular [2+2] photocycloaddition of quinolone **38** was investigated. Unfortunately, the results obtained using the homologous quinolone **38** were much less promising than those obtained employing quinolone **34**, with low enantioselectivities consistently observed (at best, 41% ee using 30 mol%

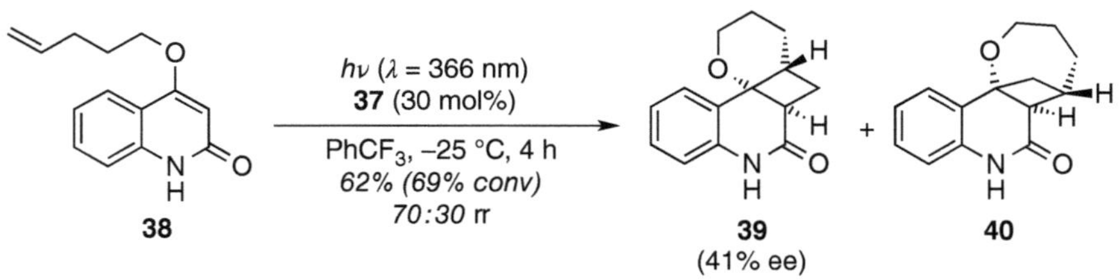

Scheme 11.19 Photochemical [2+2] cycloaddition of homologous quinolone **38**, using chiral photocatalyst **37**.

37; Scheme 11.19). Lowering the catalyst loading led to further decreases in enantioselectivity (27% ee using 10 mol% **37**; 21% ee using 5 mol% **37**).

Crucially, much higher enantioselectivity was achieved using stoichiometric amounts of chiral template **24** (87% ee), which suggested that the low enantioselectivity is not due to low enantioface differentiation in the substrate–catalyst complex, but is likely linked to the relative rates of the photocycloaddition and the dissociation of the substrate from the catalyst. Since the rate of cyclization to a six-membered ring is expected to be significantly slower than to a five-membered ring, the intramolecular [2+2] photocycloaddition of quinolone **38** is expected to be slower than that of quinolone **34**. It is suggested that dissociation of the excited **37**·**38** complex is competitive with formation of the six-membered ring, meaning that significant reaction occurs outside of the chiral environment provided by the catalyst, thus eroding enantioselectivity. On the contrary, dissociation of the excited **37**·**34** complex is *not* competitive with five-membered ring formation, such that very limited cycloaddition occurs in a nonchiral environment, leading to excellent enantioselectivity. These conclusions were adequately supported by photophysical experiments, including transient absorption spectroscopy and laser flash photolysis studies [38, 40].

Several other quinolone substrates were also studied (Scheme 11.20) [40]. In particular, the influence of the alkene substitution pattern on the product distribution was investigated, through the reactions of two isomeric quinolone substrates *E*-**41** and Z-**41**. Irradiation of either *E*-**41** or Z-**41** generated three products: the major regioisomer **42** (present as two diastereoisomers, only the major isomer is depicted) and the minor regioisomer **43** (Scheme 11.20). The influence of the starting olefin geometry had negligible effect on the product distribution, both in the racemic and enantioselective reactions. The best results were obtained at −25 °C, using 10 mol% of catalyst **37**, furnishing the corresponding photoadducts in high yields and high ee. Notably, the catalyzed reactions also benefited from increased regioselectivity and diastereoselectivity compared with the racemic reactions.

Subsequent work focused on further extending the substrate scope of the methodology [40]. Substitution on the tether chain as well as the pendant alkene was tolerated well, generating products **44** and **45** in high yields and enantioselectivities (Figure 11.3). Quinolone substrates substituted in the 3-position were found to be less soluble in trifluorotoluene than the corresponding 4-substituted quinolones; thus, the reactions were carried out at higher dilution (2.5 mM). The

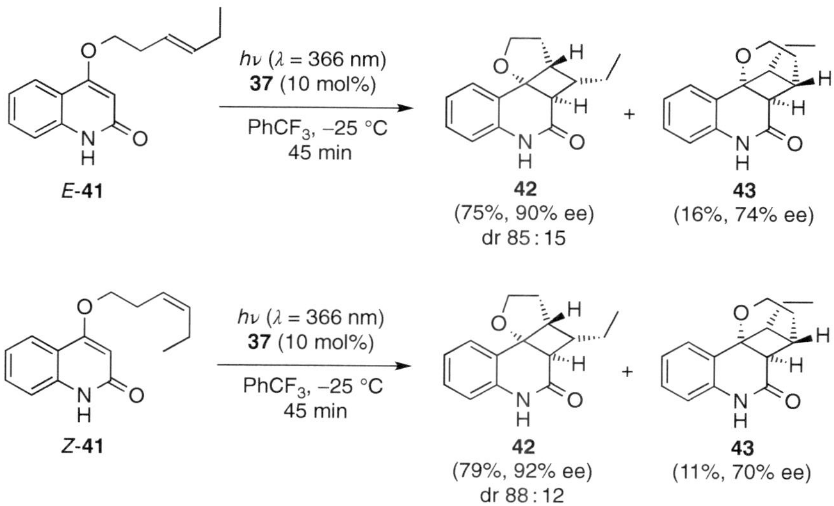

Scheme 11.20 Photochemical 2+2 cycloaddition with internal Z and E alkenes using photocatalyst **37**.

Figure 11.3 Enantioselectivities in the photochemical 2+2 cycloaddition of other quinolones containing different substituents on the alkene tether or quinolone structure.

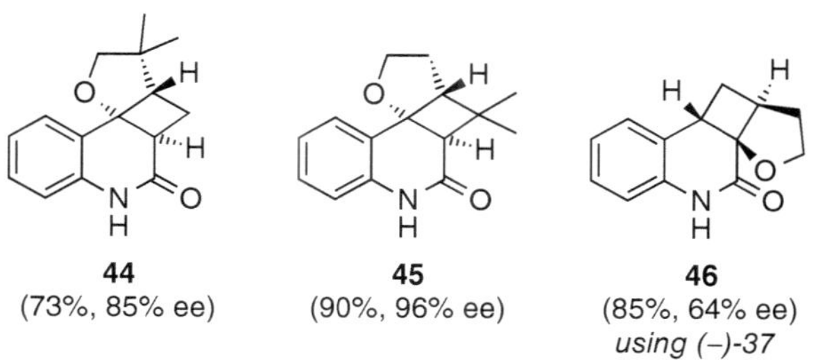

photoreactions proceeded rapidly and the desired cyclobutane products were obtained in high yields (83–88% isolated yield). However, the enantioselectivity was disappointing, and never exceeded 64% ee (Figure 11.3). The moderate enantioselectivity observed was attributed to a relatively rapid background reaction due to the increased absorption of these substrates under the irradiation conditions employed.

Very recently, Alonso and Bach reported the use of a closely related photocatalyst **47** that contains a thioxanthone photosensitizing unit (Figure 11.4) [41]. Making use of the fact that thioxanthone absorbs at longer wavelength than xanthone [39], this new catalyst was designed to allow efficient enantioselective

Figure 11.4 Chiral thioxanthone photocatalyst capable of UV-A and visible light sensitization.

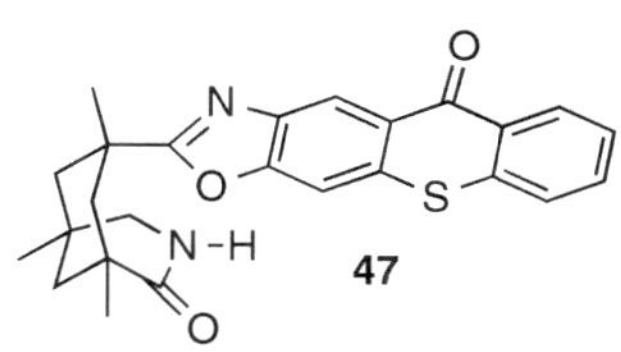

photocatalysis using *visible* light instead of the UV-A ultraviolet light required for catalyst **37**.

The new catalyst **47** is the first photocatalyst that allows the processing of visible light in a highly enantioselective [2+2] photocycloaddition, and thus represents an important breakthrough in enantioselective photocatalysis. Crucially, when irradiating with visible light, the background reactions are much slower than upon irradiation with the shorter wavelength light sources described earlier, and, in many cases, no background reaction is observable under such conditions. Thus, photoreaction takes place selectively within the substrate–catalyst complex, ensuring high levels of enantioselectivity. Catalyst **47** has been demonstrated to mediate highly enantioselective intramolecular [2+2] photocycloadditions of quinolones bearing pendant alkenes, generating the corresponding photoadducts in excellent yields and with excellent enantioselectivity. The optimum reaction conditions involved irradiation of the quinolone substrates with visible light (400–700 nm) in $PhCF_3$ for 1 h at −25 °C. Under these conditions, quinolone **48** gave cyclobutane **49** in 88% yield and 90% ee (Scheme 11.21).

Six other quinolone substrates (including heteroatom-containing alkenyl tethers) were also investigated under very similar reaction conditions, and gave comparable results both in terms of chemical yield and enantioselectivity (Figure 11.5).

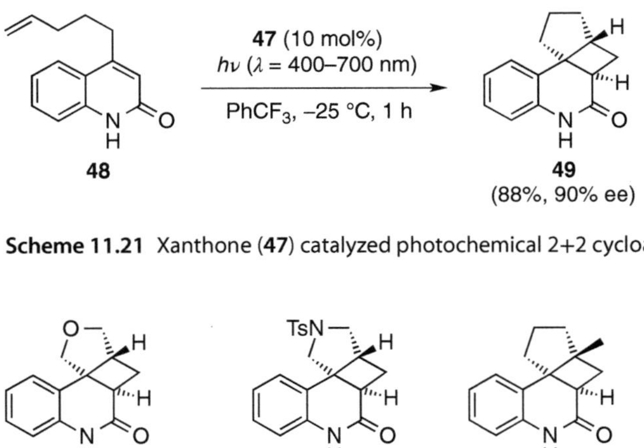

Scheme 11.21 Xanthone (**47**) catalyzed photochemical 2+2 cycloaddition.

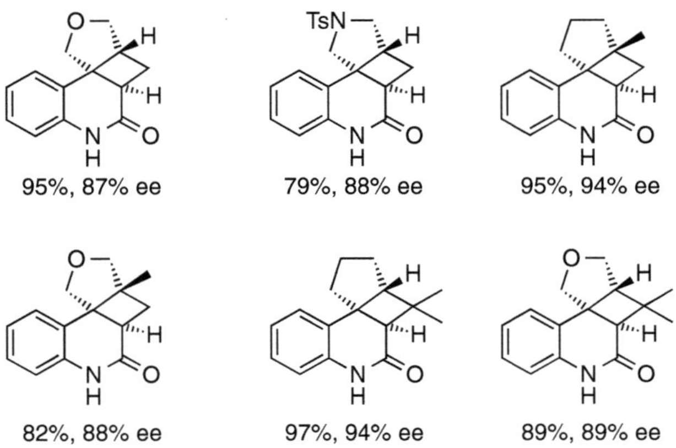

Figure 11.5 Reaction scope of photochemical 2+2 cycloadditions catalyzed by xanthone **47**.

Interestingly, thioxanthone catalyst **47** was found to be much less susceptible to degradation through hydrogen atom abstraction than the corresponding xanthone catalyst **37**, only undergoing very limited decomposition under the optimized reaction conditions.

In summary, the most effective small-molecule photocatalysts developed to date are the family of catalysts developed by the Bach group, which combine a chiral template (derived from Kemp's acid) and a photosensitizing moiety (benzophenone, xanthone, or thioxanthone units). These catalysts have been shown to enable highly enantioselective photoreactions of prochiral lactam-containing substrates, particularly the intramolecular [2+2] photoreactions of quinolone substrates. Of particular note, these reactions can now be carried out under visible light irradiation, due to the incorporation of a thioxanthone photosensitizing unit that is able to process visible light, transferring it to quinolone substrates through triplet energy transfer. Further developments in this area are expected to significantly expand the scope of enantioselective processes that can be catalyzed using this approach.

11.3.3 Lewis Acid-Mediated Photoreactions

An innovative approach to enantioselective photocatalysis that has the potential for widespread use involves the use of chiral Lewis acids as activating agents. The idea is based on the observation that upon coordination to Lewis acids, the photophysical properties of many substrates may be significantly altered. Crucially, if the substrate–Lewis acid complex can be selectively excited in the presence of noncoordinated substrate, and the Lewis acid is capable of inducing high levels of enantioselection in the subsequent photoreaction, chiral multiplication can be expected, such that substoichiometric quantities of Lewis acid could be employed to effect enantioselective photoreactions. To date, two such systems have been successfully developed, and are discussed in the next sections.

In 1983, work by Lewis and Barancyk demonstrated that the [2+2] photocycloaddition of coumarin with 2,3-dimethylbut-2-ene is promoted by Lewis acids [42], but is inefficient in the absence of such additives. Similarly, Görner and Wolff showed that the photodimerization of coumarin may be catalyzed by the addition of BF_3 [43]. In this report, it was shown that the quantum yield of intersystem crossing increased upon addition of BF_3, thus increasing the rate of dimerization through the triplet excited state. Conversely, in the absence of BF_3, coumarin is known to undergo rapid internal conversion [44], leading to slow dimerization through a singlet excited state. Building on these interesting results, Bach and coworkers disclosed the first enantioselective intramolecular [2+2] photocycloadditions of substituted coumarins bearing pendant alkenes [45, 46]. Since oxazaborolidine-based Lewis acids gave particularly good results in an initial screen, several $AlBr_3$-activated oxazaborolidine catalysts were investigated in detail. Lewis acid **51** was found to give the best performance, with the optimum results obtained by employing 50 mol% **51** and running the reaction in dichloromethane at low temperature (−75 °C; Scheme 11.22). Lowering the Lewis acid loading to 20 mol% gave a similar product yield but led to a decrease in enantioselectivity (54% ee).

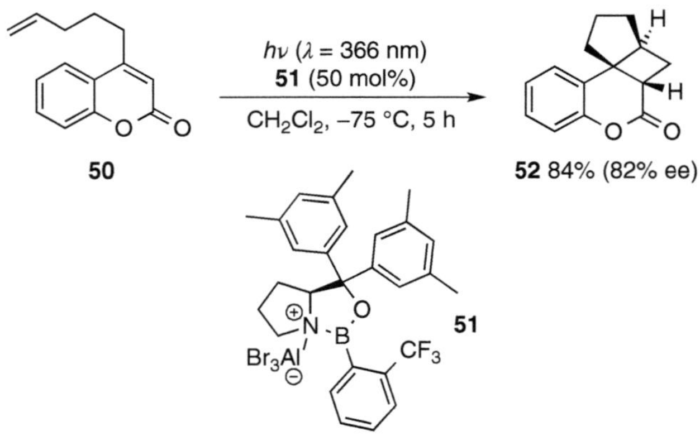

Scheme 11.22 Lewis acid-mediated photochemical 2+2 cycloaddition of alkene tethered coumarin derivatives.

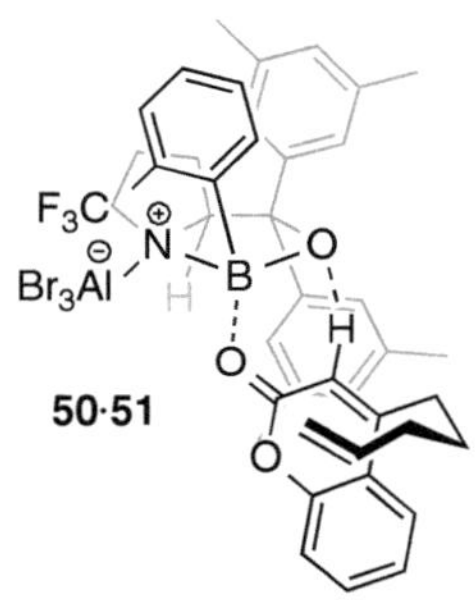

Figure 11.6 Proposed model for enantioselectivity in 2+2 photocycloaddition of **50** using Lewis acid complex **51**.

The stereochemical outcome of the reaction was attributed to the formation of a 1 : 1 substrate–catalyst complex in which the coumarin carbonyl group coordinates to the boron atom of the catalyst (Figure 11.6). Potentially, a second electrostatic interaction between the coumarin C3 hydrogen atom and an oxygen atom of the oxazaborolidine (nonclassical hydrogen bond) could also be important in complex formation.

The use of a substoichiometric amount of chiral Lewis acid clearly enables the effective suppression of the background reaction, such that high enantioselectivity is observed in these photoreactions. The mode of action of the Lewis acid catalyst is likely to be at least twofold. First, coordination of Lewis acids to coumarin **50** was shown to induce a bathochromic shift in its UV–visible absorbance spectrum, such that the absorbance of the Lewis acid–coumarin complex is greater at 366 nm than the absorbance of nonassociated coumarin. Although a slow background reaction was observed at 366 nm, irradiation at this wavelength selectively excites the complex rather than the noncomplexed coumarin. More importantly, complexation of the Lewis acid with the coumarin is thought to increase the lifetime of the singlet excited state (qualitatively demonstrated by the observation of increased fluorescence of coumarin **50** upon coordination with Lewis acids) and increase rates of intersystem crossing to the corresponding triplet excited state. Thus, complexation allows photocycloaddition to become competitive with the rapid internal conversion that occurs upon excitation of noncomplexed coumarin [44]. Support for the involvement of triplet intermediates in the Lewis acid-mediated reaction was provided by the observation that the catalyzed photoreaction of isomeric coumarin substrates is stereoconvergent. In contrast, the racemic reaction was shown to proceed through the singlet excited state.

The methodology was shown to be compatible with a number of different substituted coumarins bearing different functional groups (Scheme 11.23) [46]. For example, incorporation of heteroatom-containing alkenyl chains was tolerated, generating the corresponding photoadducts in high yields and with good enantioselectivities (Scheme 11.24, X = O, NBoc, NZ, or NTs). In contrast, incorporation of a sulfur-containing tether led to a low 22% product yield (due to a competing side reaction), and lower enantioselectivity (42% ee), which was attributed to likely partial coordination of the catalyst to the sulfur atom.

Brimioulle and Bach recently reported a much more general approach to Lewis acid-mediated enantioselective photocatalysis, focusing on the intramolecular [2+2] photocycloadditions of 5,6-dihydro-4-pyridones [47]. Such photocycloadditions have previously been employed in the total synthesis of a variety of natural products [48], and thus the development of an efficient enantioselective version

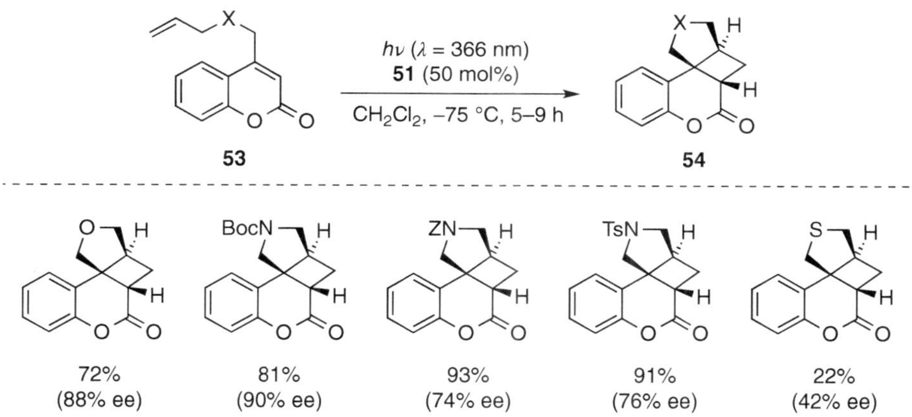

Scheme 11.23 Reaction scope of photochemical 2+2 using chiral Lewis acid promoter **51**.

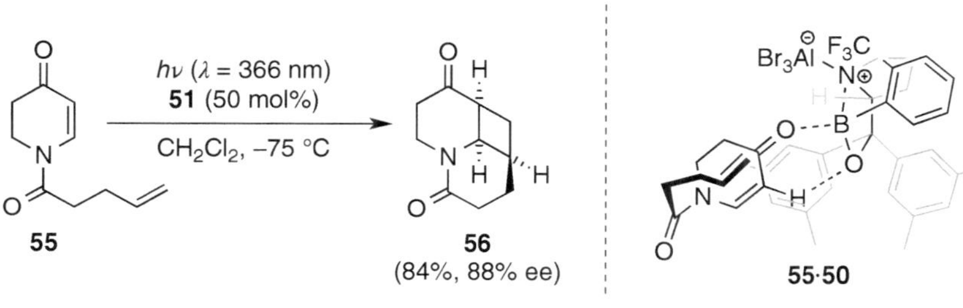

Scheme 11.24 Photochemical 2+2 cycloaddition of **55** utilizing Lewis acid complex **51** to induce enantioselectivity.

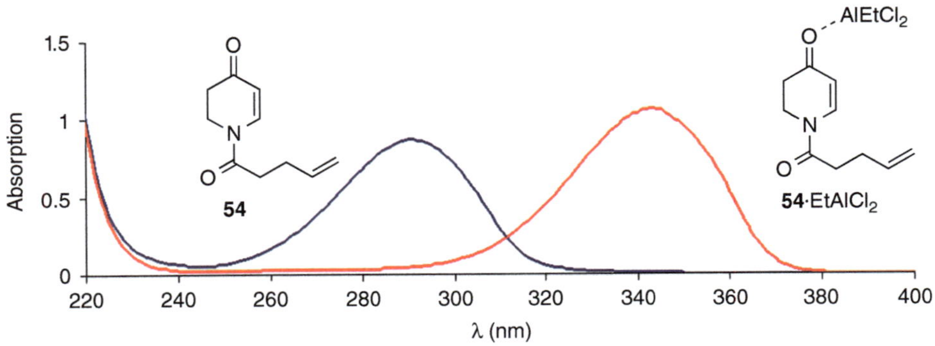

Figure 11.7 Bathochromic absorbance shift of **54** upon Lewis acid complexation.

of the reaction is highly desirable in terms of synthetic applications. Upon coordination to Lewis acids, 5,6-dihydro-4-pyridones display a large bathochromic absorbance shift; in the case of complexation with $EtAlCl_2$ or BCl_3, the shift was determined to be of more than 50 nm (Figure 11.7).

Using the same catalyst **51** as described in the previous section, selective excitation of the Lewis acid-coordinated enone was possible upon irradiation at 366 nm at low temperature (−70 °C). Thus, irradiation of substrate **55** generated the desired photoadduct as a single isomer in 84% yield and with high enantioselectivity (88% ee; Scheme 11.24). The proposed coordination mode of enone **55** to the catalyst is depicted (Scheme 11.24); the complex **55 51** ensuring the preferential approach of the pendant olefin to the *Si*-face of the enone (relative to the α-carbon atom of enone **55**).

Reducing the Lewis acid loading led to a decrease in enantioselectivity (64% ee using 40 mol% **51**; 50% ee using 30 mol% **51**), presumably due to an increase in the contribution of the racemic background reaction at lower concentrations of Lewis acid, in which a higher concentration of noncomplexed enone is present. It should be noted that these photoreactions, when carried out in the presence of Lewis acid, do not proceed faster than in the absence of Lewis acid, and thus cannot strictly be termed "catalytic" with respect to the Lewis acid. In fact, these reactions are significantly slower in the presence of Lewis acid than in its absence. However, it seems that the preferential absorption of light by the complexed substrate compared with the noncomplexed substrate is sufficient to confer high enantioselectivity in these processes. Irradiation using longer wavelength light (419 nm) was also investigated in an attempt to further suppress the racemic background reaction. However, under these conditions the reaction was found to proceed extremely slowly, with only a small amount of product observed even after extended irradiation times.

Six other substrates were also investigated under the same conditions, generating the desired products in generally high yields and with good enantioselectivities (80–90% ee; Figure 11.8). In the case of product **58**, a lower yield was obtained due to a competing side reaction. Of particular note, product **60** is a highly versatile intermediate, and was converted to the lupin alkaloid (+)-lupinine in a further

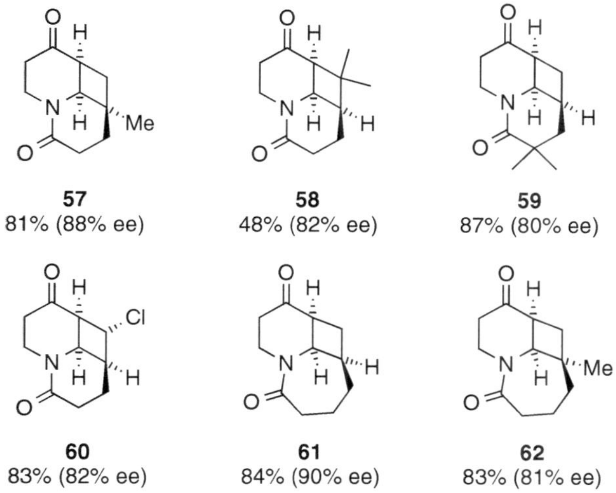

Figure 11.8 Reaction scope of intramolecular photochemical 2+2 cycloaddition utilizing Lewis acid complex **51** for enantioinduction.

five synthetic steps, as well as applied in an enantioselective formal synthesis of (+)-thermopsin.

In summary, the area of Lewis acid-mediated enantioselective photoreactions has until recently remained completely unexplored. Initial contributions in this area have established the viability of such an approach, and further study will no doubt extend the types of transformation that can be enabled using similar systems. In the case of coumarin substrates, coordination to a Lewis acid increases the rate of intersystem crossing to the triplet excited state, from which rapid photocycloaddition can occur. Although impressive results were obtained using this system, it does not serve as a general approach to enantioselective photocatalysis, since this change in mechanism upon Lewis acid coordination is specific to coumarins. In contrast, the relatively large bathochromic absorption shift observed upon coordination of Lewis acids to dihydropyridones allows selective excitation of the Lewis acid–dihydropyridone complex in the presence of noncomplexed dihydropyridone, and represents a much more general approach. Since other enone substrates were also shown to exhibit similar absorption phenomena upon coordination to Lewis acids, this method potentially allows access to a wide variety of enantioenriched products by simple application of chiral Lewis acid activators.

11.4 Conclusions and Outlook

Despite a relatively slow start, the field of enantioselective photocatalysis has matured into a fascinating research area, which is now able to offer levels of enantioselection that are competitive with ground-state catalytic asymmetric protocols. In the context of developing general enantioselective photocatalytic protocols, arguably the most promising avenues for further investigations center on chiral small-molecule photosensitizers (which are able to sensitize photochemical reactions whilst imposing a chiral environment on the substrate) and activation by chiral Lewis acids (through relatively large changes in a substrate's absorption characteristics upon complexation with a chiral Lewis acid), which emerged recently as a potentially general approach to asymmetric photocatalysis. Interest in the use of *visible* light in synthetic chemistry has increased dramatically in the past few years, and it will continue to play a key role in organic synthesis for the foreseeable future. For this reason, Alonso and Bach's recent disclosure of a photocatalyst capable of processing visible light to enable catalytic [2+2] photocycloadditions with a high degree of enantioselectivity is particularly noteworthy, and will no doubt provide much inspiration to drive forward future developments in enantioselective photochemistry.

Looking to the future, much work remains to be done. Where previous work has understandably focused on relatively specific systems and substrates, more general approaches to enantioselective photocatalysis are highly desirable, and would greatly facilitate the more widespread use of photochemical methods in organic synthesis. Moreover, while significant progress has been made in the realm of catalytic enantioselective unimolecular photoreactions, the corresponding bimolecular reactions have yet to be developed [49, 50]. The increased complexity involved renders the latter reactions particularly challenging, and generally applicable solutions will require highly creative and innovative approaches. The area of enantioselective photocatalysis remains a fascinating field of research; and with recent developments confirming that enantioselective photocatalysis is now able to compete with "classical" asymmetric synthesis, there has never been a more exciting time to be involved in this intriguing, continually evolving area.

References

1 For selected previous reviews on enantioselective photochemistry:(a) Rau, H. (1983) *Chem. Rev.*, **83**, 535–547; (b) Inoue, Y. (1992) *Chem. Rev.*, **92**, 741–570; (c) Müller, C. and Bach, T. (2008) *Aust. J. Chem.*, **61**, 557–564; (d) Yang, C. and Inoue, Y. (2014) *Chem. Soc. Rev.* doi: 10.1039/C3CS60339C

2 For books and monographs on enantioselective photochemistry:(a) Inoue, Y. and Ramamurthy, V. (2004) *Molecular and Supramolecular Photochemistry (Volume 11): Chiral Photochemistry*, Dekker, New York; (b) Austin, K.A.B. and Bach, T. (2012) Enantioselective photoreactions in solution, in *CRC Handbook of Photochemistry and Photobiology* (eds G. Axel, O. Michael, and G. Francesco), CRC Press, Boca Raton, pp. 177–200; (c) Bauer, A. and Alonso Ruiz, R. (2013) Templated enantioselective catalysis, in *Chemical Photocatalysis* (ed. B. König), de Gruyter, Berlin, pp. 67–86.

3 Nijland, A. and Harutyunyan, S.R. (2013) *Catal. Sci. Technol.*, **3**, 1180–1189.

4 (a) Iriondo-Alberdi, J. and Greaney, M.F. (2007) *Eur. J. Org. Chem.*, 4801–4815; (b) Hoffmann, N. (2008) *Chem. Rev.*, **108**, 1052–1103; (c) Bach, T. and Hehn, J.P. (2011) *Angew. Chem. Int. Ed.*, **50**, 1000–1045.

5 (a) Klán, P. and Wirz, J. (2009) *Photochemistry of Organic Compounds: From Concepts to Practice*, Wiley-Blackwell, Chichester; (b) Albini, A. and Fagnoni,

M. (eds) (2010) *Handbook of Synthetic Photochemistry*, Wiley-VCH, Weinheim; (c) Griesbeck, A., Oelgemöller, M., and Ghetti, F. (eds) (2012) *CRC Handbook of Photochemistry and Photobiology*, CRC Press, Boca Raton.
6 Feringa, B.L. and van Delden, R.A. (1999) *Angew. Chem. Int. Ed.*, **38**, 3418–3438.
7 Rau, H. (2004) Direct asymmetric photochemistry with circularly polarised light, in *Molecular and Supramolecular Photochemistry (Volume 11): Chiral Photochemistry* (eds Y. Inoue and V. Ramamurthy), Dekker, New York, pp. 1–46.
8 (a) Hammond, G.S. and Cole, R.S. (1965) *J. Am. Chem. Soc.*, **87**, 3256–3257; (b) Murov, S.L., Cole, R.S., and Hammond, G.S. (1968) *J. Am. Chem. Soc.*, **90**, 2957–2958.
9 Ouannès, C., Beugelmans, R., and Roussi, G. (1973) *J. Am. Chem. Soc.*, **95**, 8472–8474.
10 Becker, R.S., Edwards, L., Bost, R. *et al.* (1972) *J. Am. Chem. Soc.*, **94**, 6584–6592.
11 Vondenhof, M. and Mattay, J. (1990) *Chem. Ber.*, **123**, 2457–2459.
12 Inoue, Y., Shimoyama, H., Yamasaki, N., and Tai, A. (1991) *Chem. Lett.*, 593–596.
13 Balavoine, G., Jugé, S., and Kagan, H.B. (1973) *Tetrahedron Lett.*, 4159–4162.
14 Mislow, K., Axelrod, M., Rayner, D.R. *et al.* (1965) *J. Am. Chem. Soc.*, **87**, 4958–4959.
15 Demuth, M., Raghavan, P.R., Carter, C. *et al.* (1980) *Helv. Chim. Acta*, **63**, 2434–2439.
16 (a) Inoue, Y., Kunitomi, Y., Takamuku, S., and Sakurai, H. (1978) *J. Chem. Soc. Chem. Commun.*, 1024–1025; (b) Inoue, Y., Kunitomi, Y., Takamuku, S., and Sakurai, H. (1980) *J. Chem. Soc. Perkin Trans. 2*, 1672–1677.
17 (a) Inoue, Y., Yokoyama, T., Yamasaki, N., and Tai, A. (1989) *J. Am. Chem. Soc.*, **111**, 6480–6482; (b) Inoue, Y., Yokoyama, T., Yamasaki, N., and Tai, A. (1989) *Nature*, **341**, 225–226.
18 Inoue, Y., Yamasaki, N., Yokoyama, T., and Tai, A. (1992) *J. Org. Chem.*, **57**, 1332–1345.
19 Hoffmann, R. and Inoue, Y. (1999) *J. Am. Chem. Soc.*, **121**, 10702–10710.
20 (a) Kim, J.-I. and Schuster, G.B. (1990) *J. Am. Chem. Soc.*, **112**, 9635–9637; (b) Kim, J.-I. and Schuster, G.B. (1992) *J. Am. Chem. Soc.*, **114**, 9309–9317.
21 (a) Calhoun, G.C. and Schuster, G.B. (1984) *J. Am. Chem. Soc.*, **106**, 6870–6871; (b) Calhoun, G.C. and Schuster, G.B. (1986) *J. Am. Chem. Soc.*, **108**, 8021–8027.
22 (a) Tamaki, T., Kokubu, T., and Ichimura, K. (1987) *Tetrahedron*, **43**, 1485–1494; (b) Nakamura, A. and Inoue, Y. (2003) *J. Am. Chem. Soc.*, **125**, 966–972.
23 Ke, C., Yang, C., Mori, T. *et al.* (2009) *Angew. Chem. Int. Ed.*, **48**, 6675–6677.
24 Ueda, T., Kanomata, N., and Machida, H. (2005) *Org. Lett.*, **7**, 2365–2368.
25 Maeda, R., Wada, T., Mori, T. *et al.* (2011) *J. Am. Chem. Soc.*, **133**, 10379–10381.
26 Bach, T., Bergmann, H., Grosch, B. *et al.* (2001) *Synthesis*, 1395–1405.

27 (a) Bach, T. and Bergmann, H. (2000) *J. Am. Chem. Soc.*, **122**, 11525–11526; (b) Selig, P. and Bach, T. (2008) *Angew. Chem. Int. Ed.*, **47**, 5082–5084; (c) Selig, P., Herdtweck, E., and Bach, T. (2009) *Chem. Eur. J.*, **15**, 3509–3525; (d) Coote, S.C. and Bach, T. (2013) *J. Am. Chem. Soc.*, **135**, 14948–14951.

28 (a) Bach, T., Bergmann, H., and Harms, K. (2000) *Angew. Chem. Int. Ed.*, **39**, 2302–2304; (b) Albrecht, D., Vogt, F., and Bach, T. (2010) *Chem. Eur. J.*, **16**, 4284–4296; (c) Austin, K.A.B., Herdtweck, E., and Bach, T. (2011) *Angew. Chem. Int. Ed.*, **50**, 8416–8419.

29 Bach, T., Bergmann, H., and Harms, K. (2001) *Org. Lett.*, **3**, 601–603.

30 Bach, T., Grosch, B., Strassner, T., and Herdtweck, E. (2003) *J. Org. Chem.*, **68**, 1107–1116.

31 (a) Bach, T., Aechtner, T., and Neumüller, B. (2001) *Chem. Commun.*, 607–608; (b) Bach, T., Aechtner, T., and Neumüller, B. (2002) *Chem. Eur. J.*, **8**, 2464–2475.

32 (a) Grosch, B., Orlebar, C.N., Herdtweck, E. *et al.* (2003) *Angew. Chem. Int. Ed.*, **42**, 3693–3696; (b) Grosch, B., Orlebar, C.N., Herdtweck, E. *et al.* (2004) *Chem. Eur. J.*, **10**, 2179–2189.

33 Wiegand, C. and Herdtweck, E. (2012) *Chem. Commun.*, **48**, 10195–10197.

34 Bauer, A., Westkämper, F., Grimme, S., and Bach, T. (2005) *Nature*, **436**, 1139–1140.

35 For commentaries on reference 34, see: (a)a Inoue, Y. (2005) *Nature*, **436**, 1099–1100; (b) Wessig, P. (2006) *Angew. Chem. Int. Ed.*, **45**, 2168–2171.

36 Cauble, D.F., Lynch, V., and Krische, M.J. (2003) *J. Org. Chem.*, **68**, 15–21.

37 Müller, C., Bauer, A., and Bach, T. (2009) *Angew. Chem. Int. Ed.*, **48**, 6640–6642.

38 Müller, C., Bauer, A., Maturi, M.M. *et al.* (2011) *J. Am. Chem. Soc.*, **133**, 16689–16697.

39 Herkstroeter, W.G., Lamola, A.A., and Hammond, G.S. (1964) *J. Am. Chem. Soc.*, **86**, 4537–4540.

40 Maturi, M.M., Wenninger, M., Alonso, R. *et al.* (2013) *Chem. Eur. J*, **19**, 7461–7472.

41 Alonso, R. and Bach, T. (2014) *Angew. Chem. Int. Ed.*, **53**. doi: 10.1002/anie.201310997

42 (a) Lewis, F.D., Howard, D.K., and Oxman, J.D. (1983) *J. Am. Chem. Soc.*, **105**, 3344–3345; (b) Lewis, F.D. and Barancyk, S.V. (1989) *J. Am. Chem. Soc.*, **111**, 8653–8661.

43 Görner, H. and Wolff, T. (2008) *Photochem. Photobiol.*, **84**, 1224–1230.

44 (a) Hammond, G.S., Stout, C.A., and Lamola, A.A. (1964) *J. Am. Chem. Soc.*, **86**, 3103–3106; (b) Wells, P.P. and Morrison, H. (1975) *J. Am. Chem. Soc.*, **97**, 154–159; (c) Gallivan, J.B. (1970) *Mol. Photochem.*, **2**, 191–211.

45 Guo, H., Herdtweck, E., and Bach, T. (2010) *Angew. Chem. Int. Ed.*, **49**, 7782–7785.

46 Brimioulle, R., Guo, H., and Bach, T. (2012) *Chem. Eur. J.*, **18**, 7552–7560.

47 Brimioulle, R. and Bach, T. (2013) *Science*, **342**, 840–843.

48 (a) Comins, D.L. and Zheng, X. (1994) *J. Chem. Soc. Chem. Commun.*, 2681–2682; (b) Comins, D.L. and Zhang, Y.-M. (1998) *Chem. Commun.*, 2509–2510; (c) Comins, D.L., Zheng, X., and Goehring, R.R. (2002) *Org. Lett.*, **4**, 1611–1613.

49 For a rare example of an enantioselective bimolecular catalytic photoreaction, see: Asaoka, S., Wada, T., and Inoue, Y. (2003) *J. Am. Chem. Soc.*, **125**, 3008–3027.

50 For additional reading on this topic see: (a) Amador, A.G. and Yoon, T.P. (2016) A chiral metal photocatalyst architecture for highly enantioselective photoreactions. *Angew. Chem. Int. Ed.*, **55** (7), 2304–2306; (b) Mayr, F., Brimioulle, R., and Bach, T. (2016) A chiral thiourea as a template for enantioselective intramolecular [2+2] photocycloaddition reactions. *J. Org. Chem.*, **81** (16), 6965–6971; (c) Meggers, E. (2015) Asymmetric catalysis activated by visible light. *Chem. Commun.*, **51** (16), 3290–3301; (d) Wang, C. and Lu, Z. (2015) Catalytic enantioselective organic transformations via visible light photocatalysis. *Org. Chem. Front.*, **2** (2), 179–190; (e) Brimioulle, R., Lenhart, D., Maturi, M.M., and Bach, T. (2015) Enantioselective catalysis of photochemical reactions. *Angew. Chem. Int. Ed.*, **54** (13), 3872–3890; (f) Amador, A.G., Sherbrook, E.M., and Yoon, T.P. (2016) Enantioselective photocatalytic [3+2] cycloadditions of aryl cyclopropyl ketones. *J. Am. Chem. Soc.*, **138** (14), 4722–4725; (g) Yoon, T.P. (2016) Photochemical stereocontrol using tandem photoredox–chiral lewis acid catalysis. *Acc. Chem. Res.*, **49** (10), 2307–2315; (h) Blum, T.R., Miller, Z.D., Bates, D.M. *et al.* (2016) Enantioselective photochemistry through Lewis acid–catalyzed triplet energy transfer. *Science*, **354** (6318), 1391–1395; (i) Vallavoju, N., Selvakumar, S.; Jockusch, *et al.* (2014) Enantioselective Organo-Photocatalysis Mediated by Atropisomeric Thiourea Derivatives. *Angew. Chem. Int. Ed.* **53** (22), 5604-5608; (j) Maturi, M. M. and Bach, T. (2014) Enantioselective Catalysis of the Intermolecular [2+2] Photocycloaddition between 2-Pyridones and Acetylenedicarboxylates. *Angew. Chem. Int. Ed.* **53** (29), 7661-7664; (k) Brimioulle, R. and Bach, T. (2014) [2+2]-Photocycloaddition Reactions of 3-Alkenyloxycyclohex-2-enones: Enantioselective Lewis Acid Catalysis and Ring Expansion. *Angew. Chem. Int. Ed.* **53** (47), 12921–12924; (l) Coote, S. C. and Bach, T. (2015) Enantioselective Template–Directed [2+2] Photocycloadditions of Isoquinolones: Scope, Mechanism and Synthetic Applications. *Chem. Eur. J.* **21** (18), 6906-6912; (m) Brimioulle, R., Bauer, A. and Bach, T. (2015) Enantioselective Lewis Acid Catalysis in Intramolecular [2+2] Photocycloaddition Reactions: A Mechanistic Comparison between Representative Coumarin and Enone Substrates. *J. Am. Chem. Soc.* **137** (15), 5170-5176; (n) Maturi, M. M., Pöthig, A. and Bach, T. (2015) Enantioselective Photochemical Rearrangements of Spirooxindole Epoxides Catalyzed by a Chiral Bifunctional Xanthone. *Aust. J. Chem.* **68** (11), 1682-1692; (o) Tröster, A., Alonso, R., Pöthig, A. and Bach, T. (2016) Enantioselective Intermolecular [2+2] Photocycloaddition Reactions of 2(1*H*)-Quinolones Induced by Visible Light Irradiation. *J. Am. Chem. Soc.* **138** (25), 7808-7811; (p) Edtmüller, V., Pöthig, A. and Bach, T. (2017) Enantioselective Photocyclization Reactions of 2-Aryloxycyclohex-2-enones Mediated by a Chiral Copper-Bisoxazoline Complex. *Tetrahedron* **73** (33), 5038-5047; (q) Huang, X.; Quinn, T. R.; Harms, K. *et al.* Direct Visible-Light-Excited Asymmetric Lewis Acid Catalysis of Intermolecular [2+2] Photocycloadditions. *J. Am. Chem. Soc.* **139** (27), 9120-9123.

12

Photomediated Controlled Polymerizations

Nicolas J. Treat[1], Brett P. Fors[2], and Craig J. Hawker[1]

[1] *University of California, Materials Research Laboratory, Santa Barbara, CA 93106, USA*
[2] *Cornell University, Department of Chemistry, Ithaca, NY 14853, USA*

A variety of controlled chain-growth methods that enable the production of low dispersity polymers with targeted molecular weights and complex macromolecular architectures have been developed in recent years, giving nonexperts access to new functional materials. At the forefront of these methods are controlled radical and ring-opening methods, including atom transfer radical polymerization (ATRP) [1, 2], reversible addition-fragmentation chain transfer polymerization (RAFT) [3], nitroxide-mediated polymerization (NMP) [4, 5], and ring-opening metathesis polymerization (ROMP) [6].

Further inspired by nature's elegant and complex synthetic approach to macromolecules, researchers have recently focused on developing externally regulated polymerizations [7]. To this end, many groups have reported the use of allosteric, chemical, electrochemical, mechanical, and photochemical methods for activating/deactivating polymerizations [7]. Of these, light is an exceptionally attractive external stimulus as it can be easily tuned via wavelength and intensity, and it provides a noninvasive method for temporal regulation. This ability is clearly demonstrated in the broad area of photoinitiated polymerizations which have been used industrially in areas such as coatings, adhesives, inks, and microelectronics for decades [8]. These methods rely on the photoinitiation of free radical or cationic polymerizations to produce polymers with poorly defined architecture, dispersities, and molecular weights [8, 9]. Significantly, these prior developments in photolithography and photocuring provide a platform of infrastructure that may enable the rapid development of spatially and temporally controlled light-mediated strategies.

As a result, there has been a surge of interest in the field of photomediated controlled polymerizations. Typically, these reactions exhibit accurate molecular weight control, low polydispersities, living chain ends, and efficient regulation of the chain-growth process with light (Figure 12.1). Ideally, the most mild conditions should be used, including room temperature reactions, a low-intensity as well as a low-energy light source, and the ability to polymerize a

Visible Light Photocatalysis in Organic Chemistry, First Edition.
Edited by Corey R. J. Stephenson, Tehshik P. Yoon and David W. C. MacMillan.

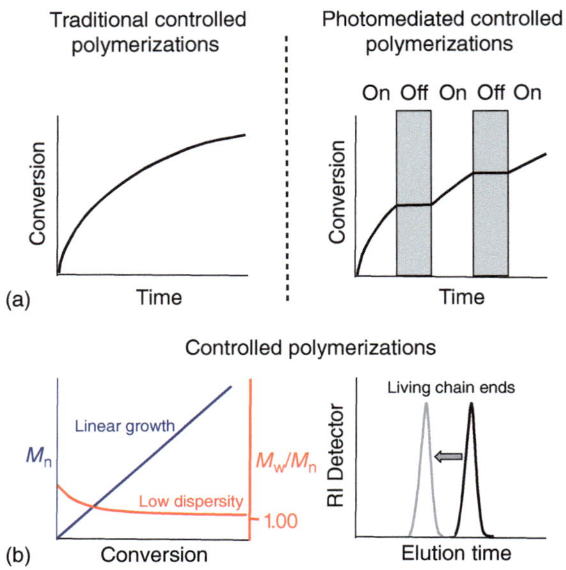

Figure 12.1 (a) Comparison of expected kinetic behavior for controlled polymerizations versus photomediated controlled polymerizations and (b) characteristics expected for typical controlled polymerizations.

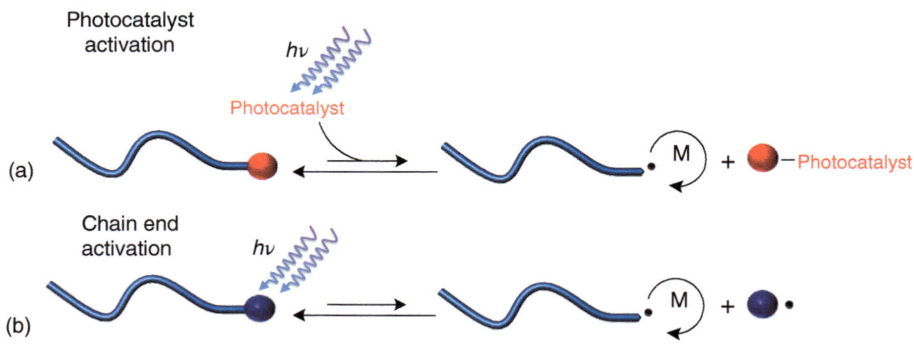

Figure 12.2 (a) Representative photocatalyst activation for photomediated polymerization and (b) chain-end-activated polymerizations.

variety of monomer families (i.e., methacrylates, acrylates, styrenics, etc.) while maintaining excellent functional group tolerance (i.e., amines, alcohols, halides). This chapter highlights recent advances in photocontrolled polymerizations with special emphasis on the aforementioned attributes, as well as the mechanisms involved in each system described.

Within the field of photomediated controlled polymerizations, two primary modes of photoregulation have been developed: photocatalyst activation and activation of a photoreactive chain end (Figure 12.2). Special emphasis is put on photocatalyst-activated systems, as they do not require specialized chain ends for polymerization to occur, and provide more robust polymers. However, chain-end photoactivatable systems are also described, as they remain an important area of photoregulated polymerizations.

12.1 Catalyst Activation by Light

12.1.1 Cu-Catalyzed Photoregulated Atom Transfer Radical Polymerizations (photoATRP)

ATRP is one of the most powerful methods for synthesizing well-defined materials, providing access to an array of functional polymers with varying architectures using commercially available catalyst-ligand systems [1]. The classical form of ATRP takes advantage of a transition-metal-mediated redox equilibrium with an alkyl bromide initiator to gain control over polymerization. Mechanistically, these reactions rely on the reversible formation of propagating radicals in sufficiently low concentration to avoid termination events (Figure 12.3).

Although ATRP has been worked on for over 20 years, it was not until 2010 that initial efforts to perform photoregulated atom transfer radical polymerization (photoATRP) were conducted by employing a dithiocarbamate as initiator in place of the traditionally used activated halide initiators [10]. The use of a dithiocarbamate was inspired by the iniferter (initiator–transfer–termination) chemistry developed in the Otsu Laboratory in the 1980s [11], wherein dithiocarbamates were cleaved photochemically to reversibly initiate and terminate chain ends. The initial photoATRP system employed CuBr to activate a dithiocarbamate upon UV irradiation at room temperature and was shown to exhibit moderate control ($M_w/M_n < 1.30$) at low conversions (<25%). Furthermore, the polymerization rate was greatly retarded (25-fold) when the light was removed from the system (Figure 12.4). However, due to residual CuBr, polymerization continued in the dark, and the low-molecular-weight distributions were only reported for low conversions. Nonetheless, this was a significant step forward in demonstrating a photomediated ATRP process.

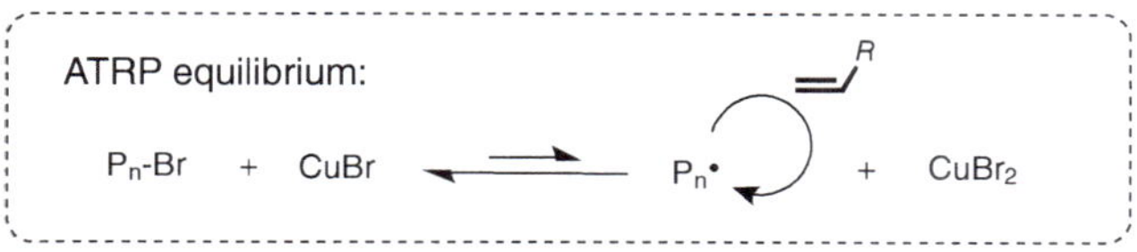

Figure 12.3 Traditional ATRP redox equilibrium.

Figure 12.4 Photocontrolled ATRP using alkyl dithiocarbamates showing rate retardation and enhancement with cycling of the light source. (Reprinted with permission from reference [10]. Copyright 2010 American Chemical Society.)

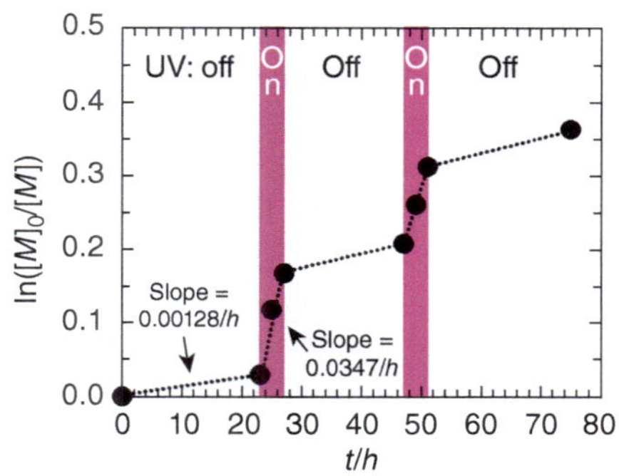

Following this, Yagci and coworkers were the first to report a photoregulated ATRP process using a typical alkyl bromide initiator [12, 13]. A 350-nm light source (3 mW cm^{-2}) was used to activate Cu(II)Br_2 with *N,N,N′,N′,N″*-pentamethyldiethylenetriamine (PMDETA) as a ligand for the polymerization of methyl methacrylate (MMA). These reactions showed good control ($M_w/M_n < 1.30$) up to moderate conversions (60–80%), with a linear increase in molecular weight with conversion and first-order kinetics. Control experiments in the absence of light showed no polymerization, and, importantly, when the reaction's exposure to light was cycled, very little polymerization occurred in the dark. Chain extension experiments were also conducted to further confirm the living nature of the polymerization. Importantly, this was the first example of photochemical formation of Cu(I)Br from Cu(II)Br_2. However, these polymerizations had an inhibition period which was attributed to the time to generate an appreciable amount of Cu(I) species; and, furthermore, these experiments needed high catalyst loadings (1 : 1 Cu(II) to initiator).

Following this report, Mosnáček *et al.* developed a photoATRP using only 50–100 ppm of Cu(II)Br_2 for the polymerization of MMA using a mercury lamp coupled with a pyrex filter ($\lambda > 350$ nm, 20 mW cm^{-2}) [14]. In this system, catalysts based on both PMDETA and tris(2-pyridylmethyl)amine (TPMA) were shown to exhibit controlled polymerization behavior even at high (60–80%) conversions. An initial proof of photoregulation was demonstrated, with little conversion observed when the light was turned off. Living chain ends were demonstrated using chain extension experiments. However, because of the intense light source used in these reactions, temperatures from 30 to 35 °C were typically observed.

Matyjaszewski and coworkers further extended the photoATRP system using 100 ppm of Cu(II)Br_2 to polymerize acrylates [15]. Various acrylate monomers were polymerized using tris((4-methoxy-3,5-dimetylpyridin-2-yl)methyl)amine (TPMA*) as ligand with blue (450 nm LEDs), violet (392 nm LEDs), and solar irradiation (Figure 12.5), where solar irradiation showed the fastest kinetics. Lower energy red light (650 nm LEDs) showed no polymerization. Importantly, this was the first example of efficient photoregulation being observed using a Cu-based catalyst under such mild light sources, with no polymerization observed in the absence of light. The conditions also enabled these polymerizations to be conducted in water, giving oligoethylene oxide-based methacrylate polymers. Analogously, Yagci and coworkers developed a photomediated ATRP in inverse microemulsions that showed good control over the polymerization of oligoethyloxide methacrylates [16].

Jordan and coworkers extended Cu-based photoATRP to the use of a standard fluorescent lamp for the polymerization of MMA [17]. Further, visible light was used to activate Cu(II)Br_2/PMDETA for surface-initiated polymerizations on silicon surfaces that were functionalized with a pre-patterned ATRP initiator. Spatial control over this process was also demonstrated using photomasks to selectively grow polymer brushes. This was the first example of using a Cu-based catalyst to achieve spatial control of polymer brush growth on surfaces.

Haddleton and coworkers also reported a photoinduced living radical polymerization of acrylates, using both sunlight and a UV light source ($\lambda_{max} = 360$ nm) [18]. Methyl acrylate was shown to be effectively polymerized using the UV light

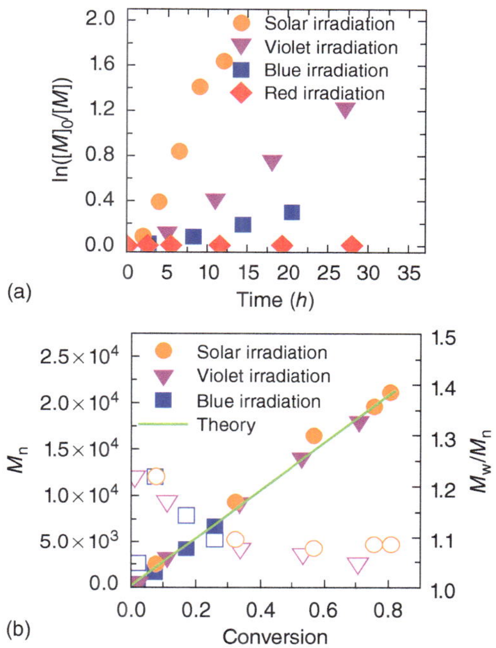

Figure 12.5 (a) Kinetics and (b) M_n and M_w/M_n evolution in polymerization of methyl acrylate using different irradiation sources under the following conditions: [MA]/[EBiB]/[$CuBr_2$]/[TPMA*] = 300 : 1 : 0.03 : 0.135 in 50 vol% DMF at room temperature. (Reprinted with permission from reference [15]. Copyright 2012 American Chemical Society.)

source and Cu(II)Br_2 in DMSO. This system gave characteristics similar to the traditional single-electron transfer living radical polymerizations that have been performed in DMSO previously, with living behavior up to very high conversions (>95%), and with low dispersities ($M_w/M_n < 1.10$). This method also allowed for one-pot block copolymer formation and was shown to polymerize a variety of acrylate monomers. This initial report was expanded to also include the photoinduced living radical polymerization of a variety of acrylates including lauryl, octadecyl, and diethylene glycol ethyl ether acrylate [19]. Further, this method was later used to synthesize decablock copolymers [20].

Haddleton and coworkers expanded on their initial work by isolating a preformed Cu(II) formate complex which could be used without any additional reducing agents or ligand to give identical photoregulated controlled radical polymerizations [21]. Control experiments showed that having equimolar amounts of Cu(II)Br_2 and Me_6TREN gave no polymerization; but in the presence of either excess ligand or sodium formate, control over the polymerization was achieved with characteristics similar to the optimized system. Ishihara and coworkers have also expanded this Cu(II)Br_2 photomediated system to using TPMA in methanol to polymerize zwitterionic monomers such as 2-methacryloyloxyethyl phosphorylcholine [22]. Importantly, this extends the scope of Cu-based photomediated ATRP to more difficult monomer types.

A vast body of mechanistic work has been conducted to better understand photoregulated Cu-catalyzed ATRP [23–27]. Perhaps the most experimentally rigorous evaluation of the mechanism came from the groups of Haddleton and Barner-Kowollik, where a combination of pulsed-laser polymerization ($\lambda = 350$ nm) and high-resolution mass spectrometry was used to study the Cu(II)/Me_6TREN/DMSO system. Through this evaluation, it was found that initiation of polymerization can be through the following: UV-light-induced C—Br bond scission of the initiator (at 350 nm), Cu(I) activation of initiators after its reduction from Cu(II) via an electron transfer from a photoexcited amine ligand, and the reduction of Cu(II) to Cu(I) via an excited copper species oxidizing the amine ligand (Figure 12.6). The combination of these processes and their relative contributions can vary based on light sources and reaction conditions. For example, a fluorescent lamp will likely not be capable of directly inducing C—Br bond scission, and complexes based on ligands such as PMDETA will likely not absorb as much light as those comprised of TPMA. Further, using high-intensity lasers to induce polymerization for these studies, likely causes some experimental differences from less intense light sources typically employed. UV–vis spectroscopy was also used to verify that the Cu(II) was being consumed under experimental conditions, showing a decrease in absorption upon irradiation both in the presence and absence of monomer (Figure 12.7). In a separate study, Matyjaszewski and coworkers reported a mechanistic analysis using a combination of experimental and theoretical data to confirm under their light source that a photoreduction of Cu(II) with excess tertiary amine-based ligand is the primary mode of generating Cu(I) for polymerization in their system [25].

12.1.2 Photomediated ATRP with Non-Copper-Based Catalyst Systems

Hawker and coworkers reported one of the earliest truly photocontrolled polymerizations, which was catalyzed by an iridium-based catalyst [28]. Fifty parts per million of a *fac*-[Ir(ppy)$_3$] (Figure 12.8a) was shown to efficiently control the polymerization of MMA in the presence of an alkyl bromide initiator. Excellent control over the polymerization was maintained even after multiple iterations of cycling the reaction's exposure to light. In addition, these polymerizations

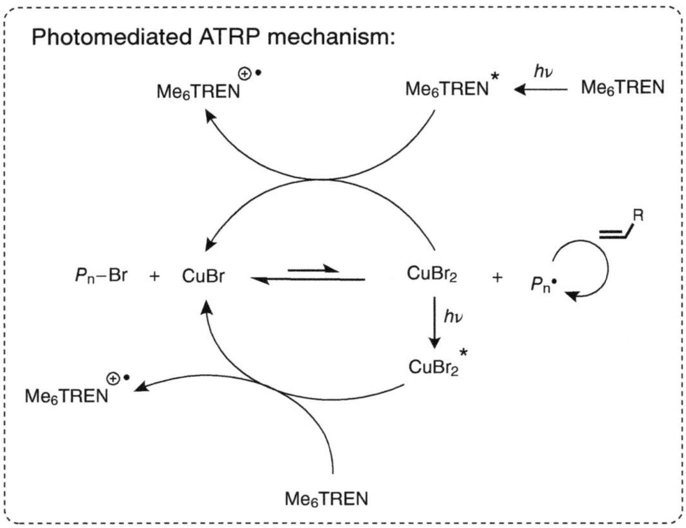

Figure 12.6 Mechanism for photomediated ATRP using Cu(II)Br$_2$ and electron-rich amine ligands.

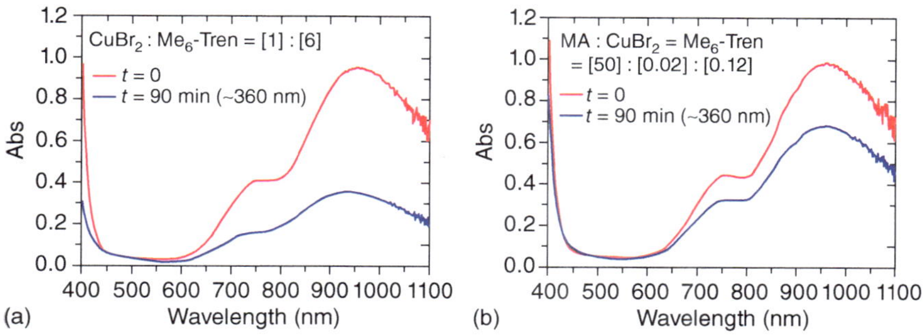

Figure 12.7 UV–vis spectroscopy giving evidence for a decrease in Cu(II) concentration in the presence of excess ligand under irradiation.

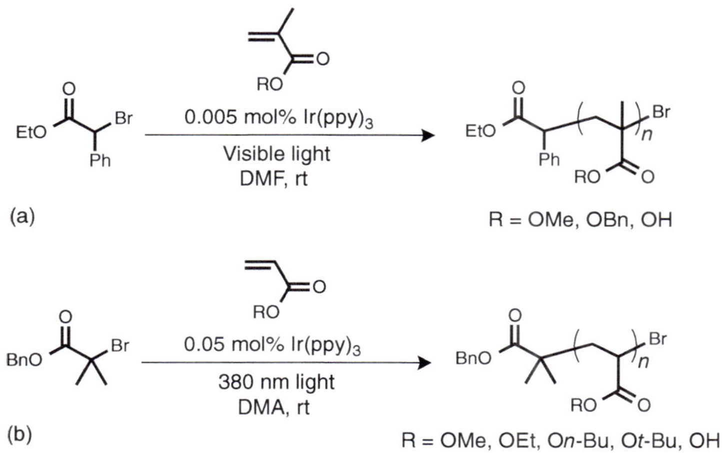

Figure 12.8 Scope of Ir-catalyzed photomediated polymerizations.

displayed living characteristics and enabled the efficient synthesis of block copolymers (PMMA $M_w/M_n = 1.28$, PMMA-*b*-PBnMA $M_w/M_n = 1.30$). Polymerization of methacrylic acid could also be performed with this system ($M_w/M_n = 1.61$), demonstrating the iridium catalyst's high functional group tolerance.

The mechanism of this process was proposed to occur through a pathway previously reported for a photomediated atom transfer radical addition [29]. Excitation of the Ir(III) complex with visible light affords a species that will reduce the alkyl bromide, resulting in the formation of an Ir(IV) complex and a propagating radical (Figure 12.10a). The highly oxidizing Ir(IV) species can then deactivate polymerization via chain-end oxidation to afford the dormant alkyl bromide. This process is fundamentally different from the previously discussed photoATRP using Cu-halide catalysts, as each propagating event is mediated by light, and does not rely on the traditional ATRP redox equilibrium.

Photomediated Ir-based polymerizations were also extended to acrylate monomers with increased catalyst loadings (500 ppm Ir(ppy)$_3$, Figure 12.8b) [30].

Well-defined polymers of methyl, *n*-butyl, and *t*-butyl acrylate were demonstrated, and acrylic acid could be randomly copolymerized with ethyl acrylate with control up to 50 mol% acid incorporation. In contrast to traditional ATRP, block copolymers were produced without any monomer sequence requirements (i.e., poly(acrylate)-*b*-(acrylate), poly(methacrylate)-*b*-(acrylate), and poly(acrylate)-*b*-(methacrylate)). This report demonstrated the ability of nontraditional Ir-based photocatalysts to be a platform for polymer synthesis that is complementary to Cu-based ATRP.

Further, a variety of other photocatalysts that have highly reducing excited states have also shown much promise to be useful for ATRP-type polymerizations. A dinuclear gold(I) complex, $[Au_2(dppm))_2]Cl_2$, was shown to very efficiently reduce alkyl bromide initiators when shining a UV light source or sunlight on the solution [31]. Various wavelengths of UV light (i.e., 300, 350, 400 nm) were used to conduct polymerization with 1.25 mol% Au photocatalyst to polymerize MMA in DMF. Although dispersities were high ($M_w/M_n > 1.5$), chain extension of PMMA homopolymers showed living behavior. However, the molecular weight with conversion was not tracked for these experiments, and on/off kinetics were not demonstrated. Thus, this system appears to have promise for the development toward a photomediated ATRP, but further work is necessary.

A niobium nanoparticle system has also been developed for the polymerization of a variety of acrylate monomers using visible light [32]. Good polymerization characteristics were observed for this system when polymerizing *N*-isopropylacrylamide (NIPAM), with a linear increase in molecular weight with conversion (up to 70%), and good on/off behavior, with no residual polymerization in the absence of light. Using niobium nanoparticles also allowed recycling of the photocatalyst through centrifugation. Further, niobium nanoparticles were shown to be tolerant to a variety of monomer systems, including MMA, acrylic acid, and a variety of acrylates.

Recently, Yang and coworkers also reported a photomediated ATRP, employing a photoredox Cu catalyst [33]. Bis(1,10-phenanthroline)copper(I) was used as it has a strong absorption band in the visible regime, allowing a ~1 W blue LED to be used for polymerization. Mechanistically, this differs from previous photoATRP results as it relies on the copper complex to enter an excited state that will reduce the alkyl bromide to undergo propagation. Controlled characteristics were observed during the polymerization, including first-order kinetics, linear increase in molecular weight with conversion (up to ~50%), and low dispersities. It was also found that the addition of triethylamine (TEA) as a reducing agent helped speed up the reaction. Further, an Ir complex base on 2-(2′-benzothienyl)pyridine ligands ($Ir(btp)_2(tmd)$) was also reported for controlled polymerization of MMA under a similar mechanism with blue LEDs [34]. In each case, these catalysts were demonstrated to efficiently stop and start polymerization in the absence and presence of light.

Hawker and coworkers more recently published a metal-free ATRP that relied on an organic photoredox catalyst, 10-phenylphenothiazine (PTH), to activate and deactivate chain ends with light (Figure 12.10b) [35]. PTH operated under a mechanism similar to that of the $Ir(ppy)_3$ catalyst, providing a metal-free alternative that offers significant advantages for numerous applications. Key to this system is that PTH is highly reducing in its excited state ($E^*_{red} = -2.1$ V vs SCE (saturated calomel electrode)) and upon oxidation forms a stable radical cation that is present for deactivating chain ends. This system was highly photocontrolled, with no loss in control even after multiple on/off cycles (Figure 12.9). It also offered broader monomer scope than the Ir-based system, giving control over dimethylaminoethyl methacrylate. Mass spectrometry was used to confirm the alpha and omega chain ends of the resulting polymers and compared with PMMA synthesized using traditional ATRP methods to validate the structure of these polymers. Finally, the PTH system showed living characteristics and was combined with other ATRP methods to produce a variety of block copolymers.

Following this work, Matyjaszewski and coworkers reported the use of phenothiazine-based photocatalysts for the controlled polymerization of acrylonitrile [36]. These polymerizations were conducted with a UV light source, and chain extension experiments of the polyacrylonitrile were conducted using MMA to confirm living behavior. Further, naphthalene and para-methoxy phenothiazine derivatives were also explored and were shown to give controlled polymerization of acrylonitrile. Although no evidence for photoregulation was given in this report, it is likely that a photomediated system is also present.

Miyake and coworkers also published a metal-free photoATRP with a perylene photocatalyst [37]. Photoregulation was demonstrated with first-order kinetics and moderately low dispersities ($M_w/M_n \sim 1.5$). However, a linear increase in molecular weight with conversion was not observed for this experiment. Mass spectrometry was used to verify polymer structure, and an initial block copolymerization experiment demonstrated some living PMMA chain ends. This study unveiled a second catalyst class for metal-free ATRP, and has potential for further development (Figure 12.10).

A mechanistically distinct Ruthenium catalyst system has been developed that is photoregulated via ligand control [38]. Irradiation at 60 °C causes ligand dissociation from the metal center, creating a 16-electron species that undergoes ATRP in the presence of an alkyl halide initiator (Figure 12.11). Light could be used to turn on and off these reactions due to the reversibility of the ligand coordination. However, broad molecular weight distributions ($M_w/M_n = 1.4–1.6$) were observed when polymerizing MMA, styrene, and butyl acrylate. A combination of ^{1}H NMR and cyclic voltammetry provided good evidence for the proposed ligand dissociation mechanism. However, ^{1}H NMR also indicated some catalyst degradation, showing 35% decomposition after 6 h of irradiation. The authors propose that the degradation is likely the formation of a catalytically inactive dimeric Ru species. This is an interesting and promising form of photopolymerization, but further development is needed to fully demonstrate a living system.

12.1.3 Iodine-Mediated Photopolymerizations

Controlled radical polymerization using alkyl iodide initiators coupled with various catalyst systems have also been well studied. Due to the lability of the iodo chain end, this reduction event is much more facile than the carbon-bromide

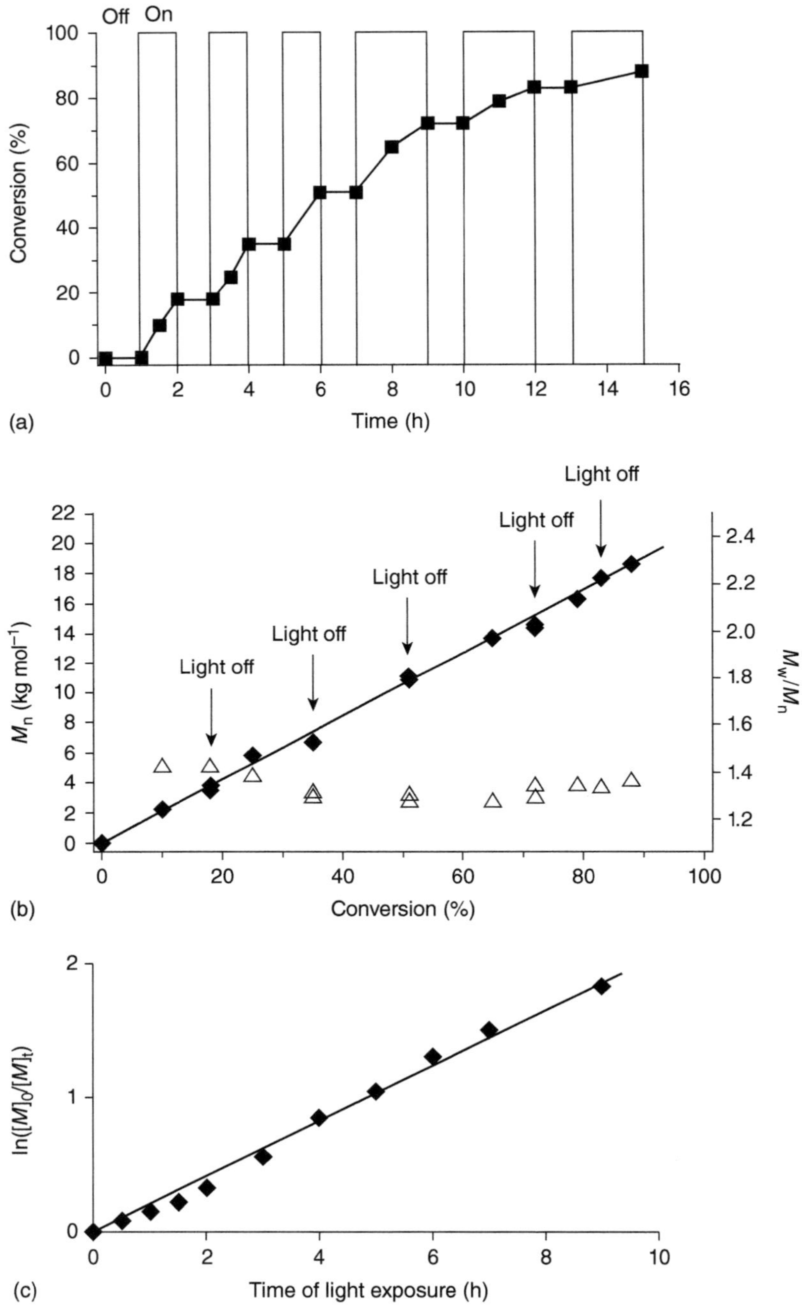

Figure 12.9 Polymerization of benzyl methacrylate using PTH with repeated "on–off" cycling of the reaction to light. (Reprinted with permission from reference [35]. Copyright 2014 American Chemical Society.)

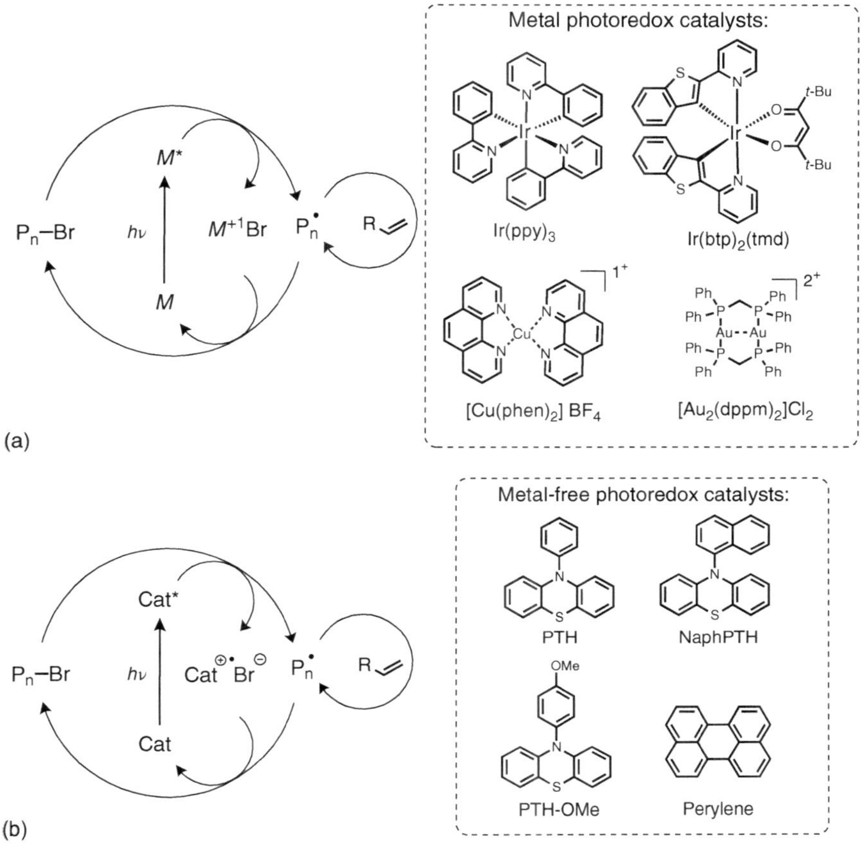

Figure 12.10 Photoredox catalysts used in photoATRP.

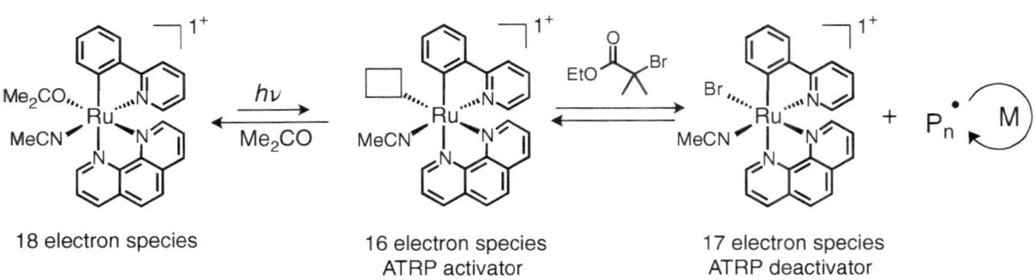

Figure 12.11 Photoregulated polymerization via ligand control.

bond that is typically utilized for ATRP. However, this also leads to difficulty in livingness, as chain-end degradation can readily occur.

The first report of light-mediated iodo-chain-end polymerization was from Koumura et al. [39–41]. A dinuclear manganese carbonyl complex [$Mn_2(CO)_{10}$] was shown to control polymerization of methyl acrylate, vinyl acetate, and styrene at 40 °C under a 27 W fluorescent bulb using an alkyl iodide initiator.

Although photoregulation was observed for vinyl acetate, a steady loss in control was observed at higher conversion (i.e., increase in polydispersity), indicating that chain ends were not efficiently capped in the dark. Styrene gave the best control for this system, with a linear increase in molecular weight with conversion and acceptable molecular weight distributions ($M_w/M_n \sim 1.3$). Mechanistically, it was proposed that initiation was via the dinuclear manganese complex homolytically cleaving to form a manganese radical that then abstracted an iodo radical from the initiator, forming an Mn-I bond and propagating chain end. The Mn-I is proposed to recap the chain end when light is turned off.

Ma and coworkers have reported the visible-light-mediated polymerization of methacrylate monomers using perfluoro-1-iodohexane as initiator with Ir(ppy)$_3$ as catalyst [42]. Mechanistically, this work was proposed to be similar to the originally reported light-mediated polymerizations of methacrylates using Ir(ppy)$_3$. This report demonstrated that the same Ir catalyst can be extended to iodo chain ends and showed control over polymerization with fluorinated and glycidyl methacrylate monomers. Good photocontrol was observed for the system, with linear increase in molecular weight versus conversion.

Wolpers and Vana have reported the use of traditional photoinitiators coupled with alkyl iodide initiators for controlled radical polymerization of butyl methacrylate in bulk when irradiating with an 8 W mercury lamp ($\lambda > 366$ nm) [43]. This study also explored thermal polymerization of iodo chain ends, and found through a combination of experimental results and simulations that UV light gave polymers with lower dispersity and better control. It was hypothesized that the increase in control was due to the light-induced C-I bond cleavage causing a nanomolar buildup of free iodine that led to both a reversible termination (i.e., ATRP) and degenerative transfer (i.e., RAFT) mechanism during polymerization. Good photocontrol was observed over the process with no reaction when the light was turned off, indicating that the chain ends were activated primarily by light. Finally, elevated temperatures were also used in combination with UV light to increase polymerization rate but retain the added level of control afforded by light. This was an important demonstration of using an exceedingly simple system to gain control over polymerization, with light as activator and no added catalysts, ligand, or solvent necessary.

Another approach to gaining light control over iodine-based living radical polymerizations has been the use of amine catalysts [44]. This process used visible light ($\lambda = 350$–600 nm, 60 W xenon lamp) with tributylamine as catalyst (0.25–1 mol%). It was demonstrated that a variety of different methacrylates could be polymerized with low dispersities and targeted molecular weights. It was also shown that using different light intensities (300, 150, and 60 W) allowed for tuning the polymerization rate. A polymerization was conducted cycling the reaction's exposure to light, but using different light intensities for each irradiation period, demonstrating disparate rates and illustrating good photoregulation. The same group also reported the use of various photocatalysts (12–25 mol%, $\lambda_{max} = 530$, 600, and 720 nm) for living radical polymerization over a range of wavelengths [45]. Further, it was shown for the first time that red light could be used in combination with a carbocyanine dye ($\lambda_{max} = 720$ nm) to induce controlled polymerization (Figure 12.12). The mechanism was proposed

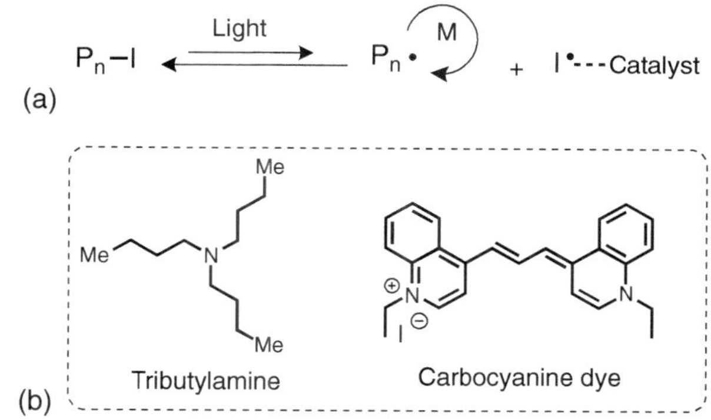

Figure 12.12 Proposed mechanism and representative catalysts for organic catalyzed iodine living radical polymerization.

to occur via energy transfer and complexation of the catalyst with the iodo radical species upon homolysis to retain control over the reaction. A variety of methacrylate monomer functionalities (i.e., hydroxyethyl, glycidyl, polyethylene glycol, 2-ethylhexyl, and dimethylamino methacrylate) were tolerated using this method, and various block copolymers were produced as well. Finally, using an initiator that was capable of initiating ring-opening as well as radical polymerization, light wavelength was modulated to conduct a one-pot polymerization of both MMA and δ-valerolactone, forming functional block copolymers.

12.1.4 Metal-Free Photomediated Ring-Opening Metathesis Polymerization

Contrary to controlled radical polymerizations, ROMP utilizes cyclic monomers that are activated via ring strain. These polymerizations typically employ a metal-based catalyst (i.e., Ru, W, Mo) that propagates via metathesis reactions on the chain end. A longstanding challenge of this field has been the development of metal-free catalysts, which necessitated the development of new mechanistic approaches to ROMP.

Boydston and coworkers have recently demonstrated a controlled ROMP with organic photocatalysts [46]. This seminal work demonstrated for the first time that a metal-free ROMP process is possible, opening the doors to broader applications for ROMP-produced polymers. The key to this process is the use of a pyrylium photoredox catalyst that, upon absorption of visible light, enters a highly oxidizing excited state that can undergo single-electron oxidation of a vinyl ether initiator (Figure 12.13). It was proposed that the oxidized vinyl ether forms a transient [2+2] complex with norbornene monomer that subsequently ring opens, undergoing an effective chain propagation. In the dark, the photocatalyst will reduce the chain end to form the stable vinyl ether and stop any propagation. A linear increase in molecular weight with conversion was demonstrated with norbornene, with dispersities from 1.3 to 1.5. Photoregulation was demonstrated, with very efficient stopping and starting of the reaction when the light was cycled on and off. The polymers prepared were compared to traditionally prepared ROMP polymers and shown to have identical properties, as evidenced by glass transition temperatures

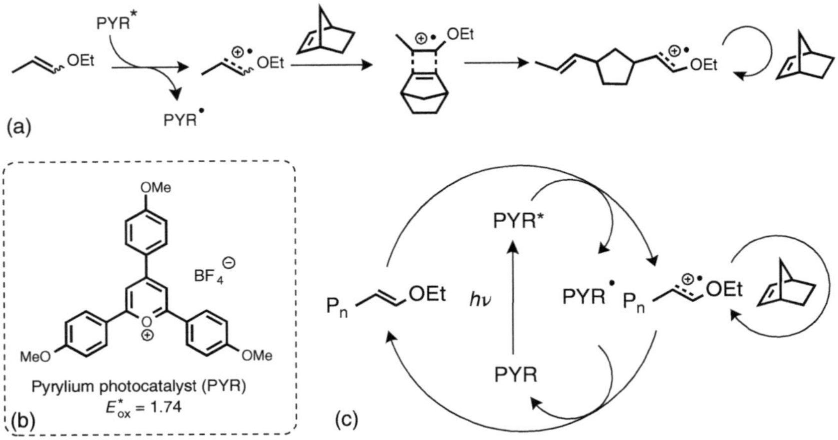

Figure 12.13 (a) Proposed mechanism for metal-free ROMP initiation and propagation. (b) Pyrylium photoredox catalyst employed with excited-state oxidation potential (vs SCE). (c) Mechanistic rational for photoregulation.

and ^{1}H NMR. Following this initial work, it was then demonstrated that this system also tolerated the polymerization of dicyclopentadiene [47]. Random copolymerizations of norbornene and dicyclopentadiene showed equivalent incorporations of the two monomers into the polymer backbone with no cross-linking. The resulting copolymers could then be subsequently cross-linked using thiol-ene click chemistry. One limitation was that homopolymerization of dicyclopentadiene stopped at low conversions (10–20%), greatly limiting the molecular weights produced (2–4 kg mol^{-1}). Mechanistic studies demonstrated that the predominant *endo*-dicyclopentadiene caused both steric congestion and intramolecular side reactions of the radical cation with the pendant alkene, both of which contributed to the lack of conversion for the polymerization. Importantly, this work demonstrated the metal-free production of cross-linked ROMP-based polymers for the first time, and showed that this process is not limited to norbornene as monomer.

Metal-free ROMP still has several limitations to overcome. No evidence for the ability to form block copolymers has been given, indicating that the polymer chain ends are not capable of being reinitiated, and the dispersities are higher than the metal ROMP processes, while the kinetics are much slower. Nonetheless, photoregulated metal-free ROMP is a significant advancement for the field, and it is anticipated that further development will lead to overcoming the aforementioned hurdles.

12.1.5 Photoregulated Reversible-Addition Fragmentation Chain Transfer Polymerizations (photoRAFT)

RAFT polymerization is one of the most commonly used techniques for producing well-defined polymers with access to advanced architecture and functionality. The mechanism of RAFT relies on a high chain transfer constant of thiocarbonyl-thio (i.e., trithiocarbamate, dithioester, etc.) species to establish an

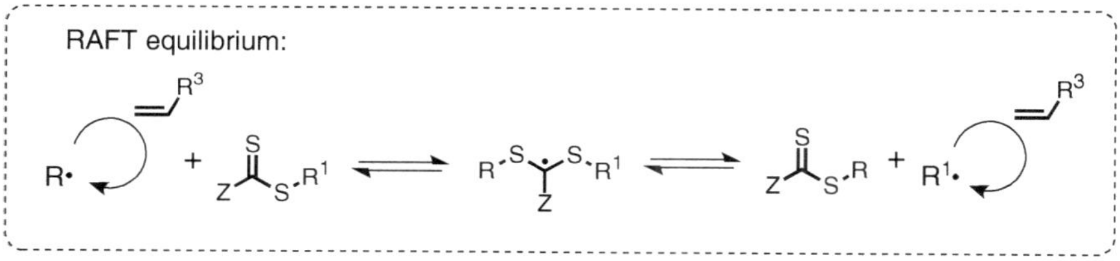

Figure 12.14 RAFT polymerization equilibrium operating via a chain (degenerative) transfer mechanism.

equilibrium of growing radical chain ends (Figure 12.14). RAFT is one of the most tolerant polymerization methods, allowing for a number of different functional groups to be polymerized with access to high conversions.

Several groups have reported the use of photoinitiated UV [48–53] and visible [54] light RAFT polymerizations. However, these polymerizations were not demonstrated to be photomediated, and thus is not the focus of this review, although the potential for photomediation in these systems remains. In 2009, Cai and coworkers were the first to report a photomediated RAFT process [55]. Polymerizations were conducted in acidic aqueous solution using a water-soluble trithiocarbamate (EDMAT, see Figure 12.15), (2,4,6-trimethylbenzoyl)diphenylphosphine oxide (TPO) as the photoinitiator, and N-(2-acryloyloxyethyl) pyrrolidone (NAP) as a water-soluble acrylic monomer. A mercury vapor lamp equipped with an optical filter ($\lambda = 405$–577 nm, Intensity $= 150\ \mu W\ cm^{-2}$) was used to irradiate the solution. A linear increase in molecular weight with conversion was demonstrated, with low dispersities and first-order kinetics. Further, block copolymers were also produced, validating living chain-end fidelity. Finally, on–off kinetics were conducted at 7 °C, with very little conversion occurring in the dark, illustrating the first attempts at a photoregulated RAFT system. However, low temperatures were required in order to achieve photoregulation, and a photoinitiator was present in the system, which may have aided in the reinitiation process when cycling the exposure to light. Later, the same group reported a photocontrolled polymerization using a visible-light-absorbing RAFT agent (CPADB, see Figure 12.15) to polymerize amino-functional methacrylamide monomers in water [56, 57]. In this case, photoregulation was demonstrated at 25 °C, but the use of a photoinitiator (TPO) was still required. In both of these initial reports, it was hypothesized that the origin of photocontrol was from a stabilized intermediate RAFT radical in the dark (Figure 12.14), although no experimental evidence was given to support this claim.

Following this, a photocontrolled RAFT process in the absence of any additional photoinitiator was demonstrated by Johnson and coworkers [58]. A trithiocarbonate (bis-norbornene TTC, Figure 12.15) was irradiated with an 8 W black light (peak emission at 352 nm) in the presence of N-isopropylacrylamide (NIPAM) at room temperature. Due to lack of an added external photoinitiator, this process relied solely on photocleavage of the C—S bond in the trithiocarbonate to initiate polymerization (Figure 12.15a). No reaction was observed in the dark, and efficient activation of the polymerization occurred upon exposure

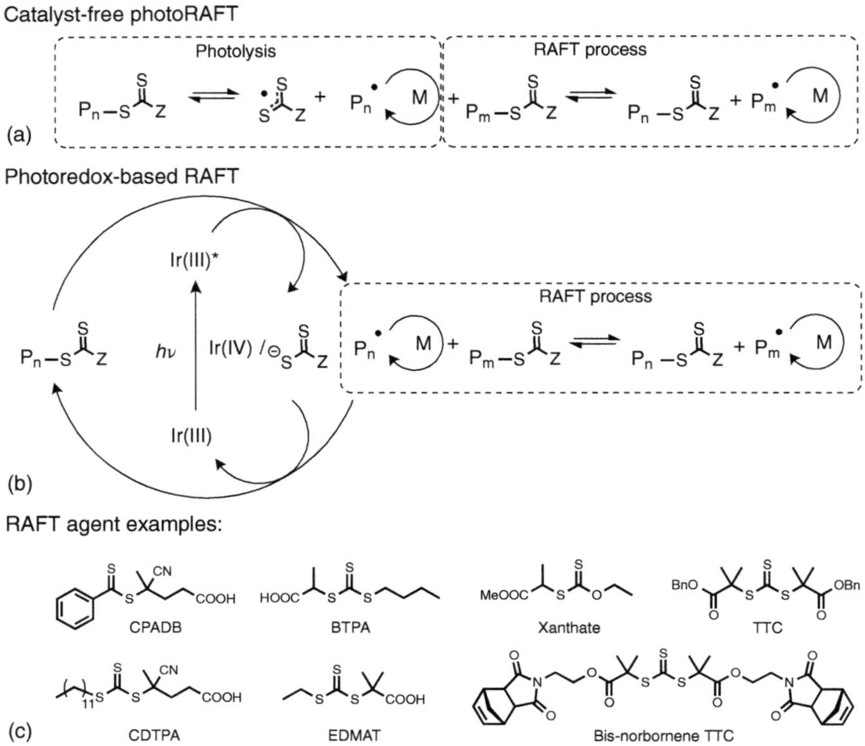

Figure 12.15 (a) Simplified mechanism of catalyst-free photomediated RAFT. (b) Proposed mechanism of RAFT performed using photoredox catalysts. (c) Examples of RAFT agents used in both catalyst-free and photoredox processes.

to light. Multiple "on"–"off" cycles were demonstrated with low dispersities ($M_w/M_n < 1.2$) observed up to 90% conversion. To verify livingness, a chain extension was carried out with NIPAM to form poly(NIPAM)-*b*-(NIPAM), with very efficient reinitiation of chain ends evidenced via the size exclusion chromatogram (SEC). These photopolymerizations were then used to synthesize a cross-linked polymer network of NIPAM utilizing the same RAFT agent. Subsequent chain extension of the network with sunlight demonstrated the novel concept for conversion of solar energy to mass in bulk materials.

A similar method was then adapted to the use of a continuous-flow setup, allowing access to rapid large-scale synthesis of polymers [59]. TTC (see Figure 12.15) was used as RAFT agent, and again irradiated using UV light (peak emission at 352 nm). The livingness as well as photoregulation was demonstrated for the flow system (Figure 12.16), with various acrylamide and acrylate monomers being tolerated, enabling the synthesis of functional triblock copolymers. Importantly, by simply increasing the collection time, 3 g of poly(dimethylacrylamide) could be produced in 400 min, demonstrating a significant increase in reaction rate for the flow versus bulk system without

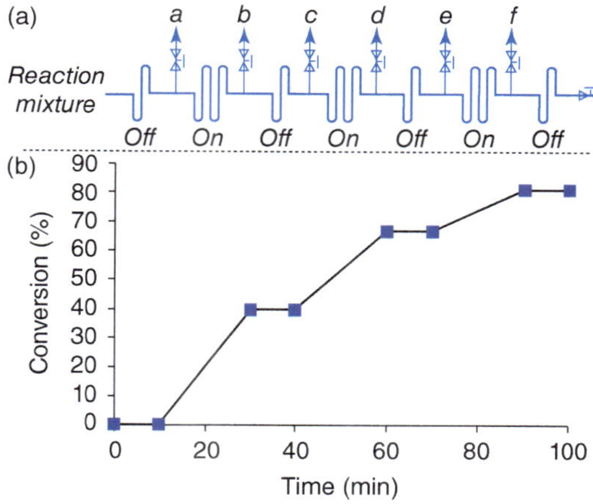

Figure 12.16 (A,B) Photoregulated RAFT under continuous flow conditions using TTC as RAFT agent to polymerize NIPAM, taking aliquots before and after each "on" period. (Reprinted with permission from reference [59]. Copyright 2015 Royal Society of Chemistry.)

sacrificing any polymer properties. Further, this development opens the door for using flow systems to synthesize increasingly complex polymer structures.

Boyer and coworkers have reported photoregulated RAFT polymerization of methacrylate monomers under visible (green) light using low-power (1–5 W) LEDs in combination with trithiocarbonate (CDTPA) RAFT agents [60]. Importantly, this was the first example of a photomediated RAFT polymerization controlling methacrylates using such mild visible-light irradiation. Good photoregulation was observed with low dispersities throughout, and various organic solvents were tested (i.e., dioxane, acetonitrile, toluene, *N,N*-dimethylformamide), all showing good control over this process. One drawback of this system is the sensitivity of the RAFT agent (CDTPA) to chain-end degradation, with the authors noting that polymers must be stored in the dark due to ambient light sensitivity. However, block copolymers are produced using this process, indicating that isolation and purification can be performed without adversely affecting chain ends.

Boyer and coworkers have also pioneered a photoredox-based RAFT system, employing small amounts (typically parts per million) of photoredox catalysts in combination with RAFT agents to undergo living photomediated polymerization and extending photoRAFT to numerous monomer families and functionalities. Initially, their work focused on the use of $Ir(ppy)_3$ as a highly reducing photoredox catalyst ($E^*_{red} = -1.7$ V vs SCE) for the reduction of a variety of RAFT agents (CPADB, $E_{red} = -0.4$ vs SCE; BTPA, $E_{red} = -0.6$ vs SCE; Xanthate, see Figure 12.15) to undergo polymerization of a large range of monomers, including methacrylates, acrylates, acrylamides, methacrylamides, styrenics, vinyl acetate, and *N*-vinyl pyrrolidinone [61]. When conducting polymerizations, very low power (1–5 W) blue LEDs were used to give reasonable reaction rates (time = 2–48 h). Further, catalyst concentrations as low as 0.1 ppm were demonstrated to give control, although a lower rate was observed. Very good photoregulation was demonstrated for MMA using 5 ppm of $Ir(ppy)_3$, with first-order kinetics during light exposure and a linear increase in molecular weight with conversion ($M_w/M_n = 1.05$–1.3). Living chain ends were verified using ^{1}H NMR and UV–vis absorbance to observe the presence of the RAFT agent in the purified polymer. In addition, a variety of block copolymers were formed under these conditions, thoroughly demonstrating the ability to synthesize more complex materials.

Interestingly, it was discovered that these polymerizations were tolerant to oxygen due to the use of a photoredox catalyst that can directly react with molecular oxygen to form an inactive oxygen species. Thus, without degassing, polymerization of acrylates and methacrylates were demonstrated to be controlled, with only a moderate inhibition period observed (~1–3 h), likely due to the time it takes for the catalyst to convert the O_2 to an unreactive species (Figure 12.17). Finally, by letting conversion approach 99% and subsequently adding additional monomers, multiblock acrylate copolymers were produced [62].

Following Boyer and coworkers' original report, a more thorough analysis was conducted for polymerizations mediated by $Ru(bpy)_3Cl_2$ [63]. Extensive evidence for a controlled photomediated system was demonstrated for both methacrylate and acrylate monomers. Further, kinetic studies demonstrated that a 1- to 2-h inhibition period was observed for polymerizations of acrylates, methacrylates, and acrylamides in the presence of oxygen. To confirm there was no catalyst degradation, the catalyst was pre-irradiated for 16 h in solution before adding monomer and RAFT agent. Upon monomer and RAFT agent addition, polymerization was observed with identical characteristics to the non-pre-irradiated solution, providing evidence that an insignificant amount of catalyst degradation occurs. In addition, a more thorough analysis of the polymerization of vinyl

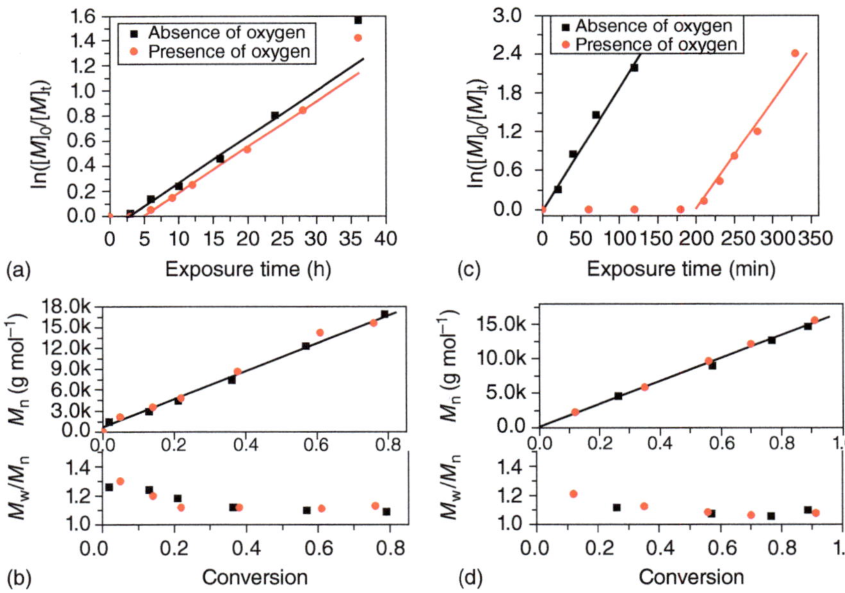

Figure 12.17 Kinetics and molecular weight control experiments conducted in the absence and presence of oxygen for (a) and (b) methyl methacrylate and (c) and (d) methyl acrylate using $Ir(ppy)_3$ in DMSO. (Reprinted with permission from reference [61]. Copyright 2014 American Chemical Society.)

acetate and *N*-vinylpyrrolidone was conducted using $Ir(ppy)_3$ [64]. Again, very similar characteristics for photoregulated polymerizations and block copolymerizations were demonstrated for unactivated monomers both in the presence and absence of oxygen. In nearly all cases, very good photoregulation and low dispersities (often < 1.2) were observed.

This system was then extended to conducting photoRAFT in the presence of water using $Ru(bpy)_3Cl_2$ as a water-soluble, biocompatible photocatalyst [65]. Typically, 1–10 ppm could be used to control the polymerization of dimethylacrylamide, with good photoregulation and living behavior. Further, in addition to extensive evidence for a controlled polymerization in water, protein polymer conjugates were synthesized using a grafting-from approach (Figure 12.18). Bovine serum albumin (BSA) was chosen as a model protein, and the BTPA-based RAFT agent was synthesized containing a disulfide bond. The RAFT agent was subsequently attached to the BSA via a pyridyl disulfide exchange. Following this, polymerization was conducted under the optimized conditions to synthesize the protein–polymer conjugate. To verify livingness, SECs of the protein–polymer conjugate were obtained, demonstrating that chain extension occurred, although some unconjugated BSA remained in the solution due to the inefficiency of the disulfide exchange reaction. The polymers were then cleaved and run on the SEC independently, verifying a low dispersity and linear increase in molecular weight with conversion for polymerizations from the protein. Finally, BSA was shown to lose no activity after being subjected to the polymerization conditions.

PhotoRAFT was then extended to using a porphyrin-based chlorophyll catalyst in combination with various RAFT agents [66]. In this case, the most widely found form of chlorophyll (Chlorophyll a, Figure 12.19) was extracted from spinach leaves and isolated via column chromatography. The naturally occurring catalyst was then employed for the polymerization of methacrylates, acrylates, and acrylamides with good control. An interesting aspect of this system is the catalyst's absorption ($\lambda_{max} = 461$ and 635 nm), which allowed the polymerization to occur under irradiation from both blue and red LED light sources. Photoregulation was observed for polymerizations under red light, and no conversion occurred when irradiating with green light due to the lack of photocatalyst absorption under that wavelength. The control of polymerization under red light indicates the ability of photoRAFT to occur under remarkably mild conditions, giving promise for future biological applications.

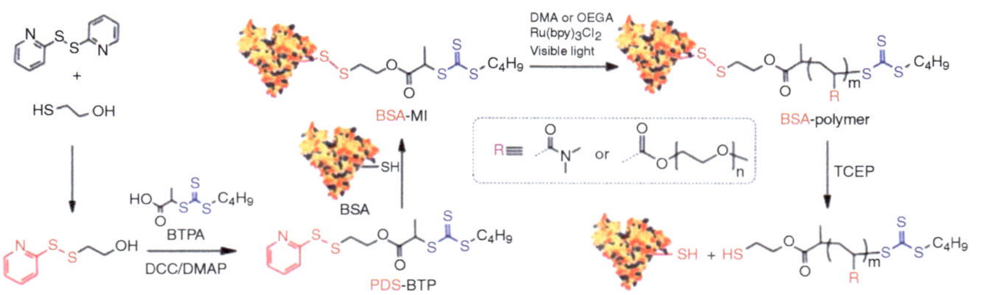

Figure 12.18 Synthetic approach for synthesis of polymer–protein conjugates using $Ru(bpy)_3Cl_2$ as photocatalyst. (Reprinted with permission from reference [65]. Copyright 2014 Royal Society of Chemistry.)

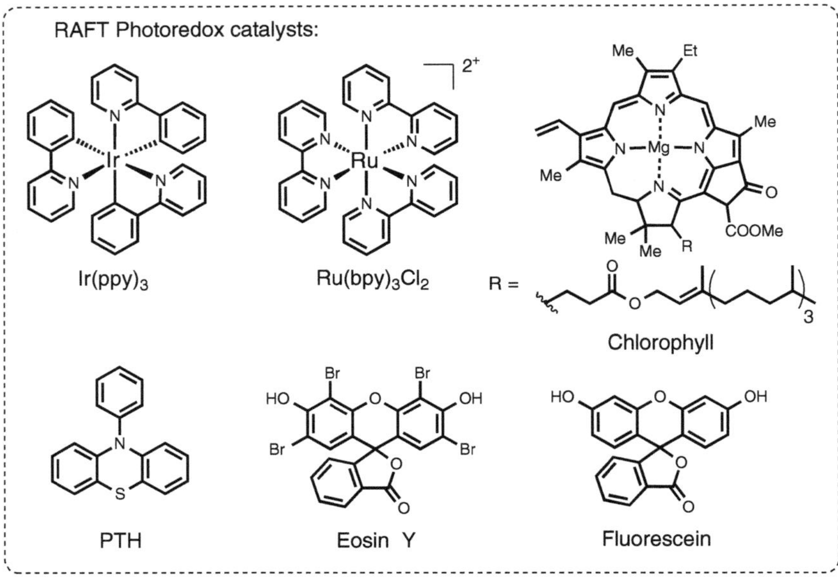

Figure 12.19 Photoredox catalysts used for photoRAFT polymerizations.

PhotoRAFT was also extended to organic photoredox catalysts, using fluorescein and eosin Y to perform visible-light-mediated controlled polymerization [67]. The excited-state redox potentials of eosin Y and fluorescein are −1.1 and −1.2 V versus SCE, respectively, rendering them reducing enough to activate the RAFT agents (−0.4 to −0.8 V vs SCE) in order to achieve controlled polymerization. Eosin Y proved to be the best organic photocatalyst. Further, TEA could be added to improve the kinetics of the polymerization, both in the presence and absence of oxygen. This is likely due to the oxidation of TEA by eosin Y*, leading to the formation of a more reducing eosin Y radical anion that will more efficiently initiate polymerization. Polymerization was well controlled from 10 to 100 ppm of eosin Y, and block copolymers were formed using this process, verifying living chain ends.

Johnson and coworkers took advantage of the highly reducing phenothiazine (PTH) catalyst used for photoregulated metal-free ATRP to conduct photoRAFT using TTC and irradiating with visible light (14 W CFL) [68]. Very good control was observed for acrylamides and acrylates, with dispersities below 1.2, and a linear increase in molecular weight with conversion. The final polymer structure was verified using MALDI-TOF mass spectrometry, with no secondary peaks observed due to side reactions.

PhotoRAFT is arguably the most general photomediated strategy to date, with a variety of organic and metal photoredox catalysts used for polymerization. This is potentially due to a heavier reliance on the RAFT mechanism for gaining control. Nonetheless, these systems are highly tolerant to numerous conditions, and exemplify how photoregulation can open up previously inaccessible levels of control over polymerization.

12.2 Chain-End Activation by Light

Another area of photoregulation has been achieved through using photosensitive chain ends. In each case, the chain end absorbs light, undergoing homolysis to form a propagating radical, and stable radical in solution that reversibly caps the growing polymer chains. It should be noted that many of the RAFT and Iodine systems previously discussed also follow a light-absorbing chain-end mechanism. Apart from these, three main systems have been reported for chain-end photoregulated polymerizations.

First, Yang and coworkers reported a cycloketyl radical mediated living polymerization of MMA and butyl acrylate using a 9,9′-bixanthene-9,9′-diol (BiXANDL) initiator (Figure 12.20) [69]. These polymerizations followed a dissociation–combination mechanism, with UV light causing the BiXANDL chain end to undergo homolysis to initiate polymerization, and the resulting stable xanthone radicals capping the growing polymer chains. Also, due to UV light activation of chain ends, the polymerization conversion could be controlled by turning the light on and off, with control observed throughout the reaction.

Another chain-end activation system takes advantage of the lability of Cobalt—carbon bonds. Most commonly light has been used as an initiation source for cobalt polymerizations [70–74], and, similar to photoRAFT, many of these systems still have potential for photoregulation. However, the only report to date of a photoregulated cobalt-mediated radical polymerization has been using a cobalt porphyrin complex (Figure 12.20) [75]. To gain control, these polymerizations required the preparation of a poly(methyl acrylate) macroinitiator. These initiators could be synthesized by combining methyl acrylate, AIBN, and the cobalt porphyrin at 60 °C and running the reaction to very low conversion. Although this is an unconventional form of an initiating species, these macroinitiators were demonstrated to polymerize *N,N*-dimethylacrylamide (DMA), *N,N*-diethylacrylamide (DEA), and *N*-acryloylmorpholine (AMO) with good control up to high conversions using visible light (500 W Xe lamp, 400–800 nm

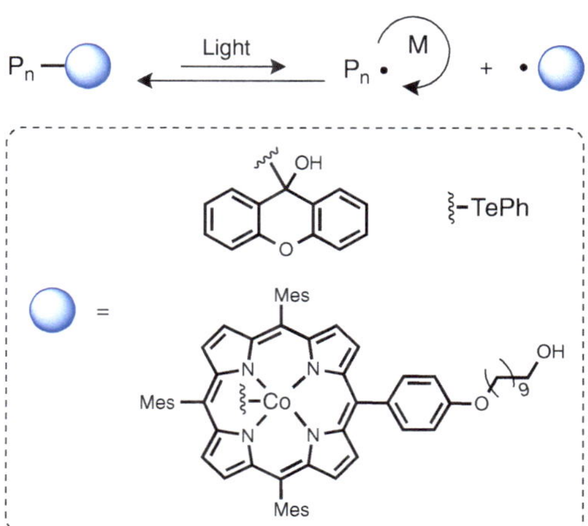

Figure 12.20 Mechanism of dissociation-combination-type polymerizations and examples of photoabsorbing chain ends.

optical filter). A kinetic experiment demonstrated first-order kinetics and a linear increase in molecular weight up to 80% conversion with low polydispersities ($M_w/M_n < 1.25$). Furthermore, photomediation was demonstrated via an "on"/"off" experiment, with efficient stopping and starting of the reaction. Triblock copolymers between the PMA macroinitiator, DMA, and AMO or DEA were synthesized using visible light (PMA-*b*-DMA-*b*-DEA $M_n = 69{,}100\,g\,mol^{-1}$, $M_w/M_n = 1.25$; PMA-*b*-DMA-*b*-AMO $M_n = 84{,}100\,g\,mol^{-1}$, $M_w/M_n = 1.28$) to demonstrate livingness. A limitation of these polymerizations is the inherent lability of the Co—C bond, making the materials difficult to handle and, therefore, requiring the preparation of an ill-defined macroinitiator. However, this system remains a powerful method for producing functional materials and represents an important step for photoregulation using cobalt chain ends.

Another impressive light-controlled polymerization is organotellerium living radical polymerization (TERP) [76]. Yamago et al. have elegantly reported the use of a organotellerium transfer reagent with an absorption maximum at 351 nm, that, when irradiated with a 500 W high-pressure mercury lamp with a short-wavelength cutoff filter (>470 nm), underwent C—Te bond homolysis. The polymerization of butyl acrylate with the 500 W Hg lamp at ~50 °C reached 98% conversion in 2 h. When using no cutoff filter from the same light source, uncontrolled polymerization was observed ($M_w/M_n = 1.87$). However, whenever using lower intensity light sources such as a 30–100 W black lamp, 6 W light emitting diode lamp [77], or even sunlight, the polymerization was controlled, implying that a steady-state equilibrium of polymer chain growth occurs with a variety of light sources. An "on–off" kinetic study was conducted to demonstrate the ability to retain chain ends through multiple cycles of stopping and starting the reaction. The polymerization was stopped and started multiple times, in each case observing no conversion in the dark, and rapid reinitiation upon reexposure to light, with good control. Higher molecular weight polymers (polybutylacrylate, $M_n = 223{,}000\,g\,mol^{-1}$, PDI = 1.18) could be synthesized under conditions identical to those previously described. Another attractive feature of this system is the modularity of the organotellerium transfer agent. A single initiator was used for the synthesis of a variety of monomer families (acrylates, acrylamides, nonconjugated *N*-vinyl monomers).

12.3 Conclusions

Light-mediated polymerizations have emerged as an importance synthetic approach for the preparation of materials for a broad audience [78]. These methods have advantages over the previously held methods, including access to metal-free systems and greater tolerance over a broad range of conditions. The extra dimension of control afforded by photomediation will surely continue to open doors to diverse multifaceted materials applications at the interface of physics, biology, polymer chemistry, and materials science.

References

1 Matyjaszewski, K. and Tsarevsky, N.V. (2014) *J. Am. Chem. Soc.*, **136**, 6513–6533.
2 Ouchi, M., Terashima, T., and Sawamoto, M. (2009) *Chem. Rev.*, **109**, 4963–5050.
3 Moad, G., Chong, Y.K., Postma, A. *et al.* (2005) *Polymer*, **46**, 8458–8468.
4 Hawker, C.J., Bosman, A.W., and Harth, E. (2001) *Chem. Rev.*, **101**, 3661–3688.
5 Nicolas, J., Guillaneuf, Y., Lefay, C. *et al.* (2013) *Prog. Polym. Sci.*, **38**, 63–235.
6 Bielawski, C.W. and Grubbs, R.H. (2007) *Prog. Polym. Sci.*, **32**, 1–29.
7 Leibfarth, F.A., Mattson, K.M., Fors, B.P. *et al.* (2012) *Angew. Chem. Int. Ed.*, **52**, 199–210.
8 Yagci, Y., Jockusch, S., and Turro, N.J. (2010) *Macromolecules*, **43**, 6245–6260.
9 Dietlin, C., Schweizer, S., Xiao, P. *et al.* (2015) *Polym. Chem.*, **6**, 3895–3912.
10 Kwak, Y. and Matyjaszewski, K. (2010) *Macromolecules*, **43**, 5180–5183.
11 Otsu, T., Yoshida, M., and Tazaki, T. (1982) *Die Makromol. Chem.*, **3**, 133–140.
12 Tasdelen, M.A., Uygun, M., and Yagci, Y. (2010) *Macromol. Rapid Commun.*, **32**, 58–62.
13 Tasdelen, M.A., Uygun, M., and Yagci, Y. (2010) *Macromol. Chem. Phys.*, **211**, 2271–2275.
14 Mosnáček, J. and Ilčíková, M. (2012) *Macromolecules*, **45**, 5859–5865.
15 Konkolewicz, D., Schröder, K., Buback, J. *et al.* (2012) *ACS Macro Lett.*, **1**, 1219–1223.
16 Ciftci, M., Tasdelen, M.A., Li, W. *et al.* (2013) *Macromolecules*, **46**, 9537–9543.
17 Zhang, T., Chen, T., Amin, I., and Jordan, R. (2014) *Polym. Chem.*, **5**, 4790.
18 Anastasaki, A., Nikolaou, V., Zhang, Q. *et al.* (2014) *J. Am. Chem. Soc.*, **136**, 1141–1149.
19 Anastasaki, A., Nikolaou, V., Simula, A. *et al.* (2014) *Macromolecules*, **47**, 3852–3859.
20 Chuang, Y.-M., Ethirajan, A., and Junkers, T. (2014) *ACS Macro Lett.*, **3**, 732–737.
21 Anastasaki, A., Nikolaou, V., Brandford-Adams, F. *et al.* (2015) *Chem. Commun.*, **51**, 5626–5629.
22 Chantasirichot, S., Inoue, Y., and Ishihara, K. (2015) *Polymer*, **61**, 55–60.
23 Frick, E., Anastasaki, A., Haddleton, D.M., and Barner-Kowollik, C. (2015) *J. Am. Chem. Soc.*, **137**, 6889–6896.
24 Ribelli, T.G., Konkolewicz, D., Pan, X., and Matyjaszewski, K. (2014) *Macromolecules*, **47**, 6316–6321.
25 Ribelli, T.G., Konkolewicz, D., Bernhard, S., and Matyjaszewski, K. (2014) *J. Am. Chem. Soc.*, **136**, 13303–13312.
26 Zhou, Y.-N. and Luo, Z.-H. (2015) *AIChE J.*, **61**, 1947–1958.

27 Mosnáček, J., Eckstein-Andicsová, A., and Borská, K. (2015) *Polym. Chem.*, **6**, 2523–2530.
28 Fors, B.P. and Hawker, C.J. (2012) *Angew. Chem. Int. Ed.*, **51**, 8850–8853.
29 Nguyen, J.D., Tucker, J.W., Konieczynska, M.D., and Stephenson, C.R.J. (2011) *J. Am. Chem. Soc.*, **133**, 4160–4163.
30 Treat, N.J., Fors, B.P., Kramer, J.W. *et al.* (2014) *ACS Macro Lett.*, **3**, 580–584.
31 Goddard, J.-P., Fensterbank, L., Ollivier, C. *et al.* (2015) *Polym. Chem.*, **6**, 4605–4611.
32 Cao, Y., Xu, Y., Zhang, J. *et al.* (2015) *Polymer*, **61**, 198–203.
33 Yang, Q., Dumur, F., Morlet-Savary, F. *et al.* (2015) *Macromolecules*, **48**, 1972–1980.
34 Telitel, S., Dumur, F., Telitel, S. *et al.* (2015) *Polym. Chem.*, **6**, 613–624.
35 Treat, N.J., Sprafke, H., Kramer, J.W. *et al.* (2014) *J. Am. Chem. Soc.*, **136**, 16096–16101.
36 Pan, X., Lamson, M., Yan, J., and Matyjaszewski, K. (2015) *ACS Macro Lett.*, **4**, 192–196.
37 Miyake, G.M. and Theriot, J.C. (2014) *Macromolecules*, **47**, 8255–8261.
38 Alfredo, N.V., Jalapa, N.E., Morales, S.L. *et al.* (2012) *Macromolecules*, **45**, 8135–8146.
39 Koumura, K., Satoh, K., and Kamigaito, M. (2008) *Macromolecules*, **41**, 7359–7367.
40 Koumura, K., Satoh, K., and Kamigaito, M. (2009) *Macromolecules*, **42**, 2497–2504.
41 Koumura, K., Satoh, K., and Kamigaito, M. (2009) *J. Polym. Sci. A Polym. Chem.*, **47**, 1343–1353.
42 Liu, Q., Liu, L., Ma, Y. *et al.* (2014) *J. Polym. Sci. A Polym. Chem.*, **52**, 3283–3291.
43 Wolpers, A. and Vana, P. (2014) *Macromolecules*, **47**, 954–963.
44 Ohtsuki, A., Goto, A., and Kaji, H. (2013) *Macromolecules*, **46**, 96–102.
45 Ohtsuki, A., Lei, L., Tanishima, M. *et al.* (2015) *J. Am. Chem. Soc.*, **137**, 5610–5617.
46 Ogawa, K.A., Goetz, A.E., and Boydston, A.J. (2015) *J. Am. Chem. Soc.*, **137**, 1400–1403.
47 Goetz, A.E. and Boydston, A.J. (2015) *J. Am. Chem. Soc.*, **137**, 7572–7575.
48 You, Y.-Z., Hong, C.-Y., Bai, R.-K. *et al.* (2002) *Macromol. Chem. Phys.*, **203**, 477–483.
49 Quinn, J.F., Barner, L., Barner-Kowollik, C. *et al.* (2002) *Macromolecules*, **35**, 7620–7627.
50 Lu, L., Yang, N., and Cai, Y. (2005) *Chem. Commun.*, 5287–5288.
51 Jiang, W., Lu, L., and Cai, Y. (2007) *Macromol. Rapid Commun.*, **28**, 725–728.
52 Muthukrishnan, S., Pan, E.H., Stenzel, M.H. *et al.* (2007) *Macromolecules*, **40**, 2978–2980.
53 Khan, M.Y., Cho, M.-S., and Kwark, Y.-J. (2014) *Macromolecules*, **47**, 1929–1934.
54 McKenzie, T.G., Fu, Q., Wong, E.H.H. *et al.* (2015) *Macromolecules*, **48**, 3864–3872.
55 Shi, Y., Liu, G., Gao, H. *et al.* (2009) *Macromolecules*, **42**, 3917–3926.

56 Shi, Y., Gao, H., Lu, L., and Cai, Y. (2009) *Chem. Commun.*, 1368.
57 Liu, G., Shi, H., Cui, Y. *et al.* (2013) *Polym. Chem.*, **4**, 1176.
58 H Zhou and JA Johnson. *Angew. Chem. Int. Ed.* 2013, **52**, 2235–2238.
59 Chen, M. and Johnson, J.A. (2015) *Chem. Commun.*, **51**, 6742–6745.
60 Xu, J.; Shanmugam, S.; Corrigan, N. A.; Boyer, C. *Controlled Radical Polymerization: Mechanisms*; American Chemical Society: Washington, D.C., 2015; **1187**, 247–267.
61 Xu, J., Jung, K., Atme, A. *et al.* (2014) *J. Am. Chem. Soc.*, **136**, 5508–5519.
62 Fu, C., Xu, J., Tao, L., and Boyer, C. (2014) *ACS Macro Lett.*, **3**, 633–638.
63 Xu, J., Jung, K., and Boyer, C. (2014) *Macromolecules*, **47**, 4217–4229.
64 Shanmugam, S., Xu, J., and Boyer, C. (2014) *Macromolecules*, **47**, 4930–4942.
65 Xu, J., Jung, K., Corrigan, N.A., and Boyer, C. (2014) *Chem. Sci.*, **5**, 3568–3575.
66 Shanmugam, S., Xu, J., and Boyer, C. (2015) *Chem. Sci.*, **6**, 1341–1349.
67 Xu, J., Shanmugam, S., Duong, H.T., and Boyer, C. (2015) *Polym. Chem.*, **6**, 5615–5624.
68 Chen, M., MacLeod, M.J., and Johnson, J.A. (2015) *ACS Macro Lett.*, **4**, 566–569.
69 Zheng, X., Yue, M., Yang, P. *et al.* (2012) *Polym. Chem.*, **3**, 1982.
70 Detrembleur, C., Versace, D.-L., Piette, Y. *et al.* (2012) *Polym. Chem.*, **3**, 1856.
71 Kermagoret, A., Wenn, B., Debuigne, A. *et al.* (2015) *Polym. Chem.*, **6**, 3847–3857.
72 Debuigne, A., Schoumacher, M., Willet, N. *et al.* (2011) *Chem. Commun.*, **47**, 12703–12705.
73 Miao, X., Zhu, W., Zhang, Z. *et al.* (2013) *Polym. Chem.*, **5**, 551–557.
74 Zhao, Y., Yu, M., Zhang, S. *et al.* (2015) *Chem. Sci.*, **6**, 2979–2988.
75 Zhao, Y., Yu, M., and Fu, X. (2013) *Chem. Commun.*, **49**, 5186–5188.
76 Yamago, S., Ukai, Y., Matsumoto, A., and Nakamura, Y. (2009) *J. Am. Chem. Soc.*, **131**, 2100–2101.
77 Nakamura, Y. and Yamago, S. (2013) *Beilstein J. Org. Chem.*, **9**, 1607–1612.
78 For additional discussion on this topic see: (a) Chen, M., Zhong, M., and Johnson, J.A. (2016) Light-controlled radical polymerization: mechanisms, methods, and applications. *Chem. Rev.*, **116** (17), 10167–10211; (b) Dadashi-Silab, S., Doran, S., and Yagci, Y. (2016) Photoinduced electron transfer reactions for macromolecular syntheses. *Chem. Rev.*, **116** (17), 10212–10275; (c) Pan, X., Tasdelen, M.A., Laun, J. *et al.* (2016) Photomediated controlled radical polymerization. *Prog. Polym. Sci.*, **62**, 73–125; (d) Boyer, C., Corrigan, N.A., Jung, K. *et al.* (2016) Copper-mediated living radical polymerization (atom transfer radical polymerization and copper(0) mediated polymerization): from fundamentals to bioapplications. *Chem. Rev.*, **116** (4), 1803–1949; (e) Trotta, J.T. and Fors, B.P. (2016) Organic catalysts for photocontrolled polymerizations. *Synlett*, **27** (5), 702–713.

13

Accelerating Visible-Light Photoredox Catalysis in Continuous-Flow Reactors

Natan J. W. Straathof and Timothy Noël

Eindhoven University of Technology, Micro Flow Chemistry & Process Technology, Department of Chemistry and Chemical Engineering, Den Dolech 2 (STO 1.37), 5612 AZ Eindhoven, The Netherlands

13.1 Introduction

The application of photochemistry in organic synthetic chemistry has gained an increasing amount of interest in the past decade. Herein, visible-light photocatalysis stands out due to its unique reactivity and mild reaction conditions, as well as its potential to provide a cleaner and more sustainable alternative compared to conventional synthetic methodologies [1]. Despite the infinite supply of solar energy, the direct application of this energy source in organic chemistry remains limited. This can be mainly attributed to the unreliable intensity and diffusive nature of the irradiation. As an alternative to sunlight, many reported examples in visible-light photocatalysis utilize conventional visible light sources (e.g., high-power LED, or compact fluorescent light bulbs) as a convenient, commercially available and inexpensive visible light source.

While typical photocatalytic operations in batch offer practical advantages, such as the use of nonspecialized equipment and noncomplex process environment, it performs poorly with regard to photon utilization (low photon flux, low quantum yield, Φ), transport phenomena (heat, mass, and photon transfer) and scalability (Bouguer–Lambert–Beer limitation). Despite the fact that these drawbacks are mostly negligible on a small batch scale, they prevent efficient scaling of the photochemical process on a pilot or industrial scale. Arguably, this represents the most significant challenge for any photochemical process and is imposed by the Bouguer–Lambert–Beer law. This law states that there is a decrease in photonic energy profile along the path length due to absorption of the photons. This limitation can be partially overcome in a number of ways, for example, increasing the amount of available photons, lowering the reactant concentrations. or decreasing the perpendicular unidirectional pathway of the reaction vessel (Figure 13.1). Increasing the energy input is often not desired due to over-irradiation of the reaction sample, which leads to the formation of undesired by-products. Furthermore, unreacted photoexcited species usually release their energy in the form of heat, which requires sufficient cooling. Decreasing the concentration of the absorbing species is also not

Visible Light Photocatalysis in Organic Chemistry, First Edition.
Edited by Corey R. J. Stephenson, Tehshik P. Yoon and David W. C. MacMillan.

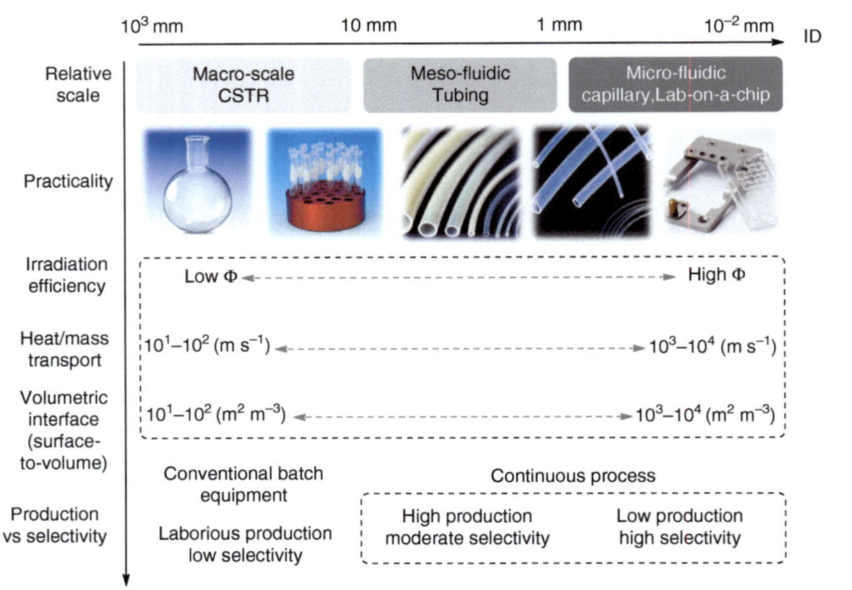

Figure 13.1 Simplified overview for the comparison of visible-light photocatalysis in conventional batch methodology and continuous-flow processing. Taking account of the different scales of reaction environments, the relative practicality (equipment), irradiation efficiency (Φ, %), energy transfer (K_L, m s^{-1}), surface-to-volume ratio (a, m^2 m^{-3}) and relative production capacity.

desired since it results in a decrease in productivity. The most efficient way to improve the efficiency of a photocatalytic process is to minimize the pathway by utilizing microreactor technology (Figure 13.1) [2]. These reactors have small characteristic dimensions, ranging from 100 to 1000 μm internal diameter (ID), and provide a homogeneous irradiation of the reaction medium. The latter allows reducing the catalyst loadings and the reaction times, and avoids the generation of by-products due to overexposure to irradiation. Due to the high surface-to-volume ratios, any heat formed during the reaction is rapidly dissipated to the environment (improved heat transfer) [3]. Microreactors are particularly useful for reactions that require rapid mixing. Among the examples of this are multiphase reaction systems (gas–liquid or solid–liquid reaction mixtures) [4]. In such cases, the reaction rate is strongly dependent on the availability (transportation) of the gaseous or heterogeneous reactants.

Another concern is the robustness of the reaction conditions, accounting for differences in mixing behavior and reaction time. In a batch process, each reaction can differ in outcome due to irregular and uncontrolled mixing, as well as uneven reaction times. However, in a continuous-flow system, the mixing efficiency is well controlled by the large and well-defined surface-to-volume ratios and the reaction times are defined by the flow rates. This results in a high reproducibility of the photocatalytic methodology in flow reactors (Figure 13.1).

As with all new technologies, many researchers wonder if continuous-flow photochemistry can really make a difference in their research. Initial hurdles to implement this technology are lack of experience with engineering fundamentals, the initial investment costs of the microreactor assembly, and microreactor clogging. The first hurdle can be tackled by adequately training chemistry students with some fundamental flow chemistry and process technology knowledge during practical courses in a laboratory environment. A comprehensive knowledge about chemical reactor engineering is of course helpful but not always required to get started with flow chemistry. To address this concern, the flow chemistry community has published a number of undergraduate and graduate books which provide easy-to-understand information about this topic [5]. The second hurdle has been addressed in recent years by commercializing microfluidic platforms which integrate reactors, pumps, and irradiation source in a convenient ready-to-use system. The increasing competition between flow companies in recent years has also resulted in lower prices, improvements in technology, and a greater flexibility of the equipment (Figure 13.2). Alternatively, the availability of cheap capillary tubing (e.g., perfluoroalkoxyalkane (PFA) and fluorinated ethylene propylene (FEP) capillary tubing for visible light photocatalysis) together with microfluidic fittings provide flexible and inexpensive alternatives for the commercial photomicroreactor platforms. Many examples described in this chapter utilize such "in-house" developed photomicroreactor

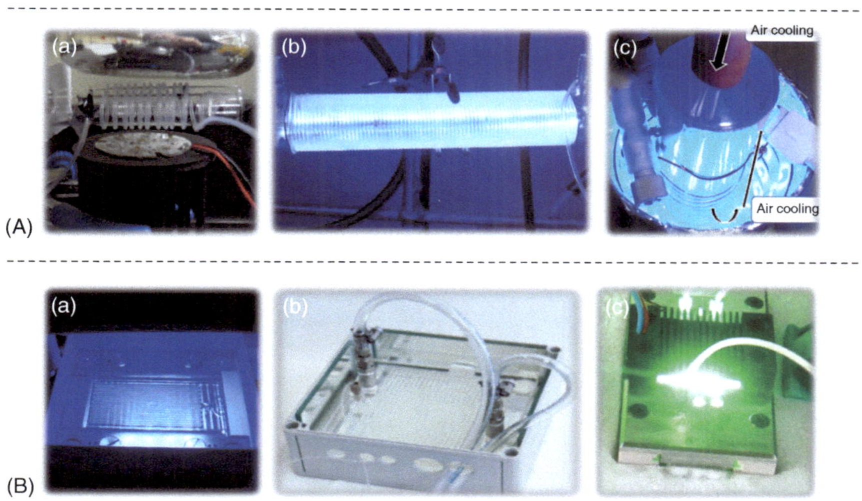

Figure 13.2 An overview of photomicroreactors used for visible-light photocatalysis: (A) Examples of self-assembled capillary microreactors using PFA or FEP capillary tubing. (a: Copyright © 2012 WILEY-VCH Verlag GmbH & Co. KGaA, Weinheim [6]; b: Copyright © 2012 WILEY-VCH Verlag GmbH & Co. KGaA, Weinheim [7]; c: Copyright © 2013 WILEY-VCH Verlag GmbH & Co. KGaA, Weinheim [8].) (B) Examples of Lab-on-a-chip photomicroreactors. (a: Copyright © 2012 WILEY-VCH Verlag GmbH & Co. KGaA, Weinheim [9]; b: Reprinted with permission from [10]. Copyright (2013) American Chemical Society. c: Reprinted with permission from [11]. Copyright (2012) American Chemical Society.)

systems (Figure 13.2A). The last hurdle concerns the irreversible blocking of microchannels due to precipitation formed during the reaction. This is arguably one of the most challenging problems associated with microreactor technology. Clogging of microchannels can be overcome using ultrasonication to break up particle agglomerates or by choosing more polar solvents [12].

This chapter gives an overview of the most important examples of visible-light photoredox catalysis in continuous-flow reactors based on the heterogeneity of the reaction mixture (homogeneous, gas–liquid, and solid–liquid). Special care is devoted to the comparison between the batch and continuous-flow protocol, thus giving guidelines where continuous-flow processing might be superior.

13.2 Homogeneous Photocatalysis in Single-Phase Flow

Photocatalysis in a homogeneous single phase can be straightforwardly converted to a continuous-flow protocol. Typically, the reaction mixture can be pumped directly in the photomicroreactor by means of a syringe pump or high-performance liquid chromatography (HPLC) pump without taking into account a lot of engineering considerations. Due to the short diffusion lengths, fast mixing can be obtained between two miscible reaction streams in a microreactor. If even faster mixing is required, one can use so-called micromixers which minimize the diffusion distance or incorporate static mixing elements.

Stephenson *et al.* demonstrated that several of their photocatalytic batch reactions could be easily adapted to a continuous-flow process (Figure 13.3) [6].

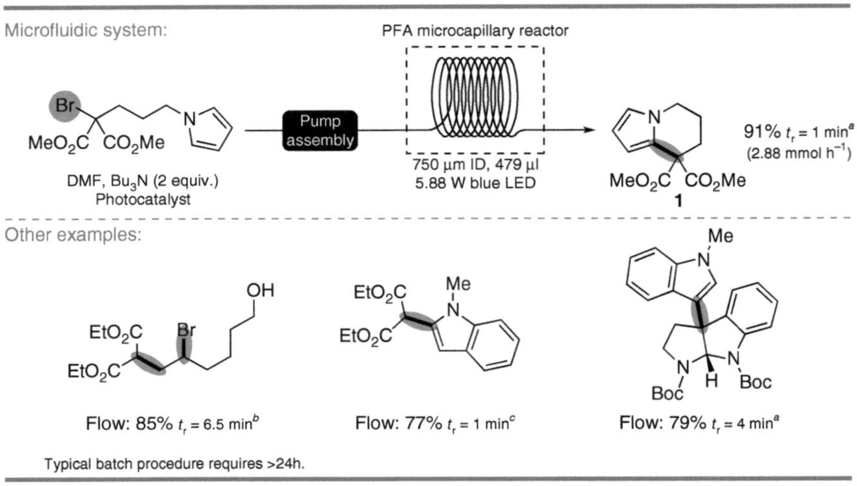

Figure 13.3 Visible-light photoredox-catalyzed dehalogenative atom-transfer radical additions in flow. [a]Reaction conditions: $Ru(bpy)_3Cl_2$ (1 mol%), Bu_3N (2 equiv.), DMF, RT. [b]Reaction conditions: $Ir(dF(CF_3)ppy)_2(dtbpy)PF_6$ (2 mol%), $LiBF_4$ (1.5 equiv.), DMF/H_2O (1 : 1), RT. [c]4-MeO-$C_6H_4NPh_2$ (2 equiv.) instead of Bu_3N. Abbreviations: dF-(CF_3)ppy = 2-(2,4-difluorophenyl)-5-(trifluoromethyl)pyridine; dtbpy = 2-(4-tert-butylphenyl)-4-tert-butylpyridine.

Hereto, an operationally simple microfluidic setup was constructed with commercially available PFA capillary tubing (750 μm ID), which was exposed to blue LED irradiation (5.88 W, λ_{max} = 447.5 nm). Carrying out the photocatalytic intramolecular cyclization in the flow reactor provided the target compound **1** in 91% yield in only a single-minute reaction time (2.88 mmol h^{-1}). A comparable large-scale batch reaction required at least 48 h albeit with a low conversion. Using the same photomicroreactor assembly, the intermolecular radical alkylation of indole derivatives could be efficiently accelerated in flow (examples given in Figure 13.3). The use of the more oxidizing $Ir(dF\text{-}(CF_3)ppy)_2(dtbpy)PF_6$ photocatalyst allowed for the coupling of terminal alkenes and alkyl bromides via an atom-transfer radical addition (ATRA) (Figure 13.2). Notably, a respectable 220-fold increase in production rate was obtained with this microfluidic protocol when compared to analogous batch protocols.

The practicality of continuous-flow technology was also demonstrated for the synthesis of symmetric anhydrides from their corresponding acids [13]. The utilization of $Ru(bpy)_3Cl_2$ in combination with CBr_4 and DMF (Figure 13.4) resulted in the formation of the Vilsmeier–Haack reagent. This reagent allowed subsequently for the formation of the anhydride. The use of continuous-flow technology provided an 11-fold increase in productivity compared to a batch reaction on a 2.5 g scale (batch: 0.04 mmol h^{-1}, flow: 0.45 mmol h^{-1}).

Removal of carbon-bound halogens by radical reductive dehalogenation is well known and offers a broad utility in synthetic organic chemistry. Nevertheless, typical radical initiators or hydrogen atom donors, such as tributyltin hydride, azobisisobutyronitrile (AIBN), or peroxides, are highly toxic or explosive in large quantities. Moreover, popular single-electron-transfer (SET) reagents, such as SmI_2, are air and moisture sensitive and thus nonpractical. To overcome these limitations, Stephenson *et al.* utilized *fac*-$Ir(ppy)_3$ (ppy = 2-phenylpyridine) to convert a wide variety of unactivated alkyl, alkenyl, and aryl iodides to the corresponding carbon-centered radical intermediates [14]. Subsequent reaction with Hantzsch ester or formic acid resulted in a hydrogen atom transfer and afforded the reduced products in good yields. Despite the broad applicability, the batch protocol requires several hours to days to reach full conversion (2.5–50 h reaction time). By employing continuous-flow microreactor technology, the reaction times could be significantly reduced from hours to minutes (Figure 13.5). It should be noted that due to the high photon flux inside the

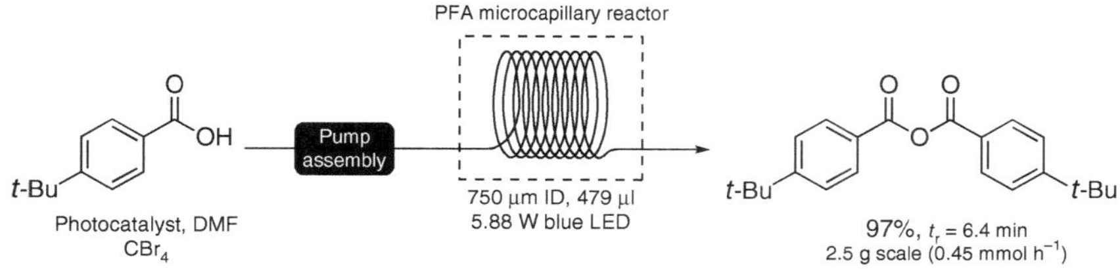

Figure 13.4 Continuous photocatalytic synthesis of symmetric anhydrides. Reagents and conditions: Ru(bpy)3Cl2 (1.0 mol%), CBr4 (2 equiv.), DMF, RT.

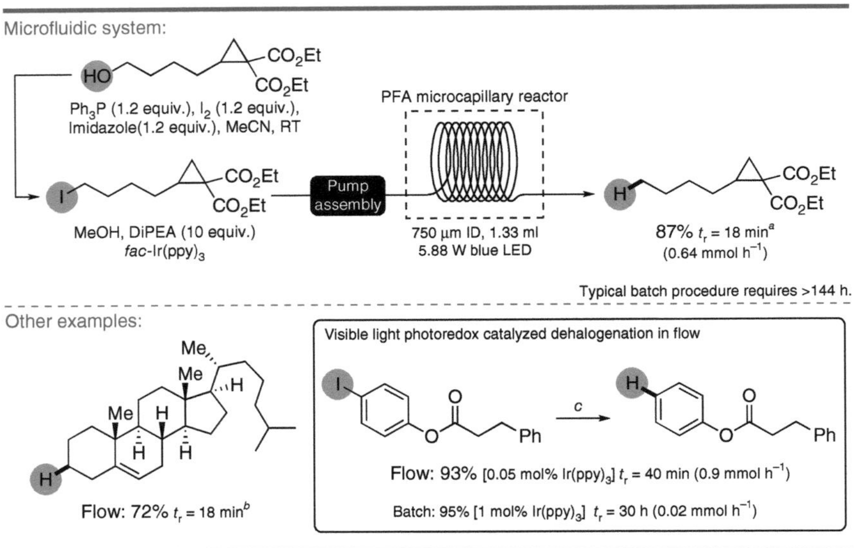

Figure 13.5 Visible-light photocatalytic reductive dehalogenation. [a]Reaction conditions: *fac*-Ir(ppy)$_3$ (0.25 mol%), DiPEA (10 equiv.), MeOH, RT. [b]Toluene instead of MeOH. [c]Reaction conditions: *fac*-Ir(ppy)$_3$ (0.05 mol%), HCO_2H (5 equiv.), Bu_3N (5 equiv.), MeCN, RT. ppy, 2-phenylpyridine; DiPEA, diisopropylethylamine.

microchannels, the catalyst loading could be reduced to 0.05 mol% (representing a TON of 1860) without significant loss of efficiency. In comparison, the batch protocol required at least 1 mol% of *fac*-Ir(ppy)$_3$ to yield comparable results.

Based on these results, the scope of this procedure was extended to deoxygenize primary and secondary alcohols. Hereto, a Garegg–Samuelsson reaction was combined with a consecutive photocatalytic dehalogenation in a flow reactor (Figure 13.5) [15]. Several optimizations were required to obtain a homogeneous mixture to avoid microreactor clogging. Hereto, the combination of Hantzsch ester and tributylamine was replaced by *N*,*N*-diisopropylethylamine (DiPEA) in methanol. With this new protocol, a 120-fold increase in productivity was obtained allowing for the preparation of 10 different primary and secondary alkyl alcohols (67–88% yield).

Bou-Hamdan and Seeberger extended the photocatalytic dehalogenation protocol to the reduction of benzylic chlorides utilizing 1 mol% of Ru(bpy)$_3$Cl$_2$ [16]. A microreactor was constructed by simply wrapping capillary FEP tubing (750 μm ID, 4.7 ml volume) around two metal rods. Irradiation was performed with two face-to-face aligned LED arrays (cold white LED, 17 W). In addition, a backpressure of 6.9 bars was applied to maintain a steady flow regime during operation. The reaction was completed in flow within 30 min, while the corresponding reaction in batch required 30 h (Figure 13.6). Using the same reaction setup, the reduction of arylazides, hydroxyl protection, and reductive epoxide opening was also achieved within reasonable time frames (Figure 13.6, selected examples).

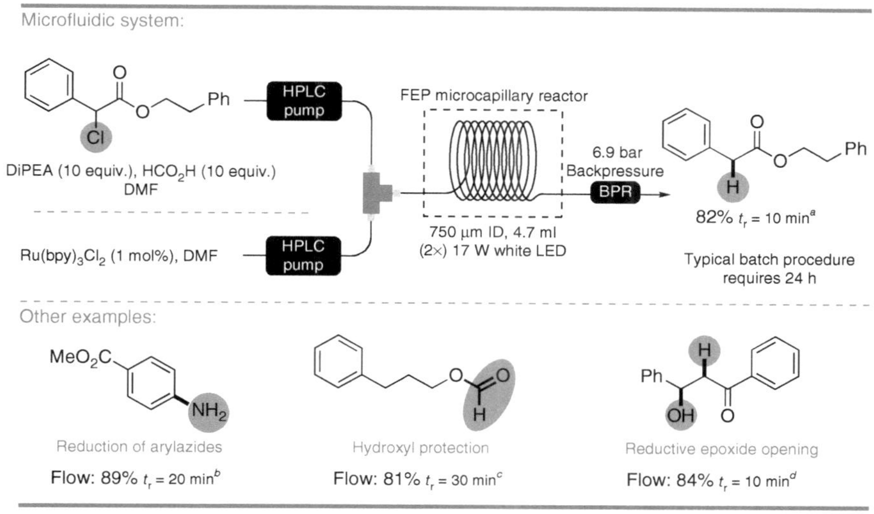

Figure 13.6 Photocatalytic dechlorination and other useful synthetic methodologies in flow. [a]Reaction conditions: Substrate, DiPEA (10 equiv.), formic acid (10 equiv.), DMF; $Ru(bpy)_3Cl_2$ (1 mol%), DMF; Mixed with a T-mixer at RT and 6.9 bar backpressure. [b]Reaction conditions: $Ru(bpy)_3Cl_2$ (1 mol%), HCO_2H (10 equiv.), DiPEA (10 equiv.), DMF. [c]Reaction conditions: $Ru(bpy)_3Cl_2$ (1 mol%), CBr_4 (2 equiv.), DMF. [d]Reaction conditions: $Ru(bpy)_3Cl_2$ (1 mol%), HCO_2H (10 equiv.), DiPEA (10 equiv.), DMF. BPR, backpressure regulator.

The intention to develop more efficient procedures for challenging synthetic procedures, such as natural product synthesis, has always been a driving force for organic synthetic chemistry. One intriguing example where photoredox catalysis was used to prepare a variety of natural products is the photocatalytic fragmentation of (+)-catharanthine [17]. Catharanthine, a commercially available alkaloid, can undergo a two-step reduction–oxidation fragmentation to afford product **2** as a versatile building block (Figure 13.7). It was proposed that the visible-light irradiation of catharanthine in the presence of $Ir(dF(CF_3)ppy)_2(dtbbpy)PF_6$ as a photocatalyst would undergo facile oxidation (catharanthine: $E_{1/2}^{\text{red}} = +0.60$ V vs SCE, Iridium catalyst: $E_{1/2}^{\text{III}*/\text{II}} = +1.21$ V vs SCE), followed by stereoselective trapping with cyanide. Consecutive reduction and protonation provides product **2** and closes the photocatalytic cycle. This system was further improved by continuous-flow processing, resulting in a decreased reaction time (hours to minutes), improved scalability (milligram scale to multi-gram scale), and increased safety due to a controlled release of HCN. With a residence time of only 2 min, the continuous-flow protocol was able to produce several grams of the fragmented product **2** in high yield (2.0 g, 96% yield, 3.84 mmol h^{-1}). This allowed for sufficient material for the preparation of valuable natural products, including (−)-pseudotabersonine, (+)-coronaridine and (−)-pseudovincadifformine. This example demonstrates one of the major advantages of continuous-flow photochemistry, which is the scalable preparation of important intermediates in total synthesis without re-optimizing the reaction parameters.

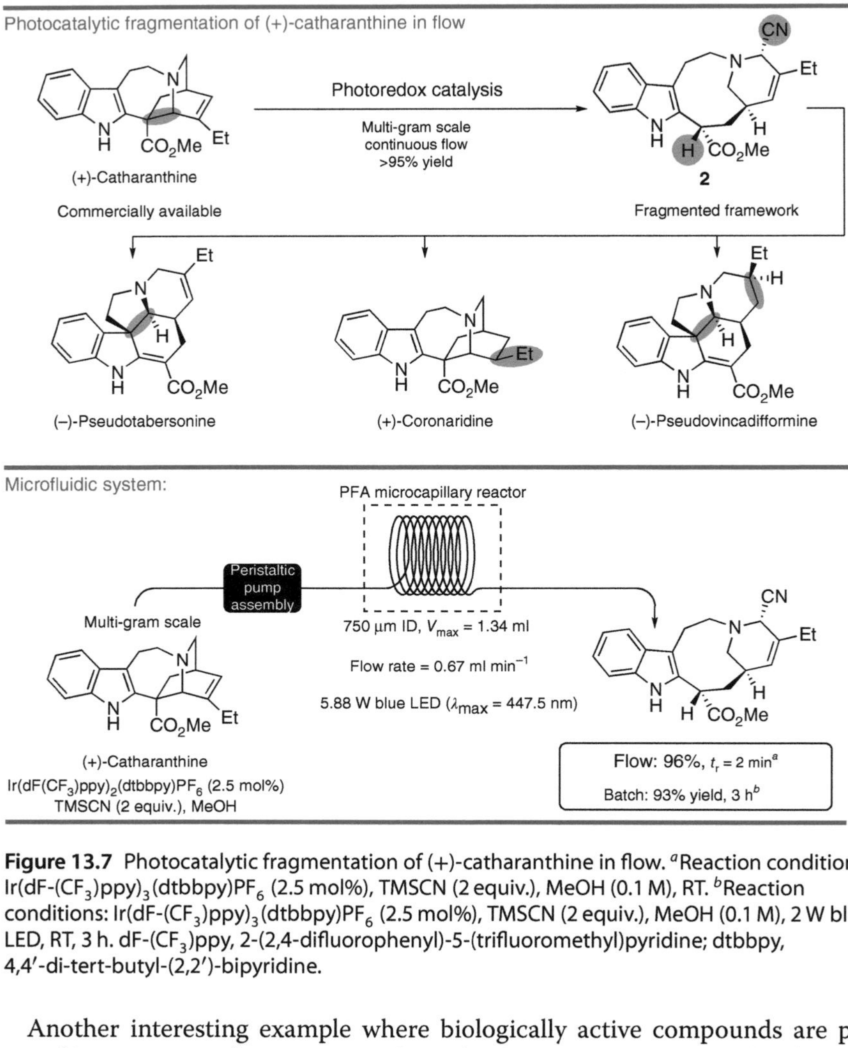

Figure 13.7 Photocatalytic fragmentation of (+)-catharanthine in flow. [a]Reaction conditions: Ir(dF-(CF_3)ppy)$_3$(dtbbpy)PF_6 (2.5 mol%), TMSCN (2 equiv.), MeOH (0.1 M), RT. [b]Reaction conditions: Ir(dF-(CF_3)ppy)$_3$(dtbbpy)PF_6 (2.5 mol%), TMSCN (2 equiv.), MeOH (0.1 M), 2 W blue LED, RT, 3 h. dF-(CF_3)ppy, 2-(2,4-difluorophenyl)-5-(trifluoromethyl)pyridine; dtbbpy, 4,4′-di-tert-butyl-(2,2′)-bipyridine.

Another interesting example where biologically active compounds are prepared in continuous flow originates from the Gagné group. Herein, a key building block for the synthesis of a number of glycoconjugates, glycosyl **3**, was prepared via a photocatalytic conjugate addition reaction between glycosyl radicals and acrolein. An operationally simple continuous-flow setup was constructed from FEP tubing and wrapped around a water-cooled tube wherein an array of blue LED was placed (Figure 13.8) [7]. This allowed for a gram-scale production of **3** in high yield (14.2 mmol in 24 h, 85% yield). Subsequent transformation of this intermediate resulted in a number of valuable *C*-glycoconjugates (two examples are displayed in Figure 13.8).

The photocatalytic generation of electrophilic iminium intermediates in a continuous-flow reactor has been demonstrated by Neumann and Zeitler [11]. A borosilicate glass microreactor (Reactor dimensions: 45.3 mm × 15.3 mm × 2.2 mm for the chip itself, 600 mm × 500 mm ID for the reaction channels,

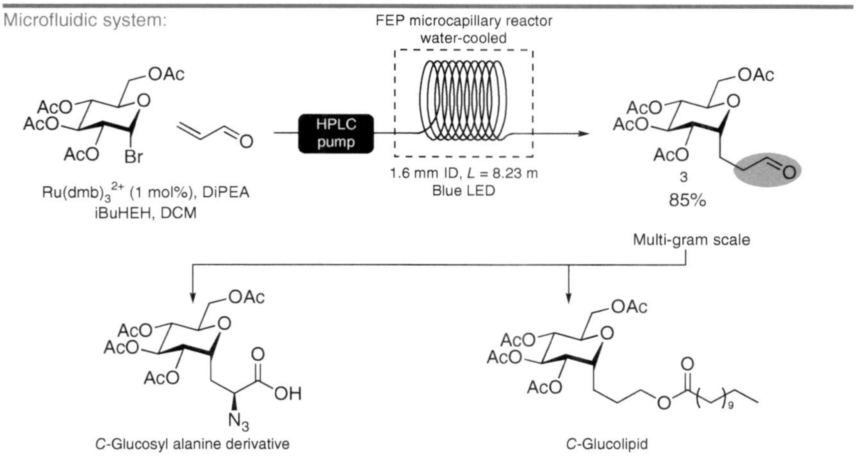

Figure 13.8 Continuouos synthesis of glucocyl aldehyde intermediate in flow. Reaction conditions: $Ru(dmb)_3(PF_6)_2$ (1 mol%), iBuHEH (2.1 equiv.), DiPEA (3 equiv.). DCM, RT. iBuHEH, diethyl 1,4-dihydro-2,6-dimethyl-3,5-pyridinedicarboxylate (Hantzsch ester); DiPEA, diisopropylethylamine; DCM, dichloromethane.

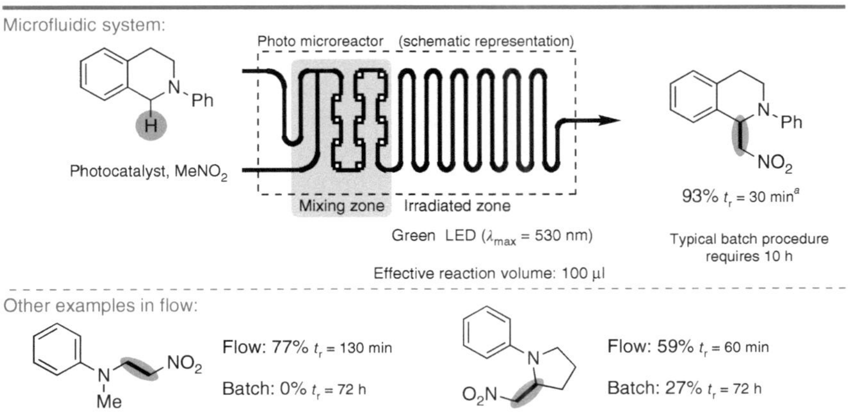

Figure 13.9 Visible-light photocatalytic aza-Henry reaction conducted in a glass microreactor. [a]Reaction conditions: $Ir(ppy)(dtbbpy)_2PF_6$ (1 mol%), nitromethane (solvent), amine (0.25 M), RT.

100 µl volume) was exposed to green light irradiation (high-power LED, $\lambda_{max} = 530$ nm) (Figure 13.9). As depicted in Figure 13.8, a clear reaction rate improvement was observed when performing the photocatalytic aza-Henry reaction in a microreactor. Notably, *N*-methyl-*N*-(2-nitroethyl)aniline could be exclusively obtained in flow in 77% yield after 130 min reaction time.

Furthermore, a comparison study revealed interesting differences for the enantioselective photocatalytic α-alkylation of *n*-octanal when different photoreactors were used, that is, standard batch reactor, a glass microreactor, and a capillary

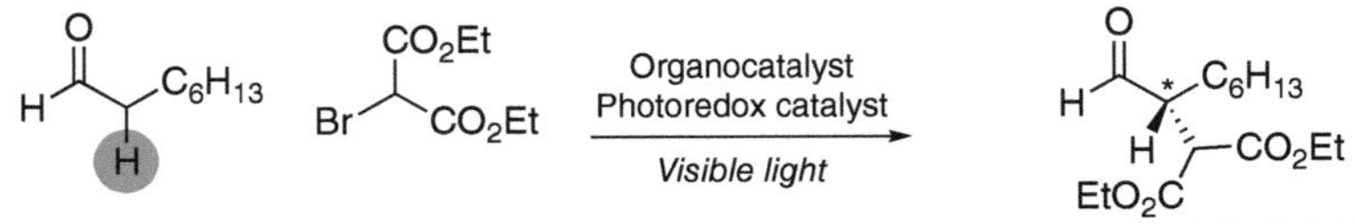

Reactor type	Yield (ee)	Productivity ($mmol\ h^{-1}$)
Batch[a]	85% (88%)	0.018
Microreactor[b]	86% (87%)	0.037
Capillary reactor[c]	92% (82%)	1.92

Figure 13.10 Organocatalytic photoredox a-alkylation of aldehydes in different reactors. Typical reaction conditions: Eosin Y (0.5 mol%), Organocatalyst (20 mol%), lutidine (2 equiv.) DMF, −5 °C. [a]0.4 mmol scale in a 5 ml snap cap vial with a 23 W CFL light bulb. [b]Borosilicate glass microreactor (reactor dimensions: 45.3 mm × 15.3 mm × 2.2 mm for the chip itself, 600 mm × 500 mm ID for the reaction channels, 100 μl volume) with a green LED light source. [c]FEP capillary microreactor (0.8 mm ID, 21 m length, 10.5 ml volume) with a 23 W CFL light bulb.

microreactor (Figure 13.10) [11]. Although the yield and enantiomeric excess are similar, a clear trend is noticeable when comparing the productivity for the different reactors (Figure 13.10). In batch, optimal results could be obtained after an irradiation time of 18 h. Conducting the reaction in a borosilicate glass microreactor revealed a twofold increase in productivity. The reason the difference is not larger can be attributed to (i) the small dimensions of the microreactor and (ii) the lower energy input of the light source. Performing the reaction in a capillary microreactor (FEP, 0.8 mm ID, $V = 10.5$ ml) greatly enhances the production rate to a respectable 1.92 $mmol\ h^{-1}$ (representing a 107-fold increase compared to the batch experiments).

Nucleophilic trapping of photocatalytically generated iminium intermediates in a capillary reactor was also reported by Rueping *et al.* [18] The researchers demonstrated the first photocatalytic multicomponent reaction of amines with isocyanides and water in a continuous process to obtain a variety of α-amino amide derivatives. The photomicroreactor design consists of FEP capillary (0.8 mm ID, $V_{max} = 9.3$ ml) wrapped around a glass tube, wherein an array of green LED was placed. The reaction mixture was pumped through the reactor by means of an HPLC pump and a backpressure regulator (BPR) (500 psi) was placed at the end of the reactor to maintain a steady flow regime (Figure 13.11). The reaction could be completed within several hours in flow instead of days for the comparative batch experiments (14 examples, 51–79% yield and 20–21 h). Notably, reactions with more challenging coupling partners provided in flow a good yield (~57% with a residence time of 21 h), while a typical batch procedure resulted in poor yield, even after days (~25%, 3 days).

Collins *et al.* reported the successful scale-up of a number of polycyclic scaffolds, including helicene and carbazoles derivatives [19, 20]. Conventional cyclization methodologies usually require UV light and show low selectivity,

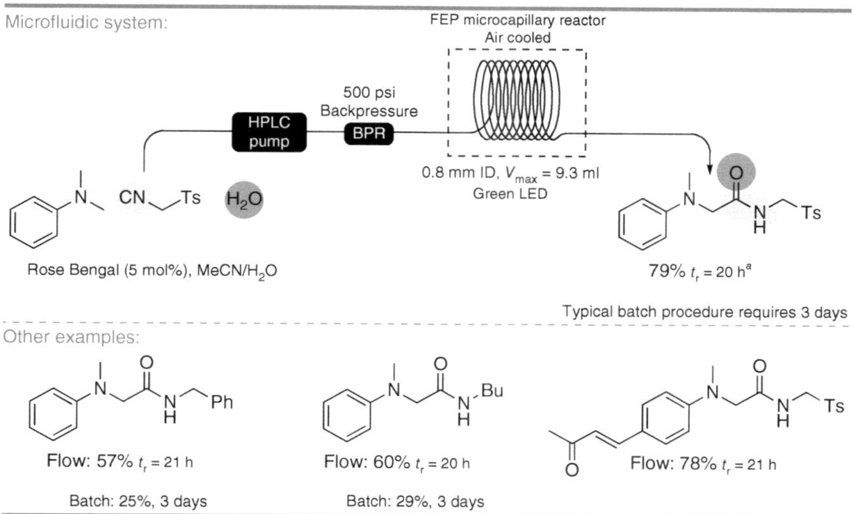

Figure 13.11 Photocatalytic multicomponent reactions in flow. [a]Reaction conditions: Rose Bengal (5 mol%), dimethyl aniline (0.05 M), isocyanide (0.05 M), MeCN/H_2O (1.2 : 1, v/v), Green LED, RT. Ts, *p*-toluenesulfonyl; Ph, phenyl; Bu, butyl.

as overoxidation is the major issue in these cases. To increase the reaction rate, a continuous-flow system was constructed with FEP tubing (1 mm ID) and irradiated with a standard 30 W CFL light bulb (Figure 13.12). While batch methodology afforded only low selectivity (25%) and typically required several days' reaction time, continuous-flow microreactor technology provided

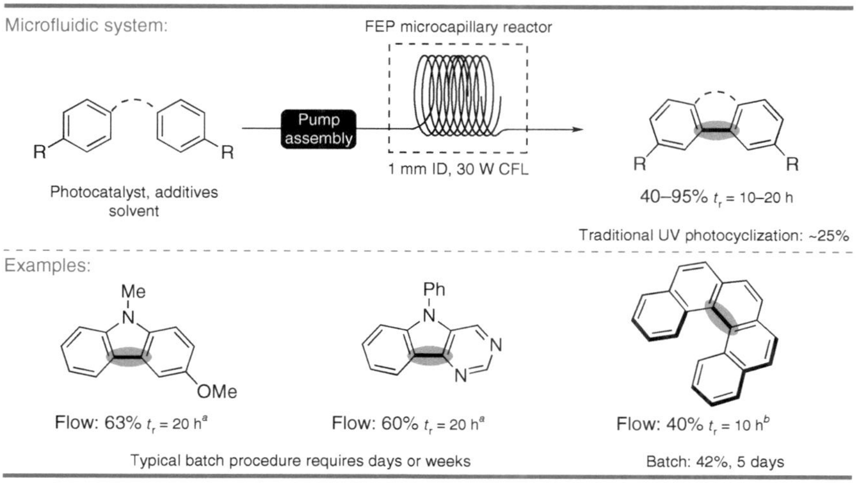

Figure 13.12 Scale-up synthesis of helicene and carbazoles derivatives in a flow reactor. [a]Reaction conditions: Cu(Xantphos)(dmp)BF_4 (5 mol%), I_2 (1 equiv.), THF/propylene oxide (56 : 1, v/v), RT. [b]Reaction conditions: Cu(Xantphos)(dmp)BF_4 (25 mol%), I_2 (1 equiv.), THF, RT. dmp, 2,9-dimethyl-1,10-phenanthroline.

a significant increase in both reaction rate and selectivity (10–20 h and up to 95% yield). Furthermore, the utilization of continuous flow allowed for a fast and reliable screening of photocatalyst, which was prepared *in situ*, and oxidant combinations. Hence, optimal reaction conditions could be obtained in a time-efficient way. The scalability of continuous-flow methodology was also demonstrated by a numbering-up strategy: a series of three photomicroreactors in parallel allowed to increase the productivity without loss of efficiency.

Due to the small dimensions of microreactors, unstable and hazardous reagents can be safely prepared on a small scale and subsequently consumed in a follow-up reaction. Hence, this multistep approach prevents storage of such hazardous compounds and allows for a safer alternative than the corresponding one-pot strategy. Noël *et al.* demonstrated this principle with the photocatalytic preparation of arylsulfides via a Stadler–Ziegler reaction [8]. This method requires the formation of explosive diazonium salts and diazosulfide intermediates. A two-step one-pot protocol was carried out in continuous flow, transforming anilines to the corresponding diazonium salts followed by photocatalytic generation of aryl radicals to afford a wide range of arylsulfides (Figure 13.13). Due to the optimal irradiation and fast mixing of the reagents, a substantial rate enhancement of this reaction was observed in microflow (15 s residence time), while a typical batch protocol required several hours (5–16 h reaction time) to reach full conversion.

The rapid introduction of trifluoromethyl groups into organic molecules is of great interest to the pharmaceutical and agrochemical industry. Carreira *et al.* developed a photocatalytic procedure for the trifluoroethylation of styrenes utilizing a cobalt-based photocatalyst. The reaction was carried out in a glass mesoscale photoreactor chip (2.2 mm ID, 8 ml volume) and exposed to light from a high-power LED array (37 W blue LED) (Figure 13.14) [10]. Typical reaction times could be decreased to several minutes in flow instead of hours or days in batch. The same flow setup was used to enable intramolecular photocatalytic

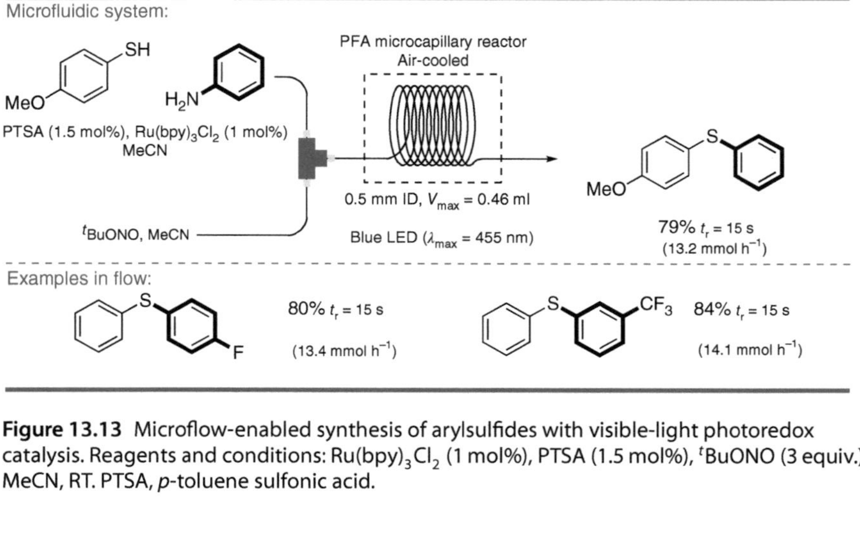

Figure 13.13 Microflow-enabled synthesis of arylsulfides with visible-light photoredox catalysis. Reagents and conditions: $Ru(bpy)_3Cl_2$ (1 mol%), PTSA (1.5 mol%), ^{t}BuONO (3 equiv.), MeCN, RT. PTSA, *p*-toluene sulfonic acid.

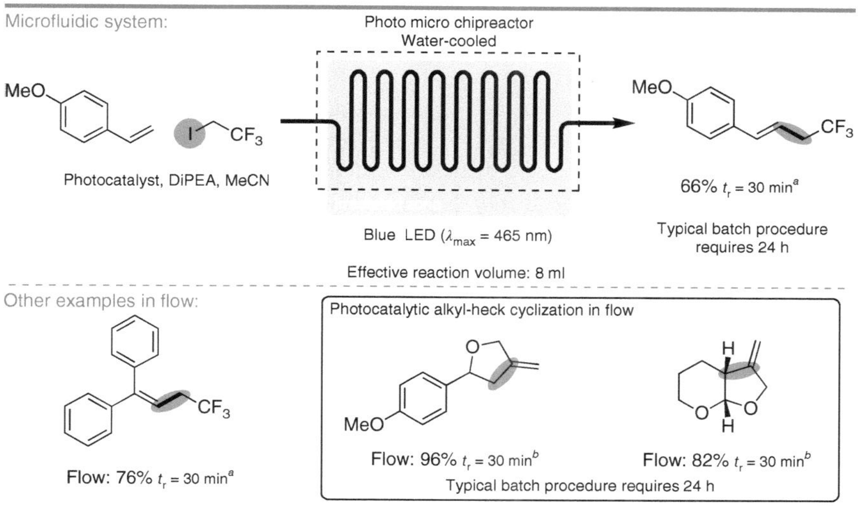

Figure 13.14 Visible-light photocatalytic perfluoroalkylation of styrenes and alkyl-Heck cyclization (inset) in flow. [a]Reagents and conditions: Co photocatalyst (20 mol%), DiPEA (2 equiv.), CF_3CH_2I (2 equiv.), MeCN/DMSO (1 : 1, v/v). [b]Co catalyst (20 mol%), DiPEA (2 equiv.), MeCN/DMSO (1 : 1, v/v).

alkyl-Heck cyclizations. Again, a significant rate enhancement was observed (30 min for flow and >24 h for batch, Figure 13.14 inset).

Kappe *et al.* have developed a two-step protocol for the α-trifluoromethylation of ketones [21]. In the first stage, the ketones were converted to the corresponding silyl enol ethers. Next, the reaction stream exiting the first reactor was merged with CF_3SO_2Cl and introduced in a photomicroreactor, which was constructed from FEP tubing (320 μm ID, V_{max} = 28 ml) (Figure 13.15). A wide variety of different α-trifluoromethylated compounds could be obtained in only 20 min overall residence time (11 examples, 56–87% yield).

13.3 Gas–liquid Photocatalysis in Flow

Reaction mixtures comprising more than one phase are very common in the chemical industry, for example, oxidations, hydrogenations, and carbonylations. Such multiphase reaction mixtures, for example, reaction mixtures comprising a gaseous reactant, warrant a different approach since the reaction rates are strongly dependent on the mass transfer rate of the reactant from one phase to the other. For gas–liquid reactions, it is important to maximize the interfacial area to avoid mass transfer limitations. In large multiphase reactors, the interfacial contact area is hard to control, leading to concentration gradients which can have an impact on the product selectivity. In microreactors, gas–liquid reaction mixtures lead to so-called segmented flow (also known as slug flow or Taylor flow) which has a well-defined and well-controlled interfacial contact area, being two to three orders of magnitude larger than in conventional batch

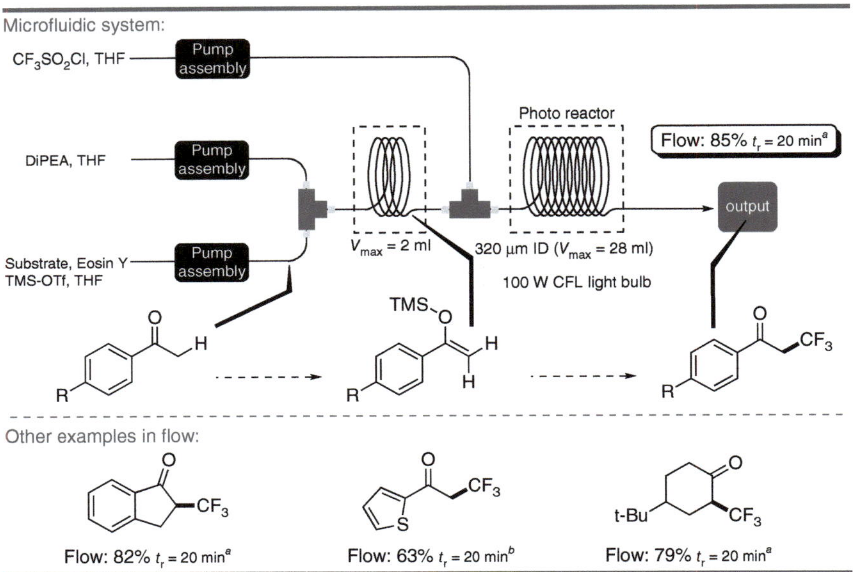

Figure 13.15 Photocatalytic α-trifluoromethylations with eosin Y in continuous flow. [a]Reagents and conditions: Eosin Y (0.5 mol%), TMS-OTf (1.5 equiv.), DiPEA (1.5 equiv.), CF_3SO_2Cl (3 equiv.), THF, RT. [b]2.0 mol% eosin Y was required. TMS-OTf, trimethylsilyl trifluoromethanesulfonate.

technology [4]. This flow pattern is characterized by elongated gas bubbles which are separated by liquid slugs. Inside the slugs, an internal circulation pattern occurs, resulting in an enhanced mixing and thus fast mass transfer from the gas phase to the liquid phase (Figure 13.16).

The sensitized generation of singlet oxygen is one of the most studied photochemical gas–liquid reactions in continuous-flow microreactors [22–25]. Several reasons can be distinguished as to why microreactor technology is advantageous for this transformation. First, due to the limited lifetime of singlet oxygen, the high mass transfer in microscale gas–liquid reactors allows for a more efficient use of the excited species. Second, the use of oxygen and the generation of peroxide intermediates and products in combination with flammable solvents requires a strict safety protocol which can be easily obtained in a microreactor. Third, reaction times can be minimized in flow reactors, thereby preventing extensive by-production formation. Fourth, gas–liquid microreactors allow for

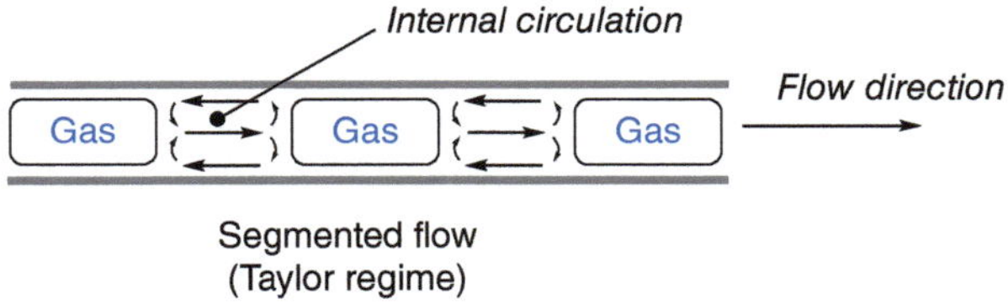

Figure 13.16 Schematic representation of gas–liquid segmented flow in a microchannel.

a straightforward scale-up of the chemistry without re-optimizing the reaction parameters.

Meyer *et al.* reported on the photocatalytic oxidation of β-citronellol in a glass microreactor (1 mm ID, 0.27 ml), using $Ru(tbpy)_3Cl_2$ as the photocatalyst (tbpy = 4,4′-*tert*-butyl-2,2′-dipyridyl) [22]. A segmented flow regime was established and irradiated with a blue LED light source. The microreactor provided for a broad range of β-citronellol concentrations, a space time yield which was one order of magnitude higher than the ones obtained with a glass Schlenck reactor (40 ml).

Most examples in flow chemistry describe microreactors with a single reaction channel leading to a segmented flow regime. Kim *et al.* have developed a membrane poly(dimethylsiloxane) (PDMS) microreactor in which two reaction channels are separated by a gas-permeable membrane (Figure 13.17) [24]. Oxygen introduced in one channel permeates through the membrane to the liquid side, which was irradiated with a white LED source. The advantage of this microreactor compared to the gas–liquid single channel microreactor is the even higher surface-to-volume ratio (area-to-volume ratios: membrane reactor = 50.9 cm^{-1}, single channel = 14.9 cm^{-1}, batch = 0.76 cm^{-1}). The methylene-blue-sensitized oxidation of (−)-citronellol resulted in 97% yield of the target compounds within 3 min. A single-channel microreactor resulted in 95% yield in 15 min reaction time, while the same reaction in batch resulted in 92% yield in 3 h.

Photochemical reactions in flow can be increased in scale by numbering-up, that is, placing several microreactors in parallel. This principle was demonstrated by deMello *et al.* for singlet oxygen generation (Figure 13.7) [25]. A thin-wall microcapillary film (MCF) made of FEP consisted of 10 parallel channels (104.2 μm ID) and was placed in a borosilicate glass encasement. The system was pressurized and oxygen was dosed in the liquid phase via diffusion through the reactor walls. The synthesis of ascaridole from α-terpinene with photosensitizer

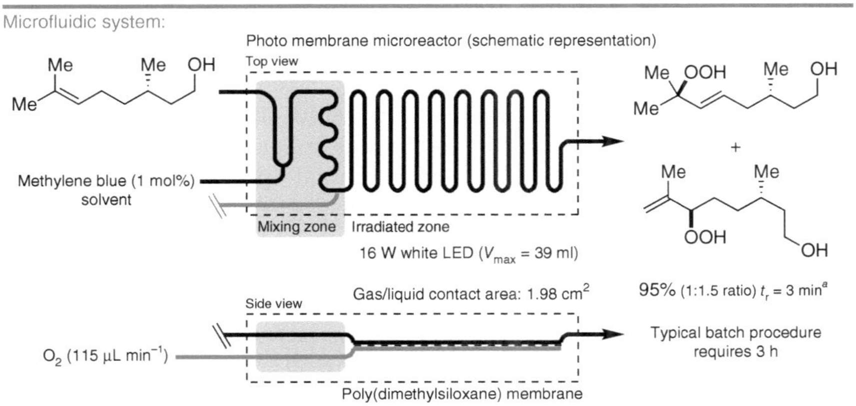

Figure 13.17 Photocatalytic membrane reactor for the oxidation of citronellol. [a]Reagents and conditions: Methylene blue (1 mol%), β-citronellol, O_2, MeCN.

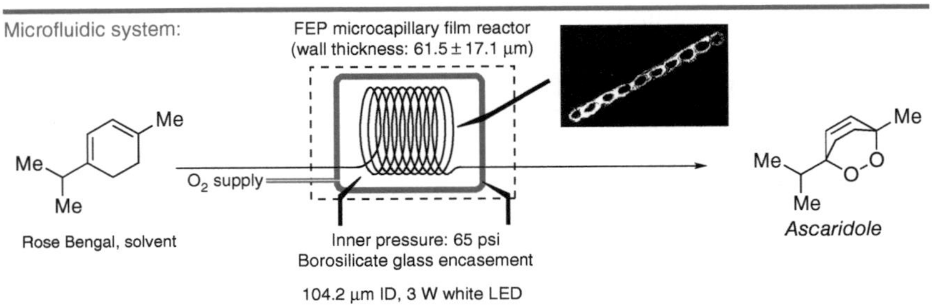

Figure 13.18 Singlet oxygen generation in oxygen-permeable FEP microcapillary film reactor. Reagents and conditions: Rose Bengal (1 mol%), α-terpinene, toluene/methanol (3.2 : 1, v/v), RT. (Reprinted with permission from [25]. Copyright (2013) American Chemical Society.)

Rose Bengal was taken as model reaction (Figure 13.18). Ascaridole could be obtained in >90% yield in 14 min residence time.

One remarkable example of singlet oxygen chemistry constitutes the continuous-flow preparation of artemisinin, one of the most potent antimalarial drugs to date [26, 27]. While conventional methods to obtain this drug depend on low-yielding extraction processes (from the plant *Artemisia Annua*) or a laborious multistep synthesis (<5% yield over 15 steps), a continuous-flow protocol developed by Seeberger *et al.* provided a high yielding process for artemisinin from dihydroartemisinic acid (DHAA) (65% yield in four steps) (Figure 13.19). DHAA can be obtained by a one-step reduction of artemisinic acid, an inexpensive and abundant biomass product. A continuous-flow system was constructed utilizing an FEP capillary photomicroreactor (750 μm ID, V_{max} = 7.5 ml) and an array of 60 high-power LEDs (12 W, λ_{max} = 420 nm). In this reactor, singlet oxygen is generated providing allylic hydroperoxide **4**. Next, a Hock cleavage and subsequent triplet-oxygen addition was carried out, resulting in the formation of intermediate **5**. Finally, several condensation steps afforded artemisinin. The overall continuous process required only 11.5 min and allowed for a production of approximately 165 g day^{-1} artemisinin (24.4 mmol h^{-1}). It should be noted that initially a 450 W UV lamp was required to enable the photochemical singlet oxidation process [28]. In a subsequent optimization of the energy efficiency of the photomicroreactor, visible-light LEDs were used since they emit a discrete spectrum which can be matched with the photocatalyst. The catalyst was changed to 9,10-dicyanoanthracene (DCA) since it is more soluble and stable under the given reaction conditions.

Singlet oxygen was also used for the oxidative generation of imine intermediates, which can be subsequently trapped with trimethylsilyl cyanide [29]. This protocol was performed and optimized in a microreactor, which was constructed with FEP capillary tubing (750 mm ID, 3.5 ml volume) and exposed to an array of blue LEDs (420 nm, 12 W) (Figure 13.20). A wide range of primary and secondary amines were able to undergo this oxidative cyanation within minutes and were obtained in high yield (11 examples, residence time: 1.5–3 min, yield: 99–71%). Despite these notable examples, it was observed that primary amines undergo

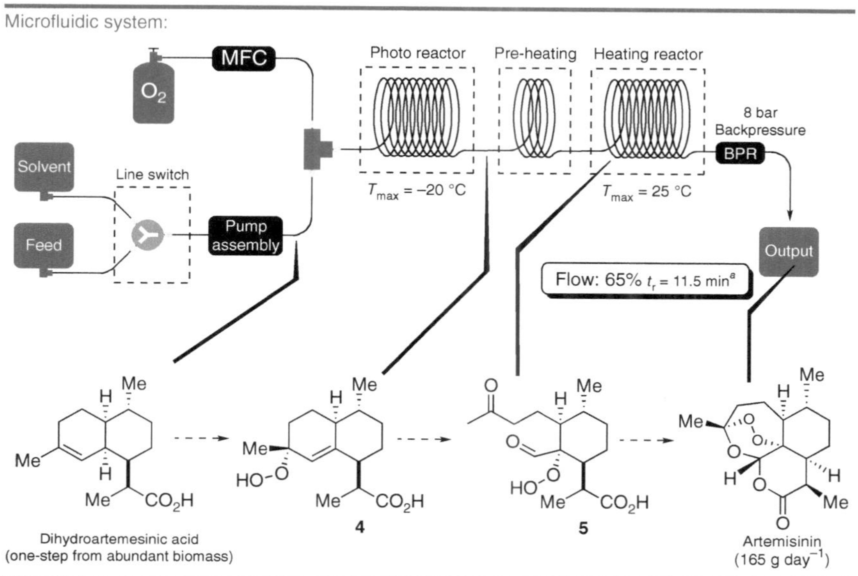

Figure 13.19 Photocatalytic continuous-flow setup for the synthesis of artemisinin. Reaction conditions: DHAA (29.5 g, 0.25 M), TFA (4.8 ml), DCA (2.5 mol%), Toluene. DHAA, dihydroartemisinic acidl; TFA, trifluoroacetic acid; DCA, 9,10-dicyanoanthracene; MFC, mass flow controller; BPR, backpressure regulator.

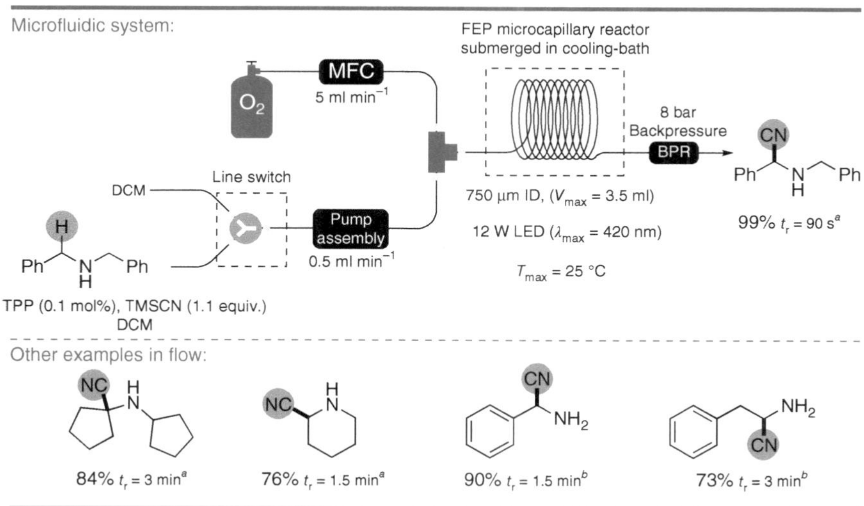

Figure 13.20 Photocatalytic oxidative cyanation of primary and secondary amines in continuous flow. [a]Reaction conditions: amine (0.3–0.5 M), TMSCN (1.1 equiv.), TPP (0.1 mol%), DCM, RT, residence time: 1.5–3 min. [b]Reaction conditions: amine (0.1 M), TMSCN (2.5 equiv.), TPP (0.1 mol%), TBAF (4 mol% based on TMSCN), THF, −50 °C, residence time: 1.5–3 min. MFC, mass flow-controller; BPR, backpressure regulator; TPP, tetrakis(pentafluorophenyl)porphyrinato.

oxidative dimerization at room temperature, prior to the cyanide trapping. Therefore, the reactor temperature was lowered to −50 °C, and small amounts of TBAF were added to increase the reactivity of TMSCN. As anticipated, this resulted predominantly in the desired α-aminonitriles without oxidative dimerization (Figure 13.20).

Noël *et al.* have developed an efficient aerobic oxidation of thiols to disulfides using eosin Y as the photosensitizer [30]. In batch, the reaction typically required a reaction time of about 16 h to reach full conversion. Transferring the chemistry to continuous flow resulted in a substantial acceleration of photocatalytic oxidation protocol (Figure 13.21). This can be attributed to the substantial increase in gas–liquid mass transfer, which has an influence on the apparent reaction rate. The apparent reaction rate constant of thiophenol was obtained for a 1-ml reaction mixture with a stirring rate of 1150 rpm, and its value was 66.6 $M^{-2}\,s^{-1}$. Despite the rather small scale and thus limited effect on mass transfer rate, it was much lower than the intrinsic reaction rate constant (252 $M^{-2}\,s^{-1}$) obtained from the continuous-flow processing using the same light source. Moreover, lower over-oxidation was observed in flow, resulting in overall higher isolated yields. This was important for the preparation of difurfuryl disulfide, which could be exclusively obtained in flow. The methodology was further extended to

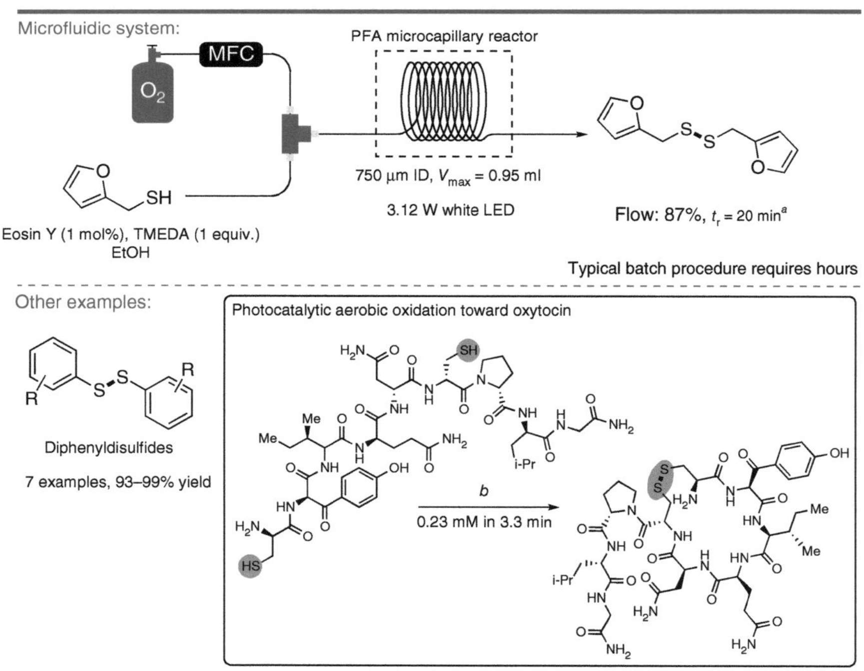

Figure 13.21 Aerobic oxidation of thiols to disulfides in continuous flow. Reagents and conditions: [a]Thiol (0.25 M), eosin Y (1 mol%), TMEDA (0.25 M), EtOH, air, 3.12 W white LED. [b]Peptide (0.23 mM), eosin Y (1 mol%), TMEDA (1 equiv.), H_2O, O_2, 3.12 W white LED. TMEDA, tetramethylethylenediamine.

the cyclization of a peptide substrate to prepare oxytocin in flow in only 200 s reaction time (Figure 13.21, inset).

The efficient generation of reactive radical species in a gas–liquid continuous-flow system was also demonstrated by Noël *et al.* for the trifluoromethylation of five-membered heterocycles [31, 32]. CF_3I gas was used as an inexpensive trifluoromethyl source which in the presence of $Ru(bpy)_3Cl_2$ and upon irradiation gave rise to CF_3 radicals (Figure 13.22). These radicals reacted subsequently with the five-membered heterocycles to yield the target compounds. The continuous-flow setup was constructed from PFA capillary tubing (0.5 mm ID, $V_{max} = 0.2$ ml) and was irradiated by blue LEDs. The high photon flux and increased mixing (Taylor regime) capacity inside the microflow reactor allows for the trifluoromethylation of a wide range of five-membered heterocycles within several minutes, including pyroles, indoles, benzofuranes, and benzothiophenes (Figure 13.22). This protocol was also applicable for the perfluoroalkylations of five-membered heterocycles (Figure 13.22) [31]. Due to the improved photon flux through the reactor, the catalyst loading could be efficiently lowered (0.05 mol%, accounting for a TOF of 2812 TOF h^{-1}). Moreover, a metal-free protocol was also developed utilizing eosin Y as photocatalyst [32].

In an extension of this work, Noël *et al.* have developed a mild and fast protocol for the trifluoromethylation of thiols by using CF_3I as the gaseous trifluoromethyl source (Figure 13.23) [33]. Similarly, CF_3 radicals were generated by irradiation of the gas–liquid reaction mixture in the presence of $Ru(bpy)_3$ as a photocatalyst.

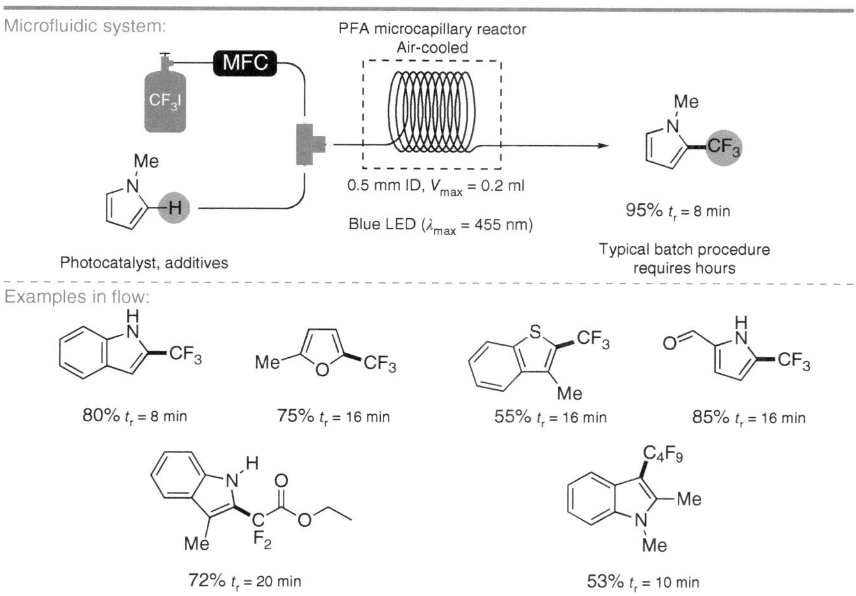

Figure 13.22 Photocatalytictrifluoromethylations and perfluoroalkylations in flow. Reaction conditions: $Ru(bpy)_3Cl_2$ (1 mol%), TMEDA (2 equiv.), MeCN, RT. MFC, mass flow controller; BPR, backpressure regulator.

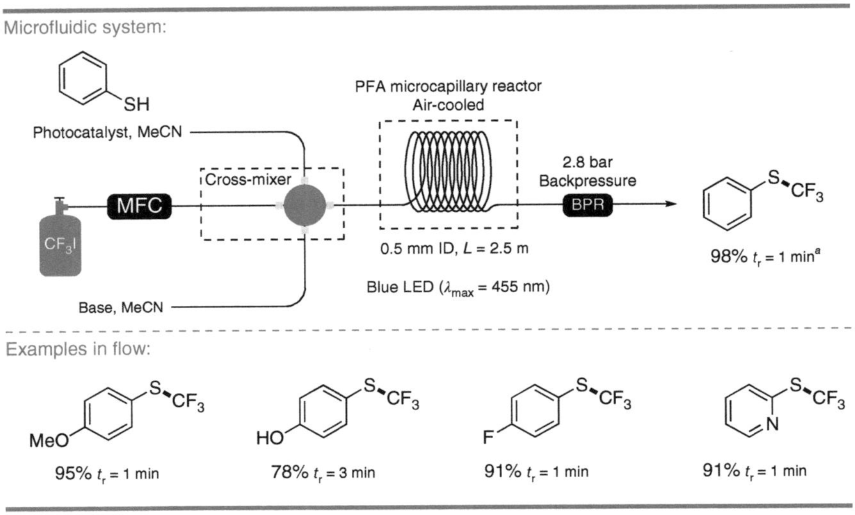

Figure 13.23 Continuous-flow setup for the synthesis of trifluoromethylarylsulfanes. Reagents and conditions: $Ru(bpy)_3Cl_2$ (1 mol%), Triethylamine (1.1 equiv.), gaseous CF_3I, MeCN, RT. MFC, mass flow-controller; BPR, back-pressure regulator.

In batch, the reaction required 4 equiv of CF_3I and reaction times of about 1 h to reach full conversion. In flow, the reaction was typically completed within 1 min residence time. Notably, due to the excellent contact between the liquid and the gas phase, the amount of CF_3I could be lowered to 1.1 equiv. A wide range of different thiophenol derivatives could be trifluoromethylated within several minutes' residence time. The methodology had a broad substrate scope, including unprotected hydroxyls and amines, as well as electron-donating, -withdrawing, and -deficient systems (Figure 13.23, selected examples).

13.4 Heterogeneous Photocatalysis in Flow

Immobilization of photocatalysts in continuous-flow reactors provides a number of advantages with regard to catalyst recuperation and reuse. It facilitates also the purification of the final product which is not contaminated with the photocatalyst. High turnover numbers can be reached for a given reaction just by directing more reactants over the catalyst. Another advantage concerns the high local loading of the photocatalyst, which allows for increased reaction rates and thus higher throughput. However, the preparation of immobilized catalysts is often labor intensive and requires a judicious choice of the reactor concept [34]. More specifically, filling the immobilized catalyst in a packed-bed microreactor is the easiest method, but only the first layer of catalyst can be effectively used due to light scattering. Immobilizing the catalyst on one side of the reactor channels is a better choice with regard to pressure drop and effective use of the photocatalyst, but it is more difficult to design and can lead to mass transfer limitations.

George, Poliakoff *et al.* reported on the successful photooxidation of α-terpinene with an immobilized photocatalyst in a continuous-flow system [35, 36]. The generation of $^1O_2$ was achieved by immobilized tetradi(2,6)chlorophenylporphyrin (TDCPP) in supercritical CO_2 ($scCO_2$) as a green solvent. Supercritical CO_2 provided excellent solubility of oxygen and high mass transfer coefficients, which are beneficial for the facilitated transport of reactants to the immobilized catalyst. The modified TDCPP was immobilized covalently on amino-polyvinyl chloride polymers and was loaded inside the reactor and exposed to high-power white LED irradiation (Figure 13.24). In addition, the microfluidic setup allowed for the safe handling of liquid O_2 in $scCO_2$. Photooxidation of α-terpinene was achieved for over 6 h with a marginal amount of photodegradation of the immobilized catalyst.

Boyle *et al.* reported on the generation of singlet oxygen in a glass microfluidic device which was functionalized with photoactive porphyrines [37]. The photooxidation of a number of compounds was achieved, including citronellol, terpinene, and cholesterol. The results were compared with that of the batch to determine the efficiency of the different photocatalytic systems. A much higher space time yield was obtained in these microfluidic devices than in conventional batch.

Blechert *et al.* have developed a continuous-flow protocol for the photocatalytic radical cyclization enabled by graphitic carbon nitride (g-C_3N_4) [38]. They filled the FEP tubing with a mixture of ground graphitic carbon nitride and silica gel and glass beads as filling material. Ground graphitic carbon nitride had a higher photocatalytic activity than non-ground, which was attributed to the increased active surface. The packed bed was next exposed to blue LEDs (420–425 nm). Bromomalonate dissolved in THF was directed over the photocatalytic bed and was converted to the corresponding cyclic product within 5–10 min residence time (Figure 13.25). The authors found that THF acted as the electron and hydrogen donor in this photocatalytic process and did not require the addition of sacrificial electron donors, such as Et_3N.

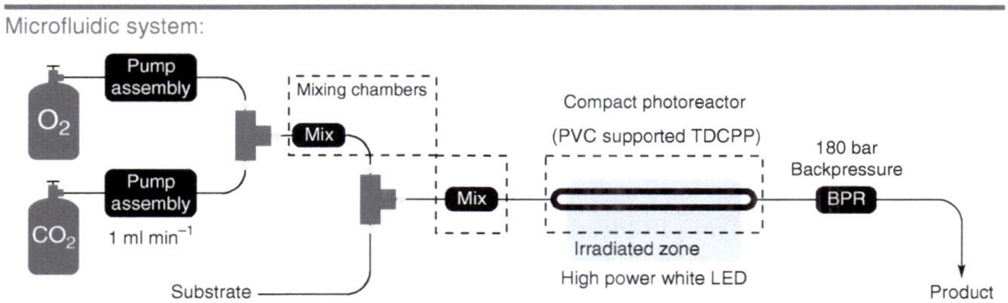

Figure 13.24 Generation of $^1O_2$ with immobilized TDCPP in a $scCO_2$ continuous-flow system. Reagents and conditions: Catalyst loading (0.3 g, glass wool matrix), substrate (0.1 ml min^{-1}), CO_2 (1 ml min^{-1}), O_2 (2 equiv.), 180 Bars, RT.

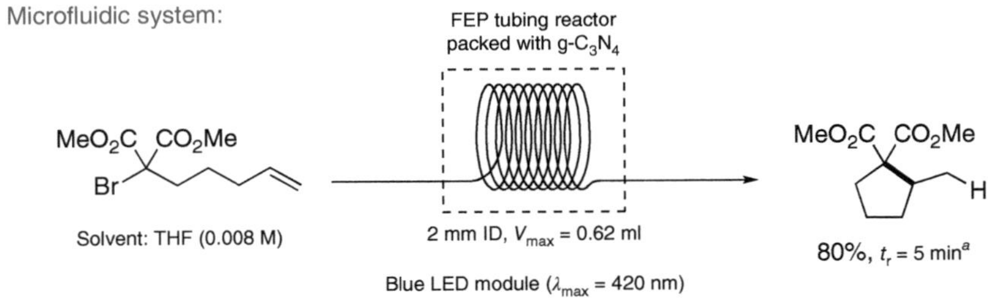

Figure 13.25 Photocatalytic radical cyclization by graphitic carbon nitride in flow. Reagents and conditions: Substrate (0.04 mmol), THF (5 ml), RT. Packed-bed microreactor: ground g-C_3N_4/silica gel (1 : 1, 50 mg, 63–200 μm particle size), 950 mg of 1 : 1 mixture of silica gel (63–200 μm particle size) and glass beads (250–500 μm).

13.5 Conclusions

In this chapter, we have reviewed the recent progress in visible-light continuous-flow photoredox catalysis. While continuous-flow chemistry, in general, is considered to be a specialist's domain, the utilization of this technology has been embraced more rapidly by the visible-light photocatalytic community. This can be attributed to the ease of implementation in a traditional chemistry laboratory. As has been shown in this chapter, photomicroreactors can be constructed out of inexpensive, commercially available microfluidic parts and light sources. Furthermore, traditional limitations of photochemistry, such as limited scalability, long reaction times, and formation of by-products, can be overcome by its use. The safe handling of hazardous intermediates, for example, singlet oxygen and diazonium salts, in microreactors has been conclusively demonstrated in this review. We expect that in the near future this technology will be increasingly used and, consequently, will be further improved [39].

Conflict of Interest

The authors have declared that there is no conflict of interest.

References

1 For selected reviews on photoredox catalysis in batch: (a) Prier, C.K., Rankic, D.A., and MacMillan, D.W.C. (2013) *Chem. Rev.*, **113**, 5322–5363; (b) Yoon, T.P. (2013) *ACS Catal.*, **3**, 895–902; (c) Ravelli, D., Fagnoni, M., and Albini, A. (2013) *Chem. Soc. Rev.*, **42**, 97–113; (d) Tucker, J.W. and Stephenson, C.R.J. (2012) *J. Org. Chem.*, **77**, 1617–1722.

2 For selected reviews on photochemistry in flow: (a) Su, Y., Straathof, N.J.W., Hessel, V., and Noël, T. (2014) *Chem. Eur. J.*, **20**, 10562–10589; (b) Garlets, Z.J., Nguyen, J.D., and Stephenson, C.R.J. (2014) *Isr. J. Chem.*, **54**, 351–360;

(c) Schuster, E.M. and Wipf, P. (2014) *Isr. J. Chem.*, **54**, 361–370; (d) Noël, T., Wang, X., and Hessel, V. (2013) *Chim. Oggi.*, **31**, 10–14; (e) Oelgemöller, M. (2012) *Chem. Eng. Technol.*, **35**, 1144–1152; (f) Knowles, J.P., Elliott, L.D., and Booker-Milburn, K.I. (2012) *Beilstein J. Org. Chem.*, **8**, 2025–2052; (g) Oelgemoeller, M. and Shvydkiv, O. (2011) *Molecules*, **16**, 7522–7550; (h) Coyle, E.E. and Oelgemoeller, M. (2008) *Photoch. Photobio. Sci.*, **7**, 1313–1322; (i) Matsushita, Y., Ichimura, T., Ohba, N. *et al.* (2007) *Pure Appl. Chem.*, **79**, 1959–1968.

3 For general reviews on continuous-flow photochemistry: (a) Hessel, V., Kralisch, D., Kockmann, N. *et al.* (2013) *ChemSusChem*, **6**, 746–789; (b) McQuade, D.T. and Seeberger, P.H. (2013) *J. Org. Chem.*, **78**, 6384–6389; (c) Pastre, J.C., Browne, D.L., and Ley, S.V. (2013) *Chem. Soc. Rev.*, **42**, 8849–8869; (d) Noël, T. and Buchwald, S.L. (2011) *Chem. Soc. Rev.*, **40**, 5010–5029; (e) Hartman, R.L., McMullen, J.P., and Jensen, K.F. (2011) *Angew. Chem. Int. Ed.*, **50**, 7502–7519; (f) Wegner, J., Ceylan, S., and Kirschning, A. (2011) *Chem. Commun.*, **47**, 4583–4592.

4 For reviews on gas–liquid reactions in flow: (a) Noël, T. and Hessel, V. (2013) *ChemSusChem*, **6**, 405–407; (b) Sobieszuk, P., Aubin, J., and Phorecki, R. (2012) *Chem. Eng. Technol.*, **35**, 1346–1358; (c) Kashid, M.N. and Kiwi-Minsker, L. (2009) *Ind. Eng. Chem. Res.*, **48**, 6465–6485; (d) Chen, G., Yue, J., and Yuan, Q. (2008) *Chin. J. Chem. Eng.*, **16**, 663–669; (e) Günther, A. and Jensen, K.F. (2006) *Lab Chip*, **6**, 1487–1503; (f) Hessel, V., Angeli, P., Gavriilidis, A., and Löwe, H. (2005) *Ind. Eng. Chem. Res.*, **44**, 9750–9769.

5 (a) Darvas, F., Dorman, G., and Hessel, V. (eds) (2014) *Flow Chemistry Volume 1: Fundamentals*, Walter de Gruyter GmbH, Berlin; (b) Darvas, F., Dorman, G., and Hessel, V. (eds) (2014) *Flow Chemistry Volume 2: Applications*, Walter de Gruyter GmbH, Berlin.

6 Tucker, J.W., Zhang, Y., Jamison, T.F., and Stephenson, C.R.J. (2012) *Angew. Chem. Int. Ed.*, **51**, 4144–4147.

7 Andrews, R.S., Becker, J.J., and Gagne, M.R. (2012) *Angew. Chem. Int. Ed.*, **51**, 4140–4143.

8 Wang, X., Cuny, G.D., and Noël, T. (2013) *Angew. Chem. Int. Ed.*, **52**, 7860–7864.

9 Cherevatskaya, M., Neumann, M., Fuldner, S. *et al.* (2012) *Angew. Chem. Int. Ed.*, **51**, 4062–4066.

10 Kreis, L.M., Krautwald, S., Pfeiffer, N. *et al.* (2013) *Org. Lett.*, **15**, 1634–1637.

11 Neumann, M. and Zeitler, K. (2012) *Org. Lett.*, **14**, 2658–2661.

12 (a) Noël, T., Naber, J.R., Hartman, R.L. *et al.* (2011) *Chem. Sci.*, **2**, 287–290; (b) Kuhn, S., Noël, T., Gu, L. *et al.* (2011) *Lab Chip*, **11**, 2488–2492; (c) Hartman, R.L. (2012) *Org. Process Res. Dev.*, **16**, 870–887; (d) Wu, K. and Kuhn, S. (2014) *Chim. Oggi*, **32**, 62–66.

13 Konieczynska, M.D., Dai, C., and Stephenson, C.R.J. (2012) *Org. Biomol. Chem.*, **10**, 4509–4511.

14 Nguyen, J.D., D'Amato, E.M., Narayanam, J.M.R., and Stephenson, C.R.J. (2012) *Nat. Chem.*, **4**, 854–859.

15 Nguyen, J.D., Reiss, B., Dai, C., and Stephenson, C.R.J. (2013) *Chem. Commun.*, **49**, 4352–4354.

16 Bou-Hamdan, F.R. and Seeberger, P.H. (2012) *Chem. Sci.*, **3**, 1612–1616.
17 Beatty, J.W. and Stephenson, C.R.J. (2014) *J. Am. Chem. Soc.*, **136**, 10270–10273.
18 Rueping, M., Vila, C., and Bootwicha, T. (2013) *ACS Catal.*, **3**, 1676–1680.
19 Hernandez-Perez, A.C., Vlassova, A., and Collins, S.K. (2012) *Org. Lett.*, **14**, 2988–2991.
20 Hernandez-Perez, A.C. and Collins, S.K. (2013) *Angew. Chem. Int. Ed.*, **52**, 12696–12700.
21 Cantillo, D., de Frutos, O., Rincon, J.A. *et al.* (2014) *Org. Lett.*, **16**, 896–899.
22 Meyer, S., Tietze, D., Rau, S. *et al.* (2007) *J. Photochem. Photobiol. A*, **186**, 248–253.
23 Wootton, R.C.R., Fortt, R., and de Mello, A.J. (2002) *Org. Process Res. Dev*, **6**, 187–189.
24 Park, C.P., Maurya, R.A., Lee, J.H., and Kim, D.-P. (2011) *Lab Chip*, **11**, 1941–1945.
25 Elvira, K.S., Wootton, R.C.R., Reis, N.M. *et al.* (2013) *ACS Sustainable Chem. Eng.*, **1**, 209–213.
26 Lévesque, F. and Seeberger, P.H. (2012) *Angew. Chem. Int. Ed.*, **51**, 1706–1709.
27 Kopetzki, D., Levesque, F., and Seeberger, P.H. (2013) *Chem. Eur. J.*, **19**, 5450–5456.
28 Holick, M.F. (2005) *J. Nutr.*, **135**, 2739S–2748S.
29 Ushakov, D.B., Gilmore, K., Kopetzki, D. *et al.* (2014) *Angew. Chem. Int. Ed.*, **53**, 557–561.
30 Talla, A., Driessen, B., Straathof, N.J.W. *et al.* (2015) *Adv. Synth. Catal.*, **357**, 2180–2186.
31 Straathof, N.J.W., Gemoets, H.P.L., Wang, X. *et al.* (2014) *ChemSusChem*, **7**, 1612–1617.
32 Straathof, N.J.W., van Osch, D.J.G.P., Schouten, A. *et al.* (2014) *J. Flow Chem*, **4**, 12–17.
33 Straathof, N.J.W., Tegelbeckers, B.J.P., Hessel, V. *et al.* (2014) *Chem. Sci.* doi: 10.1039/c4sc01982b
34 Wahlen, J., De Vos, D.E., Jacobs, P.A., and Alsters, P.L. (2004) *Adv. Synth. Catal.*, **346**, 152–164.
35 Bourne, R.A., Han, X., Poliakoff, M., and George, M.W. (2009) *Angew. Chem. Int. Ed.*, **48**, 5322–5325.
36 Han, X., Bourne, R.A., Poliakoff, M., and George, M.W. (2011) *Chem. Sci.*, **2**, 1059–1067.
37 Lumley, E.K., Dyer, C.E., Pamme, N., and Boyle, R.W. (2012) *Org. Lett.*, **14**, 5724–5727.
38 Woznica, M., Chaoui, N., Taabache, S., and Blechert, S. (2014) *Chem. Eur. J.*, **20**, 14624–14628.
39 for additional reading of primary literature on this subject matter published after the transcription of this chapter see: (a) Cambié, D., Bottecchia, C., Straathof, N.J.W. *et al.* (2016) Applications of continuous-flow photochemistry in organic synthesis, material science, and water treatment. *Chem. Rev.*, **116** (17), 10276–10341; (b) Yayla, H.G., Peng, F., Mangion, I.K. *et al.* (2016) Discovery and mechanistic study of a photocatalytic indoline dehydrogenation

for the synthesis of elbasvir. *Chem. Sci.*, **7** (3), 2066–2073; (c) Rackl, D., Kreitmeier, P., and Reiser, O. (2015) Synthesis of a polyisobutylene-tagged fac-Ir(ppy)3 complex and its application as recyclable visible-light photocatalyst in a continuous flow process. *Green Chem.*, **18** (1), 214–219; (d) Guerra, J., Cantillo, D., and Kappe, C.O. (2016) Visible-light photoredox catalysis using a macromolecular ruthenium complex: reactivity and recovery by size-exclusion nanofiltration in continuous flow. *Catal. Sci. Technol.*, **6** (13), 4695–4699; (e) Lima, F., Kabeshov, M.A., Tran, D.N. *et al.* (2016) Visible light activation of boronic esters enables efficient photoredox C(sp2)–C(sp3) cross-couplings in flow. *Angew. Chem. Int. Ed.*, **55** (45), 14085–14089; (f) Penders, I.G.T.M., Amara, Z., Horvath, R. *et al.* (2014) Photocatalytic hydroxylation of arylboronic acids using continuous flow reactors. *RSC Adv.*, **5** (9), 6501–6504; (g) Lefebvre, Q., Hoffmann, N., and Rueping, M. (2016) Photoorganocatalysed and visible light photoredox catalysed trifluoromethylation of olefins and (hetero)aromatics in batch and continuous flow. *Chem. Commun.*, **52** (12), 2493–2496; (h) Straathof, N.J.W., Su, Y., Hessel, V., and Noël, T. (2016) Accelerated gas–liquid visible light photoredox catalysis with continuous-flow photochemical microreactors. *Nat. Protoc.*, **11** (1), 10–21; (i) Su, Y., Talla, A., Hessel, V., and Noël, T. (2015) Controlled photocatalytic aerobic oxidation of thiols to disulfides in an energy-efficient photomicroreactor. *Chem. Eng. Technol.*, **38** (10), 1733–1742; (j) Seo, H., Katcher, M.H., and Jamison, T.F. (2016) Photoredox activation of carbon dioxide for amino acid synthesis in continuous flow. *Nat. Chem.* doi: 10.1038/nchem.2690 (k) Su, Y., Hessel, V., and Noël, T. (2015) A compact photomicroreactor design for kinetic studies of gas–liquid photocatalytic transformations. *AIChE J.*, **61** (7), 2215–2227.

14

The Application of Visible-Light-Mediated Reactions to the Synthesis of Pharmaceutical Compounds

James. J. Douglas

Early Chemical Development, Pharmaceutical Sciences, IMED Biotech Unit, AstraZeneca, Macclesfield, UK

14.1 Introduction

The recent rise in both the quantity and diversity of visible-light-mediated reactions has led to the establishment of this mode of catalysis as a defined research area. Key to advancement within a branch of synthetic methodology is the display of versatility and utility across the broad landscape of chemical structure. Within this setting, the successful application of visible-light-mediated processes to pharmaceutical products and biologically relevant molecules is paramount for its continued expansion. This chapter surveys such application of visible-light-mediated reactions, spanning the spectrum of the novel, previously unreported disconnection, through to the already well-established, for example, amide bond formation. Furthermore, pharmaceutical products provide a fertile testing ground for modern methodologies due to the high percentage of aromatic and heteroaromatic rings, as well as the quantity of heteroatoms, and Lewis basic sites. There has been a recent drive by Merck, among others, to concurrently develop, and screen new transformations for their robustness on representative groups of pharmaceutical compounds. This represents a welcome and direct test of the effectiveness of a visible-light-mediated process, ultimately leading to the greater uptake of these types of transformations across the pharmaceutical, agrochemical, and fine chemical industries [1].

14.2 Asymmetric Benzylation

One of the first examples highlighting the application of a visible-light-mediated process to the synthesis of a pharmaceutically relevant target was by the group of MacMillan in 2010 [2]. In this report, their pioneering merger of photoredox and organocatalysis allowed the asymmetric α-benzylation of aldehydes in generally high yield, typically >70%, and good enantioselectivities of up to 93% ee. The angiogenesis inhibitor **3**, developed for the treatment of diabetic retinopathy and tumor proliferation, was prepared in three steps and 34% overall yield from simple building blocks. The key enantioselective visible-light-mediated step

Visible Light Photocatalysis in Organic Chemistry, First Edition.
Edited by Corey R. J. Stephenson, Tehshik P. Yoon and David W. C. MacMillan.

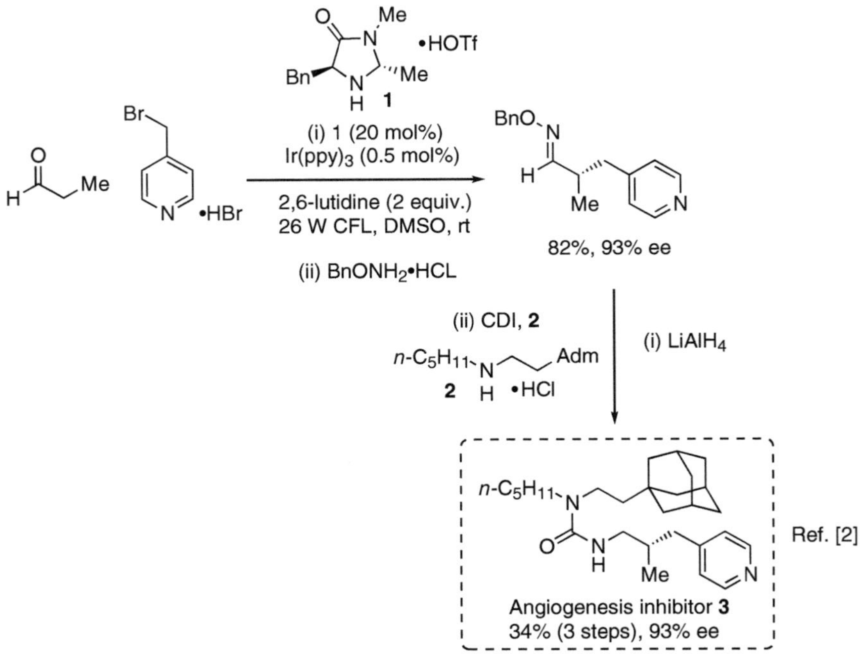

Scheme 14.1 Synthesis of angiogenesis inhibitor 3 via visible-light-mediated asymmetric benzylation.

proceeded in 82% yield and 93% ee. This example provided an early indication of the wider utility of this mode of reactivity and its ability to overcome long-standing challenges in organic synthesis (Scheme 14.1).

14.3 Amide Bond Formation

Numerous strategies have been reported for the visible-light-mediated formation of amide bonds, a highly prevalent structure found in a significant number of pharmaceutical products. Recently the groups of Kokotos and Cho have disclosed alternate strategies for the preparation of amides from aldehydes, both applying these methods to the synthesis of the antidepressant moclobemide (**6**). Cho and coworker prepared the key amide bond via *in situ* generation of the acid chloride [3]. The process likely proceeds through the acyl radical via C—H abstraction, mediated by the catalytic decomposition of *tert*-butyl hydroperoxide by $Ru(bpy)_3Cl_2$, followed by chloride atom abstraction from *N*-chlorosuccinimide (NCS) by the acyl radical. The resulting acid chloride **4** acylates the morpholino amine **5** providing moclobemide in 77% yield (Scheme 14.2A). Kokotos accesses the acyl radical via the use of phenylglyoxylic acid as an organic photoredox catalyst, postulated to proceed via abstraction of the formyl C—H bond by the excited state of the catalyst [4]. Diverting from the work of Cho, this acyl species undergoes amination by

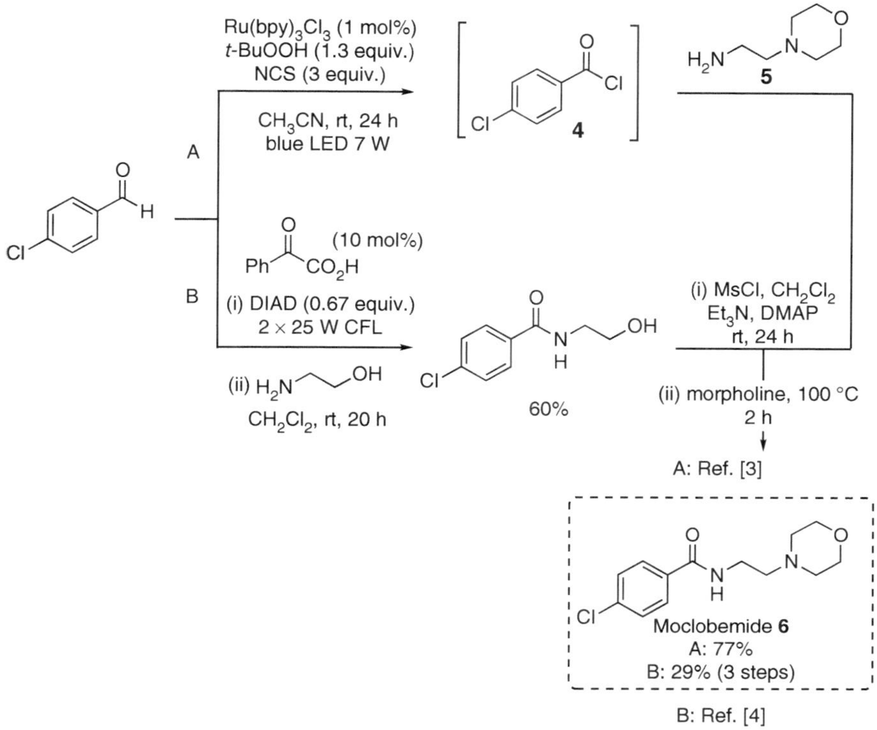

Scheme 14.2 Synthesis of moclobemide via visible-light-mediated amide bond formation.

addition to di-isopropylazodicarboxylate (DIAD), providing the intermediate carbonyl imide. This then proceeds to the final product via the addition of ethanolamine and subsequent mesylation and displacement with morpholine giving moclobemide in 29% overall yield (Scheme 14.2B).

14.4 C—H Azidation

A recent example from the research group of Greaney has applied a visible-light-mediated C—H azidation to the synthesis of the antiepileptic rufinamide (**8**) [5]. The general methodology uses Cu(dap)$_2$Cl as the photocatalyst and likely proceeds via a propagative pathway. An initially formed benzylic radical abstracts N_3 from the Zhdankin azidoiodinane reagent **7**, with the resulting iodine radical propagating the chain via benzylic C—H abstraction. The yields for the general process were typically good, >60% for a range of toluene derivatives. It was also amenable for the formation of secondary and tertiary azides, as well as a more limited number of heterocyclic analogs. Rufinamide was prepared in three overall steps with the visible-light-mediated azidation of 1,3-difluoro-2-methylbenzene, followed by copper-catalyzed triazole formation and final amide generation in 38% overall yield (Scheme 14.3).

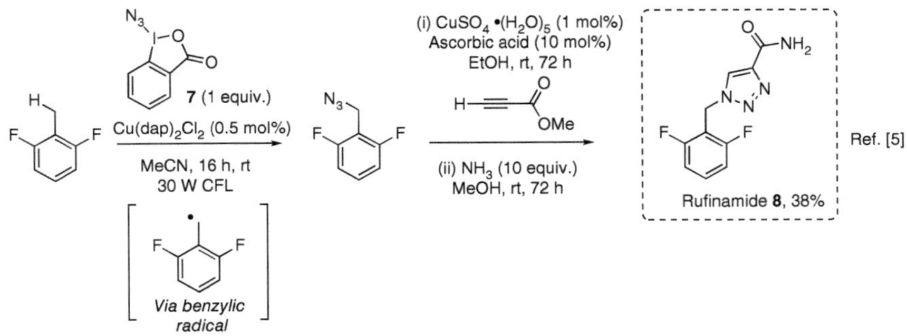

Scheme 14.3 Visible-light-mediated azidation *en route* to rufinamide.

14.5 Visible-Light-Mediated Benzothiophene Synthesis

In 2012, the group of König reported the application of their visible-light-mediated alkyne annulation of diazonium salts to the synthesis of the benzothiophene core of raloxifene (**12**) [6]. The general methodology employing the organic photocatalyst eosin Y was applicable to a range of substituted *ortho*-methylthiobenzenediazonium salts and a broad set of terminal alkynes in typically good yields, 60–80%. The key step for the synthesis of raloxifene proceeded in 70% yield to give benzothiophene **11** as a single regioisomer (Scheme 14.4). The reaction mechanism likely operates via single-electron reduction of the diazonium salt to give the aryl radical. Following this, addition to the alkyne, cyclization, and oxidative demethylation provides the benzothiophene **11**.

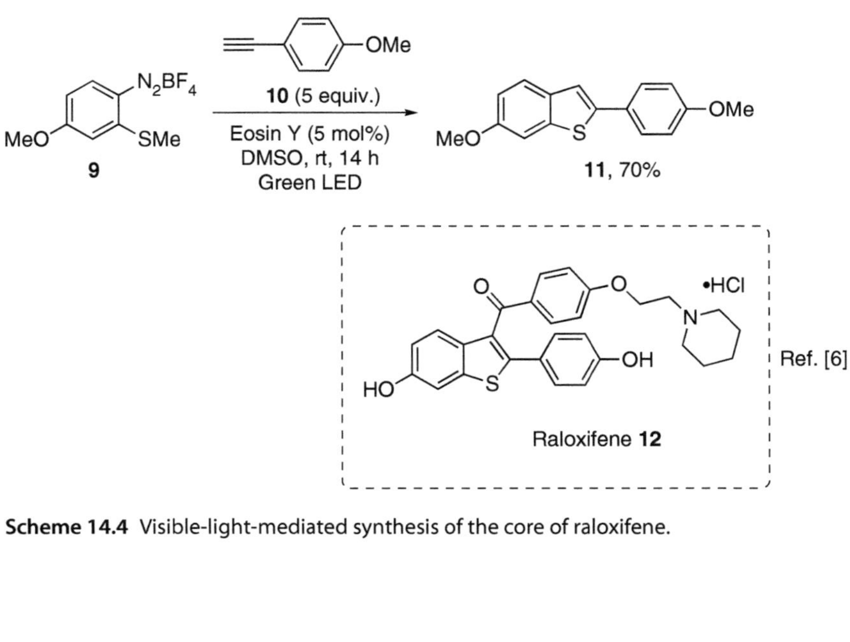

Scheme 14.4 Visible-light-mediated synthesis of the core of raloxifene.

14.6 α-Amino Radical Functionalization

In 2014, a collaboration between the Stephenson group and Eli Lilly reported their in-depth investigation into a visible-light-mediated α-amino radical addition to a complex imidazopyridazine **13** [7]. This heterocyclic core is the key intermediate of a JAK2-V617F inhibitor **14** that has recently undergone clinical trials for the treatment of several myeloproliferative disorders. After extensive optimization, the reaction proceeded in 56% yield using $Ir(ppy)_3$ as the photocatalyst, oxygen as the terminal oxidant, and an excess of *N*-methyl moroholine (NMM). This specific transformation operated best at 0 °C in a mix of DMA:H_2O with NaOAc as the base, and although a range of other *N*-methyl tertiary amines provided moderate yield, the reaction is not currently applicable to other heterocyclic cores (Scheme 14.5).

In 2014 MacMillan *et al.*, along with independent research from Koike and Akita, disclosed the generation and use of α-amino radicals *en route* to two related pharmaceuticals. Methodology developed by MacMillan uses a carboxylic acid as a "traceless activating group," allowing the regiospecific formation of an α-amino radical under mild conditions [8]. The general method employing $[Ir(dF(CF_3)ppy)_2(dtbbpy)]PF_6$ as the photocatalyst showcases a range of carboxylic acids and is also effective for secondary and tertiary alkyl radical formation. The generated α-amino radicals add efficiently to various Michael acceptors to give the products in typically good yield, >80%. Using *N*-Boc glycine **15** and Michael acceptor **16** leads to **17**, an intermediate for the synthesis of the

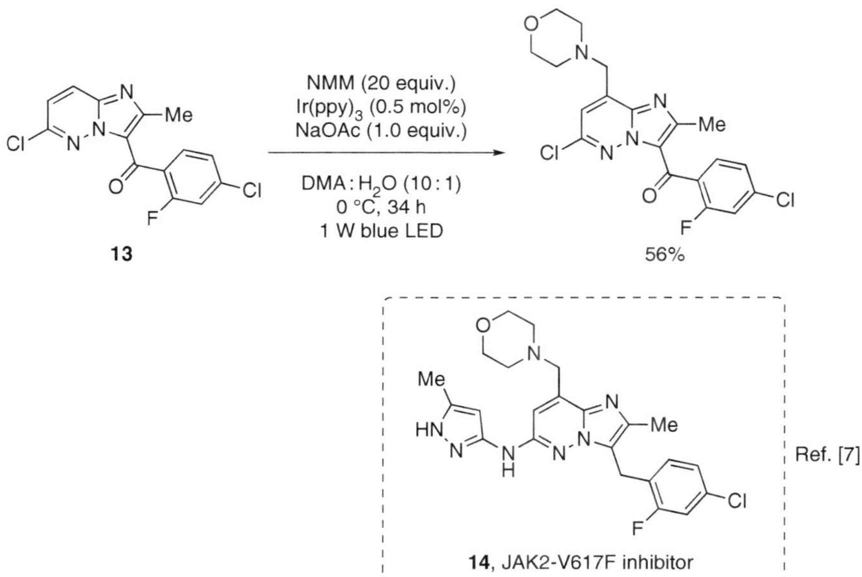

Scheme 14.5 Synthesis of the imidazopyridazine core of a JAK2 inhibitor via *N*-methyl moroholine α-amino radical addition.

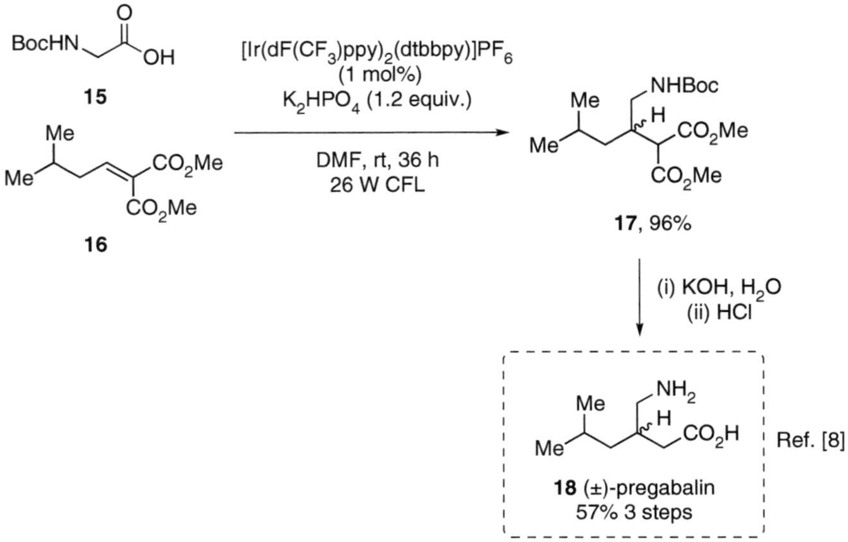

Scheme 14.6 Synthesis of pregabalin via α-amino radical conjugate addition.

anticonvulsant pregabalin (**18**) in 96% yield, with two further steps completing the racemic synthesis in 57% overall yield (Scheme 14.6).

The method disclosed by Koike and Akita [9] generates reactive α-amino radical intermediates via oxidative fragmentation of aminotrifluoroborates, mediated by the structurally related photocatalyst $[Ir(dF(CF_3)ppy)_2(bpy)]PF_6$. Their method operates in good yield for a range of activated alkene coupling partners, 10 examples >60%; however, it is limited in the scope of the amine to **19**. Although limited to formal aminomethylation, the utility of this method is demonstrated via the synthesis of baclofen, a muscle relaxant and antispastic agent. The key step operates in 89% yield, with an acid-mediated hydrolysis and decarboxylation providing racemic baclofen **20** in 87% overall (Scheme 14.7).

Another recent example reported by Sparling and Lovett from within Amgen [10] disclosed the synthesis of encainide **23**, a discontinued class 1c antiarrhythmic agent. The visible-light-mediated oxidative decarboxylation of *N*-methylpiperidine carboxylic acid **21** by $[Ir(dF(CF_3)ppy)_2(dtbbpy)]PF_6$ gives an α-amino radical that adds to styrene **22** and ultimately provides racemic encainide in 70% yield (Scheme 14.8). The general methodology was found to be selective for the β-addition to the alkene terminus, with the generation of the benzylic radical favored when methyl cinnamate was used (5.7 : 1 ratio).

In 2016, Ji *et al* reported the α-heteroarylation of amides and ethers via a visible-light-mediated transformation employing benzaldehyde and ammonium persulfate $(NH_4)_2S_2O_8$ [11]. Photoexcited benzaldehyde was proposed by the authors to catalyze the decomposition of ammonium persulfate to the sulfate radical. Abstraction of the weakest C—H bond of either the amide or ether substrate by the sulfate radical generates the desired electron-rich radical which undergoes a Minisci-type addition to an electron-deficient heteroarene,

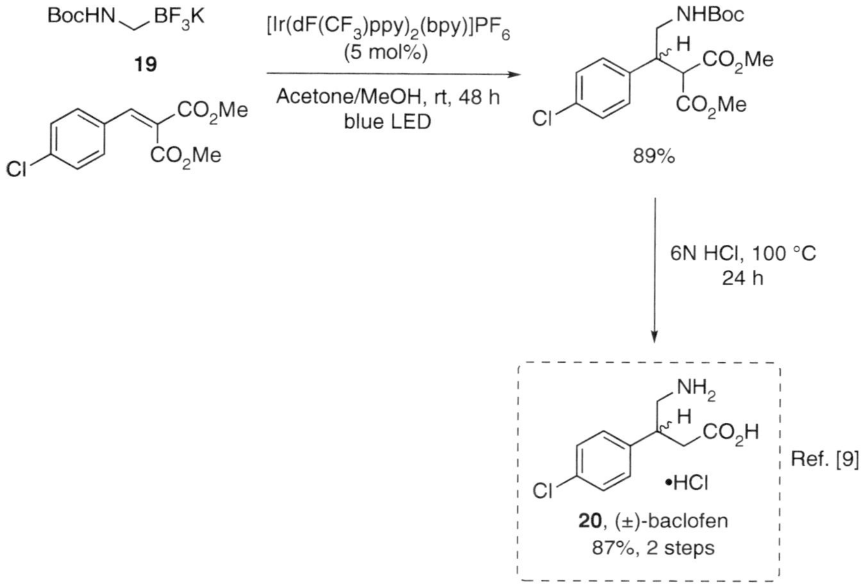

Scheme 14.7 Synthesis of baclofen via α-amino radical conjugate addition.

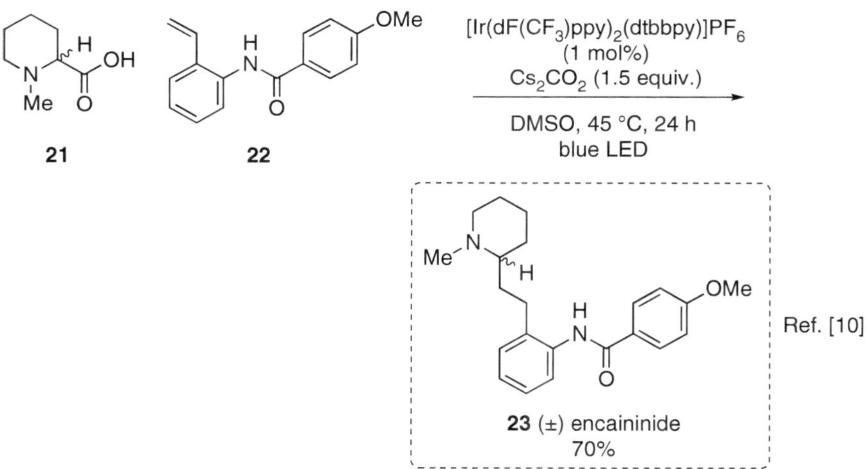

Scheme 14.8 Synthesis of encainide via α-amino radical conjugate addition.

leading to the product. The method was found to be applicable for a range of heteroaromatics when employing formamide (via abstraction of the formyl C—H bond), giving the amide products in typically moderate to good yield, 60–85%. Using dimethylacetimide (DMA) leads to C_{sp3}—H abstraction from the amide *N*-methyl group with addition to benzothiazole to form the alkylated product **24** in 87% yield. Subsequent de-acetylation provides the amine HCL salt **25**, which is an intermediate in the antidiabetic agent **26** (Scheme 14.9).

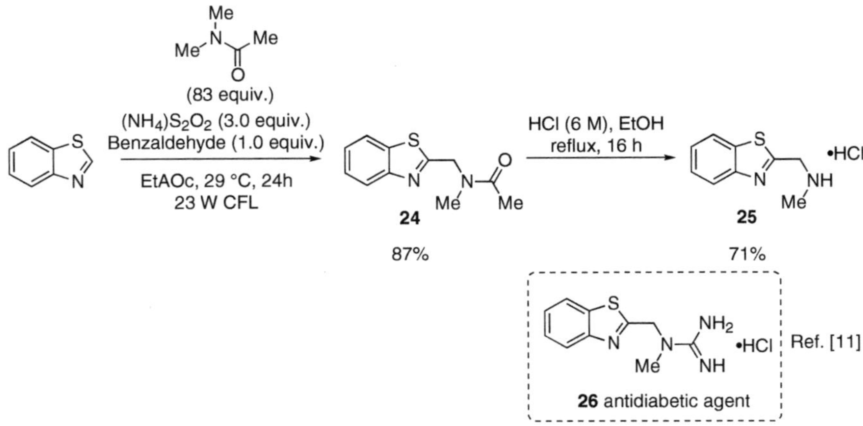

Scheme 14.9 Alkylation of benzothiazole leading to the synthesis of antidiabetic agent **26**.

14.7 Visible-Light-Mediated Radical Smiles Rearrangement

Further research carried out by the collaboration between Eli Lilly and the Stephenson group has led to the visible-light-mediated synthesis of a fluorinated intermediate *en route* to an opioid receptor-like 1 antagonist (ORL-1) **29** [12, 13]. This method details a radical Smiles rearrangement from sulfonate **27** to yield the benzylic difluorothiophene **28** that likely proceeds via an initial C—Br

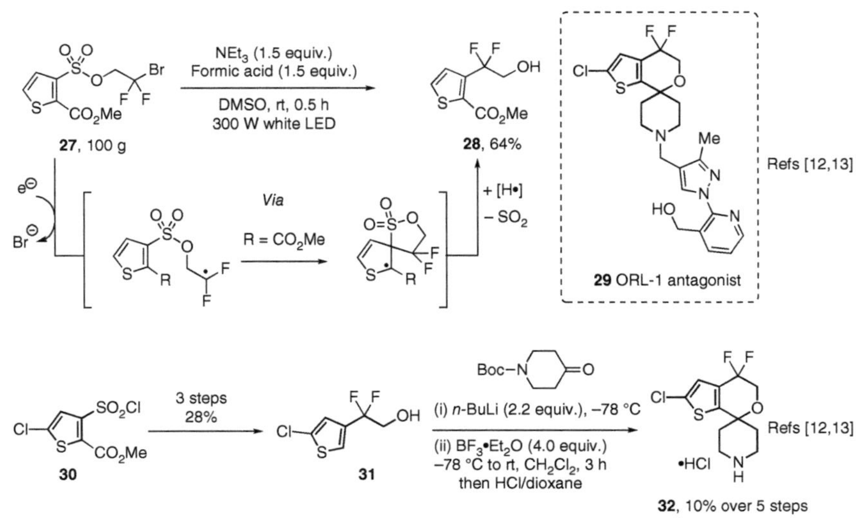

Scheme 14.10 A visible-light-mediated radical Smiles rearrangement and its application to the synthesis of spirocycle **31**.

reduction. Further mechanistic investigation showed that a radical chain process is the dominant pathway, and that the reaction could be initiated both with visible light and with thermal activation. The visible-light-mediated reaction in the absence of a photocatalyst was demonstrated on 100 g scale in 64% yield, which also proved effective in continuous flow. Application of the methodology to a chlorinated thiophene **30** provided **32** in a five-step sequence. The radical Smiles product **31** underwent lithiation, addition to *N*-Boc-4-piperidinone, followed by Lewis acid-mediated spirocyclization/Boc deprotection to afford **32** in 10% overall yield (Scheme 14.10). The radical Smiles rearrangement method was also general to other heterocycles appended with radical stabilizing groups, or extended aromatic systems, giving moderate to good yields of 24–72%.

14.8 Photoredox and Nickel Dual Catalysis

The recent advances in the combination of photoredox and nickel catalysis have provided a wide range of new synthetic methods with applications to the synthesis of pharmaceuticals. The addition of a single-electron species to a nickel complex and subsequent reductive elimination from Ni(III) has provided a valuable mechanistic alternative to traditional two-electron transmetalation manifolds, with seminal publications from Molander [14] and MacMillan/Doyle, respectively [15]. The rise of this mode of activation has led to both new bond disconnections and solutions to long-standing problems, such as nickel-mediated aryl C—O bond formation. A recent report of MacMillan has overcome the barrier to reductive elimination from a Ni(II) alkoxide complex via the visible-light-mediated generation of a Ni(III) alkoxide [16]. This allowed the facile generation of a range of aryl- and heteroaryl ethers at room temperature from the requisite alcohols and aryl bromides. The reactions proceeded in typically good yields, 70–90%, using the iridium photocatalyst $[Ir(dF(CF3)ppy)_2(dtbbpy)]PF_6$ in concert with a Ni(II) precatalyst. Fluoxetine is an antidepressant marketed under the trade name Prozac. This methodology was successfully extended to the preparation of *N*-Boc fluoxetine **34** in 82% yield from alcohol **33** and 1-bromo-4-(trifluoromethyl)benzene, demonstrating its potential utility (Scheme 14.11).

A further report of MacMillan has detailed the visible-light-mediated synthesis of ketones via a decarboxylative nickel coupling of aryl halides and ketoacids. The authors propose a visible-light-mediated decarboxylation of the keto acid fragment by $[Ir(dF(CF_3)ppy)_2(dtbbpy)]PF_6$ to generate an acyl radical [17]. This intercepts the aryl-nickel complex formed from the requisite aryl iodide, with reductive elimination generating the desired arylketone. The methodology was demonstrated on a variety of aryl- and heteroaryl halides as well as diverse selection of alkyl and aryl ketoacids in typically good yield, 70–90%. Furthermore, an efficient three-step synthesis of the cholesterol-modulating compound fenofibrate (**37**) was showcased with the key visible-light-mediated step operating in 71% yield between ketoacid **35** and aryl iodide **36** (Scheme 14.12).

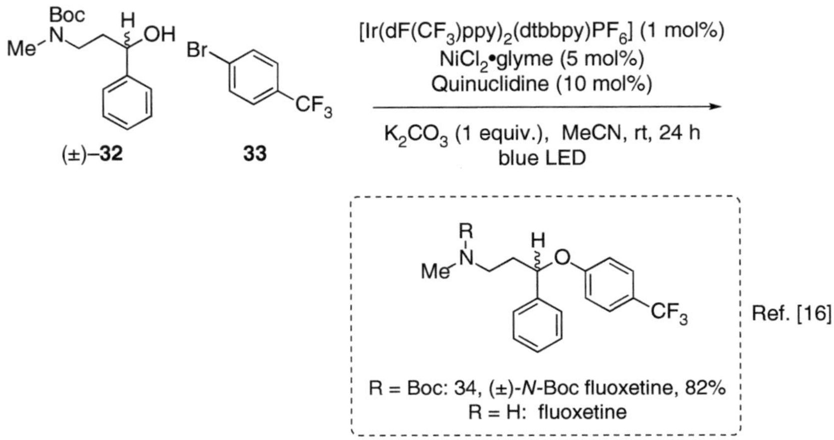

Scheme 14.11 Synthesis of *N*-Boc fluoxetine via a visible-light-mediated C—O bond formation.

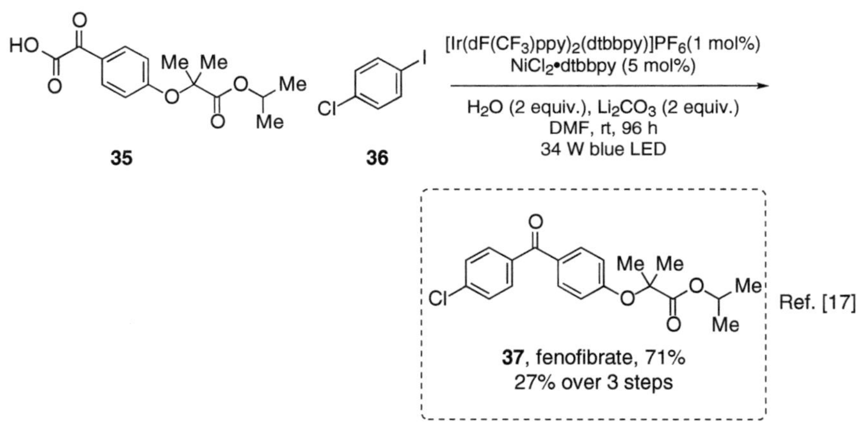

Scheme 14.12 Synthesis of fenofibrate via a visible-light-mediated aryl ketone formation.

A related method was reported in 2015 by MacMillan and coworkers to access ketones via *in situ* generated anhydrides [18]. In this visible-light-mediated process, the photocatalyst serves to modulate the oxidation state of the nickel co-catalyst, which undergoes an insertion-decarboxylation-recombination to ultimately provide the desired ketone product. A wide range of carboxylic acids, predominately with a radical stabilizing group such as an α-hetero atom, are effectively coupled with a diverse set of acid chlorides. This methodology operates at room temperature and in typically moderate to good yield, 50–80% for 25 examples. The utility of this disconnection was illustrated via the racemic synthesis of edivoxetine (**41**), a selective norepinephrine reuptake inhibitor

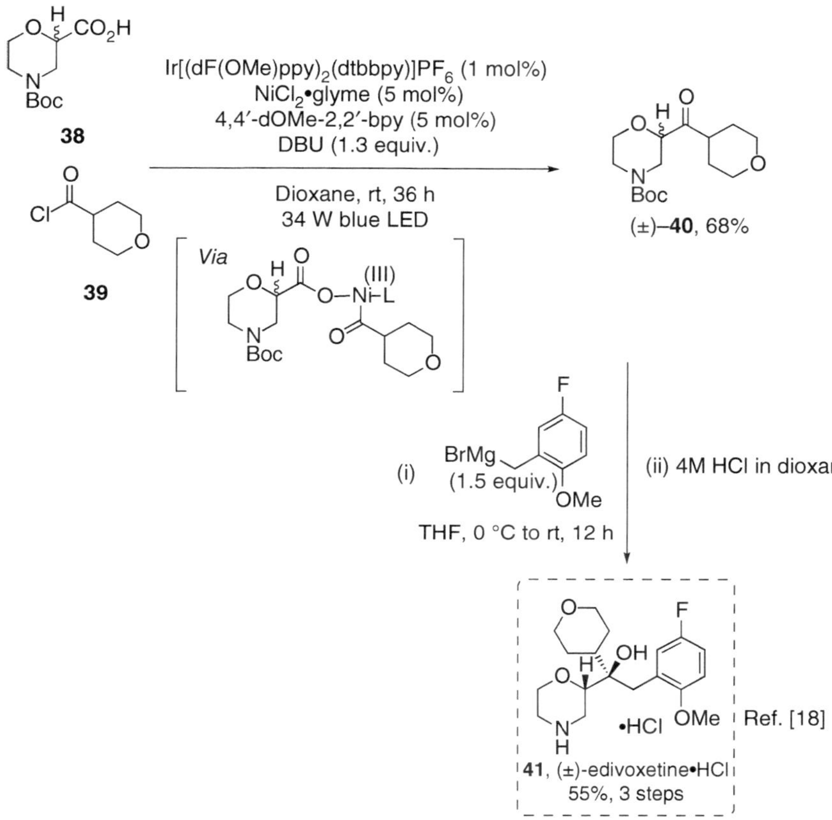

Scheme 14.13 Synthesis of (±)-edivoxetine via a visible-light-mediated aryl ketone formation.

previously in clinical trials for depression. The visible-light-mediated coupling of carboxylic acid **38** and acid chloride **39** provided ketone **40** in 68% yield, a further two steps furnished racemic edivoxetine in 55% overall yield (Scheme 14.13).

In 2016, the collaboration between the groups of MacMillan and Fu disclosed a method for the enantioconvergent decarboxylative arylation of α-amino acids [19]. This process elegantly merges previous individual reports of visible-light-mediated decarboxylative arylation [20] and enantioconvergent nickel-catalyzed cross coupling, respectively [21]. A wide range of racemic C_{sp3}-carboxylic acid containing α-amino acids are effectively cross coupled with a range of aryl- and heteroaryl bromides in typically >70% yield and >80% ee. The wider utility of this transformation was showcased via the synthesis of the northern fragment of a PEG_2 receptor antagonist **44** from the cross coupling of *N*-Boc-leucine **42** with 1-bromo-3,5-bis(trifluoromethyl)benzene **43** in 76% yield and 96% ee (Scheme 14.14).

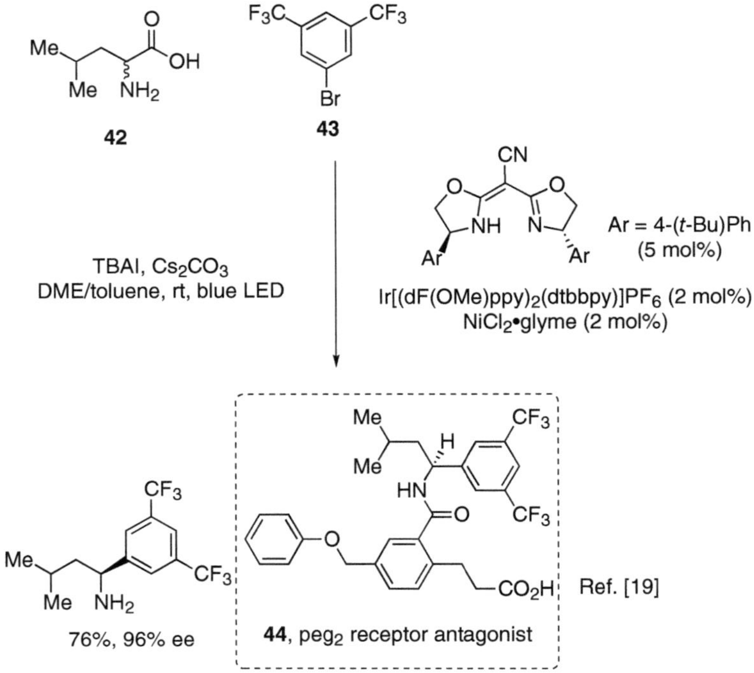

Scheme 14.14 Visible-light-mediated asymmetric decarboxylative arylation.

14.9 The Scale-Up of Visible-Light-Mediated Reactions Via Continuous Processing

Research scientists at Merck have been leaders in the application of visible-light-mediated reactions to pharmaceutical compounds [22], particularly with regard to the challenges associated with scale-up. The most attractive avenue for operation at scale appears to be via continuous flow [23], with reported reactors similar in design to those pioneered by Berry and Booker-Milburn for UV photochemistry [24]. These overcome the challenges associated with sufficient light penetration into solution by conducting the reaction in a small diameter tube, thus ensuring optimal control over reaction temperature and irradiation.

A collaboration between Merck and the Britton group have reported a method for visible-light-mediated C—H fluorination, with application to the synthesis of the cathepsin inhibitor odanacatib **47** [25]. This process uses sodium decatungstate, which upon visible-light excitation is capable of selective C—H abstraction from the leucine methyl ester **45**. Subsequent fluorine atom abstraction from *N*-fluorobenzenesulfonimide (NSFI) by the generated tertiary alkyl radical provides (*S*)-γ-fluoroleucine **46** directly, without racemization. This method was benchmarked against other routes, providing both the shortest step count and the highest yield (1 step, 90% yield). Furthermore, a flow reactor

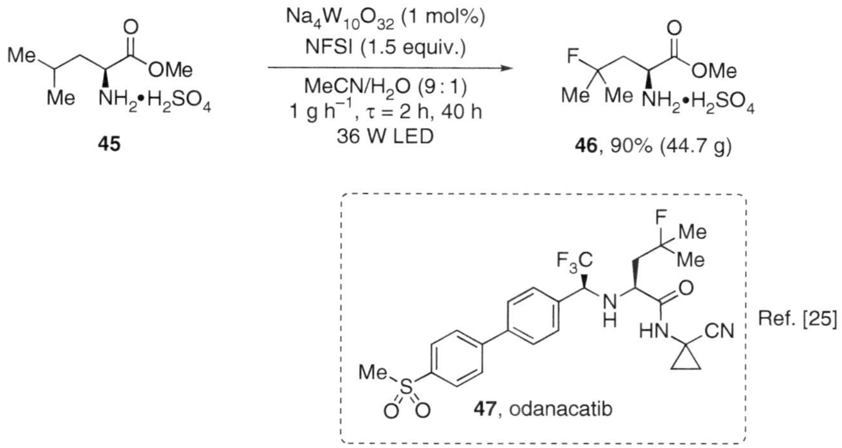

Scheme 14.15 The synthesis of odanacatib via a visible-light-mediated C—H fluorination.

designed and built in-house, was capable of producing approximately 1 g h^{-1} with a residence time (τ) of 2 h (Scheme 14.15).

Merck have also applied a visible-light-mediated process to the synthesis of elbasvir (**51**), used for the treatment of hepatitis C [26]. A detailed study in collaboration with Knowles *et al* documents the use of an iridium photocatalyst $[Ir(dF(CF3)ppy)_2(dtbbpy)]PF_6$ in combination with a perester terminal oxidant for the challenging oxidation of indoline **48**. Other oxidants tested in this reaction led to epimerization of the hemiacetal, presumably via initial *N*-centred oxidation. Extensive mechanistic investigation indicates the visible-light-mediated process operated via C—H abstraction to generate the α-amino radical **49**, thereby avoiding pathways to epimerization. The final visible-light-mediated conditions were optimized via high-throughput experimentation providing **50** in 85% yield and >99% ee (Scheme 14.16). Making use of a continuous flow reactor similar to that used for odanacatib, the reaction was demonstrated on a 100-g scale with residence time (τ) of 60 min and a productivity of 20 g h^{-1}.

In 2015, Stephenson *et al.* reported a method for visible-light-mediated radical trifluoromethylation using trifluoroacetic acid (TFAA) as a low-cost source of $•CF_3$ [27, 28]. This process is thought to operate via the single-electron reduction of *in situ* formed acylated pyridinium *N*-oxide complex **53**, generating the reactive trifluoromethyl radical, pyridine, and CO_2. The trifluoromethyl radical was found to add to a limited range of electron-rich aromatics in moderate yield, typically 30–60%. The reaction was successfully applied to continuous flow to overcome problems with light penetration on scale, producing 23 g of CF_3 pyrrole at 3.3 g h^{-1}. The utility of this method for trifluoromethylation was demonstrated via the preparation of a key pyridine motif used in the synthesis of anti-infective agents such as **55**. While the reaction does not proceed on pyridines directly, trifluoromethylation of *N*-methyl pyridone **52** and chlorination leads to the desired product in 25% overall yield (Scheme 14.17).

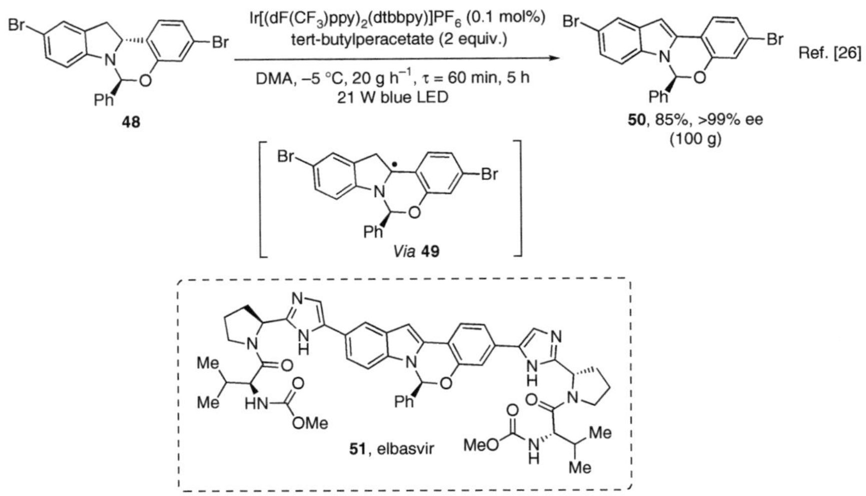

Scheme 14.16 A method for indoline oxidation via C—H abstraction.

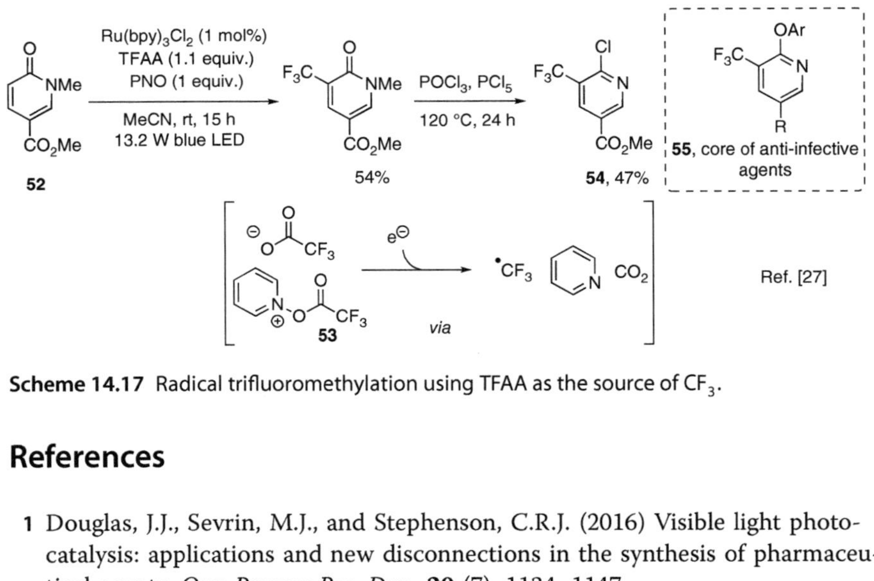

Scheme 14.17 Radical trifluoromethylation using TFAA as the source of CF_3.

References

1 Douglas, J.J., Sevrin, M.J., and Stephenson, C.R.J. (2016) Visible light photocatalysis: applications and new disconnections in the synthesis of pharmaceutical agents. *Org. Process Res. Dev.*, **20** (7), 1134–1147.

2 Shih, H.-W., Vander Wal, M.V., Grange, R.L., and MacMillan, D.W.C. (2010) Enantioselective α-benzylation of aldehydes via photoredox organocatalysis. *J. Am. Chem. Soc.*, **132** (39), 13600–13603.

3 Iqbal, N. and Cho, E.J. (2016) Visible-light-mediated synthesis of amides from aldehydes and amines via in situ acid chloride formation. *J. Org. Chem.*, **81**, 1905–1911.

4 Papadopoulos, G.N. and Kokotos, C.G. (2016) One-pot amide bond formation from aldehydes and amines via a photoorganocatalytic activation of aldehydes. *J. Org. Chem.*, **81** (16), 7023–7028.

5 Rabet, P.T.G., Fumagalli, G., Boyd, S., and Greaney, M.F. (2016) Benzylic C—H azidation using the Zhdankin reagent and a copper photoredox catalyst. *Org. Lett.*, **18** (7), 1646–1649.
6 Hari, D.P., Hering, T., and König, B. (2012) Visible light photocatalytic synthesis of benzothiophenes. *Org. Lett.*, **14**, 5334–5337.
7 Douglas, J.J., Cole, K.P., and Stephenson, C.R.J. (2014) Photoredox catalysis in a complex pharmaceutical setting: toward the preparation of JAK2 inhibitor LY2784544. *J. Org. Chem.*, **79**, 11631–11643.
8 Chu, L., Ohta, C., Zuo, Z., and MacMillan, D.W.C. (2014) Carboxylic acids as a traceless activation group for conjugate additions: a three-step synthesis of (±)-pregabalin. *J. Am. Chem. Soc.*, **136**, 10886–10889.
9 Miyazawa, K., Koike, T., and Akita, M. (2014) Hydroaminomethylation of olefins with aminomethyltrifluoroborate by photoredox catalysis. *Adv. Synth. Catal.*, **356** (13), 2749–2755.
10 Lovett, G.H. and Sparling, B.A. (2016) Decarboxylative anti-Michael addition to olefins mediated by photoredox catalysis. *Org. Lett.*, **18** (14), 3494–3497.
11 Zhang, Y., Teuscher, K.B., and Ji, H. (2016) Direct α-heteroarylation of amides (α to nitrogen) and ethers through a benzaldehyde-mediated photoredox reaction. *Chem. Sci.*, 7, 2111–2188.
12 Douglas, J.J., Albright, H., Sevrin, M.J. *et al.* (2015) A visible-light-mediated radical smiles rearrangement and its application to the synthesis of a difluoro-substituted spirocyclic ORL-1 antagonist. *Angew. Chem. Int. Ed.*, **54**, 14898–14902.
13 Douglas, J.J., Sevrin, M.J., Cole, K.P., and Stephenson, C.R.J. (2016) Preparative scale demonstration and mechanistic investigation of a visible light-mediated radical smiles rearrangement. *Org. Process Res. Dev.*, **20** (7), 1148–1155.
14 Tellis, J.C., Primer, D.N., and Molander, G.A. (2014) Single-electron transmetalation in organoboron cross-coupling by photoredox/nickel dual catalysis. *Science*, **345**, 433–436.
15 Zuo, Z., Ahneman, D. T., Chu, L., Terrett, J. A., Doyle, A. G. MacMillan, D. W. C. Merging photoredox with nickel catalysis: coupling of α-carboxyl sp3-carbons with aryl halides. *Science*, **345**, 437–440.
16 Terrett, J.A., Cuthbertson, J.D., Shurtleff, V.W., and MacMillan, D.W.C. (2015) Switching on elusive organometallic mechanisms with photoredox catalysis. *Nature*, **524**, 330–334.
17 Chu, L., Lipshultz, J.M., and MacMillan, D.W.C. (2015) Merging photoredox and nickel catalysis: the direct synthesis of ketones by the decarboxylative arylation of a-oxo acids. *Angew. Chem. Int. Ed.*, **54**, 7929–7933.
18 Le, C. and MacMillan, D.W.C. (2015) Fragment couplings via CO2 extrusion–recombination: expansion of a classic bond-forming strategy via metallaphotoredox. *J. Am. Chem. Soc.*, **137**, 11938–11941.
19 Zuo, Z., Cong, H., Li, W. *et al.* (2016) Enantioselective decarboxylative arylation of α-amino acids via the merger of photoredox and nickel catalysis. *J. Am. Chem. Soc.*, **138** (6), 1832–1835.

20 Noble, A., McCarver, S.J., and MacMillan, D.W.C. (2015) Merging photoredox and nickel catalysis: decarboxylative cross-coupling of carboxylic acids with vinyl halides. *J. Am. Chem. Soc.*, **137**, 624–627.
21 For a recent example see;Cordier, C.J., Lundgren, R.J., and Fu, G.C. (2013) Enantioconvergent cross-couplings of racemic alkylmetal reagents with unactivated secondary alkyl electrophiles: catalytic asymmetric negishi α-alkylations of N-Boc-pyrrolidine. *J. Am. Chem. Soc.*, **135** (30), 10946–10949.
22 For a pioneering example see:DiRocco, D.A., Dykstra, K., Krska, S. *et al.* (2014) Late-stage functionalization of biologically active heterocycles through photoredox catalysis. *Angew. Chem. Int. Ed.*, **53**, 4802–4806.
23 For a recent review see:Cambié, D., Bottecchia, C., Straathor, N.J.W. *et al.* (2016) Applications of continuous-flow photochemistry in organic synthesis, material science, and water treatment. *Chem. Rev.*, **116** (17), 10276–10341.
24 Hook, B.D.A., Dohle, W., Hirst, P.R. *et al.* (2005) Practical flow reactor for continuous organic photochemistry. *J. Org. Chem.*, **70** (19), 7558–7564.
25 Halperin, S.D., Kwon, D., Holmes, M. *et al.* (2015) Development of a direct photocatalytic C—H fluorination for the preparative synthesis of odanacatib. *Org. Lett.*, **17** (21), 5200–5203.
26 Yayla, H.G., Peng, F., Mangion, I.K. *et al.* (2016) Discovery and mechanistic study of a photocatalytic indoline dehydrogenation for the synthesis of elbasvir. *Chem. Sci.*, **7**, 2066.
27 Beatty, J.W., Douglas, J.J., Cole, K.P., and Stephenson, C.R.J. (2015) A scalable and operationally simple radical trifluoromethylation. *Nat. Commun.*, **6**, 7919.
28 Beatty, J.W., Douglas, J.J., Miller, R. *et al.* (2016) Photochemical perfluoroalkylation with pyridine N-oxides: mechanistic insights and performance on a kilogram scale. *Chem*, **1** (3), 456–472.

Index

Visible Light Photocatalysis in Organic Chemistry, First Edition.
Edited by Corey R. J. Stephenson, Tehshik P. Yoon and David W. C. MacMillan.